T0255012

LONDON MATHEMATICAL SOCIETY LECTURE NOTE SERIES

Managing Editor: Professor N.J. Hitchin, Mathematical Institute,
University of Oxford, 24–29 St Giles, Oxford OX1 3LB, United Kingdom

The titles below are available from booksellers, or, in case of difficulty, from Cambridge University Press.

88 FPF ring theory, C. FAITH & S. PAGE
90 Polytopes and symmetry, S.A. ROBERTSON
96 Diophantine equations over function fields, R.C. MASON
97 Varieties of constructive mathematics, D.S. BRIDGES & F. RICHMAN
99 Methods of differential geometry in algebraic topology, M. KAROUBI & C. LERUSTE
100 Stopping time techniques for analysts and probabilists, L. EGGHE
104 Elliptic structures on 3-manifolds, C.B. THOMAS
105 A local spectral theory for closed operators, I. ERDELYI & WANG SHENGWANG
107 Compactification of Siegel moduli schemes, C.-L. CHAI
109 Diophantine analysis, J. LOXTON & A. VAN DER POORTEN (eds)
113 Lectures on the asymptotic theory of ideals, D. REES
116 Representations of algebras, P.J. WEBB (ed)
119 Triangulated categories in the representation theory of finite-dimensional algebras, D. HAPPEL
121 Proceedings of *Groups - St Andrews 1985*, E. ROBERTSON & C. CAMPBELL (eds)
128 Descriptive set theory and the structure of sets of uniqueness, A.S. KECHRIS & A. LOUVEAU
130 Model theory and modules, M. PREST
131 Algebraic, extremal & metric combinatorics, M.-M. DEZA, P. FRANKL & I.G. ROSENBERG (eds)
138 Analysis at Urbana, II, E. BERKSON, T. PECK, & J. UHL (eds)
139 Advances in homotopy theory, S. SALAMON, B. STEER & W. SUTHERLAND (eds)
140 Geometric aspects of Banach spaces, E.M. PEINADOR & A. RODES (eds)
141 Surveys in combinatorics 1989, J. SIEMONS (ed)
144 Introduction to uniform spaces, I.M. JAMES
146 Cohen-Macaulay modules over Cohen-Macaulay rings, Y. YOSHINO
148 Helices and vector bundles, A.N. RUDAKOV *et al*
149 Solitons, nonlinear evolution equations and inverse scattering, M. ABLOWITZ & P. CLARKSON
150 Geometry of low-dimensional manifolds 1, S. DONALDSON & C.B. THOMAS (eds)
151 Geometry of low-dimensional manifolds 2, S. DONALDSON & C.B. THOMAS (eds)
152 Oligomorphic permutation groups, P. CAMERON
153 L-functions and arithmetic, J. COATES & M.J. TAYLOR (eds)
155 Classification theories of polarized varieties, TAKAO FUJITA
158 Geometry of Banach spaces, P.F.X. MÜLLER & W. SCHACHERMAYER (eds)
159 Groups St Andrews 1989 volume 1, C.M. CAMPBELL & E.F. ROBERTSON (eds)
160 Groups St Andrews 1989 volume 2, C.M. CAMPBELL & E.F. ROBERTSON (eds)
161 Lectures on block theory, BURKHARD KÜLSHAMMER
163 Topics in varieties of group representations, S.M. VOVSI
164 Quasi-symmetric designs, M.S. SHRIKANDE & S.S. SANE
166 Surveys in combinatorics, 1991, A.D. KEEDWELL (ed)
168 Representations of algebras, H. TACHIKAWA & S. BRENNER (eds)
169 Boolean function complexity, M.S. PATERSON (ed)
170 Manifolds with singularities and the Adams-Novikov spectral sequence, B. BOTVINNIK
171 Squares, A.R. RAJWADE
172 Algebraic varieties, GEORGE R. KEMPF
173 Discrete groups and geometry, W.J. HARVEY & C. MACLACHLAN (eds)
174 Lectures on mechanics, J.E. MARSDEN
175 Adams memorial symposium on algebraic topology 1, N. RAY & G. WALKER (eds)
176 Adams memorial symposium on algebraic topology 2, N. RAY & G. WALKER (eds)
177 Applications of categories in computer science, M. FOURMAN, P. JOHNSTONE & A. PITTS (eds)
178 Lower K- and L-theory, A. RANICKI
179 Complex projective geometry, G. ELLINGSRUD *et al*
180 Lectures on ergodic theory and Pesin theory on compact manifolds, M. POLLICOTT
181 Geometric group theory I, G.A. NIBLO & M.A. ROLLER (eds)
182 Geometric group theory II, G.A. NIBLO & M.A. ROLLER (eds)
183 Shintani zeta functions, A. YUKIE
184 Arithmetical functions, W. SCHWARZ & J. SPILKER
185 Representations of solvable groups, O. MANZ & T.R. WOLF
186 Complexity: knots, colourings and counting, D.J.A. WELSH
187 Surveys in combinatorics, 1993, K. WALKER (ed)
188 Local analysis for the odd order theorem, H. BENDER & G. GLAUBERMAN
189 Locally presentable and accessible categories, J. ADAMEK & J. ROSICKY
190 Polynomial invariants of finite groups, D.J. BENSON
191 Finite geometry and combinatorics, F. DE CLERCK *et al*
192 Symplectic geometry, D. SALAMON (ed)
194 Independent random variables and rearrangement invariant spaces, M. BRAVERMAN
195 Arithmetic of blowup algebras, WOLMER VASCONCELOS
196 Microlocal analysis for differential operators, A. GRIGIS & J. SJÖSTRAND
197 Two-dimensional homotopy and combinatorial group theory, C. HOG-ANGELONI *et al*
198 The algebraic characterization of geometric 4-manifolds, J.A. HILLMAN
199 Invariant potential theory in the unit ball of C^n, MANFRED STOLL
200 The Grothendieck theory of dessins d'enfant, L. SCHNEPS (ed)
201 Singularities, JEAN-PAUL BRASSELET (ed)
202 The technique of pseudodifferential operators, H.O. CORDES
203 Hochschild cohomology of von Neumann algebras, A. SINCLAIR & R. SMITH
204 Combinatorial and geometric group theory, A.J. DUNCAN, N.D. GILBERT & J. HOWIE (eds)
205 Ergodic theory and its connections with harmonic analysis, K. PETERSEN & I. SALAMA (eds)
207 Groups of Lie type and their geometries, W.M. KANTOR & L. DI MARTINO (eds)
208 Vector bundles in algebraic geometry, N.J. HITCHIN, P. NEWSTEAD & W.M. OXBURY (eds)

209 Arithmetic of diagonal hypersurfaces over finite fields, F.Q. GOUVÊA & N. YUI
210 Hilbert C*-modules, E.C. LANCE
211 Groups 93 Galway / St Andrews I, C.M. CAMPBELL *et al* (eds)
212 Groups 93 Galway / St Andrews II, C.M. CAMPBELL *et al* (eds)
214 Generalised Euler-Jacobi inversion formula and asymptotics beyond all orders, V. KOWALENKO *et al*
215 Number theory 1992–93, S. DAVID (ed)
216 Stochastic partial differential equations, A. ETHERIDGE (ed)
217 Quadratic forms with applications to algebraic geometry and topology, A. PFISTER
218 Surveys in combinatorics, 1995, PETER ROWLINSON (ed)
220 Algebraic set theory, A. JOYAL & I. MOERDIJK
221 Harmonic approximation, S.J. GARDINER
222 Advances in linear logic, J.-Y. GIRARD, Y. LAFONT & L. REGNIER (eds)
223 Analytic semigroups and semilinear initial boundary value problems, KAZUAKI TAIRA
224 Computability, enumerability, unsolvability, S.B. COOPER, T.A. SLAMAN & S.S. WAINER (eds)
225 A mathematical introduction to string theory, S. ALBEVERIO, J. JOST, S. PAYCHA, S. SCARLATTI
226 Novikov conjectures, index theorems and rigidity I, S. FERRY, A. RANICKI & J. ROSENBERG (eds)
227 Novikov conjectures, index theorems and rigidity II, S. FERRY, A. RANICKI & J. ROSENBERG (eds)
228 Ergodic theory of $\boldsymbol{Z}^d$ actions, M. POLLICOTT & K. SCHMIDT (eds)
229 Ergodicity for infinite dimensional systems, G. DA PRATO & J. ZABCZYK
230 Prolegomena to a middlebrow arithmetic of curves of genus 2, J.W.S. CASSELS & E.V. FLYNN
231 Semigroup theory and its applications, K.H. HOFMANN & M.W. MISLOVE (eds)
232 The descriptive set theory of Polish group actions, H. BECKER & A.S. KECHRIS
233 Finite fields and applications, S. COHEN & H. NIEDERREITER (eds)
234 Introduction to subfactors, V. JONES & V.S. SUNDER
235 Number theory 1993–94, S. DAVID (ed)
236 The James forest, H. FETTER & B. GAMBOA DE BUEN
237 Sieve methods, exponential sums, and their applications in number theory, G.R.H. GREAVES *et al*
238 Representation theory and algebraic geometry, A. MARTSINKOVSKY & G. TODOROV (eds)
239 Clifford algebras and spinors, P. LOUNESTO
240 Stable groups, FRANK O. WAGNER
241 Surveys in combinatorics, 1997, R.A. BAILEY (ed)
242 Geometric Galois actions I, L. SCHNEPS & P. LOCHAK (eds)
243 Geometric Galois actions II, L. SCHNEPS & P. LOCHAK (eds)
244 Model theory of groups and automorphism groups, D. EVANS (ed)
245 Geometry, combinatorial designs and related structures, J.W.P. HIRSCHFELD *et al*
246 *p*-Automorphisms of finite *p*-groups, E.I. KHUKHRO
247 Analytic number theory, Y. MOTOHASHI (ed)
248 Tame topology and o-minimal structures, LOU VAN DEN DRIES
249 The atlas of finite groups: ten years on, ROBERT CURTIS & ROBERT WILSON (eds)
250 Characters and blocks of finite groups, G. NAVARRO
251 Gröbner bases and applications, B. BUCHBERGER & F. WINKLER (eds)
252 Geometry and cohomology in group theory, P. KROPHOLLER, G. NIBLO, R. STÖHR (eds)
253 The *q*-Schur algebra, S. DONKIN
254 Galois representations in arithmetic algebraic geometry, A.J. SCHOLL & R.L. TAYLOR (eds)
255 Symmetries and integrability of difference equations, P.A. CLARKSON & F.W. NIJHOFF (eds)
256 Aspects of Galois theory, HELMUT VÖLKLEIN *et al*
257 An introduction to noncommutative differential geometry and its physical applications 2ed, J. MADORE
258 Sets and proofs, S.B. COOPER & J. TRUSS (eds)
259 Models and computability, S.B. COOPER & J. TRUSS (eds)
260 Groups St Andrews 1997 in Bath, I, C.M. CAMPBELL *et al*
261 Groups St Andrews 1997 in Bath, II, C.M. CAMPBELL *et al*
263 Singularity theory, BILL BRUCE & DAVID MOND (eds)
264 New trends in algebraic geometry, K. HULEK, F. CATANESE, C. PETERS & M. REID (eds)
265 Elliptic curves in cryptography, I. BLAKE, G. SEROUSSI & N. SMART
267 Surveys in combinatorics, 1999, J.D. LAMB & D.A. PREECE (eds)
268 Spectral asymptotics in the semi-classical limit, M. DIMASSI & J. SJÖSTRAND
269 Ergodic theory and topological dynamics, M.B. BEKKA & M. MAYER
270 Analysis on Lie Groups, N.T. VAROPOULOS & S. MUSTAPHA
271 Singular perturbations of differential operators, S. ALBEVERIO & P. KURASOV
272 Character theory for the odd order function, T. PETERFALVI
273 Spectral theory and geometry, E.B. DAVIES & Y. SAFAROV (eds)
274 The Mandelbrot set, theme and variations, TAN LEI (ed)
275 Computatoinal and geometric aspects of modern algebra, M. D. ATKINSON *et al* (eds)
276 Singularities of plane curves, E. CASAS-ALVERO
277 Descriptive set theory and dynamical systems, M. FOREMAN *et al* (eds)
278 Global attractors in abstract parabolic problems, J.W. CHOLEWA & T. DLOTKO
279 Topics in symbolic dynamics and applications, F. BLANCHARD, A. MAASS & A. NOGUEIRA (eds)
280 Characters and automorphism groups of compact Riemann surfaces, T. BREUER
281 Explicit birational geometry of 3-folds, ALESSIO CORTI & MILES REID (eds)
282 Auslander-Buchweitz approximations of equivariant modules, M. HASHIMOTO
283 Nonlinear elasticity, R. OGDEN & Y. FU (eds)
284 Foundations of computational mathematics, R. DEVORE, A. ISERLES & E. SULI (eds)
285 Rational points on curves over finite fields: theory and applications, H. NIEDERREITER & C. XING
286 Clifford algebras and spinors 2nd edn, P. LOUNESTO
287 Topics on Riemann surfaces and Fuchsian groups, E. BUJALANCE, A.F. COSTA & E. MARTINEZ (eds)
288 Surveys in combinatorics, 2001, J.W.P. HIRSCHFELD (ed)
289 Aspects of Sobolev-type inequalities, L. SALOFF-COSTE
290 Quantum groups and Lie theory, A. PRESSLEY
291 Tits buildings and the model theory of groups, K. TENT
292 A quantum groups primer, S. MAJID
293 Second order partial differential equations in Hilbert spaces, G. DA PRATO & J. ZABCZYK
294 Introduction to operator space theory, G. PISIER
296 Lectures on invariant theory, I. DOLGACHEV
297 The homotopy category of simply connected 4-manifolds, H.-J. BAUES

London Mathematical Society Lecture Note Series. 307

Surveys in Combinatorics 2003

Edited by

C. D. Wensley
University of Wales, Bangor

CAMBRIDGE
UNIVERSITY PRESS

University Printing House, Cambridge CB2 8BS, United Kingdom

One Liberty Plaza, 20th Floor, New York, NY 10006, USA

477 Williamstown Road, Port Melbourne, VIC 3207, Australia

314-321, 3rd Floor, Plot 3, Splendor Forum, Jasola District Centre, New Delhi - 110025, India

103 Penang Road, #05-06/07, Visioncrest Commercial, Singapore 238467

Cambridge University Press is part of the University of Cambridge.

It furthers the University's mission by disseminating knowledge in the pursuit of education, learning and research at the highest international levels of excellence.

www.cambridge.org
Information on this title: www.cambridge.org/9780521540124

First published 2003

A catalogue record for this publication is available from the British Library

ISBN 978-0-521-54012-4 Paperback

Contents

Preface

The British Combinatorial Committee invited nine distinguished combinatorial mathematicians to give survey talks on the latest developments in their own fields at the 19th British Combinatorial Conference held at the University of Wales, Bangor, in July 2003. This volume contains the survey articles on which these talks were based, together with a memoir of Bill Tutte.

The conference was held in Bangor to honour the 100th anniversary of the birth of Professor D.E. Littlewood, Professor of Mathematics at Bangor from 1948 to 1970. The talk by Arun Ram, which dealt with recent developments in generalized Hall–Littlewood and Kostka–Foulkes polynomials, was designated the *Littlewood Anniversary Lecture.*

The customary *Rado Lecture*, this year given by Imre Leader, described recent developments in Ramsey Theory for infinite systems. The nine articles in this volume provide a splendid source of reference to the current state of the art in combinatorics.

In coordinating this volume I have been greatly assisted by previous local organisers, by members of the British Combinatorial Committee, and by Jonathan Walthoe of Cambridge University Press. The delivery on schedule of this volume was made possible by the careful attention to detail, and by the prompt response to requests, of all the authors and referees.

The conference is grateful for financial support from the London Mathematical Society.

Christopher D. Wensley
Bangor, March 2003
c.d.wensley@bangor.ac.uk

Preface

W.T. Tutte 1917-2002

Norman Biggs

Bill Tutte will take his place in history for two reasons. First, as part of the now-famous codebreaking team at Bletchley Park, he made a significant individual contribution to the outcome of the Second World War. Secondly, he formulated and proved many of the theorems that form the foundations of Graph Theory. In both cases his achievements resulted from very deep insights into matters that, at first sight, might be thought simple. In 1977 Paul Erdős recalled [13]:

> I first heard about Tutte in early September 1939 .... But of the real powers I only learned later. T. Gallai and I as freshmen in 1930 took the course on graph theory by D. Kőnig – he mentioned the conjecture of Tait and the extension of Petersen's theorem on factorisation of graphs as important outstanding problems – we tried both unsuccessfully. As is well known, Tutte settled both problems – and many others.

William Thomas Tutte was born on May 14, 1917 at Fitzroy House, a horseracing establishment in Newmarket, England. Around 1921 the family settled in Chevely, a village near Newmarket, where his father was gardener at the Rutland Arms Hotel. He attended the village school until, at the age of 11, he won a scholarship to the Cambridge and County High School. The school was 15 miles from his home and the daily journey was difficult, but well worth the effort. He won many prizes, and in the school library he discovered W.W. Rouse Ball's book of *Mathematical Recreations and Essays*, in which he read about the Five-Colour Theorem and Petersen's Theorem. Both these results were to figure largely in his life's work.

In 1935 he went up to Trinity College, Cambridge, supported by a State Scholarship and a College Scholarship. He read Natural Sciences, specialising in Chemistry, and getting a First Class Honours degree. He also joined the Trinity Mathematical Society, where he met R.L. Brooks, C.A.B. Smith, and A.H. Stone. Together they worked on the problem of dividing a square into squares of different sizes. The story of their work on this problem has been told many times, including by Tutte himself in the *Scientific American* [6]. It is always a surprise to find just how many important ideas arose first in this work on a problem in 'recreational' mathematics. The ideas about spanning trees can be traced back to Kirchhoff, but many of their algebraic results were new, as were the insights into planarity and duality.

By the time that the paper on squaring the square was published [1], Tutte had started his research in Chemistry, and had produced two papers on his experimental results. His progress was interrupted when he was called up for national service in the Second World War, and after initial training he arrived at Bletchley Park, the British cryptographic HQ, in 1941. He was one of many who regarded signing the Official Secrets Act as a lifelong obligation, and when stories of the great deeds done at Bletchley began to leak out, often in a garbled fashion, he did not immediately leap on the bandwagon. It was probably a relief to him when, in the 1990s, it became clear that at least some of the secrets were no longer official. At his eightieth birthday celebrations in 1997 he felt able to talk informally about some of the details, and in 1998 he was persuaded to give a talk at the opening of the Center for Applied Cryptographic Research at the University of Waterloo.

In this talk [10], entitled *Fish and I*, he tells how, others having failed, he was asked to work on the cipher system known at Bletchley as Tunny. This was one of the 'Fish' codes used by the German High Command. He had an idea and, although not optimistic, he 'thought it best to seem busy'. So he copied out some ciphertext onto sheets of squared paper, using chunks of various lengths, noticed certain patterns, and was able to infer the structure of the system. Indeed, he achieved a virtual reconstruction of an extremely complex machine using only scraps of information – an amazing feat that must rank as one of his greatest intellectual achievements.

The success of Bletchley as an institution was partly due to the fact that the powers-that-be were not stupid, and soon many people were helping to work out the implications of Tutte's discovery. This continued throughout 1942 and 1943, with regular upgrading of the techniques to deal with improvements in the system. Eventually it became necessary to use a form of number-crunching statistical analysis, and Tutte saw how this could be done. He reported his ideas and, in his own words, 'there were rapid developments'. The outcome was that the famous Colossus computer was deployed on these problems.

At the end of the War, Trinity College elected Tutte to a Research Fellowship in Mathematics. Although less prestigious in the public eye than the awards given to other civil servants, it was probably more appreciated by the recipient. Exactly how it came about is unclear. C.A.B. Smith recalled being stopped in the street by a Fellow of Trinity, who said 'we've just elected Tutte to a Fellowship but we don't know what he has done or where he lives' [14].

The period at Trinity was a highly productive one. His Ph.D. thesis on 'An Algebraic Theory of Graphs' contained many seminal ideas, and these were published in papers that quickly established graph theory as a significant area of mathematics, with Tutte as its master builder. Among the papers that were published at that time there are several classics. In a paper published in 1946 [2] he disproved Tait's conjecture by constructing a planar cubic graph that has no Hamilton cycle. His paper on the symmetry of cubic graphs [3] contains a truly unexpected bound on the order of a vertex-stabilizer, a fact that was to resurface twenty years later in the work of permutation-group-theorists. Perhaps the most influential paper is the one on factorization of graphs [4], in which he obtains the canonical form of the basic result on this topic, with Petersen's theorem as a simple corollary.

Much later, in his book *Graph Theory As I Have Known It* [11], he gave a fascinating account of how he arrived at some of these fundamental results. Perhaps not surprisingly, it was often by a process that offered an intellectual challenge rather than a guarantee of success. The use of Pfaffians in the proof of the factorization theorem was marvellous, even if it was later shown to be unnecessary. As well as graph theory, his thesis also contained important results about matroids, a subject that had been inaugurated by Hassler Whitney. Many of these results were published about ten years later [7], but their significance was not fully recognized until they appeared in a series of lectures in 1965 [8].

In 1948 he took up a post at the University of Toronto. Here he continued to produce a stream of new ideas and, rather unexpectedly, for he was a very shy man, he got married. His wife Dorothea would bemoan the fact that weekends had to be spent on research, because Bill feared that mathematical inspiration would dry up before he was 40 (at least, that's what he told her). Some of his Toronto papers discussed aspects of the chromatic polynomial and its two-variable generalization, now known (justifiably) as the Tutte polynomial [5]. Several famous conjectures, such as the conjecture that every bridgeless graph has a 5-flow, also appeared in print at this time.

In 1962 he was persuaded to move to the newly-established University of Waterloo. By this time he had been appointed a Fellow of the Royal Society of Canada, and his eminence was being recognised internationally. The university created around him a world-famous Department of Combinatorics and Optimisation, and it was instrumental in the foundation of the *Journal of Combinatorial Theory.* He himself was not an administrator, but he supported and encouraged people whose talents lay in that direction, and his placid temperament helped to calm the troubled waters that sometimes threatened his Department.

By the 1970s the growth of air travel meant that Bill and Dorothea were able to travel extensively, and they returned to England on several occasions. In 1971 he was the principal guest at a small meeting held in Royal Holloway College. The success of that meeting led to the establishment of the continuing series of British Combinatorial Conferences, and Bill spoke at the first one to be organised on a regular basis, the Fourth BCC in Aberystwyth (1973).

His work at this period centred on the enumeration of planar graphs, and specifically four-colourable planar graphs. There was a slight chance that the four-colour conjecture could be settled 'asymptotically', but he did not have great hopes for the method. He greeted the Appel-Haken resolution of the conjecture enthusiastically, agreeing that the strategy was sound, even if the calculations could not be checked by hand [9].

He retired formally in 1985, but continued to be active in mathematics. In his quiet way he enjoyed the recognition that accompanied the growth in popularity and status of Graph Theory, the subject he had built. Outstanding mathematicians were attracted to work in this field, many of them inspired by Tutte's earlier results.

After Dorothea's death in 1994 he lived in England for a while, but he did not settle, and eventually returned to his adopted home in Ontario. It was proper that his eightieth birthday should be marked by a celebration in Waterloo where he was able to talk about his work to an audience that fully appreciated what he had achieved. In Britain we were fortunate to have him as the Rado Lecturer at the BCC in Canterbury in 1999, where he spoke about *The Coming of the Matroids* [12]. In this talk he explained how some of his work at Bletchley had helped him to understand the properties of linear dependence, and how this led to some of the fundamental theorems of matroid theory.

In 2001 his eminence was recognised by the award of the Order of Canada, which he received with characteristic humour and humility. At that time he was in good health, but in March 2002 he was diagnosed with a serious medical condition, and he died on May 2, in his 85th year.

Author's Note This is an extended version of the obituary published in *The Independent* on 9 May 2002. I am grateful to several people, in particular Dan Younger, for additional information. A full appreciation of Tutte's mathematical work is planned to appear in the *Bulletin of the London Mathematical Society.*

References

Books and Papers by W.T.Tutte:

[1] (with R.L. Brooks, C.A.B. Smith and A.H. Stone), The dissection of rectangles into squares, *Duke Math. J.* **7** (1940), 312-340.

[2] On Hamiltonian circuits, *J. London Math. Soc.* **21** (1946), 97–101.

[3] A family of cubical graphs, *J. London Math. Soc.* **43** (1947), 459–474.

[4] The factorization of linear graphs, *J. London Math. Soc.* **22** (1947), 107–111.

[5] A contribution to the theory of chromatic polynomials, *Canad. J. Math.* **6** (1954), 347–352.

[6] Squaring the Square, *Scientific American* (November 1958), reprinted in *More Mathematical Puzzles and Diversions* (ed. M. Gardner), Penguin, London (1961), pp. Chapter 17.

[7] A homotopy theorem for matroids I, II, *Trans. Amer. Math. Soc.* **88** (1958), 144–160, 161–174.

[8] Lectures on Matroids, *J. Res. Natl. Bur. Stand.* **B49** (1965), 1–47.

[9] Colouring problems, *Math. Intelligencer* **1** (1978), 72–75.

[10] Fish and I, lecture given on 19 June 1998 at the University of Waterloo, `http://www.math.uwaterloo.ca/CandO_Dept/corr98-39.pdf`.

[11] *Graph Theory As I Have Known It*, Clarendon Press, Oxford (1998).

[12] The coming of the matroids, in *Surveys in Combinatorics, 1999* (eds. J.D. Lamb & D.A. Preece), *London Math. Soc. Lecture Note Ser.*, 267, Cambridge Univ. Press, Cambridge (1999), pp. 3–14.

Other items:

[13] P. Erdős, A Tribute, in *Graph Theory and Related Topics* (eds. J.A. Bondy & U.S.R. Murty), Academic Press, New York (1979), p. xxiii.

[14] C.A.B. Smith, Early Reminiscences, in *Graph Theory and Related Topics* (eds. J.A. Bondy & U.S.R. Murty), Academic Press, New York (1979), pp. xix–xxi.

Department of Mathematics
London School of Economics
Houghton Street
London, WC2A 2AE
United Kingdom
`n.l.biggs@lse.ac.uk`

Decompositions of complete graphs: embedding partial edge-colourings and the method of amalgamations

L.D. Andersen and C.A. Rodger

Abstract

We consider decompositions of the edges of complete graphs, mainly with each part inducing a spanning subgraph with specified properties. We give special attention to embedding questions, where a decomposition of the edges of a complete subgraph of a complete graph is given, and the problem is to extend the decomposition to a full decomposition with the desired properties. One way to obtain decomposition and embedding results simultaneously is by the method of amalgamations of vertices, to which we devote a large part of the paper.

1 Introduction

1.1 Some sample questions

As a starting point, consider the edge-coloured complete graph K_6 in Figure 1. Its edge-colouring (indicated by numbers) is proper, meaning that each colour occurs on at most one edge incident with each vertex, so thinking of the graph as a subgraph of a K_{11} it makes sense to ask

1. Can the edge-colouring be completed to a proper edge-colouring of K_{11} with the least possible number of colours, 11?

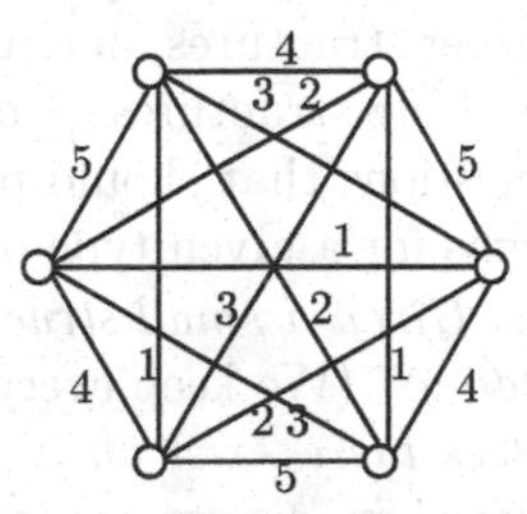

Figure 1

On the other hand, in K_{11} the colouring could easily be part of a colouring of the edges which is not necessarily proper, but which may have other nice properties, such as each colour class inducing a given subgraph or a subgraph with some given property. In this spirit, one could ask

2. Can the edge-colouring be completed to an edge-colouring of K_{11} in which every colour class induces a spanning connected subgraph?

3. Can the edge-colouring be completed to an edge-colouring of K_{11} in which every colour class induces a Hamiltonian cycle?

4. Can the edge-colouring be completed to an edge-colouring of K_{11} in which every colour class induces the disjoint union of a 6-cycle and a 5-cycle?

1.2 Embeddings

Many questions can be asked about completing partial structures such as edge-colourings. In this paper, we focus on situations where the prescribed part is concentrated in a given part of the larger structure, and consequently we shall mainly talk about *embeddings* rather than completions (not that this distinction matters much).

Proper edge-colourings are models of timetables for pairwise meetings and, in the case of complete graphs, round-robin tournaments. In such connections, it is natural to wish to extend smaller schedules to larger ones, giving rise to questions like sample question 1 above. In many cases the derived problems, for example *'When can a proper edge-colouring of K_r be embedded in a proper edge-colouring of K_n with the minimum number of colours?'* are not hard, but they can be extremely challenging with small additional conditions such as requiring a given set of mutually independent edges in $K_n \setminus K_r$ to have prescribed colours. This kind of condition arises naturally when the embedding result is developed as a tool for solving other combinatorial problems.

Quite generally, considering embedding problems for combinatorial structures is both a natural and a fruitful exercise. Thus embedding problems have been considered for a variety of such structures: partial quasigroups (latin squares), edge-coloured graphs, partial balanced incomplete block designs, and many more. The insight gained can have many facets: can smaller structures be simple building blocks in larger structures, does the embedding method give an algorithm for construction of big structures — or maybe even an existence proof? There are two basic questions that should perhaps be asked first when considering embedding problems for a given type of combinatorial structure.

The first basic question is: *'Given a small structure of the type considered, is there always a finite embedding?'* (We keep everything finite in this paper.) Again, a question such as *'Does there exist an n such that any proper edge-colouring of K_r can be embedded in a proper edge-colouring of K_n with the minimum number of colours?'* is not too hard, but other more difficult, yet analogous questions can be asked.

We usually require the target structure to be complete in some sense, such as a latin square (not partial), a proper edge-colouring with the minimum number of colours, and so on. The second natural question applies when the smaller structure is complete as well; for example *'What is the smallest larger latin square into which a latin square can be embedded?'* Here we have a simple, clean version of the general problem, where one would expect the answer to be more easily found than in the general case. In sample question 1, we actually have this situation, as the graph of Figure 1 is properly coloured with the minimum number of colours, 5. And it is easy to see that it cannot be embedded in the desired way in a complete graph with fewer than 11 vertices.

The embedding problems related to sample question 1 have a tight relationship to the theory of latin squares, as this is where the most obvious previous analogues are to be found, and also where some of the questions demanding more subtle embedding results arose. We survey some of the latin square analogues in Section 2, and then devote Section 3 to the treatment of sample question 1 and extensions of it.

1.3 Amalgamations

For the remaining three sample questions, or rather for the vast class of problems for which they are just simple examples, we describe the *method of amalgamations*. Designed to address more general questions, this method — when it works — almost always produces an embedding result as well. It was first introduced in connection with latin squares, but in 1984 Hilton [27] showed how to construct Hamiltonian decompositions of complete graphs with it.

When used on graphs, the point of the method is to find a clever description of what the target edge-colouring (such as a Hamiltonian decomposition with a colour for each Hamiltonian cycle) looks like when subsets of vertices are each identified, that is *amalgamated* to form a single vertex. It is then often easier to recognise such an amalgamated colouring in a smaller graph, and the strength of the method is in situations where it can be shown that, given what might be an amalgamated colouring, the vertices can indeed be pulled apart again to give the desired result. We elaborate on this in Section 4. It is important to note that the method involves both multiple edges and loops, so *multiple edges and loops are allowed in the graphs of this paper.*

The embedding corollaries are obtained from the method of amalgamations when all vertices outside the graph to be embedded are amalgamated into one single vertex.

Since Hilton's paper, itself covering sample question 3, the method has been extended and used to address questions such as sample question 2. Sample question 4 is an example of an embedding problem derived from the *Oberwolfach problem.* This problem asks whether the complete graph K_{2n+1} can be decomposed into isomorphic spanning subgraphs, each being the union of disjoint cycles of specified lengths. (The cycle lengths correspond to hypothetical table sizes, in terms of numbers of persons seated at the table, at the mathematical research institute in Oberwolfach in Germany; the vertices of the complete graph correspond to the participants in a given meeting; and the decomposition is a seating schedule ensuring that, over a number of meals, each participant sits next to each other participant exactly once.) The Oberwolfach problem is not solved, and the ultimate goal of using the method of amalgamations on it would be to obtain a solution. So here it is the existence problem that is in focus, not the embedding.

1.4 Flashback

Still, we begin by looking at embedding results, old and new, in the style of sample question 1. The inspiration comes very much from latin squares, so we take a look at these first.

By the way, the answer to all four sample questions in Subsection 1.1 is yes ....

2 The heritage from latin squares

2.1 Latin square terminology

A *latin square* of side n is an $n \times n$ matrix with entries from an n-set, in which each symbol occurs exactly once in each row and exactly once in each column. We shall define two partial structures: a *partial latin square* of side n is an $n \times n$ matrix in which some cells may be empty and the rest have entries from an n-set in such a way that each symbol occurs at most once in each row and at most once in each column, and a *latin rectangle* of size $r \times s$ on n symbols is an $r \times s$ matrix with entries from an n-set, where each symbol occurs at most once in each row and at most once in each column.

Thus latin rectangles have no empty cells. The theorems mentioned in the following are mainly concerned with latin rectangles, sometimes with a few elements outside a latin rectangle also prescribed.

We note that a latin square of side n is equivalent to a proper edge-colouring of the complete bipartite graph $K_{n,n}$ with n colours (one such equivalence is obtained by taking rows as one bipartition class of vertices, columns as the other, and symbols as colours, but in fact the roles of rows, columns and symbols can be permuted).

A latin square, a partial latin square or a latin rectangle is said to be *symmetric* if cell (i, j) always contains the same entry as cell (j, i).

A symmetric latin square of side n corresponds to a proper edge-colouring with n colours of the complete graph K_n with a loop on each vertex, the loops corresponding to the diagonal cells.

Remark In a symmetric latin square of even side, each symbol occurs an even number of times on the diagonal. In a symmetric latin square of odd side, each symbol occurs exactly once on the diagonal.

2.2 The first embedding theorem

Although this whole excursion into latin squares is meant to be brief, we dwell a little on the following theorem, proved by Hall [24] in 1945.

Theorem 2.1 *Every latin rectangle of size $r \times n$ on n symbols can be embedded in a latin square of side n.*

A consequence of this simple theorem is a very simple way of constructing latin squares: it can be done row by row. Write down any permutation of the n symbols, and you have a latin rectangle of size $1 \times n$, and by the theorem it can be completed to a latin square of side n. Find, in any way, a second permutation, with no repetition in any column when used as the second row, and so on. Thus the theorem is also an existence proof for latin squares! (Although the existence problem never was very hard . . . .)

The result can be interpreted as stating that a complete bipartite graph $K_{n,n}$ can be properly edge-coloured by repeatedly giving a 1-factor a new colour and then deleting it, and so it is generalised by Kőnig's theorem that a bipartite graph with maximum degree Δ has a proper edge-colouring with Δ colours. This theorem also gives an easy proof of Theorem 2.1 (in Section 4.2, we present a generalisation of Kőnig's theorem needed for the method of amalgamations). Another proof of Hall's theorem, echoing more the row-by-row completion, is obtained by always constructing a new row by finding a system of distinct representatives for the sets of symbols not yet appearing in each column.

Both these techniques can be used to prove the latin square embedding results of the next subsection as well.

2.3 Ryser type embeddings

If R is a latin rectangle, we let $R(\sigma)$ denote the number of cells of R occupied by the symbol σ.

Ryser proved the following generalisation of Theorem 2.1 in 1951.

Theorem 2.2 ([55]) *A latin rectangle R of size $r \times s$ on n symbols can be embedded in a latin square of side n if and only if*

$$R(\sigma) \geqslant r + s - n$$

for all symbols σ.

In an influential paper [21] from 1960, Evans formulated the obvious corollary.

Corollary 2.3 *A latin rectangle of size $r \times r$ on n symbols can be embedded in a latin square of side n for all $n \geqslant 2r$, and this bound is best possible.*

The graph theoretic translation of these results deals with complete bipartite graphs, and it is not our intention to pursue the connection in this paper. However, in 1974, Cruse [17] proved analogues of both for symmetric latin squares.

Theorem 2.4 (i) *A symmetric latin rectangle R of size $r \times r$ on n symbols can be embedded in a symmetric latin square of side n if and only if*

$$R(\sigma) \geqslant 2r - n \quad \text{for all symbols } \sigma\text{, and}$$

$$R(\sigma) \equiv n \pmod 2 \quad \text{for at least } r \text{ symbols } \sigma.$$

(ii) *A symmetric latin rectangle R of size $r \times r$ can be embedded in a symmetric latin square of any even order $n \geqslant 2r$, and it can be embedded in a symmetric latin square of any odd order $n > 2r$ if and only if the diagonal elements of R are distinct. Both bounds are best possible.*

As we shall see in Section 3, Theorem 2.4 has direct analogues concerning edge-colourings of complete graphs.

2.4 Quasigroups and embeddings with prescribed diagonal

In [21], Evans suggested what has become known as the Evans conjecture (although other people seem to have thought of it at approximately the same time): *Every partial latin square of side n with at most $n - 1$ non-empty cells can be completed to a latin square of side n.* This was first proved by Smetaniuk [56] in 1981. An independent proof by Andersen and Hilton [7] is of interest here, because it relies heavily on embedding latin rectangles with elements of the diagonal outside the rectangle also specified. This is just one example that such embeddings are a useful tool in combinatorics — there are several others. The topic of completing partial latin squares, rather than embedding latin rectangles, is very interesting but outside the scope of this paper.

More than anyone else, the person who took over after Evans and promoted embedding problems for latin squares, was Lindner. (Later he inspired several others such as Hilton and — directly and through Hilton — the present authors.) Lindner considered latin squares as *quasigroups* and, by requiring certain *quasigroup identities* to be satisfied in the larger latin square, he listed a whole class of embedding problems. For all of these it has been shown that finite embedding is possible, and good upper bounds on the size of the containing quasigroups have been found. Lindner has written a number of surveys on the topic, the most recent being [46]. The early notes [47] (one part written by Lindner, the other by Evans) are interesting, because they demonstrate the relationship between this kind of embedding question and universal algebra. Evans wrote in these notes that his original motivation for studying finite completion of partial algebras was the fact that the embedding problem for a variety is solvable if and only of the word problem is solvable for the variety.

We mention here that requiring the so-called Steiner identities gives the embedding problem for Steiner triple systems, to which we shall return. This is a case where the best known upper bound can probably be improved.

Requiring solely the identity $x^2 = x$ corresponds to the embedding problem for idempotent latin squares, where the diagonal is required to contain distinct elements (the order required in the quasigroup formulation can always be achieved with permutations).

So there was more than one reason why embeddings with prescribed diagonal received attention. Although a necessary Ryser type condition (given as condition (a) of Theorem 2.6 below) can easily be deduced, it is not sufficient, and the number of occurrences of each symbol is not enough to determine whether embedding is possible (see [12]). Still, the best possible bound for the idempotent case was proved by Hilton and the present authors [11].

Theorem 2.5 *An idempotent latin rectangle R of size $r \times r$ on n symbols can be embedded in an idempotent latin square of side n for all $n \geqslant 2r+1$.*

The difficulties occurring for $n \leqslant 2r$ seem similar to those encountered in Section 3 when embedding proper edge-colourings with additional prescribed colours on mutually independent edges outside the complete graph to be embedded. They are, however, still unresolved, in contrast to the graph case.

Finally we state a generalisation of Theorem 2.5 due to Rodger [51].

Theorem 2.6 *Let R be a latin rectangle of size $r \times r$ on n symbols, with $r \geqslant 10$ and $n \geqslant 2r+1$, and let f be a non-negative integral-valued function on the symbols such that $\sum_\sigma f(\sigma) = n - r$. Then R can be embedded in a latin square of side n on the same symbols in which each symbol σ occurs $f(\sigma)$ times on the diagonal outside R, if and only if the following hold:*

(a) *$R(\sigma) \geqslant 2r - n + f(\sigma)$ for all symbols σ,*

(b) *$R(\sigma) = n \Rightarrow f(\sigma) \neq n - r - 1$ for all symbols σ, and*

(c) *if R is a latin square on symbols $\sigma_1, \ldots, \sigma_r$ and $n = 2r+1$, then $\sum_{i=1}^{r} f(\sigma_i) \neq 1$.*

Interestingly, and with an analogous phenomenon occurring for embeddings of properly edge-coloured graphs, the Ryser condition (a) alone is sufficient for all $n \geqslant r$ if at least one cell on the diagonal is left unprescribed [3].

As for *symmetric* latin rectangles with prescribed diagonal, the remark from subsection 2.1 must be taken into account. If the diagonal is admissible with respect to this, the Ryser condition is sufficient. This was proved independently by Andersen [2] and Hoffman [35].

2.5 Generalised latin rectangles

In [5] and [6], Andersen and Hilton introduced *generalised latin rectangles*, where each cell can contain more than one symbol, and each symbol can occur more than once in each row and column. With their definitions, it was possible to view some rectangles as obtained from others by merging sets of rows and sets of columns (as well as sets of symbols), and to study when the reverse process was feasible.

The merging results of these two papers dealt with very uniform amalgamation, where the set of columns was partitioned into subsets of the same size, and the columns of each subset were then merged to form one new column, and similarly for rows. But for the hardest embedding result, the best proof the authors found actually operated by having simple columns and just one big amalgamated column, much like the main application of the method of amalgamations today. Thus this seems to be the origin of the method. In the survey [4] written after, but published before [5] and [6], Andersen and Hilton presented general amalgamations for latin squares, where different amalgamated columns could originate from a different number of original columns, and likewise for rows.

3 Embedding proper edge-colourings into proper edge-colourings

3.1 Embedding a properly edge-coloured K_r into a properly edge-coloured K_n

A direct graph-theoretic translation of Theorem 2.4 concerns complete graphs with loops. We denote the complete graph with n vertices with one loop on each vertex by K_n^l and call a colouring of some or all of its loops *admissible* if, when n is odd, all coloured loops have distinct colours, and, when n is even, the number of uncoloured loops is at least as great as the number of colours each occurring on an odd number of loops. Taking into account the generalisation due to Andersen and to Hoffman, we can state the following graph theoretic result.

Theorem 3.1 *In K_n^l, let a subgraph K_r^l have a proper edge-colouring with n colours, and assume that some loops outside K_r^l are coloured, so that the loop-colouring is admissible. Then the edge-colouring can be completed to a proper edge-colouring of K_n^l if and only if each colour c occurs on at least $2r-n+l(c)$ vertices of K_r^l, where $l(c)$ is the number of loops outside K_r^l precoloured with colour c.*

In this paper we wish to focus on results about complete graphs without loops, and here a distinction between complete graphs of even and odd order is necessary.

If n is even, the minimum number of colours in a proper edge-colouring of K_n is $n-1$, with every colour occurring at every vertex in such a colouring. The colouring is a 1-factorisation of the complete graph.

If n is odd the minimum number of colours is n, with every colour being absent from exactly one vertex. Here the colouring is a near-1-factorisation of the graph.

Because of the greater regularity of the colouring for even order, and because this usually makes it the harder case, there is a tendency to give this more attention than odd order. We shall, however, consider both in the following.

We obtain the following corollary from Theorem 3.1.

Corollary 3.2 (i) *A properly edge-coloured K_r with $2m-1$ colours can be embedded in a properly edge-coloured K_{2m} with $2m-1$ colours if and only if each colour occurs on at least $r-m$ edges of K_r.*
(ii) *A properly edge-coloured K_r with $2m-1$ colours can be embedded in a properly edge-coloured K_{2m-1} with $2m-1$ colours if and only if each colour occurs on at least $r-m$ edges of K_r.*

Proof (i) We use Theorem 3.1, requiring all loops to have the same new colour, making the total number of colours $2m$, as required. The loop-colouring is clearly admissible, the embeddings of the theorem and the corollary are equivalent, and the conditions on the $2m-1$ old colours are the same. The condition of the theorem on the new colour is satisfied, as it occurs on r loops, must occur on $2m-r$ further loops, and $r = 2r - 2m + (2m-r)$.
(ii) This follows from statement (i) by observing that embedding into K_{2m-1} is equivalent to embedding into K_{2m}, as the colour missing at each vertex in the first case can be used on an edge from the vertex to the extra vertex of K_{2m}. □

The second part of Corollary 3.2 readily provides an affirmative answer to sample question 1 of Subsection 1.1.

When considering ordinary complete graphs, without loops, there is no obvious analogue to the diagonal of latin squares. It turns out, however, that the problem of embedding latin squares (not symmetric!) with additional prescribed cells on the diagonal has a very close analogue in the problem of embedding properly edge-coloured complete graphs with additional prescribed colours on mutually independent edges outside the given complete graph.

The analogy does not involve any direct translation from one problem to the other — after all, the edge-coloured graphs would correspond to symmetric squares, where the diagonal is easy to handle — but manifests itself on two other accounts. The first is the usefulness of the embedding in solving other problems. In [8], Andersen and Hilton proved a symmetric analogue of the Evans conjecture, and although they work with graphs with loops, they also consider ordinary graphs, and it is clear that additional independent edges play an important part (see also [9]). The second is the kind of difficulty that is encountered, especially when embedding into a complete graph of even order. For example, the Ryser type condition is not sufficient in general, but it it is sufficient if one diagonal cell is left unprescribed, and if one vertex is left with no prescribed edge incident with it.

3.2 Embedding with free edges

Given a K_r as a subgraph of K_n, let us call a set of edges that are mutually independent and belong to $K_n \setminus K_r$ *free edges*. We are concerned with embedding a properly edge-coloured K_r with some free edges having prescribed colours.

Suppose for a while that we embed into a complete graph K_{2m} of even order. Then each colour must occur on every vertex. We deduce the necessary Ryser type conditions. Assume that the colour c is prescribed on $e(c)$ free edges. Then, in the embedding, it can occur on at most $2m - r - 2e(c)$ edges joining $K_n \setminus K_r$ to K_r, and so it must occur on at least $r - (2m - r - 2e(c)) = 2r - 2m + 2e(c)$ vertices in the edge-colouring of K_r, and so on at least $r - m + e(c)$ edges of K_r.

Andersen and Hilton [9] proved that these conditions are sufficient, if the graph consisting of K_r and the precoloured free edges does not include all vertices of K_{2m}.

Theorem 3.3 *Let G be a subgraph of K_{2m} consisting of a K_r, possibly some free edges, and at least one isolated vertex, and assume that G has a proper edge-colouring with $2m - 1$ colours. For each colour c, let $e(c)$ be the number of free edges of colour c. Then the edge-colouring of G can be completed to an edge-colouring of K_{2m} with $2m - 1$ colours if and only if each colour c occurs on at least $r - m + e(c)$ edges of K_r.*

The way to prove this theorem, and also Theorem 3.5 below, is to extend the edge-colouring of K_r to one of K_{r+2}, including one of the original free edges, also satisfying the conditions, and thus proceed by induction. Naturally, it must be checked that this results in a situation (where the whole graph is coloured, or where there are no more free edges, or a similar case) in which the conditions imply that the desired embedding has been obtained or is readily obtainable. The extension is found by considering a bipartite graph with the vertices of K_r as one bipartition class, and a vertex for each colour in the other, a colour being joined to a K_r-vertex if and only if it does not occur at that vertex in the edge-colouring of K_r. In this graph we must find two disjoint matchings M_1 and M_2, each incident with all K_r-vertices. Then we can colour the edge from one of the new vertices v_1 to each vertex u of K_r by the colour joined to u in M_1, and colour the edges incident with the other new vertex v_2 in the same way, using M_2. The difficulty is that M_1 and M_2 must satisfy additional conditions; an obvious one is that neither can include the vertex corresponding to the colour of the added free edge, as this would make that colour occur twice at one of the new vertices.

A lucky corollary of Theorem 3.3 is that the case of embedding into complete graphs of odd order is solved completely: the Ryser condition is necessary and sufficient. There is no requirement that the graph has an isolated vertex.

Corollary 3.4 *Let G be a subgraph of K_{2m-1} consisting of a K_r and possibly some free edges, and assume that G has a proper edge-colouring with $2m-1$ colours. For each colour c, let $e(c)$ be the number of free edges of colour c. Then the edge-colouring of G can be completed to an edge-colouring of K_{2m-1} with $2m-1$ colours if and only if each colour c occurs on at least $r-m+e(c)$ edges of K_r.*

Proof Add an isolated vertex to G and note that the desired embedding is equivalent to an embedding of the extended G into the K_{2m} spanned by K_{2m-1} and the new vertex. Now apply Theorem 3.3. □

If, in Theorem 3.3, the condition that the graph has an isolated vertex is dropped, the theorem becomes false. A simple counterexample is obtained if m is even by letting $r=m$, giving K_m an edge-colouring where some colour c_1 occurs at every vertex, some colour c_2 occurs on at least one edge, with no further requirement on the colouring, and then having $e(c_1)=\frac{m}{2}-1$, $e(c_2)=1$, and $e(c)=0$ for all other colours c. Then the Ryser condition is satisfied (it is void except for c_1 and c_2). But the edge-colouring cannot be completed, because the colour c_1 occurs everywhere except at the two end-vertices of the free edge to be coloured c_2, and so it cannot be made to occur at these.

Put differently, if the K_r and the free edges are allowed to span all of the K_{2m}, some additional requirement must be added to the Ryser condition.

Dugdale and Hilton found a further necessary condition for this case and suggested that — together with the Ryser condition — it was also sufficient. Recently, this was confirmed by Henderson and Hilton.

Theorem 3.5 ([25]) *Let r be even and let G be a subgraph of K_{2m} consisting of a K_r and $\frac{1}{2}(2m-r)$ free edges, and assume that G has a proper edge-colouring with $2m-1$ colours. For each colour c, let $e(c)$ be the number of free edges of colour c, and let $R(c)$ be the number of edges of K_r of colour c. Then the edge-colouring of G can be completed to an edge-colouring of K_{2m} with $2m-1$ colours if and only if*

(i) *$R(c) \geqslant r-m+e(c)$ for each colour c, and*

(ii) *if $2m-r \geqslant 4$ there is no set A of colours such that*

(ii.1) *$\sum_{c\in A} e(c) = \frac{1}{2}(2m-r-2)$,*

(ii.2) *$\sum_{c\in A}(r-2R(c)) = (2m-r)(|A|-1)$, and*

(ii.3) *$|J(A)| = |A|-1$,*

where $J(A)$ denotes the set of vertices of K_r that are not incident with an edge of every colour $c \in A$.

Condition (ii.2) can also be stated as follows: each colour in A has equality in condition (i), except for one colour c which has $R(c)$ exceeding the bound by one.

In the example described before the theorem, the set $A = \{c_1\}$ would violate condition (ii).

To see the necessity of condition (ii), suppose that the condition fails so that a set A as described exists, but that completion is possible, and consider the K_{r+2} spanned by K_r and the free edge, e say, of colour not in A (e exists by (ii.1)) in such a completion. Since condition (i), with $r+2$ in place of r, must be satisfied in this K_{r+2}, and by the remark above regarding (ii.2), all colours of A but one must each occur on two edges joining K_r to the end-vertices of e, and the last colour of A must occur on one such edge. But this requires at least $|A|$ vertices of K_r available for a new edge with a colour from A, which is impossible by (ii.3).

Before Henderson and Hilton obtained Theorem 3.5, but after Dugdale and Hilton had conjectured it, Allen [1] proved it true in the case where each colour either occurred on at least two free edges, or on no free edges at all.

Henderson's and Hilton's theorem is quite a big achievement in this area, and it gives hope that some of the remaining questions concerning embeddings of latin squares and edge-coloured graphs are within reach. Two such questions are embeddings of latin rectangles of size $r \times r$ into latin squares of side n with prescribed diagonal for $n \leqslant 2r$, and embeddings of graphs with loops in the case where a set of mutually independent edges *and* loops outside K_r have prescribed colours.

The question about embedding proper edge-colourings with free edges is one case that so far has not lent itself to treatment with the otherwise very powerful method of amalgamations. We now turn to situations where this method works well.

4 The method of amalgamations

4.1 The basic idea

As described informally in Section 1.3, amalgamations are graph homomorphisms applied to graph decompositions.

The technique begins by imagining a partition P of the edges of some target graph G so that the subgraph induced by each element of P has some desirable characteristic, such as being connected (sample question 2) or being a Hamiltonian cycle (sample question 3). The partition P is usually described by an edge-colouring: two edges receive the same colour if and only if they are in the same element of P. The target graph is usually taken to be the complete graph K_n, or perhaps a complete multipartite graph. The aim is to move beyond imagining and actually construct such a graph decomposition.

To proceed further, a formal definition is in order. Given two graphs G and H, $f : V(G) \to V(H)$ is said to be an *amalgamating* function of G if f is surjective and if there exists a bijection $\phi : E(G) \to E(H)$ that satisfies:

(a) if $v, w \in V(G)$ and if e joins v and w then $\phi(e)$ is an edge joining $f(v)$ to $f(w)$ in H (so if $f(v) = f(w)$ then $\phi(\{v, w\})$ is actually a loop in H); and

(b) if e is a loop on $v \in V(G)$ then $\phi(e)$ is a loop on $f(v)$ in H.

H is said to be an *amalgamation* of G, and G is said to be a *disentanglement* of H. So one can dynamically envision the amalgamation H as being formed by partitioning the vertices of G, then amalgamating the vertices occurring in the same element of the partition into a single vertex, the edge ends clinging to their incident vertices as the fusing of the vertices takes place. This makes it clear that H naturally inherits an edge-colouring from an edge-colouring of G (formally this is so since ϕ is a bijection). It is also worth defining the amalgamation number $\eta(h)$ for each vertex $h \in V(H)$ to be $|f^{-1}(h)|$; so continuing with our visual description, $\eta(h)$ is the number of vertices in G that are amalgamated together to form the vertex h in H.

Now the idea is to try to describe the edge-coloured H without having the specific edge-coloured graph G at hand. It may even be that G does not exist, but that does not stop us trying to characterise H!

For example, suppose that the graph decomposition we have in mind is a k-factorisation of K_n. So $G = K_n$, and its edges are coloured so that each colour class induces a k-factor; so certainly k must divide $n - 1$. Then without seeing G surely we could make the following observations about H:

(i) each vertex h is incident with $\binom{\eta(h)}{2}$ loops,

(ii) each pair of distinct vertices h_1 and h_2 is joined by exactly $\eta(h_1)\eta(h_2)$ edges,

(iii) each vertex h is incident with exactly $k\eta(h)$ edges of each colour when we count loops twice, and

(iv) $\sum_{h \in V(H)} \eta(h) = n$.

One could consider these four properties as defining an outline sketch of what an amalgamation of G would look like; so any graph satisfying (i)–(iv) is said to be an *outline* k-factorisation of K_n. Then one can prove the following result [6, 7, 54]; see also Subsection 4.2, and see [19] for a generalisation.

Theorem 4.1 *Every outline k-factorisation is the amalgamation of some k-factorisation of K_n.*

With Theorem 4.1 in hand, it is now possible to establish the existence of a k-factorisation of K_n by describing *any* outline k-factorisation, H. This is most simply done by letting H have a single vertex h, defining $\eta(h) = n = kx + 1$ for some x, and placing exactly $kn/2$ loops of each of x colours on h (it follows from (iii) and (iv) that kn must be even).

Of course, directly finding a k-factorisation of K_n is easy to do; nevertheless this example demonstrates the general use of amalgamations in settling the existence of certain graph decompositions:

(Step 1) find some properties that must be satisfied by all amalgamations of the edge-coloured target graph;

(Step 2) show that any "outline" graph that satisfies these properties is in fact the amalgamation of some target graph with the desired edge-decomposition; and

(Step 3) find one example of such an outline graph.

Obviously there is some work involved in establishing Step 2! So the gain over simply finding the desired graph decomposition directly is that hopefully establishing Step 3, finding the outline graph, is *much* simpler.

As mentioned before, there is also a second use for which this technique works well, namely the embeddings of partial graph decompositions. To see this, we continue examining k-factorisations of K_n. Suppose that you have been given H_1, an edge-coloured copy of K_v; is it possible to construct a k-factorisation of K_n (colouring its edges to describe the k-factorisations) with the property that it contains a subgraph isomorphic to H_1 (with the *same* edge-colouring)? This kind of question lends itself beautifully to an answer obtained by using amalgamations. We can very easily make use of properties (i)–(iv) to complete Step 1, and use Theorem 4.1 to complete Step 2. So it remains to find an outline graph, H.

Before describing H, we turn to one more issue that arose in finding k-factorisations of K_n, namely that sometimes it is not possible! So at some stage (like now!) one must find conditions that are necessary for the desired embedding to exist. In this case it is clear that

(N1) each vertex in H_1 is incident with at most k edges of each colour, since H_1 is to be a subgraph of K_n in which each vertex is incident with exactly k edges of each colour.

Also we need that

(N2) if n is odd then k is even, since each k-factor must have an even number of vertices of odd degree.

Finally, mirroring condition (a), we need to make sure that a colour class in H_1 is not so small that it forces one of the added vertices to have degree more than k. To be precise, if the smallest colour class has ϵ edges, then there must be $kv - 2\epsilon$ edges joining vertices in H_1 to the "new" $n - v$ vertices, so it must be the case that

(N3) $kv - 2\epsilon \leqslant k(n - v)$.

Knowing that we can assume such conditions is crucial in establishing that our outline graph H can indeed be formed, as we will now see.

Again, taking the simplest approach to form H works well. Start with H_1, which we know has to be in our K_n anyway, and for each vertex h in H_1 let $\eta(h) = 1$ – these are vertices that we know are already in their "final state". Now add one more vertex α with $\eta(\alpha) = n - v$; so α "contains" the remaining vertices. What are the added edges? The good news is that we really have no choice! Focusing mainly on property (iii), each vertex H_1 must be joined with enough additional edges of each colour so that in H it is incident with exactly k edges of each colour; necessary condition (N1) makes this possible, ensuring that the number of edges to be added is non-negative. Next, if there are ϵ_i edges coloured i in H_1, then we just added $kv - 2\epsilon_i$ additional edges coloured i joining vertices in H_1 to α, so we need to place $(k(n - v) - (kv - 2\epsilon_i))/2$ loops coloured i on α to boost the number of edges coloured i incident with α to $k\eta(\alpha)$ (recalling that each loop contributes 2 to this number). The fact that this number of loops is an integer follows from (N2), and that it is non-negative from (N3).

It is now easy to check that conditions (i)–(iv) are satisfied by H, and so by Theorem 4.1 we have necessary conditions that are also sufficient to guarantee that the desired embedding of H_1 does actually exist.

4.2 Using edge-colourings

In this section we discuss in more detail just how Step 2 is often accomplished. That is, given an outline decomposition of some sort, how do we use its existence to establish the existence of the decomposition itself? The answer is to disentangle the outline structure! This is accomplished with an inductive proof, pulling the vertices out of each amalgamated vertex v one by one, at each step guaranteeing that the resulting edge-coloured graph also satisfies the defining properties of an outline decomposition. Of course, when a vertex is disentangled from v, the amalgamation number of v is decreased by 1, and the new vertex is assigned an amalgamation number of 1. This process is continued until a graph is reached in which each vertex has an amalgamation number of 1. The defining properties of the outline decomposition of interest must be chosen in such a way that *every* outline decomposition in which each vertex has an amalgamation number of 1 is a decomposition of the desired type. Then clearly this process ensures that Step 2 can be completed.

The most common method for achieving one inductive disentangling step is to make use of the following neat result, proved by de Werra [58] without property (3), which can be added using methods of McDiarmid [49]. Hilton and Rodger [32] give a proof of this.

Theorem 4.2 *For any bipartite graph B and for any integer $k \geqslant 1$, there exists an edge-colouring of B with colours $1, \ldots, k$ in which for $1 \leqslant i < j \leqslant k$ and for each $u, v \in V(B)$:*

(1) $|c_i(v) - c_j(v)| \leqslant 1$,

(2) $|c_i(u,v) - c_j(u,v)| \leqslant 1$, *and*

(3) $|c_i(B) - c_j(B)| \leqslant 1$,

where $c_i(v)$ is the number of edges coloured i incident with v, $c_i(u,v)$ is the number of edges coloured i that join u and v, and $c_i(B)$ is the number of edges coloured i in B.

Edge-colourings that satisfy property (1) are said to be *equitable*, those that satisfy (1) and (2) are said to be *balanced*, and those that satisfy (3) are said to be *equalised*.

By letting k be the maximum degree of the graph, we can obtain from Property (1) Kőnig's well known result, mentioned in Section 2.2, about proper edge-colourings of bipartite graphs.

To see how to make use of such edge-colourings, we return to outline k-factorisations, defined by properties (i)–(iv) in Section 4.1. Let G be an outline k-factorisation of K_n, and let v be a vertex with amalgamation number $\eta(v) > 1$. To disentangle one vertex from v, form a bipartite graph B with bipartition $\{\{\ell\} \cup V(G) \setminus \{v\}, \{c_1, \ldots, c_k\}\}$ by joining each vertex $w \in V(G) \setminus v$ to c_i for $1 \leqslant i \leqslant k$ with x edges if and only if there are x edges coloured i joining v and w in G, and by joining vertex ℓ to c_i for $1 \leqslant i \leqslant k$ with $2x$ edges if and only if there are x loops coloured i incident with v in G. Then properties (i)–(iv) tell us the degrees of the vertices in B:

(a) $d_B(\ell) = \eta(v)(\eta(v) - 1)$ (by (i)),

(b) $d_B(w) = \eta(v)\eta(w)$ for all $w \in V(G) \setminus \{v\}$ (by (ii)), and

(c) $d_b(c_i) = k\eta(v)$ for all i, $1 \leqslant i \leqslant k$ (by (iii)).

Now let $f : E(B) \to \{1, 2, \ldots, \eta(v)\}$ be a balanced edge-colouring of B with $\eta(v)$ colours, and let B_1 be the subgraph of B induced by the edges coloured 1. Then since f is equitable we know from (a)–(c) that

(a′) $d_{B_1}(\ell) = \eta(v) - 1$,

(b′) $d_{B_1}(w) = \eta(w)$, and

(c′) $d_{B_1}(c_i) = k$.

Furthermore, since f is balanced and since $\eta(v) \geqslant 2$, property (2) of Theorem 4.2 ensures that at most half of the edges joining ℓ to c_i are coloured 1 in B.

This is a crucial point since it guarantees that the number of edges joining ℓ to c_i in B_1 is at most the number of loops coloured i on v in G; recall that each loop on v in G is represented by *two* edges incident with ℓ in B.

So now we disentangle one vertex v_1 from v in G to form G_1, with $V(G_1) = V(G) \cup \{v_1\}$, as follows.

(a$''$) If there are x edges joining ℓ to c_i in B_1 then remove x loops coloured i on v in G and join v to v_1 with x edges coloured i.

(b$''$) If there are x edges joining $w \in V(G) \setminus \{v\}$ to c_i in B_1 then remove x edges coloured i joining v to w in G and add x edges coloured i joining v_1 to w.

We should now check that this procedure does indeed produce an outline k-factorisation of K_n with an associated amalgamation number function η_1 defined by

$$\eta_1(h) = \begin{cases} \eta(h) - 1 & \text{if } h = v, \\ 1 & \text{if } h = v_1, \text{ and} \\ \eta(h) & \text{otherwise.} \end{cases}$$

So we check properties (i)–(iv) of an outline k-factorisation in turn. Since in G_1 the number of edges and loops of each colour incident with each vertex other than v and v_1 is exactly the same as in G, we need consider only v and v_1. Vertex v is incident with $\binom{\eta(v)}{2} - (\eta(v) - 1) = \binom{\eta(v)-1}{2} = \binom{\eta_1(v)}{2}$ loops (by (a$'$) and (i)) and $\eta(v_1)$ is incident with $0 = \binom{\eta_1(v_1)}{2}$ loops; so G_1 satisfies (i). Vertices v and v_1 are joined by $\eta(v) - 1 = \eta_1(v)\eta_1(v_1)$ edges (by (a$'$)), vertices v_1 and $w \in V(G) \setminus \{v_1\}$ are joined by $\eta(w) = \eta_1(w)\eta_1(v_1)$ edges (by (b$'$)), and v and w are joined by $\eta(v)\eta(w) - \eta(w) = \eta(w)(\eta(v) - 1) = \eta_1(w)\eta_1(v)$ edges (by (ii) and (b$'$)); so (ii) is satisfied by G_1. By (c$'$), v_1 is incident with k edges of each colour, and in G_1, v is incident with $k\eta(v) - k = k\eta_1(v)$ edges of each colour (by (c$'$) and (iii)); so G_1 satisfies (iii). Since also (iv) is clearly satisfied by G_1, G_1 is indeed an outline k-factorisation, and so Step 2 is complete.

The main point here is that to disentangle G, the requirement is to find a subset of edges of G that are selected "fairly" in all sorts of senses: a fair selection of edges at each vertex, of loops at each vertex, of colours on the selected edges at each vertex. Often balanced edge-colourings are the perfect tool for achieving such aims, but we will see in Subsection 4.5 that sometimes more powerful tools are needed.

We close this subsection by mentioning another edge-colouring result that has been crucial in several of the theorems surveyed in this paper. It was proved by Häggkvist and Johansson [23] in 1994 after being conjectured by Hilton and Rodger [32].

Theorem 4.3 *Let G be a connected graph where all vertices have degree $2k$ or $2k-2$, for some $k > 1$. Then G admits an equitable and equalised edge-colouring with k colours if and only if the number of vertices of degree $2k-2$ is either 0 or at least 2, and not an odd number if $k = 2$.*

Henderson and Hilton used this result in their proof of Theorem 3.5.

4.3 Connectivity issues

One could easily argue that the solution of the embedding problem for partial triple systems of even index described in Section 5 is the prized result obtained using amalgamations. But probably the most effective use of amalgamations has come when studying graph decompositions that involve Hamiltonian cycles, and it is this advance that we consider in this subsection.

In view of the results concerning k-factorisations that are described in Subsection 4.1, it is clear that attacking problems involving Hamiltonian cycles is certainly within our sights, since such cycles are just 2-factors that are also *connected.* It was this insight and a beautiful way of dealing with the connectivity issue that Hilton brought into play when he extended Theorem 4.1 in [27]. Clearly the disentangling of any outline k-factorisation of K_n will never increase the edge-connectivity of any colour class (that is, the subgraph induced by the edges of a specified colour). Indeed, the disentangling process could easily lower the edge-connectivity of any colour class, and even disconnect it! So if the intent is to disentangle an outline 2-factorisation of K_n and thereby produce a Hamiltonian decomposition of K_n (that is, a 2-factorisation in which each 2-factor is a Hamiltonian cycle), then of course we need to add the necessary condition

(N4) each colour class of the outline 2-factorisation is connected

to conditions (N1)–(N3) described in Subsection 4.1. Define an outline Hamiltonian decomposition of K_n to be an outline 2-factorisation that also satisfies condition (N4). So now, armed with the new condition (N4) we at least have a chance to accomplish our goal, but then as we peel off vertices one by one from a vertex v with $\eta(v) > 1$, we need to make sure that each colour class remains connected. To deal with this issue, it is worth thinking about how we could possibly go astray!

We focus first on one colour class: let $\mathcal{C}(G)$ be the graph induced by the edges coloured c in the outline Hamiltonian decomposition G of K_n. First notice that each vertex v has even degree; namely $d_{\mathcal{C}(G)}(v) = 2\eta(v)$ using condition (iii). This means that the size of any edge-cut $\mathcal{E}$ in $\mathcal{C}(G)$ must be *even* because the sum of the degrees of the vertices in each component of $\mathcal{C}(G) - \mathcal{E}$ must be even. So in particular, if we are attempting to peel v_1 out from v to form G_1, each component of $\mathcal{C}(G) - v$ is joined to v by either 2 edges or at least 4 edges. Therefore, since v_1 is incident with *exactly* 2 edges coloured c in G_1,

> the only way that $\mathcal{C}(G)$ could be disconnected is if the 2 edges moved from v to v_1 when disentangling G actually form an edge-cut of size 2 in $\mathcal{C}(G)$.

With this observation in mind, we can now see that one way to ensure that each colour class remains connected in G_1 would be to adopt the following strategy:

> when choosing the edges coloured c to be moved from v to v_1 make sure that *at most half* of the edges joining each component of $\mathcal{C}(G) - v$ to v are selected.

This may seem like a strange way to approach the problem. However, it turns out that this approach works well even in the more general setting of k-factorisations, and, as is seen below, guarantees that our objective can be achieved.

How do we make sure that the two edges coloured c chosen to move from v to v_1 do not form a cut-set in $\mathcal{C}(G)$? It turns out that we can avoid this possibility by considering the bipartite graph B_2 formed by the edges induced by *two* colour classes in B, say the edges coloured either 1 or 2. Then, since f is equitable, we have that

(a‴) $d_{B_2}(\ell) = 2(\eta(v) - 1)$,

(b‴) $d_{B_2}(w) = 2\eta(w)$, and

(c‴) $d_{B_2}(c_i) = 4$.

We now make the crucial step of forming a bipartite graph B_3 by disentangling each vertex c_i in B_2 into two vertices each of degree 2; this is done so that if two edges incident with c_i in B_2 correspond to edges that form an edge-cut in $\mathcal{C}_i(G)$, the graph induced by the edges of colour c_i in G, then they are still adjacent in B_3. So now what happens if B_3 is given an equitable 2-edge-colouring? If we let B_1 be the subgraph of B_3 induced by the edges coloured 1 then clearly it satisfies conditions (a′), (b′) and (c′). Furthermore, if an edge-cut of size 2 in $\mathcal{C}_i(G)$ corresponds to edges e_1 and e_2 in B_3, then by our construction e_1 and e_2 are adjacent in B_3, and so receive different colours in the 2-edge-colouring of B_3. Therefore at most one of e_1 and e_2 occurs in B_1; thus we have avoided disconnecting $\mathcal{C}_i(G)$ in forming G_1. Using this method, Hilton [27] proved the following theorem.

Theorem 4.4 *Every outline Hamiltonian decomposition of K_n is the amalgamation of some Hamiltonian decomposition of K_n.*

Following this result, Hilton and Rodger extended it to Hamiltonian decompositions of complete multipartite graphs [31].

It is also natural to see if this approach can be extended to consider the connectivity of each k-factor in a k-factorisation of K_n for values of $k > 2$. An outline k-factorisation is said to be m-edge-connected if each colour class is m-edge-connected. It turns out that a much more complicated argument, but one which uses the same tools and the same method, can be used to ensure that each colour class begins and remains 2-edge-connected throughout the disentangling process. So Rodger and Wantland eventually obtained a generalisation of Theorem 4.4.

Theorem 4.5 ([54]) *Every outline 2-edge-connected k-factorisation of K_n is the amalgamation of some 2-edge-connected k-factorisation of K_n.*

In our minds, these results are shown in their best light when they are applied to the embedding problem. Thus Theorem 4.4 immediately yields the following beautiful corollary.

Corollary 4.6 ([27]) *Let G be any edge-coloured copy of K_n that uses at most k colours. Then G can be embedded in a k-edge-coloured copy of K_{2k+1} in which each colour class is a Hamiltonian cycle if and only if*

(a) *each component of each of the k colour classes is a path (possibly of length 0), and*

(b) *each colour class has at most $2k+1-n$ components.*

From this corollary follows the affirmative answer to sample question 3 of Subsection 1.1 (and hence to sample question 2 as well).

A companion corollary giving necessary and sufficient conditions for an edge-coloured K_n to be embedded in a 2-edge-connected k-factorisation of K_x follows similarly from Theorem 4.5, and can be found in [54].

4.4 Multipartite graphs

The method of amalgamations has been extremely useful when considering graph decompositions of complete multipartite graphs. Let $K(a_1, \ldots, a_p)$ denote the simple graph with vertex set being the disjoint union of p *parts* $A_1 \cup \cdots \cup A_p$, where $|A_i| = a_i$, and where two vertices u and w are joined if and only if they occur in different parts. Let K_a^p denote this graph when $a_1 = \cdots = a_p = a$. In this setting most progress has been made by considering outline decompositions that have *no loops.* Clearly this is different from the complete graph setting! Such outline graphs can be obtained by amalgamating vertices that occur in the same part. Restricting one's attention to outline decompositions of this sort has the advantage of making Step 2 much simpler than if the general situation was considered. To date, it is turning out to be difficult to prove that, for example, outline k-factorisations of complete multipartite graphs that contain loops are indeed amalgamations of k-factorisations of K_a^p. Step 2 has been successfully addressed for such outline decompositions in the case where $a = 2$, but more work remains to be done here.

Of course, the difficulty in restricting Step 2 to loopless outline decompositions is that it makes Step 3 more difficult to achieve. To see this, consider the problem of finding maximal sets of Hamiltonian cycles in K_a^p:

> for which values of m does there exist a set S of m edge-disjoint Hamiltonian cycles in K_a^p with the property that if the edges in the cycles in S are deleted from K_a^p then the resulting graph is non-Hamiltonian?

This question has been addressed in several papers [14, 18, 39], in each case making use of the following result.

Theorem 4.7 *Every loopless outline Hamiltonian decomposition of a subgraph of K_a^p is the amalgamation of a Hamiltonian decomposition of some subgraph of K_a^p.*

Maximal sets of Hamiltonian cycles in K_a^p are then obtained by finding an outline Hamiltonian decomposition G of K_a^p of a very special type. Almost always it is possible to form such a graph on the $2p$ vertices $u_1, \ldots, u_p, v_1, \ldots, v_p$ in which u_i and v_j are joined by exactly $\eta(u_i)\eta(v_j)$ edges when $i \neq j$, and by 0 edges if $i = j$ (so the vertices in the ith part of the copy H of K_a^p eventually produced after disentangling G all come from u_i or v_i). The point of this additional property is that once G is disentangled, it *must* be the case that no matter how this is done, *all* the edges in H joining vertices disentangled from $u_1, \ldots, u_p$ to those disentangled from $v_1, \ldots, v_p$ occur in Hamiltonian cycles. So if the edges in these Hamiltonian cycles are removed from K_a^p then the resulting subgraph of K_a^p is disconnected, and thus is clearly non-Hamiltonian.

Constructing this edge-coloured graph G with each vertex v having degree $2m\eta(v)$ ($2\eta(v)$ edges of each of the m colours) in the myriad of cases to consider is no easy feat. Indeed the problem is yet to be completely solved.

Theorem 4.8 ([14, 18, 39, 48]) *There exists a maximal set of m Hamiltonian cycles in K_a^p if and only if*

(a) $\lceil a(p-1)/4 \rceil \leqslant m \leqslant \lceil a(p-1)/2 \rceil$, *and*

(b) $m > a(p-1)/4$ *if*

  (i) a *is odd and* $p \equiv 1 \pmod 4$, *or*

  (ii) $p = 2$, *or*

  (iii) $a = 1$,

except possibly if all three of the following conditions hold:

$$a \geqslant 3 \quad \textit{is odd,} \quad p \quad \textit{is odd,} \quad \textit{and} \quad m \leqslant ((a+1)(p-1)-2)/4\,.$$

This result is hard to prove since Step 3 is made difficult by the severe limitation Theorem 4.7 places by requiring G to be loopless. In just one case, namely when $a = 2$, proving Step 2 when G has loops has been accomplished (see [48]). This increase in the power of the result obtained in Step 2 allows us to accomplish Step 3 easily by finding G with just 2 vertices; so Step 3 suddenly becomes essentially trivial.

Briefly, the difficulty in generalising Theorem 4.7, by allowing G to have loops, lies in automating the process of recognizing, as v_1 is disentangled from v, where the other vertices that are in the same part as v_1 currently lie. Of course, v_1 cannot be adjacent to any such vertices, so this is an important consideration.

There is one further captivating result that must be mentioned here. Not only is the result memorable, but at this point it is not at all clear how it could be proved in any other way; so its proof using amalgamations is a fine testament to the power of this technique. It also reinforces the theme, prevalent in this section, of sharing edges fairly among the various colour classes. A Hamiltonian decomposition of K_a^p is said to be *fair* if for every pair of parts A_i and A_j, the edges joining vertices in A_i to vertices in A_j are shared out as evenly as possible among the colour classes. Then one can prove the following result of Leach and Rodger.

Theorem 4.9 ([43]) *For all $a, p \geqslant 1$, there exists a fair Hamiltonian decomposition of K_a^p if and only if $a(p-1)$ is even.*

As one might expect by now, the proof follows Steps (1)–(3), with Step 2 essentially being Theorem 4.7, so the work required is the completion of Step 3. In this case, obtaining the required fair outline Hamiltonian decomposition of K_a^p is really quite neat and worth the read!

4.5 More on connectivity

As described in Subsection 4.3, existence and embedding results for 2-edge-connected k-factorisations have been successfully obtained using amalgamations. If connectivity is ignored, then the situation has been addressed in Subsections 4.1 and 4.2. But the reader is probably wondering why is there a jump from 0-edge-connected to 2-edge-connected? What happens if we are looking for necessary and sufficient conditions for an outline k-factorisation to be the amalgamation of some k-factorisation of K_n in which each k-factor is connected (or 1-edge-connected if you prefer).

The answer is that this turns out to be far more complicated than the 2-edge-connected case. The reason for this is mainly that in the 2-edge-connected case, for each colour class c of G in turn we can always focus on the edges joining v to each of the components in $\mathcal{C}(G) - v$ one by one. What happens to one such component is largely independent of what happens to the other components.

This is far from the case when we look for connected k-factors. We should note, however, that when k is even a k-regular graph can have no cut-edge, and so 2-edge-connectedness is equivalent to connectedness. But when k is odd, it is quite possible for $\mathcal{C}(G) - v$ to have many components that are each joined to v by a single edge; that is, a cut-edge in $\mathcal{C}(G)$. If we mistakenly select k such edges to move to v_1 when forming G_1 then we would disconnect $\mathcal{C}(G_1)$. So we may have to be extremely efficient about how we use the edges joining components of $\mathcal{C}(G) - v$ to v with more than one edge; call these the *gluing* edges! For example, it would be most efficient to move $k - 1$ cut-edges in $\mathcal{C}(G)$ to v_1 together with 1 gluing edge. It could even be that $\mathcal{C}(G)$ may have so many cut-edges that such efficiency would be *required*! This approach is enough to guarantee that $\mathcal{C}(G_1)$ is connected.

After many years of tackling this issue, a very complicated argument was found that *bundled* edges together in careful ways, and then fair distributions of edges within such bundles amongst the various colour classes achieved the desired result. But then a jewel of a result was brought to the scene. This idea of bundling edges was inherent in the following powerful result of Nash-Williams. A family $\mathcal{F}$ of sets is said to be a *laminar set* if for all $X, Y \in \mathcal{F}$, we have $X \subseteq Y$, $Y \subseteq X$ or $X \cap Y = \emptyset$.

Theorem 4.10 ([50]) *If $\mathcal{F}_1$ and $\mathcal{F}_2$ are laminar sets of subsets of a finite set M and if h is a positive integer then there exists a set $J \subseteq M$ such that*

$$\lfloor |X|/h \rfloor \leqslant |J \cap X| \leqslant \lceil |X|/h \rceil$$

for every $X \in \mathcal{F}_1 \cup \mathcal{F}_2$.

Informally, this again addresses the notion of fair distribution by finding a set J that contains its fair share of the elements of M in *every* set in the *two* laminar sets.

To apply this result, let M be the set of edges in B, the bipartite graph defined in Subsection 4.2. The set $\mathcal{F}_1$ is constructed by letting the edges incident with each $w \in V(G) \setminus \{v\}$ form an element of $\mathcal{F}_1$, and taking the edges incident with ℓ to form another set in $\mathcal{F}_1$. Clearly $\mathcal{F}_1$ is laminar since its elements actually partition $M = E(B)$. Similarly, the set of edges C_i incident with c_i for $1 \leqslant i \leqslant k$ form an element of $\mathcal{F}_2$, as do the set of edges $C_{i,j}$ in B that correspond to edges joining the jth component of $\mathcal{C}_i(G) - v$ to v; so clearly $C_{i,j} \subseteq C_i$. Finally $\mathcal{F}_2$ also contains the sets S_i of edges in B that correspond to loops coloured i on v in G, or to edges incident with v coming from components in $\mathcal{C}_i - v$ that are joined to v by more than one edge. So $\mathcal{F}_2$ is also laminar. The edges in S_i are the ones that play the crucial gluing role, being efficiently used to tie the cut-edges of $\mathcal{C}_i(G) - v$ together. This efficient use is guaranteed by the Nash-Williams result because we choose $h = \eta(v)$, so that any step only the small portion $1/\eta(v)$ of these gluing edges are ever used.

Finally the following result was proved. It involves a technical requirement that is more than just connectedness, namely the property of being *tree-connected*, a definition of which can be found in [29]. (Briefly, an outline factorisation H is tree-connected if each colour class induces a connected graph, and if, for each colour c and each vertex v, the number of edges and loops of colour c incident with v exceeds the number of components of $\mathcal{C}(H)-v$ by at least the amalgamation number of v minus 1.) It also requires all vertices but one to have an amalgamation number of 1. This is enough to prove the corresponding embedding result, but it would be good to remove this restriction.

Theorem 4.11 *Let H be an outline k-factorisation of K_{kn+1} in which $\eta(v) = 1$ for each vertex except possibly one. Then H is the amalgamation of some connected k-factorisation of K_{kn+1} if and only if it is tree-connected.*

This result is just what is needed to obtain the corresponding embedding result. So in [29] one can find necessary and sufficient conditions for an edge-coloured copy of K_n to be embedded in a k-factorisation of K_v in which each k-factor is connected.

The above was further generalised by Johnson, introducing at the same time a very promising new technique. In his thesis [41], he considered decompositions of K_n into factors $F_1, \ldots, F_t$, where, for a given triple $(t, K = (k_1, \ldots, k_t), L = (l_1, \ldots, l_t))$ with $\sum_{i=1}^{t} k_t = n-1$ and $0 \leqslant l_i \leqslant k_i$ for all i, F_i is k_i-regular and l_i-edge-connected. Such factorisations can be constructed by other means, which Johnson also mentioned, but he also employed amalgamations and obtained a very neat embedding theorem.

One of the novel features of Johnson's work is that *he did not use disentangling* of the amalgamated vertices to prove that any outline structure is in fact an amalgamation of a target structure. (He got around the requirement of tree-connectedness by basically having it included in his definition of outline factorisation.) Instead, he considered each colour class by itself and chose some subgraph of the target graph that could give rise to the given edge-set in the outline graph. The chosen graphs were subsequently manipulated so that together they form a decomposition of the target graph, while keeping the desired degree and connectivity properties.

The thesis [41] contains a similar treatment of complete multipartite graphs K_a^p. It will be interesting to follow the future development and application of Johnson's method.

5 Further uses of amalgamations

In this section we briefly touch upon some other uses of amalgamations and refer the reader to the literature for details.

5.1 Triple systems

One of the classic embedding problems that still remains unsolved is the Lindner Conjecture [45] on partial triple systems. A (partial) triple system of index λ and order n is a partition of (a subset of) the edges of λK_n (the complete multigraph with λ edges between each pair of vertices) into sets of size 3, each of which induces a copy of K_3. A Steiner triple system is a triple system of index 1. Lindner conjectured that

> every partial triple system of index λ and of order n can be embedded in a triple system of index λ and of order v for all admissible values of $v \geqslant 2n+1$.

Here, 'admissible' means that λ and v are chosen so that each vertex has even degree (the copies of K_3 partition edges incident with a vertex v into pairs) and the total number of edges is divisible by 3. It should also be noted that there do exist partial triple systems of order n that cannot be embedded in any triple system of order v with $v \leqslant 2n$, so the lower bound is best possible in this sense.

It was this conjecture that Hilton was studying using the amalgamation technique when he made the insightful observation that the method worked admirably on Hamiltonian cycles, thereby producing Corollary 4.6. One can think of K_3 as being a 3-cycle, so in this sense Hamiltonian cycles are at the other extreme, being n-cycles in K_n. It would be a big breakthrough for amalgamations if somebody could extend the method in a flexible way to consider graph decompositions that involve cycles of other lengths (see Section 6).

Even dealing with 3-cycle decompositions required much more effort in using edge-colourings. The method of attack can best be described by thinking of an amalgamation G of K_n with exactly one vertex v that has an amalgamation number exceeding 1. The copies of K_3 will each consist of one of the following: 3 loops on v; one loop on v joined by a pair of edges to some other vertex w; a K_3, one vertex of which is v; and, a K_3 in which no vertex is v. The third of these cases really points to one of the difficulties when considering cycles of length less than n, namely that at some stage one must "finish off" the cycle by selecting both edges that are still incident with v to move to v_1. This is definitely a new wrinkle that never emerges when dealing with Hamiltonian cycles.

Progress towards settling Lindner's Conjecture was made by Hilton and Rodger [32] when they considered the graph H formed from $G-v$ by deleting all copies of K_3 of the last type, then appending a subdivided loop to each vertex w involved in a triple of the second type. Then H has an equitable $\eta(v)$-edge-colouring, since each original edge can be coloured with the name of the third vertex in its triple, a vertex which is currently amalgamated into v; and two edges on each subdivided loop can be coloured with the names of the other two vertices in their corresponding triple. In fact, H has vertices of degree $\lambda\eta(v)$ and 2; and each vertex of degree $\lambda\eta(v)$ is incident with exactly λ edges of each of the $\eta(v)$ colours.

If we were to start with an outline triple system, then we should be able to form this graph H, and then in some way be able to come up with such an edge-colouring, or inductively guarantee it exists after v_1 is disentangled from v. Such edge-colourings are very close in nature to edge-colourings of r-regular graphs with r colours; edge-colouring such graphs is known to be very difficult!

To some extent this problem was avoided in the two results that have settled half of Lindner's Conjecture. If attention is focused on the case where λ is even, then the problem becomes one that is closer to finding 2-factors than to finding 1-factors, a significant decrease in difficulty. Also, H is Eulerian, and this fact can be exploited to find equitable edge-colourings, something which will be much harder to do when attempting to solve the case where λ is odd.

Using this approach, Hilton and Rodger were able to confirm Lindner's Conjecture in the case that 4 divides λ. Much of the work in [32] was written for the more general case where λ is even, but a strengthening of one edge-colouring result was missing. This was later obtained with Theorem 4.3, and using this, Johansson managed to use their approach to complete the solution for all even λ.

Theorem 5.1 ([32, 40]) *Let λ be even. Every partial triple system of order n and index λ can be embedded in a triple system of order v and index λ for any admissible $v \geqslant 2n+1$.*

It is also worth noting that the proof of this theorem did rely on a disentangling result that required there to be at most one vertex that has an amalgamation number that is more than 1. Recently Ferencak and Hilton have removed this restriction, proving the following more general result. Again, Theorem 4.3 was used in the proof.

Theorem 5.2 ([22]) *Every outline triple system of order n and of even index λ is the amalgamation of some triple system of order n and index λ.*

For odd λ, including the case of Steiner triple systems, the best known result towards Lindner's Conjecture is that embedding is possible into a triple system of order v for all admissible $v \geqslant 4n+1$ [10, 53].

5.2 Colourings and 1-factors in complete multipartite graphs

Amalgamations proved to be the right tool to use in considering the problems of finding both the chromatic index and the total chromatic number of $K(a_1, \ldots, a_p)$. The *chromatic index* of a graph is the least integer k for which it has a proper edge-colouring with k colours, and the *total chromatic number* is the least integer ℓ for which the graph has a colouring of both vertices and edges with ℓ colours, such that the vertex- and the edge-colouring are proper, and no edge incident with a vertex has the same colour as the vertex. In both cases, the results were basically obtained by amalgamating all the vertices in one of the parts, colouring the rest of the graph in some way, then disentangling the final part.

A graph G is said to be Class 1 if its chromatic index equals its maximum degree, $\Delta(G)$ ($\Delta(G)$ is an obvious lower bound on the chromatic index of G). A graph is said to be *overfull* if $|E(G)| > \Delta(G)\lfloor |V(G)|/2 \rfloor$. It is easy to see that if a graph is Class 1 then it cannot be overfull.

Theorem 5.3 ([38]) *$K(a_1, \ldots, a_p)$ is Class 1 if and only if it is not overfull.*

We used this result on complete graphs ($p = 1$) in Section 3 — a complete graph of odd order is overfull.

So this problem is completely solved, but the total chromatic number of $K(a_1, \ldots, a_p)$ is still in doubt. Using amalgamations one can settle the existence problem for total colourings of $K(a_1, \ldots, a_p)$ of a certain type, known as *biased* total colourings. This seems to oppose all the *fair* colourings considered so far, but in a way this is fair to every colour except one! A total colouring of $K(a_1, \ldots, a_p)$ is biased if all vertices in one part receive the same colour, but no other colour appears on more than one vertex. Yap [57] has shown that $K(a_1, \ldots, a_p)$ has total chromatic number at most $\Delta+2$, but there remains the classification problem which asks one to characterise the complete multipartite graphs with total chromatic number $\Delta+1$ ($\Delta(G)+1$ is an obvious lower bound on the total chromatic number of G). Necessary and sufficient conditions for a biased $(\Delta+1)$-total colouring of $K(a_1, \ldots, a_p)$ are given in [36]. It turns out that this result nearly settles the classification problem, leaving in doubt only those graphs which are very close to being regular. The deficiency def(G) of a graph G is defined to be $\sum_{v \in V(G)}(\Delta(G) - d_G(v))$; so def($G$) is a measure of how far G is from being regular. Hoffman and Rodger proved the following.

Theorem 5.4 ([36]) *Let $a_1 \leqslant \cdots \leqslant a_p$. The graph $K = K(a_1, \ldots, a_p)$ has total chromatic number $\Delta(K) + 1$ if*

(a) *$|V(K)|$ is odd, or*

(b) *$|V(K)|$ is even and*

$$\text{def}(K) \geqslant \begin{cases} |V(H)| - a_1 & \text{if } p = 2 \text{ or} \\ & \text{if } p \text{ is even, } a_1 \text{ is odd and } a_1 = a_{p-1}, \\ |V(K)| - a_p & \text{otherwise.} \end{cases}$$

Hoffman and Rodger went on to make the following conjecture in [36].

Conjecture 5.5 *$K = K(a_1, \dots, a_p)$ has total chromatic number $\Delta(K) + 2$ if and only if*

(a) *$p = 2$ and K is regular, or*

(b) *$|V(K)|$ is even and $\sum_{v \in V(K)} (\Delta(K) - d(v))$ is less than the number of parts of odd size.*

Lastly in this subsection, we point the reader to an interesting graph parameter, somewhat related to the chromatic index, namely the maximum number of edge-disjoint 1-factors in a graph G. So if G happens to be k-regular, then deciding if G is Class 1 or not is equivalent to deciding whether or not G has k edge-disjoint 1-factors. Using a technique that follows the method used for the previous two problems in this section, but with the addition of some more sophisticated results involving conditions that ensure a graph is Class 1, Hoffman and Rodger [37] made some progress towards finding this parameter for the complete multipartite graphs. Among other results, they managed to find the maximum number k of edge-disjoint 1-factors in $K(a_1, ..., a_p)$ in the case that $k \geqslant \delta - a_2$, where δ is the minimum degree of the graph, and where the assumption is that the sizes of the parts are in non-decreasing order (so a_2 is the size of the second smallest part).

5.3 Other cycle lengths

One of the difficulties facing amalgamations lies in finding a way to use them when looking for graph decompositions of K_n that involve cycle lengths other than 3 or n. To date, only one idea has surfaced to attack this barrier. In his Ph.D. thesis, Buchanan [15] observed that if one started with a graph decomposition of K_n into Hamiltonian cycles and one 2-factor F, then amalgamating the vertices occurring in the same component of F into one vertex resulted in a graph for which the disentangling techniques described in Subsection 4.2 would still work! The reason for this is that disentangling the last vertex in each component is just like disentangling the last vertex in a Hamiltonian cycle — it closes itself off automatically. This enabled him to prove the following neat result.

Theorem 5.6 ([15]) *Let n be odd and let F be any 2-factor of K_n. Then there exists a Hamiltonian decomposition of $K_n - E(F)$.*

Recently, this result was proved by Rodger [52] using difference methods, and was extended by Bryant [13] who, with a similar proof, showed that, for $n \geqslant 11$, given any three 2-regular graphs on n vertices there exists a 2-factorisation of K_n in which three of the factors are isomorphic to the given 2-regular graphs, and the rest are Hamiltonian cycles. These proofs do not use amalgamations.

An analogue of Theorem 5.6 was proved by Leach and Rodger who considered decompositions of complete bipartite graphs.

Theorem 5.7 ([42]) *Let n be even and let F be any 2-factor F of $K_{n,n}$. Then there exists a Hamiltonian decomposition of $K_{n,n} - E(F)$ except if $n = 4$ and F is a pair of 4-cycles, in which case such a decomposition does not exist.*

The same authors have had some success extending these results to complete multipartite graphs [44], and recently a solution for regular such graphs was obtained by Rodger [52] using difference methods.

5.4 As fair as you can be

One of the main recurring themes in the use of amalgamations is that of fairness in distributing edges amongst colour classes. This idea has been taken to the limit in a neat result that currently appears in the thesis of Clark [16]. To understand this result, for the purposes of this survey we continue to describe it in terms of edge-colourings of complete graphs, though it is easily possible to be more general (essentially by having a dummy colour class). So suppose that G is an outline decomposition of K_n. Let G be edge-coloured in any way. The theorem appearing in Clark's thesis shows that there does exist an edge-coloured copy H of K_n that is a disentanglement of G, where the edge-colouring of H satisfies *four* notions of balance simultaneously. To describe these four notions, notice that G induces a natural partition $\mathcal{P}$ of the vertex set of H, where two vertices are in the same element of $\mathcal{P}$ if and only if they are amalgamated into the same vertex in G. For each $P, Q \in \mathcal{P}$, H satisfies the following balance properties, where $H[P]$ is the subgraph of H induced by P, and $H[P,Q]$ is the bipartite subgraph of H induced by the edges joining vertices in P to vertices in Q:

(1) in $H[P]$, the edges in each colour class are distributed as evenly as possible among the vertices in P;

(2) in $H[P,Q]$, the edges in each colour class are distributed as evenly as possible among the vertices in P;

(3) in $H[P, V(H) - P]$, the edges in each colour class are distributed as evenly as possible among the vertices in P; and

(4) in H, the edges in each colour class are distributed as evenly as possible among the vertices in P.

5.5 Algebraic structures

After the first ideas about amalgamations had occurred to Andersen and Hilton in connection with their work on generalised latin squares (see Subsection 2.5), the method was soon generalised to other latin-square-like matrices such as those arising in school timetabling and experimental design (for example in Hilton's paper [26], which introduces the now standard terminology of *outline* structures), and to match-tables ([30] by Hilton and Rodger). Already in 1979 Andersen and Hilton [4] had made the natural interpretation of amalgamations of latin squares as *set-multiplication tables for quasigroups.*

This more algebraic way of looking at things was taken further by Hilton and Wojciechowski [33], and further still by Dugdale and the same two authors in [20], who introduced concepts such as *fractional latin squares* and *simplex algebras.* We refer to these papers for more details.

Hilton and Wojciechowski also looked at amalgamations of infinite latin squares [34] (this remark being the exception to the rule that everything in this paper is finite!).

Recently, Glebsky and Gordon have used similar extended amalgamation methods to prove that a locally compact group can be approximated by finite quasigroups if and only if it is unimodular (personal communication with Hilton).

6 The future

Probably the most outstanding challenge, relevant to the solution of many problems, is to find a way to disentangle outline decompositions of K_n so that each colour class is a cycle. This would, for example, most likely lead to a solution of a conjecture of Alspach, namely that if $l_1, \cdots, l_x$ are each integers at least 3, and if they add to $n(n-1)/2$, then there exists a partition of the edges of K_n into sets $S_1, \cdots, S_x$, where for $1 \leqslant i \leqslant x$ the set S_i induces a cycle of length l_i in K_n. This would be an astounding result to prove, so may well be worth the effort. The difficulty in obtaining such a result has been alluded to several times in previous sections: how do you ensure that when v_1 is disentangled from v, at a time when colour class i has only 2 edges incident with v, you either select *both* these edges coloured i or you select *neither.* One can see that this requirement is vastly different from that described in Subsection 4.3, where it was sufficient to ensure that *at most one* of these two edges is selected.

Another question, involving similar difficulties if attacked by the method of amalgamations, is the Oberwolfach problem mentioned in Subsection 1.3 and illustrated in its embedding form in sample question 4 in the beginning of the paper. A recent paper by Hilton and Johnson [28] solves a number of new cases of the problem, and part of their proof can be said to be a (small) step towards employing amalgamation or, rather, embedding techniques to answer this attractive question.

A third, though less stunning advance, would be to prove an analogue of Theorem 4.4 for complete multipartite graphs. Of course, Theorem 4.7 is progress towards such a result, but the restriction that the outline graphs be loopless makes it difficult to use for solving other problems.

References

[1] S. Allen, *Extending edge-colourings of graphs*, Ph.D. thesis, University of Reading, UK, 1996.

[2] L.D. Andersen, Embedding latin squares with prescribed diagonal, *Annals of Discrete Mathematics*, 15, North-Holland, Amsterdam (1982), pp. 9–26.

[3] L.D. Andersen, R. Häggkvist, A.J.W. Hilton and W.B. Poucher, Embedding incomplete latin squares in latin squares whose diagonal is almost completely prescribed, *European J. Combin.* **1** (1980), 5–7.

[4] L.D. Andersen and A.J.W. Hilton, Generalized latin rectangles, in *Graph theory and combinatorics* (ed. R.J. Wilson), *Research Notes in Mathematics*, 34, Pitman, London (1979), pp. 1–17.

[5] L.D. Andersen and A.J.W. Hilton, Generalized latin rectangles I: construction and decomposition, *Discrete Math.* **31** (1980), 125–152.

[6] L.D. Andersen and A.J.W. Hilton, Generalized latin rectangles II: embedding, *Discrete Math.* **31** (1980), 235-260.

[7] L.D. Andersen and A.J.W. Hilton, Thank Evans!, *Proc. London Math. Soc. (3)* **47** (1983), 507–522.

[8] L.D. Andersen and A.J.W. Hilton, Symmetric latin square and complete graph analogues of the Evans conjecture, *J. Combin. Des.* **2** (1994), 197–252.

[9] L.D. Andersen and A.J.W. Hilton, Extending edge-colourings of complete graphs and independent edges, in *Graph Theory and its applications: east and west, proceedings of the first China-USA international graph theory conference* (1989), pp. 30–41.

[10] L.D. Andersen, A.J.W. Hilton and E. Mendelsohn, Embedding partial Steiner triple systems, *Proc. London Math. Soc. (3)* **41** (1980), 557–576.

[11] L.D. Andersen, A.J.W. Hilton and C.A. Rodger, A solution to the embedding problem for partial idempotent latin squares, *J. London Math. Soc. (2)* **26** (1982), 21–27.

[12] L.D. Andersen, A.J.W. Hilton and C.A. Rodger, Small embeddings of incomplete idempotent latin squares, *Annals of Discrete Mathematics*, 17, North-Holland, Amsterdam (1983), pp. 19–31.

[13] D.E. Bryant, Hamilton cycle rich two-factorisations of complete graphs, manuscript, University of Queensland, Australia, 2003.

[14] D.E. Bryant, S. El-Zanati and C.A. Rodger, Maximal sets of hamilton cycles in $K_{n,n}$, *J. Graph Theory* **33** (2000), 25–31.

[15] H. Buchanan, Graph factors and Hamiltonian decompositions, Ph.D. thesis, University of West Virginia, USA, 1997.

[16] S.A. Clark, Edge-color balance with respect to a partition of the vertices of K_n, Ph.D. thesis, University of Auburn, USA, 1997.

[17] A. Cruse, On embedding incomplete symmetric latin squares, *J. Combin. Theory Ser. A* **16** (1974), 18–27.

[18] M. Daven, J.A. MacDougall and C.A. Rodger, Maximal sets of hamilton cycles in complete multipartite graphs, *J. Graph Theory* , in press.

[19] J.K. Dugdale and A.J.W. Hilton, Amalgamated factorizations of complete graphs, *Combin. Probab. Comput.* **3** (1994), 215–231.

[20] J.K. Dugdale, A.J.W. Hilton and J. Wojciechowski, Fractional latin squares, simplex algebras, and generalized quotients, *J. Statist. Plann. Inference* **86** (2000), 457–504.

[21] T. Evans, Embedding incomplete latin squares, *Amer. Math. Monthly* **67** (1960), 958–961.

[22] M.N. Ferencak and A.J.W. Hilton, Outline and amalgamated triple systems of even index, *Proc. London Math. Soc. (3)* **84** (1992), 1–34.

[23] R. Häggkvist and A. Johansson, (1, 2)-factorizations of general Eulerian nearly regular graphs, *Combin. Probab. Comput.* **3** (1994), 87–95.

[24] M. Hall, An existence theorem for latin squares, *Bull. Amer. Math. Soc.* **51** (1945), 387–388.

[25] M.J. Henderson and A.J.W. Hilton, Completing the edge-colouring of a complete graph, manuscript, University of Reading, UK, 2002.

[26] A.J.W. Hilton, The reconstruction of latin squares with applications to school timetabling and to experimental design, *Mathematical Programming Study* **13** (1980), 68–77.

[27] A.J.W. Hilton, Hamiltonian decompositions of complete graphs, *J. Combin. Theory Ser. B* **36** (1984), 125–134.

[28] A.J.W. Hilton and M. Johnson, Some results on the Oberwolfach problem, *J. London Math. Soc. (2)* **64** (2001), 513–522.

[29] A.J.W. Hilton M. Johnson, C.A. Rodger and E.B. Wantland, Amalgamations of connected k-factorizations, *J. Combin. Theory Ser. B* , in press.

[30] A.J.W. Hilton and C.A. Rodger, Matchtables, *Annals of Discrete Mathematics*, 15, North-Holland, Amsterdam (1982), pp. 239–251.

[31] A.J.W. Hilton and C.A. Rodger, Hamiltonian decompositions of complete regular s-partite graphs, *Discrete Math.* **58** (1986), 63–78.

[32] A.J.W. Hilton and C.A. Rodger, The embedding of partial triple systems when 4 divides λ, *J. Combin. Theory Ser. A* **56** (1991), 109–137.

[33] A.J.W. Hilton and J. Wojciechowski, Weighted quasigroups, in *Surveys in Combinatorics, 1993* (ed. K. Walker), *London Math. Soc. Lecture Notes Ser.*, 187, (1993), pp. 137–171.

[34] A.J.W. Hilton and J. Wojciechowski, Amalgamating infinite latin squares, manuscript, University of Reading, UK, 1994.

[35] D.G. Hoffman, Completing incomplete commutative latin squares with prescribed diagonals, *European J. Combin.* **4** (1983), 33–35.

[36] D.G. Hoffman and C.A. Rodger, The total chromatic number of complete multipartite graphs, *Festschrift for C.St.J.A. Nash-Williams*, *Congr. Numer.* **113** (1996), 205-220.

[37] D.G. Hoffman and C.A. Rodger, On the number of edge-disjoint 1-factors in complete multipartite graphs, *Discrete Math.* **160** (1996), 25–31.

[38] D.G. Hoffman and C.A. Rodger, The chromatic index of complete multipartite graphs, *J. Graph Theory* **16** (1992), 159–163.

[39] D.G. Hoffman, C.A. Rodger and A. Rosa, Maximal sets of 2-factors and hamilton cycles, *J. Combin. Theory Ser. B* **57** (1993), 69–76.

[40] A. Johansson, A note on embedding partial triple systems, Preprint, University of Umeå, Sweden, 1997.

[41] M. Johnson, *Some problems on graphs and designs*, Ph.D. thesis, University of Reading, UK, 2002.

[42] C.D. Leach and C.A. Rodger, Non-disconnecting disentanglements of complete multipartite graphs, *J. Combin. Theory Ser. B* **85** (2002), 290–296.

[43] C.D. Leach and C.A. Rodger, Fair hamilton decompositions of complete multipartite graphs, *J. Combin. Des.* **9** (2001), 460–467.

[44] C.D. Leach and C.A. Rodger, Hamilton decompositions of complete multipartite graphs with any 2-factor leave, preprint, Auburn University, USA, 2002.

[45] C.C. Lindner, A partial Steiner triple system of order n can be embedded in a Steiner triple system of order $6n+3$, *J. Combin. Theory Ser. A* **18** (1975), 349–351.

[46] C.C. Lindner, Embedding theorems for partial latin squares, *Latin squares. New developments in the theory and applications*, (eds. J. Dénes and A.D. Keedwell), *Annals of Discrete Mathematics*, 46, North-Holland, Amsterdam (1991), pp. 217–265.

[47] C.C. Lindner & T. Evans, *Finite embedding theorems for partial designs and algebras*, *Collection seminaire de mathématiques supérieures*, 56, Les presses de l'Université de Montréal, Montreal (1977).

[48] S. Logan and C.A. Rodger, Maximal sets of hamilton cycles in $K_n - F$, Preprint, Auburn University, USA, 2002.

[49] C.J.H. McDiarmid, The solution of a timetabling problem, *J. Inst. Math. Appl.* **9** (1972), 23–34.

[50] C.St.J.A. Nash-Williams, Amalgamations of almost regular edge-colourings of simple graphs, *J. Combin. Theory Ser. B* **43** (1987), 322–342.

[51] C.A. Rodger, Embedding an incomplete latin square in a latin square with a prescribed diagonal, *Discrete Math.* **51** (1984), 73–89.

[52] C.A. Rodger, Hamilton decomposable graphs with specified leaves, manuscript, Auburn University, USA, 2003.

[53] C.A. Rodger and S.J. Stubbs, Embedding partial triple systems, *J. Combin. Theory Ser. A* **44** (1987), 241–252.

[54] C.A. Rodger and E.B. Wantland, Embedding edge-colorings into 2-edge-connected k-factorizations of K_{kn+1}, *J. Graph Theory* **19** (1995), 169–185.

[55] H.J. Ryser, A combinatorial theorem with an application to latin rectangles, *Proc. Amer. Math. Soc.* **2** (1951), 550–552.

[56] B. Smetaniuk, A new construction on latin squares - I: A proof of the Evans conjecture, *Ars Combin.* **9** (1981), 155–172.

[57] H.P. Yap, Total colorings of graphs, *Bull. Lond. Math. Soc.* **21** (1989), 159–163.

[58] D. de Werra, Equitable colorations of graphs, *Rev. Fran. Rech. Oper.* **5** (1971), 3–8.

Department of Mathematical Sciences
Aalborg University
Fredrik Bajers Vej 7G
9220 Aalborg Ø, Denmark
lda@math.auc.dk

Department of Discrete and Statistical Sciences
225 Allison Lab
Auburn University
Al 36849-5307, USA
rodgec1@auburn.edu

Combinatorial schemes for protecting digital content

Simon R. Blackburn

Abstract

When digital information is widely distributed in some fashion, the distributor would often like to trace the source of pirate copies of the information. The paper surveys some of the schemes that have been proposed to achieve this aim, especially emphasising the underlying combinatorics involved. In particular, codes with the identifiable parent property, frameproof codes, secure frameproof codes, traceability schemes and related concepts are surveyed.

1 Introduction

Nowadays a vast amount of data is distributed in digital fashion (music, film and software being three important examples). Purchasers of the data want easy access, and data providers want to make sure that their data is not pirated. This paper concentrates on one specialised problem in this huge research area, where combinatorial methods have been especially useful. A rough description of the problem is as follows. A *traitor* is a user who is allowed access to the data (because, for example, they have paid a subscription fee) but who passes on data to an unauthorised third party, a *pirate*. There could be several traitors, all with different versions of the data, and all giving the information they know to the pirate. The pirate then creates a new version of the data to share with other unauthorised users. How can the data provider identify a traitor, if the provider comes into possession of the pirate's new version?

The solutions to the problem that we will be most interested in arose from the papers of Boneh and Shaw [13] (who introduced the notions of frameproof code and collusion secure code) and Chor, Fiat and Naor [14] (who introduced the concept of traitor tracing; see their paper with Pinkas [15]).

The rest of the paper is structured as follows. Section 2 defines one of the main combinatorial problems we will be concerned with — the problem of constructing good codes with the k-identifiable parent property. Section 3 motivates this problem by discussing three related applications for these codes. Section 4 returns to combinatorics: known 'coding theoretic' results relating to the applications discussed in Section 3 are surveyed. Section 5 considers other related models. Section 6 is a short conclusion.

All logarithms are to the base 2 and all sets are finite, unless specified otherwise.

2 A Taster: The Identifiable Parent Property

This section introduces the notion of a k-IPP code from the perspective of abstract combinatorics. This notion was introduced in the special case when $k = 2$ by Hollmann, van Lint, Linnartz and Tolhuizen [26], and in full generality by Staddon, Stinson and Wei [44]. We postpone a discussion of any applications until Section 3, and any general results until Section 4.

Let q and ℓ be positive integers. To avoid trivialities, we assume that $q \geqslant 2$ and $\ell \geqslant 2$. Let F be a finite set of cardinality q.

Let X be a set of words of length ℓ over F. So $X \subseteq F^\ell$. For $x \in F^\ell$, we write x_i for the ith component of x. We define the *set of descendants* of X to be the subset $\mathrm{desc}(X) \subseteq F^\ell$ given by

$$\mathrm{desc}(X) = \{d \in F^\ell \mid \forall\, i \in \{1, 2, \ldots, \ell\}\ \exists\, x \in X : d_i = x_i\}. \tag{2.1}$$

So, for example, if $X = \{111, 123\}$ then

$$\mathrm{desc}(X) = \{111, 113, 121, 123\}.$$

If $|X| = k$, then $|\mathrm{desc}(X)| \leqslant k^\ell$, with equality if and only if for all $i \in \{1, 2, \ldots, \ell\}$ the ith components of the words in X are all distinct. Note that $X \subseteq \mathrm{desc}(X)$ for any set X. Moreover $\mathrm{desc}(\emptyset) = \emptyset$, and $\mathrm{desc}(\{x\}) = \{x\}$ for any word x.

The terminology 'descendant' comes from thinking of the words in X as the strings of DNA in an isolated population of living organisms. Then $\mathrm{desc}(X)$ is the set of possible DNA strings of the population after some time, assuming no mutations occur.

Let C be a length ℓ code over F. (We are using the terminology 'code' to mean a set of words over a finite alphabet of fixed length; so we just mean $C \subseteq F^\ell$ here. The code C is not necessarily an error correcting code, and C has nothing to do with any cipher!) Let k be a positive integer. We define the *k-descendant set* $\mathrm{desc}_k(C)$ of C by

$$\mathrm{desc}_k(C) = \bigcup_{X \subseteq C,\ |X| \leqslant k} \mathrm{desc}(X). \tag{2.2}$$

For example, suppose $C = \{111, 222, 333, 123\}$. Clearly $\mathrm{desc}_1(C) = C$. To calculate $\mathrm{desc}_2(C)$, note that we have

$$\begin{aligned}
\mathrm{desc}(\{111, 123\}) &= \{111, 113, 121, 123\} \\
\mathrm{desc}(\{123, 222\}) &= \{122, 123, 222, 223\} \\
\mathrm{desc}(\{123, 333\}) &= \{123, 133, 323, 333\} \\
\mathrm{desc}(\{111, 222\}) &= \{111, 112, 121, 122, 211, 212, 221, 222\} \\
\mathrm{desc}(\{111, 333\}) &= \{111, 113, 131, 133, 311, 313, 331, 333\} \\
\mathrm{desc}(\{222, 333\}) &= \{222, 223, 232, 233, 322, 323, 332, 333\}.
\end{aligned}$$

So $\mathrm{desc}_2(C)$ is the union of these sets:

$$\begin{aligned}\mathrm{desc}_2(C) &= \{111, 112, 113, 121, 122, 123, 131, 133, 211, 212, 221, 222, 223,\\ &\quad 232, 233, 311, 313, 322, 323, 331, 332, 333\}.\end{aligned}$$

It is easy to see that $\mathrm{desc}_k(C) = \{1, 2, 3\}^3$ for $k \geqslant 3$ in this example.

We now leave our example, and return to the general case. If $d \in \mathrm{desc}_k(C)$ then there is a set $X \subseteq C$ such that $|X| = k$ and $d \in \mathrm{desc}(X)$. In this case, we say that X is a *possible (k, C)-parent set* of d. (This terminology makes most sense in the case when $k = 2$!) The code C and the integer k will usually be obvious from the context, and so we will just refer to X as a *parent set* of d. Note that parent sets are not in general unique. For example, whenever $d \in C$ we find that any set $X \subseteq C$ such that $|X| \leqslant k$ and $d \in X$ is a parent set. For $d \in F^\ell$, we define $\mathcal{P}_{k,C}(d)$ by

$$\mathcal{P}_{k,C}(d) = \{X \subseteq C : |X| \leqslant k \text{ and } d \in \mathrm{desc}(X)\}. \tag{2.3}$$

So $\mathcal{P}_{k,C}(d)$ is the set of parent sets of d. Note that $\mathcal{P}_{k,C}(d) \neq \emptyset$ if and only if $d \in \mathrm{desc}_k(C)$.

Suppose that you and I both know a code C. You pick a subset X of codewords such that $|X| \leqslant k$. You choose $d \in \mathrm{desc}(X)$ and give this to me. What property should the code C have so that I can always tell you an element from X? (Since you might not have used all the elements of X when creating d, it is hopeless to require that I always tell you the whole set X.) The property we require can be succinctly expressed as: Given a k-descendant of C, at least one of its parents can be identified.

More precisely, a code C has the *k-identifiable parent property* (is a *k-IPP code*) if for all $d \in \mathrm{desc}_k(C)$ we have that

$$\bigcap_{X \in \mathcal{P}_{k,C}(d)} X \neq \emptyset. \tag{2.4}$$

If a codeword x lies in the left hand side of (2.4), then x is contained in every parent set for d. So whichever set X is the 'true' parent set giving rise to d, we must have $x \in X$. Thus at least one parent of any descendant can be identified.

The ternary length 3 code C given above is an example of a 2-IPP code. Suppose, for example, $d = 113$. Then $\mathcal{P}_{k,C}(113) = \{\{111, 333\}, \{111, 123\}\}$. Since both members of $\mathcal{P}_{k,C}(113)$ contain 111, we may identify 111 as a parent. In fact, there is an easy algorithm to identify a parent from a descendant of this code. If a descendant contains two (or three) components equal to a fixed value a, then aaa is a parent. If this is not the case, then 123 is a parent. Note that C is not a 3-IPP code, since $\{111, 222, 333\}$ and $\{123\}$ both have 123 as a descendant.

Another, and very beautiful, example of a 2-IPP code is the ternary Hamming code of length 4. This is the example given by Hollmann *et al.* [26]. Thinking of the alphabet as $\{0,1,2\}$, this code has 9 codewords:

$$C = \{0000, 0111, 0222, 1012, 1120, 1201, 2021, 2102, 2210\}.$$

It is easy to show that $\text{desc}_2(C) = \{0,1,2\}^4$. To identify a parent of a descendant d, just choose a codeword x that agrees with d in 3 or more positions.

The combinatorial problem can now be stated as follows. For fixed parameters q, ℓ and k, what is the largest number n of codewords in a q-ary k-IPP code of length ℓ? We will return to this subject in Section 4, after we have discussed some applications.

3 Applications

This section describes three application areas, all related to the IPP codes of Section 2.

3.1 Digital Fingerprinting

Consider a digital copy of a film (on DVD, for example). If the least significant bits of the brightness of some of the pixels in the film are changed, the resulting marks will be imperceptible to a viewer. The same potential for adding imperceptible marks exists in many digital media. Digital fingerprinting is a method whereby every user of some data is given a copy that is marked differently. These marks can be used to trace a copy of the data back to a specific user. This general idea goes back a long way. For example (an example given by Boneh and Shaw [13]), the makers of logarithm tables used to alter the least significant digit of certain entries for this purpose. Map makers add 'false cul-de-sacs' to their street maps for similar reasons.

We will consider the following situation (see Figure 1). A data provider transmits some data to each of n authorised users. Each user's copy of the data is marked in a different way. Some of the users are traitors (users 2 and 3 in Figure 1) and pass their data to a third party, the pirate. The pirate creates a new copy of the data (the *pirate version*). The pirate version is then intercepted by the data provider. The question we address is: How can the data provider mark their data, in such a way that the provider can identify at least one traitor that contributed to the pirate version?

The above question can only be answered once we specify how data can be marked, and which operations the pirate can carry out to modify the marks on a set of data. This depends on the marking model used. Many methods have been proposed — see Cox, Kilian, Leighton and Shamoon [17] for one such — but we are not going to delve into the details here. Normally, we would hope that a mark is undetectable to a user (even if the user sets out to look for them). It might be possible to remove undetected marks, for example by adding noise

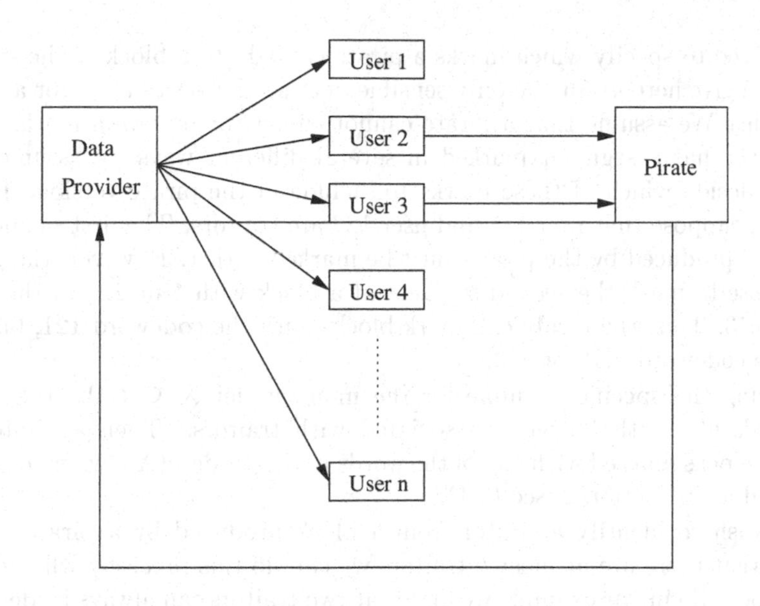

Figure 1: An overview of the model

to the data, but we assume that undetected marks cannot be removed without degrading the data to an unacceptable level. This is known as the *Marking Assumption* [13]. An aside: Marks that satisfy the Marking Assumption can be used in digital watermarking as well as digital fingerprinting. (For the purposes of this paper we define watermarking to be methods for identifying the data provider; fingerprinting identifies the user who obtained the data.) And a warning: The Marking Assumption is currently unrealistic; see Swanson, Kobayashi and Tewfik [50] for a discussion of these issues.

We assume that the data to be protected is divided up into blocks of a standard size. (A block could be 1 second of a film, for example.) Each block is divided into ℓ segments (each segment being a frame of the film, perhaps). Each segment can be marked in one of q ways; we identify the q different marks with the elements of an alphabet F of cardinality q. So a block can be marked in q^ℓ ways, and we think of a marked block as corresponding to a word x of length ℓ over F. We abuse terminology, and speak of a block being *marked by the word* x.

The data provider labels users with distinct words taken from a code C, where $C \subseteq F^\ell$. The provider marks blocks sent to user $x \in C$ with x.

As an example, suppose that each block consists of 3 segments, and each segment can be marked in one of 3 ways. We can use the length 3 code given in Section 2 as our code. So let $F = \{1, 2, 3\}$ and $C = \{111, 222, 333, 123\}$. There are 4 users, associated with the codewords 111, 222, 333 and 123. Any block sent to user 123 (for example) is marked with 123: the first segment is marked 1, the second is marked 2, and the third 3.

We need to specify which marks a pirate can add to a block. (The specification we give here is not the only sensible one: see Subsection 5.3 for another approach.) We assume that a pirate cannot change or remove a mark. However, if she has a segment marked in several different ways we assume that she can decide which of these marks to include in the pirate version. In our example, suppose that user 111 and user 123 are traitors. The first segment of any block produced by the pirate must be marked with 1. However, the pirate can choose to mark the second segment of a block with 1 or 2, and the third with 1 or 3. Thus the pirate can mark blocks with the codeword 121, but not with the codewords 112 or 223.

Leaving the specific example for the moment, let $X \subseteq C$ be the set of codewords of length ℓ over F associated with traitors. Then a pirate can produce blocks marked with any of the words in the set $\mathrm{desc}(X)$, where $\mathrm{desc}(X)$ is defined as in Section 2; see (2.1).

We wish to identify a traitor from a block produced by a pirate. If we assume that there are at most k traitors we can do this precisely when C is a k-IPP code. So in our example we find that two traitors can always be detected from a block produced by the pirate. For example, if a pirate block is marked with the word 113, the list of sets of descendants given in the previous section shows that the set of traitors must be one of $\{111, 123\}$ or $\{111, 333\}$. Since 111 lies in both sets, we know that 111 is a traitor. Similarly, if the pirate block is marked with the word 123, we know that the set of traitors is one of $\{123\}, \{111, 123\}, \{123, 222\}$ or $\{123, 333\}$, and 123 is therefore a traitor. Note that if 3 (or 4) traitors are possible then we can no longer identify a traitor using this code, since C is not a 3-IPP code. For suppose the pirate outputs a block corresponding to 123. Any subset X of 3 codewords has 123 as a descendant (as either $123 \in X$, which gives us that $123 \in \mathrm{desc}(X)$ immediately, or $X = \{111, 222, 333\}$ which clearly has 123 as a descendant) and so any single user might be innocent.

So a k-IPP code is exactly what is needed in this application. We now turn to another example.

3.2 Fingerprinting in a Broadcast Setting

The model discussed in Subsection 3.1 is useful when the data provider is able to provide each user with different data. This is not always the case. For example, in subscription T.V. broadcasting every user gets the same data from the satellite or cable connection, and data encryption is used to bar unauthorised users from access.

For readers unfamiliar with the terminology of modern cryptography, a good introduction for the mathematician is Stinson [46]. For our purposes, we only need a high level description of a (symmetric) cipher, as depicted in Figure 2. A cipher consists of two efficient algorithms, labelled 'Encrypt' and 'Decrypt'. The 'Encrypt' algorithm takes as input the raw data (the *plaintext*)

p and a *key* K (which is typically a binary string of a fixed length). It outputs a *ciphertext* $E_K(p)$ (the encrypted data). The 'Decrypt' algorithm takes $E_K(p)$ and K as input, and returns the value p. It is assumed that everyone knows the two algorithms, but that the key K is kept secret. It should be infeasible to determine the plaintext p from $E_K(p)$ unless K is also known. (As a technical remark for cryptographers, we will be assuming that the cipher is ideal.)

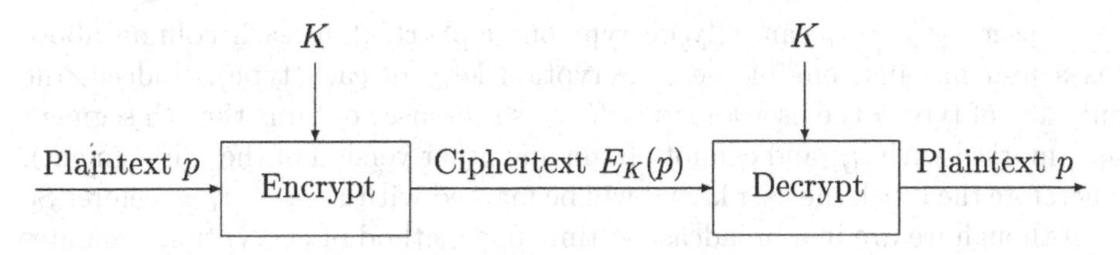

Figure 2: A symmetric cipher

Returning to the subscription T.V. setting, the data provider could use a cipher in a classical manner as follows. He chooses a key K at random, and gives K to all authorised users. Rather than broadcasting a segment p, he broadcasts $E_K(p)$. Since authorised users know K, they can apply the decryption algorithm to $E_K(p)$ and thus obtain the segment p. Unauthorised users do not know K, so cannot obtain p. The problem with this system is that a traitor can give their key to unauthorised users without fear of detection.

But it is possible for the data provider to use encryption in such a way that the model in Section 3.1 applies. One obvious way would be to give each user x a randomly chosen key K_x, and broadcast a block marked with x encrypted under key K_x. But this is an unrealistic approach — if there are n users, the amount of information broadcast would rise by a factor of n. There is a much better system, which goes as follows. Firstly, the data provider picks $q\ell$ keys uniformly and independently at random; we write the keys as $K_{i,j}$, where $1 \leqslant i \leqslant q$ and $1 \leqslant j \leqslant \ell$. Think of the keys as being arranged in a $q \times \ell$ array:

$$\begin{array}{cccc} K_{1,1} & K_{1,2} & \cdots & K_{1,\ell} \\ K_{2,1} & K_{2,2} & \cdots & K_{2,\ell} \\ \vdots & \vdots & \ddots & \vdots \\ K_{q,1} & K_{q,2} & \cdots & K_{q,\ell} \end{array}$$

For $j \in \{1, 2, \ldots, \ell\}$, we say that the q keys of the form $K_{i,j}$ (those in the jth column above) are of *type* j. A user gets one key of each type — if the user is labelled with the codeword $x_1x_2\cdots x_\ell$ with $x_j \in \{1, 2, \ldots q\}$ he gets the ℓ keys of the form $K_{x_j,j}$, one from each column.

The data provider now broadcasts the data as follows. For the jth segment p_j of a block, the provider marks the segment in all q possible ways. Let $p_{i,j}$ be segment j marked with the ith mark. He encrypts $p_{i,j}$ using key $K_{i,j}$. He then broadcasts all q resulting ciphertexts $E_{K_{i,j}}(p_{i,j})$:

$$\begin{array}{cccc} E_{K_{1,1}}(p_{1,1}) & E_{K_{1,2}}(p_{1,2}) & \cdots & E_{K_{1,\ell}}(p_{1,\ell}) \\ E_{K_{2,1}}(p_{2,1}) & E_{K_{2,2}}(p_{2,2}) & \cdots & E_{K_{2,\ell}}(p_{2,\ell}) \\ \vdots & \vdots & \ddots & \vdots \\ E_{K_{q,1}}(p_{q,1}) & E_{K_{q,2}}(p_{q,2}) & \cdots & E_{K_{q,\ell}}(p_{q,\ell}) \end{array}$$

A user $x_1 x_2 \cdots x_\ell$ can only decrypt one ciphertext in each column above (as a user has just one of the q decryption keys of each type). Indeed, the only key of type j the user knows is $K_{x_j,j}$, so the user obtains the jth segment $p_{x_j,j}$ marked with x_j (and cannot obtain any other version of the jth segment). Therefore the block the user knows will be marked with $x_1 x_2 \cdots x_\ell$ as before. So even though we are in a broadcast setting, our method of encryption simulates the situation of the subsection above, where every user gets a copy of the data marked in a different way.

The combinatorics behind this scheme is exactly the same as that in the previous subsection. So the concept of k-IPP code is useful even in the setting where the data provider is a broadcaster. Note however that implementing this scheme has a heavy price: the amount of data the provider needs to broadcast is q times as much as normal. So q should be as small as possible in this application.

3.3 Pirated Keys

Another realistic setting can be described as follows. (This is an application suggested by Chor, Fiat and Naor [15].) Access to the digital content is provided through a decoder box. The data is broadcast in encrypted form. Users who currently subscribe to the service are given a smartcard, a device which plugs into the decoder box and contains the keys needed to decrypt. The pirate is interested in distributing smartcards to provide access to the service, rather than in rebroadcasting the data itself (or in manufacturing new decoder boxes).

A system that could work in this setting is as follows. The data provider chooses keys $K_{i,j}$ as in the previous scheme. The user $x_1 x_2 \cdots x_\ell$ receives a smartcard containing the keys $K_{x_j,j}$, one key of each type, as before. The data provider then encrypts and broadcasts the data in such a way that a smartcard must contain a valid key of every type in order to work. If this is done, the only working smartcards the pirate can construct are those associated with words in $\mathrm{desc}(X)$, where X is the set of traitor codewords, and a traitor can always be identified from the smartcard if a k-IPP code is used.

Here is one way for the data provider to ensure a key of every type is needed to decrypt the data. The broadcast of each block of the data is preceded by a (short) *enabling block* that can be used to decrypt the data block that follows it. For each data block, the provider chooses quantities s_j for $1 \leqslant j \leqslant \ell$ uniformly and independently at random from the set of possible keys. Define a *session key* s by $s = s_1 \oplus s_2 \oplus \cdots s_\ell$, where $\oplus$ is some group operation on the set of keys. (In practice, keys are binary strings of a fixed length, and the operation $\oplus$ is XOR.) The enabling block consists of the values $E_{K_{i,j}}(s_j)$ for all i and j. Thus s_j is encrypted under every key of type j. The associated data block p is encrypted using the key s, to produce $E_s(p)$. For a user to decrypt the block $E_s(p)$, they need to know the session key s, and so need to know all the quantities $s_1, s_2, \ldots, s_\ell$. But in order to determine the value of s_j, a key $K_{i,j}$ of type j is needed. So a smartcard must contain a key of every type, as required.

4 Codes

This section is divided into three subsections, corresponding to three 'coding theoretic' problems based around the idea of a descendant of a set of codewords. Subsection 4.1 considers k-IPP codes as defined in Section 2; Subsections 4.2 and 4.3 consider the related notions of frameproof codes and secure frameproof codes. The area is a large one, so we are not able to deal with all results in full detail here. But we draw out some common themes. More precisely: all three types of codes can be constructed from error correcting codes of large minimum distance; all have a 'local' property that allows constructions from perfect hash families; this local property allows probabilistic existence results and upper bounds to be proved, that are comparable at least when the alphabet size is large. In addition, pointers to other results in the literature are given.

We will use some of the notation given in Section 2. So F is an alphabet of cardinality q, and the code C is a q-ary code of length ℓ with n codewords. Recall the definitions of $\mathrm{desc}(X)$ and $\mathrm{desc}_k(C)$ given by (2.1) and (2.2). Also recall the definition of the set $\mathcal{P}_{k,C}(d)$ of parent sets of a descendant d given by (2.3).

4.1 k-IPP Codes and Traceability Codes

4.1.1 Explicit Constructions Most known explicit constructions for k-IPP codes are of a stronger object: traceability codes. To define such codes, we need a notion of distance between a pair of words. Let $x, y \in F^\ell$. Recall that the *(Hamming) distance* $\mathrm{dist}(x, y)$ is defined by

$$\mathrm{dist}(x, y) = |\{i \in \{1, 2, \ldots, \ell\} : x_i \neq y_i\}|.$$

Let C be a q-ary length ℓ code, and let $d \in \mathrm{desc}_k(C)$. Let P_d be the set of codewords that are as close as possible to d. So defining $m = \min\{\mathrm{dist}(x, d) : x \in C\}$ we have that $P_d = \{x \in C : \mathrm{dist}(x, d) = m\}$. A code C is a k-*traceability code* if $P_d \subseteq X$ for all $X \in \mathcal{P}_{k,C}(d)$. Of course, this means that

$$\bigcap_{X \in \mathcal{P}_{k,C}(d)} X \supseteq P_d.$$

Since P_d is non-empty (by definition), this shows that every k-traceability code is a k-IPP code.

The notion of a traceability code predates that of a k-IPP code, going back to the original traitor tracing paper of Chor, Fiat and Naor [14]. The following theorem rephrases a result from that paper. Recall that an *error correcting code of minimum distance* δ is a code C such that $\mathrm{dist}(x, y) \geqslant \delta$ for all distinct $x, y \in C$. For an introduction to the theory of error correcting codes, see van Lint [30] for example.

Theorem 4.1 *Let C be a length ℓ error correcting code with minimum distance δ. If $\delta > (1 - 1/k^2)\ell$, then C is a k-traceability code.*

Proof Note that our condition on C is equivalent to the following. Let $x, y \in C$. If x and y agree in ℓ/k^2 (or more) positions, then $x = y$.

Let $d \in \mathrm{desc}_k(C)$, and let $x \in P_d$ be a codeword as close as possible to d. We must show that every parent set of d contains x.

We claim that d and x agree in at least ℓ/k positions. We prove this claim as follows. Since $d \in \mathrm{desc}_k(C)$, we have that $d \in \mathrm{desc}(X)$ for some subset $X \subseteq C$ where $|X| \leqslant k$. But for each $i \in \{1, 2, \ldots, \ell\}$ (by the definition of a descendant) the ith component of d agrees with the ith component of some codeword in X. Since X contains at most k codewords, there is a codeword in X that agrees with d in ℓ/k or more positions. Hence x (being a codeword as close as possible to d) agrees with d in ℓ/k or more positions. So the claim follows.

Let Y be a parent set of d. To prove the theorem, we need to show that $x \in Y$. Define $I = \{i \in \{1, 2, \ldots, \ell\} : x_i = d_i\}$. For all $i \in I$, d_i agrees with the ith component of some codeword in Y. Since $|Y| \leqslant k$, there exists $y \in Y$ such that $y_i = d_i$ for at least $|I|/k$ components $i \in I$. Since $d_i = x_i$ for all $i \in I$, we find that x and y agree in at least $|I|/k$ positions. But $|I| \geqslant \ell/k$ by the claim proved above. So x and y agree in at least ℓ/k^2 positions. Hence $x = y$ by the first paragraph of the proof, and so $x = y \in Y$ as required. $\square$

We remark that the proof of the theorem shows that we may soften our condition on the minimum distance δ slightly, to $\delta > \ell - \lceil \ell/k^2 \rceil$. We also remark that Hollmann *et al.* [26] independently constructed 2-IPP codes from error correcting codes using essentially the same arguments as the proof of Theorem 4.1.

There are many possibilities for error correcting codes that can be used with Theorem 4.1. Algebraic geometric codes are one possibility that is often mentioned — the codes of Garcia and Stichtenoth [23] are one explicit example. Standard probabilistic existence results for error correcting codes can be used if explicit constructions are not required. One often quoted possibility is a Reed–Solomon code, which may be described as follows. Suppose that q is a prime power such that $\ell \leqslant q$. Identify F with the finite field of order q, and let $\alpha_1, \alpha_2, \ldots, \alpha_\ell \in F$ be distinct. We identify C with the set of polynomials over

F of degree less than a fixed bound r, where a polynomial f corresponds to the codeword $(f(\alpha_1), f(\alpha_2), \ldots, f(\alpha_\ell))$. So (provided that $r \leqslant \ell$) there are q^r codewords. Note that (by interpolation) if two codewords agree in r positions then they are equal. So C has minimum distance greater than $n - r$. (In fact, C has minimum distance exactly $n - r + 1$.) In particular, if we choose $r = \lceil \ell/k^2 \rceil$, Theorem 4.1 and the remark after it shows that there exists a q-ary k-traceability code of length ℓ with $q^{\lceil \ell/k^2 \rceil}$ codewords.

Staddon, Stinson and Wei [44] observe that the much of the proof of Theorem 4.1 does need the full strength of the error correcting code condition. They make the following definitions. For $x \in F^\ell$ and $Y \subseteq F^\ell$, define

$$I(x, Y) = \{i \in \{1, 2, \ldots, \ell\} : x_i = y_i \text{ for some } y \in Y\}.$$

A code C is said to be (k, α)-*cover free* if for all $x \in C$ and all k-element subsets $Y \subseteq C$ such that $x \notin Y$ we have that $|I(x, Y)| < (1 - \alpha)\ell$. Staddon *et al.* prove that every error correcting code of minimum distance greater than $(1 - 1/k^2)\ell$ is a $(k, 1 - 1/k)$-cover free code (essentially using the argument in the first and last paragraphs of the proof of Theorem 4.1). They then prove that any $(k, 1 - 1/k)$-cover free code is a k-traceability code (essentially using the remainder of the proof of Theorem 4.1).

We now turn to another class of explicit constructions, of k-IPP codes from perfect hash families. An $(\ell; n, q, t)$-*perfect hash family* is a sequence $f_1, f_2, \ldots, f_\ell$ of functions from a set C of size n to a set F of size q, with the following property. For all $D \subseteq C$ such that $|D| \leqslant t$, there exists $i \in \{1, 2, \ldots, \ell\}$ such that f_i is injective when restricted to D. (We say that f_i *separates* D.) See Blackburn [8] for a survey of perfect hash families from a combinatorial point of view. The notation here suggests an obvious way of deriving a code from a perfect hash family: associate each $c \in C$ with the codeword $(f_1(c), f_2(c), \ldots, f_\ell(c))$. The following theorem is due to Staddon *et al.* [44].

Theorem 4.2 *Let $f_1, f_2, \ldots, f_c$ be an $(\ell; n, q, t)$-perfect hash family. Let C be code associated with this family, as above. Then C is a k-IPP code whenever $t \geqslant \lfloor (k+2)^2/4 \rfloor$.*

The next result, which explains where the quantity $\lfloor (k+2)^2/4 \rfloor$ in Theorem 4.2 comes from, is crucial to Staddon *et al.*'s proof.

Theorem 4.3 *A code C is a k-IPP code if and only if every subset $D \subseteq C$ such that $|D| \leqslant \lfloor (k+2)^2/4 \rfloor$ is a k-IPP code.*

Proof Clearly if C is a k-IPP code, then every subset D of C is a k-IPP code. So assume that C is not a k-IPP code. To prove the theorem it is sufficient to show that there exists a subset $D \subseteq C$ such that $|D| \leqslant \lfloor (k+2)^2/4 \rfloor$ and such that D is not a k-IPP code.

Since C is not a k-IPP code, there exists $d \in \mathrm{desc}_k(C)$ such that

$$\bigcap_{X \in \mathcal{P}_{k,C}(d)} X = \emptyset.$$

Let $\mathcal{M}$ be a subset of $\mathcal{P}_{k,C}(d)$, minimal with respect to the property that $\cap_{X \in \mathcal{M}} X = \emptyset$. Define $D = \cup_{X \in \mathcal{M}} X$. Note that $d \in \mathrm{desc}_k(D)$, since $d \in \mathrm{desc}(X)$ for all $X \in \mathcal{M}$. Moreover, since $\mathcal{P}_{k,D}(d) \supseteq \mathcal{M}$ we have that $\cap_{X \in \mathcal{P}_{k,D}(d)} X = \emptyset$ and so D is not a k-IPP code.

It remains to show that $|D| \leqslant \lfloor (k+2)^2/4 \rfloor$. Define $m = |\mathcal{M}|$ and write $\mathcal{M} = \{X_1, X_2, \ldots, X_m\}$. For $i \in \{1, 2, \ldots, m\}$, choose $x_i \in \cap_{j \in \{1,2,\ldots,m\} \setminus \{i\}} X_j$; we may do this since the right hand side is non-empty, by the minimality of $\mathcal{M}$. The elements x_i are distinct. For if $x_i = x_j$ when $i \neq j$ then x_i would be contained in all the subsets in $\mathcal{M}$, contradicting the fact that $\cap_{X \in \mathcal{M}} X = \emptyset$. Each set X_i contains exactly $m-1$ of the elements $x_1, x_2, \ldots, x_m$, and so at most $k - (m-1)$ other elements. Hence

$$|D| = \left| \cup_{i \in \{1,2,\ldots m\}} X_i \right| \leqslant m + m(k - (m-1)).$$

The right hand side is maximised when $m = \lceil k/2 \rceil + 1$, at $\lfloor (k+2)^2/4 \rfloor$. So $|D| \leqslant \lfloor (k+2)^2/4 \rfloor$ and the theorem is proved. □

Proof of Theorem 4.2 Suppose, for a contradiction, that C is not a k-IPP code. Then, by Theorem 4.3, there exists a subset $D \subseteq C$ such that D is not a k-IPP code, and $|D| \leqslant \lfloor (k+2)^2/4 \rfloor$. We derive our contradiction by showing that D is a k-IPP code.

Since $t \geqslant \lfloor (k+2)^2/4 \rfloor$, the perfect hash family condition implies that there exists $i \in \{1, 2, \ldots, \ell\}$ such that f_i is injective when restricted to D. This means that the ith components of the codewords in D are distinct. Let $d \in \mathrm{desc}_k(D)$. By the definition of descendant, any parent set $X \subseteq D$ of d must contain a codeword $x \in D$ such that $x_i = d_i$. But since the ith components of codewords in D are distinct, there is a unique codeword $x \in D$ with this property and hence $\cap_{X \in \mathcal{P}_{k,D}(d)} X \supseteq \{x\} \neq \emptyset$. Thus D is a k-IPP code, and this contradiction proves the theorem. □

Note that perfect hash family constructions cannot work when $q < \lfloor (k+2)^2/4 \rfloor$, as the corresponding perfect hash families do not exist.

We mention another explicit construction of k-IPP codes. The paper of Hollmann *et al.* [26] proves that an equidistant code of distance δ is a 2-IPP code whenever δ is odd, or δ is even and $\delta > (2/3)\ell$.

4.1.2 Probabilistic Results We now turn away from explicit constructions, and towards probabilistic existence results. There is a naive method that gives results: choose n codewords $c_1, \ldots, c_n \in F^\ell$ uniformly and independently at random, and estimate the probability that the resulting code is a k-IPP code. This idea was used by Chor *et al.* [15]. (This is a classical method: Mehlhorn [31] in the early 1980's used this approach to show the existence of perfect hash families.) But there are techniques that give better results than the naive method. Hollmann *et al.* [26] used Lovász's Local Lemma (see Spencer [43]) to prove the existence of good 2-IPP codes. This approach has been generalised k-IPP codes by Yemane [56]. But Barg, Cohen, Encheva, Kabatiansky and Zémor [5] were the first to give good probabilistic results for k-IPP codes for general values of k. Their results work for all feasible alphabet sizes. They make the following definition. Let u be an integer such that $k \leqslant u$. A code C is (k,u)*-partially hashing* if for all $D \subseteq C$ with $|D| = u$, and all $X \subseteq D$ with $|X| = k$ there exists $i \in \{1, 2, \ldots, \ell\}$ such that for any $y \in X$ and any $z \in D \setminus \{y\}$ we have that $y_i \neq z_i$. A code associated with an $(\ell; n, q, u)$-perfect hash family is (u, u)-partially hashing. Even though partially hashing codes are in general a weaker notion than codes derived from perfect hash families, an analogue of Theorem 4.2 holds.

Theorem 4.4 *Define* $u = \lfloor (k+2)^2/4 \rfloor$. *A* (k,u)*-partially hashing code with* u *or more codewords is a* k*-IPP code.*

Proof Let C be a (k,u)-hashing code. To prove that C is a k-IPP code, it suffices to prove that every subset $D \subseteq C$ such that $|D| = u$ is a k-IPP code, by Theorem 4.3.

Let $D \subseteq C$ with $|D| \leqslant u$. Let $d \in \mathrm{desc}_k(D)$. So there exists $X \subseteq D$ such that $|X| = k$ and $d \in \mathrm{desc}(X)$. By the partially hashing property of C, there exists $i \in \{1, 2, \ldots, \ell\}$ such that for any $y \in X$ and any $z \in D \setminus \{y\}$ we have that $y_i \neq z_i$. Since $d \in \mathrm{desc}(X)$, there exists $x \in X$ such that $x_i = d_i$. The partial hashing property implies that no other codeword $z \in D$ has $z_i = x_i = d_i$. Since any possible (k, D)-parent set of d must contain a codeword whose ith component is equal to d_i, we find that x is contained in any possible (k, D)-parent set of d. This implies that D is a k-IPP code, and so the theorem follows. □

Unlike perfect hash families, codes with the partial hashing property have not been extensively studied and so Theorem 4.4 does not currently give rise to new k-IPP codes. (Though a recent paper of Alon, Cohen, Krivelevich and Litsyn [1] does study partially hashing codes. They consider the case when $q = k+1$. They provide probabilistic existence results, and give upper bounds inspired by techniques due to Nilli [35] for perfect hash families.) As remarked above, Barg *et al.* [5] use Theorem 4.4 to provide a good probabilistic existence result for k-IPP codes.

Theorem 4.5 *Suppose that* $q \geqslant k+1$. *Define* $u = \lfloor (k+2)^2/4 \rfloor$. *A* q*-ary* k*-IPP code of length* ℓ *with* n *codewords exists whenever*

$$\binom{2n}{u}\binom{u}{k}\left(1 - \frac{q(q-1)\cdots(q-(k-1))(q-k)^{u-k}}{q^u}\right)^{\ell} \leqslant n. \tag{4.1}$$

Proof Define $N = 2n$. We choose N codewords $c_1, c_2, \ldots, c_N \in F^\ell$ uniformly and independently at random. We aim to remove n (or fewer) of these words, so that the code that remains has the (k,u)-partially hashing property.

For $I \subseteq \{1, 2, \ldots, N\}$ such that $|I| = u$, let $D = \{c_i : i \in I\}$. Let $J \subseteq I$ be such that $|J| = k$ and define $X = \{c_i : i \in J\}$. The probability that the partially hashing property fails for the subsets X and D is at most

$$\left(1 - \frac{q(q-1)\cdots(q-(k-1))(q-k)^{u-k}}{q^u}\right)^\ell. \tag{4.2}$$

(Note that this probability includes the failing situation when two of the words c_i in X are equal.) So the expected number E of pairs (X, D) of subsets where partially hashing property fails is at most $\binom{N}{u}\binom{u}{k}$ times the value in (4.2). If we disregard one codeword in X for every such pair, the remaining code is (k,u)-partially hashing, and there is a non-zero probability that the remaining code has at least n codewords provided that

$$E \leqslant \binom{N}{u}\binom{u}{k}\left(1 - \frac{q(q-1)\cdots(q-(k-1))(q-k)^{u-k}}{q^u}\right)^\ell \leqslant n.$$

By Theorem 4.4, the result now follows. □

We remark that the paper of Alon *et al.* [1] improves on Theorem 4.5 (in a good many cases) by using a non-uniform distribution to choose each coordinate of the codewords c_i.

So that we can see what Theorem 4.5 is saying more clearly, we let the parameters of the problem vary in two ways. Firstly, we let $\ell \to \infty$ with q and k fixed. The result shows that whenever $q \geqslant k+1$, there are infinite sequences of k-IPP codes with $q^{R\ell}$ or more codewords, where R is any positive real number such that

$$R < \frac{1}{u-1} \log_q \frac{(q-k)!q^u}{(q-k)!q^u - q!(q-k)^{u-k}}.$$

Secondly, suppose that $q \to \infty$ with ℓ and k fixed. In this case, the large bracket in (4.1) is of the order of $q^{-\ell}$, and so the theorem shows that codes having of the order of $q^{\ell/(u-1)}$ codewords exist. This is significantly better than the explicit construction based on Reed–Solomon codes above, which produces codes of size $q^{\lceil \ell/k^2 \rceil}$. Of course, the Reed–Solomon code constructs a k-traceability code.

Open Problem 1 *Is the Reed–Solomon code construction of a k-traceability code optimal, for large alphabet sizes? More generally, are there families of k-traceability codes that have rate significantly better than those constructed by Theorem 4.1?*

Open Problem 2 *Fix q and k, with $q \leqslant k^2$. Are there classes of 'large' k-traceability codes for infinitely many lengths? By 'large' is meant* $(\log_q n)/\ell$ *tends to a non-zero value as $\ell \to \infty$.*

The above open problem is due to Barg and Kabatiansky [6]. Note that a recent paper of van Trung and Martirosyan [53] constructs k-traceability codes with $n > q$ for small alphabet sizes, answering an earlier question of Staddon *et al.* [44], but these codes are not large enough to settle the above open problem.

4.1.3 Bounds on the Code Size We now turn to 'negative' results concerning k-IPP codes. The next theorem shows that the condition $q \geqslant k+1$ above is a natural one.

Theorem 4.6 *Suppose that C is a q-ary k-IPP code. If $q \leqslant k$, then C has at most q codewords.*

The repetition code C defined by $C = \{11\cdots 1, 22\cdots 2, \ldots, qq\cdots q\}$ is a k-IPP code for all k. Theorem 4.6 asserts that when $q \leqslant k$ this trivial example is the best one could hope for.

Proof of Theorem 4.6 Suppose that C contains $q+1$ or more codewords. Let $D \subseteq C$ be such that $|D| = k+1$. Since $k+1 > q$, the pigeonhole principle implies that for all $i \in \{1, 2, \ldots, \ell\}$ there is a value d_i taken by the ith components of two or more codewords in D. Define the word d by $d = d_1 d_2 \cdots d_\ell$. Then d is a descendant of any k-element subset of D. Since the intersection of all k-element subsets of D is empty, the k-IPP property is violated and so C is not a k-IPP code. □

There are several other bounds on k-IPP codes. For example, Theorem 4.14 below applies to k-IPP codes, and a bound on the size of certain set systems due to Erdős, Frankl and Füredi [19] can be used to provide bounds on k-IPP codes; see Staddon *et al.* [44]. Blackburn [10] proves the following theorem (see [10] for a proof).

Theorem 4.7 *Let $u = \lfloor (k+2)^2/4 \rfloor$. A q-ary k-IPP code of length ℓ contains at most $\frac{1}{2}u(u-1)q^{\lceil \ell/(u-1) \rceil}$ codewords.*

This generalises a bound of Hollmann *et al.* [26], who prove a similar bound in the case when $k = 2$. There is some evidence that this bound has the right order of magnitude when q is large. Theorem 4.5 shows that k-IPP codes with of the order of $q^{\ell/(u-1)}$ codewords exist when q is large, and so the bound is of the right form when $u-1$ divides ℓ. And Alon, Fischer and Szegedy [2] have used probabilistic techniques to show the following. Let ϵ be a real number such that $\epsilon > 0$. Then there exists a constant q_ϵ such that for all $q > q_\epsilon$ there exist q-ary 2-IPP codes of length 4 with more than $q^{2-\epsilon}$ codewords. This again shows that the bound is the right order of magnitude in this case. Finally, we mention a recent preprint of Alon and Stav [3], which contains an analogous bound to Theorem 4.7 but with a better constant term.

Open Problem 3 *Is the bound of Theorem 4.7 of the right order of magnitude for large q? More precisely, for fixed ℓ and k is it the case that for all positive ϵ and for all sufficiently large q there exist q-ary k-IPP codes of length ℓ with more than $q^{\lceil \ell/(u-1)\rceil-\epsilon}$ codewords? Here, as before, $u = \lfloor (k+2)^2/4 \rfloor$.*

4.1.4 Identifying Parents Efficiently Using a k-IPP code means that we can identify at least one parent of any descendant d. But the concept of a k-IPP code does not say anything about how this identification might be carried out in practice. There is an obvious method — try all k-element subsets of C in turn, determine which are parent sets of d, then form their intersection — that takes about n^k operations and so is inefficient when k is not 2 or 3. A general k-traceability code is better in this regard. It takes roughly n operations to find a nearest codeword to d, and this might well be practical. However, there are much better algorithms that work for specific classes of codes. Silverberg, Staddon and Walker [42] use a recent list decoding algorithm of Guruswami and Sudan [25] to improve the parent identification algorithm of those traceability codes based on Reed–Solomon codes (and other algebraic geometry codes). These algorithms run in time polynomial in $k \log n$. They credit Zane [57] as being the first to use such list decoding techniques in the context of watermark security. In addition, Silverberg *et al.* discuss the related problem of finding a list of all possible parent sets of d (which is a byproduct of the naive IPP algorithm), and give an algorithm that improves on the naive approach for certain classes of IPP codes. Barg and Kabatiansky [6] also construct IPP codes where parents may be traced efficiently. They also use Guruswami and Sudan decoding for algebraic-geometry codes in their construction, but concatenate these codes with linear codes of high minimum distance to keep the alphabet size of their codes small. Classes of k-IPP codes with efficient parent identification (and a reasonable number of codewords) are constructed whenever $q \geqslant k+1$. Finally, a recent preprint of van Trung and Martirosyan [52] constructs a class of IPP codes in a recursive fashion, using concatenation repeatedly. Because of their recursive nature, the resulting IPP codes have decoding algorithms that are linear in n — better than the naive algorithm for IPP codes.

4.2 Frameproof Codes

It might be the case that, rather than requiring that a traitor be identified from a pirate version of some data, we might just require the following weaker property: a pirate should be unable to create a version that is identical to that of a user who is not a traitor. In other words, we require that a pirate should not be able to 'frame' an innocent user. This idea is captured in the definition of a frameproof code. A code C is *k-frameproof* if for all $X \subseteq C$ with $|X| \leqslant k$, we have that

$$\text{desc}(X) \cap C = X.$$

Frameproof codes were first introduced by Boneh and Shaw [13]. (However, their definition of a descendant was different, at least in the non-binary case. See the discussion in Section 5.) The following simple example is from Blackburn [9]. Let $F = \{0, 1, 2, \ldots, q-1\}$, and let C consist of all weight 1 words (words with exactly one non-zero component) in F^ℓ. It is easy to see that C is a k-frameproof code for any k. For let $X \subseteq C$, and let $d \in \text{desc}(X) \cap C$. Since d is a codeword, it has a unique non-zero component: the ith, say. Since $d \in desc(X)$, we have that $d_i = x_i$ for some $x \in X$. But a codeword is determined by the value and position of its non-zero component, and so $d = x$. We have shown that $\text{desc}(X) \cap C \subseteq X$. But $X \subseteq \text{desc}(X) \cap C$ holds for any code, and so C is k-frameproof.

It can be shown [9] that the above, rather trivial, example is optimal whenever $\ell \leqslant q$. However as the length rises other, larger, codes can be constructed. Indeed, any k-IPP code is a k-frameproof code (as was noted by Staddon *et al.* [44]). For suppose $X \subseteq C$ is such that $|X| \leqslant k$ and there exists $d \in \text{desc}(X) \cap C$ such that $d \notin X$. Then $\{d\}$ and X are disjoint parent sets for d, and so no parent can be identified. Frameproof codes generally behave rather like IPP codes (but the proofs are often easier). For example, we have an analogue to Theorem 4.1 (the core idea behind this theorem is due to Boneh and Shaw [13]).

Theorem 4.8 *Let C be a length ℓ error correcting code with minimum distance δ. If $\delta > (1 - 1/k)\ell$, then C is a k-frameproof code.*

Proof Our condition on C is equivalent to the property that whenever two codewords $x, y \in C$ agree in ℓ/k or more positions, then $x = y$.

Let $X \subseteq C$ be such that $|X| \leqslant k$. Let $d \in \text{desc}(X) \cap C$. Since $d \in \text{desc}(X)$, every component of d agrees with one of the corresponding components of some codeword in X. There are at most k parents, and so there is an element $x \in X$ that agrees with d in ℓ/k or more positions. Since $x, d \in C$ this implies that $d = x$. Hence C is k-frameproof. □

In fact, the proof shows that we may weaken the condition on δ in the theorem to $\delta > \ell - \lceil \ell/k \rceil$. Theorem 4.8 implies, for example, that there exists a k-frameproof code with $q^{\lceil \ell/k \rceil}$ codewords whenever q is a prime power and $q \geqslant \ell$. (This is using a Reed–Solomon code, as in the remarks after Theorem 4.1.)

One of the themes running through Subsection 4.1 was that the combinatorics was affected by the number of codewords involved in a minimal counterexample to the k-identifiable parent property (namely $\lfloor (k+2)^2/4 \rfloor$). The same is true for k-frameproof codes.

Theorem 4.9 *A code C is a k-frameproof code if and only if every subset $D \subseteq C$ such that $|D| = k+1$ is k-frameproof.*

We leave the details of the proof to the reader. But the main point of the proof is that a counterexample to the k-frameproof property involves at most $k+1$ codewords (the set X and a codeword $d \in (\mathrm{desc}(X) \cap C) \setminus X$).

It is not difficult to show that a code is k-frameproof if for all subsets of $k+1$ codewords there exists $i \in \{1, 2, \ldots, \ell\}$ such that the ith components of these codewords are distinct. This observation gives rise to the analogue of Theorem 4.2 (due to Staddon, Stinson and Wei [44]).

Theorem 4.10 *Let $f_1, f_2, \ldots, f_c$ be an $(\ell; n, q, t)$-perfect hash family. Let C be the code associated with this family. Then C is a k-frameproof code whenever $t \geqslant k+1$.*

There have been other constructions of frameproof codes, concentrating on the binary case. Stinson and Wei [48] relate binary k-frameproof codes to certain set systems, and use this approach to construct frameproof codes from t-designs and packing designs. (These constructions perform well when parameter sizes are small.) Stinson and Wei also prove the following nice characterisation of 2-frameproof codes.

Theorem 4.11 *A binary code is 2-frameproof if and only if for all $x, y, z \in C$*

$$\mathrm{dist}(x, y) < \mathrm{dist}(x, z) + \mathrm{dist}(z, y).$$

Proof Let C be a binary code. For binary codewords $x, y, z \in C$, we have that $z \in \mathrm{desc}(\{x, y\})$ if and only if whenever $i \in \{1, 2, \ldots, \ell\}$ is such that $x_i = y_i$ we have that $z_i = x_i$. So C fails to be a binary 2-frameproof code if and only if there exist distinct codewords x, y and z such that whenever $x_i = y_i$ we have that $x_i = z_i$.

Let x, y and z be distinct binary codewords of length ℓ. Define the integer r by $r = \mathrm{dist}(x, y)$. Suppose, without loss of generality, that x and y agree in their first $\ell - r$ positions and disagree on their final r positions.

Let s_1 be the number of disagreements between x and z in the first $\ell - r$ positions, let s_2 be the number of disagreements in the last r positions. So y and z disagree in s_1 of the first $\ell - r$ positions, and (since the words are binary) they disagree in $r - s_2$ of the last r positions. Thus

$$\mathrm{dist}(x, z) + \mathrm{dist}(z, y) = (s_1 + s_2) + (s_1 + r - s_2) = 2s_1 + r = 2s_1 + \mathrm{dist}(x, y).$$

Hence $\mathrm{dist}(x, y) = \mathrm{dist}(x, z) + \mathrm{dist}(z, y)$ if and only if $s_1 = 0$. But $s_1 = 0$ means that $x_i = z_i$ whenever $x_i = y_i$, and the first paragraph of the proof shows that this situation occurs for distinct $x, y, z \in C$ if and only if C is not 2-frameproof. The theorem now follows. □

Xing [55] has constructed frameproof codes from algebraic curves, resulting in better parameters than an approach via error correcting codes. Finally, a paper of Safavi-Naini and Wang [37] uses a beautiful construction of constant weight codes due to Graham and Sloane [24] to construct binary k-frameproof codes. Their main result can be summarised as follows.

Theorem 4.12 *Suppose that ℓ is a prime power. Let w be an integer such that $1 \leqslant w \leqslant q$, and let r be an integer such that $r > (1 - \frac{1}{k})w$. Provided that $r \geqslant 3$, there exist binary k-frameproof codes of length ℓ with at least*

$$\frac{1}{\ell^r}\binom{\ell}{w}$$

codewords.

We now turn towards probabilistic results. Given Theorem 4.10, an analogue of Theorem 4.5 is easy to prove:

Theorem 4.13 *Suppose that $q \geqslant k + 1$. A q-ary k-frameproof code of length ℓ with n codewords exists whenever*

$$\binom{2n}{k+1}\binom{k+1}{k}\left(1 - \frac{q(q-1)\cdots(q-k)}{q^{k+1}}\right)^{\ell} \leqslant n. \qquad (4.3)$$

Proof An $(\ell; n, q, k+1)$-perfect hash family is equivalent to a q-ary $(k, k+1)$-partially hashing code of length ℓ with n codewords (just take the associated code). The proof of Theorem 4.5 (with $u = k+1$) shows that a q-ary $(k, k+1)$-partially hashing code C of length ℓ with n codewords exists whenever (4.3) holds. Theorem 4.10 now implies that C is a k-frameproof code. □

When $q \to \infty$ with ℓ and k fixed, Theorem 4.13 shows that there are k-frameproof codes with of the order of $q^{\ell/k}$ codewords. In contrast to the k-IPP case, for large alphabet sizes the probabilistic results are no better than the coding theoretic constructions, especially when k divides ℓ.

We now turn to upper bounds. There are large binary k-frameproof codes, and so there is no equivalent of Theorem 4.6. The following theorem is due to Staddon *et al.* [44].

Theorem 4.14 *Let C be a q-ary k-frameproof code of length ℓ with n codewords. Then*

$$n \leqslant kq^{\lceil \ell/k \rceil}.$$

Proof For $I \subseteq \{1, 2, \ldots, \ell\}$, define U_I to be the set of codewords uniquely defined by their components in I:

$$U_I = \{x \in C : \forall y \in C \setminus \{x\}\ \exists i \in I \text{ such that } x_i \neq y_i\}.$$

There are at most $q^{|I|}$ choices for the ith components of a codeword for $i \in I$, and (by definition of U_I) each codeword in U_I is uniquely specified by such a choice. Hence $|U_I| \leqslant q^{|I|}$.

Suppose that $I_1, I_2, \ldots, I_k \subseteq \{1, 2, \ldots, \ell\}$ cover $\{1, 2, \ldots, \ell\}$ (in other words $I_1 \cup I_2 \cup \cdots \cup I_k = \{1, 2, \ldots, \ell\}$). We claim that

$$C = U_{I_1} \cup U_{I_2} \cup \cdots \cup U_{I_k}.$$

For suppose not. Let $d \in C \setminus (U_{I_1} \cup U_{I_2} \cup \cdots \cup U_{I_k})$. Since $d \notin U_{I_j}$, there exists a distinct codeword x^j that agrees with d on those components in I. Define $X = \{x^1, x^2, \ldots, x^k\}$. Since the I_j cover $\{1, 2, \ldots, \ell\}$, we have that $d \in \mathrm{desc}(X)$. Moreover, $d \notin X$ and so the k-frameproof property is violated. This establishes our claim.

Choose the subsets I_j to cover $\{1, 2, \ldots, \ell\}$ and have that property that $|I_j| = \lceil \ell/k \rceil$ for all j. Then

$$|C| \leqslant \sum_{j=1}^{k} |U_{I_j}| \leqslant kq^{\lceil \ell/k \rceil},$$

as required. □

The Reed–Solomon code example shows that the upper bound is of the right order of magnitude when q is large.

The constant k at the front of the upper bound can be reduced in many cases. Indeed, Blackburn [9] gives an upper bound with an improved constant, and relates this constant to a question in the theory of set systems. (A paper of Sarkar and Stinson [39] generalises an argument in an earlier version of [9], but gets weaker results than the final version of the paper.)

Open Problem 4 *Define $M(q, k, \ell)$ to be the largest number of codewords of a q-ary k-frameproof code of length ℓ. For k and ℓ fixed, what is*

$$\lim_{q \to \infty} M(q, k, \ell)/q^{\lceil \ell/k \rceil},$$

(if this limit exists)?

4.3 Secure Frameproof Codes

In this section, we consider one last combinatorial structure that is associated with the idea of descendant. In this application, we do not want to identify a traitor: we just want to find a set of k users, at least one of whom is a traitor. (Theorem 4.6 shows that large q-ary k-IPP codes do not exist when $q \leqslant k$. If the application nevertheless dictates that $q \leqslant k$, then we must be satisfied with a weaker form of traitor identification. This motivated the definition of secure frameproof code.)

A code C is *k-secure frameproof* if for all $d \in \text{desc}_k(C)$, any two parent sets $X, Y \in \mathcal{P}_{k,C}(d)$ intersect non-trivially. The notion of a secure frameproof code was introduced by Stinson, van Trung and Wei [47]. Secure frameproof codes lie somewhere between k-IPP codes and k-frameproof codes. A k-IPP code is clearly a k-secure frameproof code (the intersection of all parent sets being nonempty implies that any pair of parent sets intersect nontrivially). Moreover, k-secure frameproof codes are k-frameproof. To see this, suppose that C is a k-secure frameproof code. If $d \in \text{desc}(X) \cap C$ for some $X \subseteq C$ with $|X| \leqslant k$, then $\{d\}$ and X are both parent sets of d and so intersect non-trivially. Hence $d \in X$, and therefore C is k-frameproof.

The following simple example of a k-secure frameproof code C is given in Stinson *et al.* [47]. The code C is binary of length $\ell = \binom{2k-1}{k-1}$, and has $2k$ codewords. Let $B_1, B_2, \ldots, B_\ell$ be the k-element subsets of $\{1, 2, \ldots, 2k\}$ containing 1. The ith codeword $c^i \in C$ is defined to be $c_{i1}c_{i2} \cdots c_{i\ell}$, where $c_{ij} = 1$ when $i \in B_j$ and $c_{ij} = 0$ otherwise. So in the case $k = 2$, listing the 2-element subsets of $\{1, 2, 3, 4\}$ containing 1 as $\{1, 2\}, \{1, 3\}, \{1, 4\}$ we find that

$$C = \{111, 100, 010, 001\}.$$

To see why the code is k-secure frameproof, consider disjoint subsets $X, Y \subseteq C$ where $|X| \leqslant k$ and $Y \leqslant k$. We must show that $\text{desc}(X) \cap \text{desc}(Y) = \emptyset$. Define $I, J \subseteq \{1, 2, \ldots, 2k\}$ so that $X = \{c^i : i \in I\}$ and $Y = \{c^j : j \in J\}$. As I and J are disjoint, there exists a k-element subset K of $\{1, 2, \ldots, 2k\}$ containing one of I and J and disjoint from the other. By replacing K by its complement if necessary, we may assume that $1 \in K$, and so $K = B_j$ for some $j \in \{1, 2, \ldots, \ell\}$. Without loss of generality, assume that $I \subseteq B_j$ and $J \cap B_j = \emptyset$. The definition of C implies that the jth component of every codeword in X is 1, and so the same is true of any element of $\text{desc}(X)$. Similarly, the jth component of every element of $\text{desc}(Y)$ is 0. Hence $\text{desc}(X) \cap \text{desc}(Y) = \emptyset$, as required.

Secure frameproof codes can be constructed from error correcting codes. See Encheva and Cohen [18], for example. Interestingly, it seems that the best construction of k-secure frameproof codes from error correcting codes is via Theorem 4.1. (We do not seem to be able to weaken the condition on δ in Theorem 4.1 if we only require C to be k-secure frameproof rather than a k-traceability code.) However, the constructions from perfect hash families and the probabilistic methods give much better results when compared to k-IPP codes. The following three theorems are analogous to theorems given earlier for IPP and frameproof codes (but I have not been able to find these results in the literature). Their proofs turn on the fact that there are $2k$ codewords involved in a minimal counterexample to the k-secure frameproof property.

Theorem 4.15 *Let $f_1, f_2, \ldots, f_c$ be an $(\ell; n, q, t)$-perfect hash family. Let C be code associated with this family. Then C is a k-secure frameproof code whenever $t \geqslant 2k$.*

Theorem 4.16 *Suppose that $q \geqslant k+1$. A q-ary k-secure frameproof code of length ℓ with n codewords exists whenever*

$$\frac{1}{2}\binom{2n}{2k}\binom{2k}{k}\left(1-\frac{q^k(q-k)^k}{q^{2k}}\right)^\ell \leqslant n. \tag{4.4}$$

(Let two sequences $s_1, s_2, \ldots, s_k$ and $t_1, t_2, \ldots, t_k$ over F be chosen at random. So the terms of the sequences are picked uniformly and independently from the alphabet F. The expression $\frac{q^k(q-k)^k}{q^{2k}}$ in Theorem 4.16 may be replaced by the probability that $s_i \neq t_j$ for all $i, j \in \{1, 2, \ldots, k\}$.)

Theorem 4.16 shows that (for large q) there are k-secure frameproof codes with about $q^{\ell/(2k-1)}$ codewords. There is a corresponding upper bound (which is proved in a similar way to Theorem 4.7).

Theorem 4.17 *Let C be a q-ary k-secure frameproof code of length ℓ with n codewords. Then*

$$n \leqslant 2k^2 q^{\lceil \ell/(2k-1) \rceil}.$$

As Stinson *et al.* [47] pointed out, binary k-secure frameproof codes are related to (i, j)-separating systems, as defined by Graham, Friedman and Ullman [22]. Indeed, a binary k-secure frameproof code is equivalent to a (k, k)-separating system. This problem (and q-ary generalisations) has been much studied in the context of the theory of automata; we refer to Sagalovich [38] for a survey of the area. Also see Cohen, Encheva and Schaathun [16] and Barg, Blakley and Kabatiansky [4] and the references there. We should also mention that Sarkar and Stinson [39] have studied codes where any r parent sets should have non-empty intersection, where r is an extra parameter. Such codes range between k-IPP codes (the case $r \geqslant k+1$)and k-secure frameproof codes (the case $r = 2$). However, these codes do not have applications at present.

5 Other Models

Section 4 concentrated on codes with properties related to the notion of descendant defined in Section 2. This section considers generalisations and variations of the coding theoretic model.

5.1 Dynamic Models

The model we have been using up to now does not allow the data provider to vary the data they broadcast depending on information from the pirate. (The fingerprinting marks, or the assignments of keys to smartcards, are determined before the traitors pass any information on to the pirate.) In this sense, the model we have been using is static. This subsection considers dynamic models, which are useful in the following situation. Suppose a subscription internet data provider detects that its data is being broadcast from a new unauthorised source. Suppose it has the facility to transmit different versions of the data to different authorised users. The provider could partition the set of users into q parts, sending a different version of the data to each part. If version i appears in the unauthorised broadcast, the provider knows that a user in part i of the partition is a traitor. (As before, we assume that a pirate in possession of several versions of the same segment has no choice but to broadcast one of the versions she has.) After obtaining this information, the provider can decide how to partition the users into q parts when sending the next segment of data. This process carries on until a traitor is identified. One this is done, we assume that the traitor can be disconnected from the system (his keys could be revoked in some way, for example) and the whole process starts again with a smaller number of traitors. In this way, every traitor can eventually be detected and removed.

This model, under the name 'dynamic traitor tracing' was introduced by Fiat and Tassa [21]. The main question is: assuming there are at most k traitors, what is the most efficient method of tracing all traitors for a fixed alphabet size q and a fixed number n of users? By efficient, we mean that the number ℓ of segments needed should be as small as possible in the worst case. The number of possible sets of traitors is equal to the number of non-trivial subsets of an n-element subset of size at most k. Since each segment received from the pirate contains at most $\log q$ bits of information, we need that

$$\ell \log q \geqslant \log\left(\sum_{i=1}^{k}\binom{n}{i}\right). \tag{5.1}$$

For reasonable parameter sizes (n large and k small) the bracketed expression on the right hand side of (5.1) is about n^k, and so we would expect that a good method for detecting all traitors would require of the order of $k\log_q n$ segments.

It is not difficult to show that dynamic traitor tracing cannot work when $q \leqslant k$. (As in Theorem 4.6, the pirate broadcasts any version of a segment that two or more traitors receive.) So we assume that $q \geqslant k+1$.

There is a natural algorithm, due to Fiat and Tassa [21], that works when $q = 2k+1$ (or larger). It can be described as follows. At each stage, the data provider's knowledge can be summarised as a collection of pairwise disjoint sets $C_1, C_2, \ldots, C_t, I$ that partition the set C of users. So $C_1 \cup C_2 \cup \cdots \cup C_t \cup I = C$. At any stage, each C_i is known to contain at least one traitor (so $t \leqslant k$). The set I consists of those users not in any of the sets C_i. Initially $t = 1$, $C_1 = C$ and $I = \emptyset$. If at any stage we have that $|C_i| = 1$, a traitor has been identified and can be removed (along with the set C_i) and so we may assume that $|C_i| \geqslant 2$ for $i \in \{1, 2, \ldots, t\}$. Before broadcasting a segment, the provider splits each C_i into two (almost) equal parts: $C_i = L_i \cup R_i$, where $0 \leqslant |L_i| - |R_i| \leqslant 1$ and $L_i \cap R_i = \emptyset$. The provider then broadcasts a different version of a segment to each of the $2t+1$ sets I, L_i and R_i. Since $2t+1 \leqslant 2k+1 \leqslant q$, the provider is able to do this. If the pirate broadcasts a version given to members of some L_i, the provider has narrowed down the location of one of the traitors. He sets $C_i = L_i$ and replaces I by $I \cup R_i$. Similarly, if a version given to a member of R_i is received, he sets $C_i = R_i$ and $I = I \cup L_i$. Finally, if the pirate version is the one given to members of I, the provider is now aware of the existence of another traitor. He increases t by one, then sets $C_t = I$ and finally $I = \emptyset$. It is not difficult to see that this algorithm eventually identifies all traitors. (The algorithm also has the nice property that the value of k does not need to be known in advance.) The value of t is increased at most $k-1$ times. There are at most k sets C_i, each is of size at most n when it is created and its size is halved each time it is modified. So each C_i is modified at most $\log n$ times. These observations show that this algorithm takes of the order of $k \log n$ segments to identify all traitors.

Fiat and Tassa give an algorithm that works when $q = k+1$, but their algorithm is exponential in k and so impractical in most situations. In a beautiful paper, Berkman, Parnas and Sgall [7] give two practical algorithms and one impractical but asymptotically good algorithm for dynamic traitor tracing in the case when $q = k+1$. (These algorithms again have the nice property that the value of k does not need to be known in advance.) The two practical algorithms use of the order of $k^2 \log n$ segments. The impractical algorithm uses of the order of $k^2 + k \log n$ segments (which is proved to be the best possible behaviour). However, this last algorithm is very complicated, and is impractical due to large hidden constants. Berkman *et al.* ask:

Open Problem 5 *Does there exist a simple and practical algorithm to solve the dynamic traitor tracing problem in the case when $q = k+1$ using of the order of $k^2 + k \log n$ segments?*

A restriction of the dynamic traitor tracing model, called sequential traitor tracing, was introduced by Safavi-Naini and Wang [36]; the full version of their paper may be obtained from the first author. Their model is closer to the coding theoretic models of Section 4. The idea is to remove the dynamic character of the broadcasts, but to keep one element of the dynamic nature of the problem: once a traitor has been identified, they are automatically disabled and can no longer be used by the pirate. The motivation behind this definition is to reduce computation by the broadcaster, and more importantly to reduce the impact of delayed broadcasting by the pirate on the efficiency of the tracing process.

An outline of the model is as follows. Users are identified with q-ary codewords taken from a code C. Let $X \subseteq C$ be a set of traitors. At a given time, the provider has identified (and removed) a subset $U \subseteq X$ of traitors. At time i, the data provider broadcasts to each user $x \in C \setminus U$ a segment marked with $x_i \in F$. By examining the mark on the next segment broadcast by the pirate, the data provider obtains an element $d_i \in \{x_i : x \in X \setminus U\}$. After receiving this information, the provider tries to identify some traitors in $X \setminus U$ (by using some algorithm $\mathcal{A}$ of his choice). He then adds these traitors to U. The code C is a *sequential k-IPP code* if there exists an algorithm $\mathcal{A}$ such that for all $X \subseteq C$ with $|X| \leqslant k$ this process always terminates when $i \leqslant \ell$ in the state where $U = X$. We note that Safavi-Naini and Wang use the terminology 'sequential k-traceability scheme' rather than 'sequential k-IPP code'. We use the latter terminology, since (as Safavi-Naini and Wang observe) every sequential k-IPP code is in fact a k-IPP code.

Safavi-Naini and Wang give examples of sequential k-IPP codes based on error correcting codes; these examples are similar to the traceability code constructions in Subsection 4.1. In fact, rather surprisingly, a k-IPP code constructed from an error correcting code as in Theorem 4.1 is also a sequential k-IPP code. The following theorem is a minor improvement on a result in the current version of the paper of Safavi-Naini and Wang.

Theorem 5.1 *Let C be a length ℓ error correcting code with minimum distance δ. If $\delta > (1 - 1/k^2)\ell$, then C is a sequential k-traceability code.*

Proof Our condition on δ means that any two codewords agreeing in ℓ/k^2 or more positions must, in fact, be equal. The following algorithm will be used as algorithm $\mathcal{A}$. In the algorithm, the integer t is the number of traitors yet to be identified, the integer i is the position that is currently being examined, and the subset U is the set of traitors that have already been identified. At some points, the algorithm adds a mark to a position $i \in \{1, 2, \ldots, \ell\}$; all positions are initially unmarked. A mark is added to a position i when d_i could have been produced by an identified traitor. Note that it is quite possible for the algorithm to identify more than one traitor when examining a fixed value of i. Also note that when there are less than k traitors, the algorithm finishes early: another symbol d_i is requested at line 2(a), but none is forthcoming since all traitors have been disabled. In this case, the algorithm terminates successfully.

1. Set $t = k$, $i = 0$ and $U = \emptyset$.
2. While $t \neq 0$:
   (a) Set $i = i + 1$. Receive a new, symbol $d_i \in \{x_i : x \in X \setminus U\}$.
   (b) While there exists $u \in C \setminus U$ whose initial i components match $d_1 d_2 \cdots d_i$ in $t\ell/k^2$ (or more) unmarked positions:
      (i) Let $u \in C \setminus U$ match $d_1 d_2 \cdots d_i$ in $t\ell/k^2$ (or more) unmarked positions.
      (ii) Set $t = t - 1$ and $U = U \cup \{u\}$.
      (iii) Mark any unmarked position j such that $u_j = d_j$.
      (iv) Endwhile.
   (c) Endwhile.
3. Output U.

Consider the algorithm above. For all $X \subseteq C$ with $|X| \leqslant k$, we must prove that $U \subseteq X$ at all times, and that the algorithm terminates before i becomes too large.

Position j is marked by the algorithm precisely when there is a codeword in U that could have given rise to d_j, so if position j is unmarked then $d_j \in \{x_j : x \in X \setminus U\}$. We now use an argument from the proof of Theorem 4.1 to show that any word u added to U lies in X. By line 2(b)(i) of the algorithm, u agrees with $d_1 d_2 \cdots d_i$ in more than $t\ell/k^2$ unmarked positions. There are t unidentified traitors, and so an unidentified traitor, say $x \in X$, agrees with u in ℓ/k^2 or more positions. But our condition on the minimum distance of C now shows that $u = x \in X$ and so u is a traitor as required.

We now show that the algorithm always terminates when $i \leqslant \ell$. We first observe that the algorithm always identifies at least one traitor. To see this, note that (since X has at most k elements) there is an element $u^1 \in X$ agreeing with $d_1 d_2 \cdots d_\ell$ in at least ℓ/k positions. Since $\ell/k = k\ell/k^2$, the codeword u^1 is identified as a traitor by the algorithm. Note that exactly $\lceil \ell/k \rceil$ positions are marked when u^1 is identified.

At the moment u^1 is identified as a traitor, all other codewords agree with $d_1 d_2 \cdots d_i$ in less than $\lceil \ell/k \rceil$ unmarked positions. Now, the expression $t\ell/k^2$ drops by more than 1 when t is decremented (as we must have $\ell \geqslant k^2$ for C to be non-trivial). So the inequality in steps (b) and (b)(i) of the algorithm becomes easier to satisfy after u^1 is added to U. Hence each traitor that is identified after u^1 causes at most $\lceil \ell/k \rceil - 1$ positions to be marked.

Suppose, for a contradiction, that the algorithm has passed through the outer while loop ℓ times, and not all traitors have been identified. By the previous paragraph, the number of marked positions is at most

$$\lceil \ell/k \rceil + (k - t - 1)(\lceil \ell/k \rceil - 1).$$

No codeword in $C \setminus U$ can agree with $d_1 d_2 \cdots d_\ell$ in $t\ell/k^2$ or more unmarked positions, otherwise it would have been identified as a traitor. But each element d_i must agree with the ith component of some traitor, and so the number of unmarked positions is at most $t(\lceil t\ell/k^2 \rceil - 1)$. Every position is either marked or unmarked, so

$$\lceil \ell/k \rceil + (k - t - 1)(\lceil \ell/k \rceil - 1) + t(\lceil t\ell/k^2 \rceil - 1) \geqslant \ell.$$

But this inequality is never satisfied (use the facts that $\lceil t\ell/k^2 \rceil < \lceil \ell/k \rceil$ and that $\lceil \ell/k \rceil \leqslant \ell/k + (k-1)/k$ to derive the contradiction). This contradiction establishes the theorem. □

It is interesting that the sequential property of codes based on error correcting codes of high minimum distance comes essentially for free (the bounds on δ in Theorems 4.1 and 5.1 are the same). In the case of (non-sequential) k-IPP codes, there are constructions not based on error correcting codes that have better parameters than the known families of k-traceability codes. Is this also the case in the sequential setting?

Open Problem 6 *In the case of a sequential q-ary k-IPP code of length ℓ with n codewords, what is the maximum value that $(1/\ell)\log_q n$ that can be obtained? In particular, what happens when $q \to \infty$ with ℓ and k fixed? Or when $\ell \to \infty$ with q and k fixed?*

Open Problem 7 *Are there upper bounds on sequential k-IPP codes, better than those coming from the fact that every sequential k-IPP code is a (non-sequential) k-IPP code?*

5.2 Set Systems

In Subsection 3.3 each user gets a subset of keys, one of each type. Session keys are generated and encrypted in such a way that a pirate smartcard must have at least one key of each type to be able to decrypt. This is just one instance of the following more general approach. Let $V = \{1, 2, \dots, v\}$, where v is some positive integer. The data provider generates v keys $K_1, K_2, \dots, K_v$. User x is associated with a subset B_x of V of size ℓ. The provider gives the subset $\{K_i : i \in B_x\}$ of keys to user x.

The session key is encrypted in such a way that any pirate smartcard must contain ℓ or more of the keys K_i in order to decrypt. There might be other restrictions on which subsets of ℓ keys enable a pirate to decrypt — dividing keys into types as in Subsection 3.3 and insisting that there be a key of each type is just one possible restriction. (If there are no restrictions, so any ℓ keys suffice to reconstruct the session key, then an ℓ out of v threshold secret sharing scheme could be used in the session key process. We do not give the details here, but see Shamir [41] who introduced the concept of a secret sharing scheme, and see Stinson [45] for an introduction to the area of secret sharing.)

Let X be a set of traitors. The pirate knows keys in the set $\{K_i : i \in \bigcup_{x \in X} B_x\}$, and a pirate smartcard contains ℓ of these: say the keys $\{K_i : i \in D\}$, where D is an ℓ-element subset of $\cup_{x \in X} B_x$ that can be used to decrypt. We require that for all such subsets D, we must be able to determine at least one of the traitors in X (under the assumption that $|X| \leqslant k$).

A *traitor tracing scheme* (more precisely an open k-resilient traitor tracing scheme) is a collection of three components: an algorithm to generate the keys K_i and distribute them to users; a method for generating session keys, and transmitting this information to users so that they can decrypt; and an algorithm to identify a traitor from the keys in a pirate smartcard (provided that there are at most k traitors). The notion of a traitor tracing scheme is due to Chor, Fiat and Naor [14]; see their paper with Pinkas [15]. Their paper constructs a scheme which is equivalent to a binary k-traceability code (they call this their 'open one-level scheme'). They also consider a more complex scheme (their 'open two-level scheme') based on a concatenation construction, which reduces the size of an enabling block.

All traitor tracing schemes must have the following property. There do not exist distinct users $x_0, x_1, x_2, \ldots, x_k$ such that

$$B_{x_0} \subseteq \bigcup_{i=1}^{k} B_{x_i}.$$

In other words, no B_x is contained in the union of any k others. If such users exist, the traitors $x_1, x_2, \ldots, x_k$ could create a pirate smartcard that was the same as that given to x_0, and so could not be traced. (This is an analogue of the k-frameproof property in Section 4.) Set systems with this property have been studied by Erdős, Frankl and Fűredi [19], and Chor *et al.* applied these in the traitor tracing scheme setting:

Theorem 5.2 *In any (open k-resilient) traitor tracing scheme with n users, in which every user obtains ℓ keys from a total of v keys, we have that*

$$n \leqslant \frac{\binom{v}{t}}{\binom{\ell-1}{t-1}}$$

where $t = \lceil \ell/k \rceil$.

We mention a class of set systems that has been studied in the traitor tracing context. A *k-traceability scheme* is a set $\mathcal{B}$ of ℓ-element subsets of a v-element set having the following property. Let $X \subseteq \mathcal{B}$ be such that $|X| \leqslant k$, and let D be an ℓ-element subset of $\cup_{B \in X} B$. Let B be an element of $\mathcal{B}$ such that $|B \cap D|$ is as large as possible. Then $B \in X$.

Stinson and Wei [48] introduced k-traceability schemes. These schemes can be used to construct traitor tracing schemes. (Session keys are split into v shares using an ℓ out of v threshold secret sharing scheme, then each share is encrypted under a different key K_i before being broadcast.) The k-traceability condition allows a traitor to be identified from a pirate smartcard containing ℓ keys: any user with the maximum number of keys in common with the smartcard is a traitor.

Stinson and Wei [48] constructed traceability schemes from t-designs and packing designs. Stinson and Wei [49] have also shown the following result.

Theorem 5.3 *Let $\mathcal{B}$ be a set of ℓ-element subsets of a v-element set. Let μ be an integer such that $|X \cap Y| \leqslant \mu$ for all $X, Y \in \mathcal{B}$. Whenever $\ell > k^2\mu$, we have that $\mathcal{B}$ is a k-tracability scheme.*

Safavi-Naini and Wang [37] use this theorem to construct traceability schemes from constant weight codes.

We end this subsection by mentioning the concept of threshold traitor tracing. The notion was introduced by Naor and Pinkas [34]. Their setting is similar to that of the traitor tracing schemes above. In the standard traitor tracing model, any pirate smartcard that decrypts with non-negligible probability can be used to trace a traitor. But in the threshold traitor tracing model, the condition is weakened: any smartcard that decrypts with probability greater than some threshold p can be used to trace a traitor. It is assumed that p is small, but not negligible. See Chor *et al.* [15] for details of the model, and for constructions of threshold traitor tracing schemes.

5.3 Another Definition of Descendant

There is another widely studied definition of descendant, going back to the early paper by Boneh and Shaw [13]. In this model, a descendant is a length ℓ word over the larger alphabet $F \cup \{?\}$. Here '?' is a formal symbol disjoint from F, representing the case when a mark on a segment has been removed, or corrupted beyond recognition. The Boneh–Shaw definition of descendant is given as follows. Let X be a set of length ℓ words over F. The set $\mathrm{Desc}(X)$ of *Boneh–Shaw descendants* of X is defined by

$$\begin{aligned}\mathrm{Desc}(X) &= \{d \in (F \cup \{?\})^\ell : d_i = a \text{ whenever} \\ &\quad x_i = a \text{ for all } x \in X \text{ and some } a \in F\}.\end{aligned}$$

We use $\mathrm{Desc}(X)$ rather than $\mathrm{desc}(X)$ to distinguish the Boneh–Shaw notion of a descendant from the notion used in the rest of this survey. As an example, suppose $F = \{1,2,3\}$ and $X = \{111,123\}$. Then

$$\begin{aligned}\mathrm{Desc}(X) &= \{111,112,113,11?,121,122,123,12?,131,132,133,13?,1?1,1?2,\\ &\quad 1?3,1??\}.\end{aligned}$$

This set is much larger than $\mathrm{desc}(X)$ (which was calculated in Section 2). Indeed, it is clear that $\mathrm{Desc}(X)$ always strictly contains $\mathrm{desc}(X)$. This is one reason why the problem of constructing analogues of the codes in Section 4 for Boneh–Shaw descendants is often more difficult than the original problem.

Boneh and Shaw were the first to define a k-frameproof code, but using their definition of descendant. So we define a *k-Boneh–Shaw frameproof code*, or k-BSF code, to be a code C such that

$$\mathrm{Desc}(X) \cap C = X$$

for all $X \subseteq C$ with $|X| \leqslant k$. In fact, it is not difficult to check that a binary code C is k-frameproof if and only if it is a k-BSF code, and so we already have a fund of examples of k-BSF codes. In particular, the set of all length ℓ binary words of weight 1 is an ℓ-BSF, by the example of a k-frameproof code given in Subsection 4.2. This construction is a special case of a more general construction. If C is a q-ary k-frameproof code of length ℓ, we may construct a binary k-BSF code of length $k\ell$ by replacing the ith symbol in F by the weight 1 length k binary string with a 1 in its ith position. (This construction was pointed out by Boneh and Shaw, in the case when C is an error correcting code of minimum distance δ satisfying $\delta > (1-1/k)\ell$.) So the length ℓ binary

ℓ-BSF we just mentioned is the case when $C = F$, a trivial frameproof code of length 1.

Boneh and Shaw define a *totally k-secure code* in the same way as a k-IPP code is defined in Section 2, but using the Boneh–Shaw notion of descendant. Sadly, they prove that no non-trivial totally k-secure codes exist (non-trivial meaning that $n \geqslant 3$ and $k \geqslant 2$). To show this, it is sufficient to consider the case $k = 2$. But suppose that C is a totally 2-secure code containing three distinct codewords x, y, z. Defining

$$d_i = \begin{cases} ? & \text{if all of } x_i, y_i, z_i \text{ are distinct, and} \\ a & \text{if two or more of } x_i, y_i, z_i \text{ are equal to a common value } a \end{cases}$$

we find that d is a Boneh–Shaw descendant of any two of x, y and z. This proves that no non-trivial totally 2-secure codes exist.

Because non-trivial totally k-secure codes do not exist, Boneh and Shaw introduced the notion of a k-secure code with ϵ error. There is an extra random element introduced into the code — the codeword sent to user x depends on some random choices made by the provider in addition to x itself. The resulting scheme is *k-secure with ϵ error* if there exists an algorithm that, when given $d \in \mathrm{Desc}(X)$ generated by a subset $X \subseteq C$ with $|X| \leqslant k$, returns a member of X with probability greater than $1 - \epsilon$. The probability is calculated over the random choices made by the distributor and any random choices made by the pirate. There is not enough space to give details here, but see the paper of Boneh and Shaw [13] for the precise model. Barg, Blakley and Kabatiansky [4] give another construction, which has an efficient traitor identification algorithm. See also Sebé and Domingo-Ferrer [40] and Tô, Safavi-Naini and Wang [51].

5.4 Public Key Techniques

This subsection lists public key schemes that can be used for traitor tracing. We do not give many details of the schemes themselves — they are not combinatorial in nature, and so they are outside the scope of this paper. However, we should emphasise that public key techniques tend to produce efficient schemes, and so the schemes in this section are very important.

The earliest public key traitor tracing scheme is due to Kurosawa and Desmedt [28]. The idea is based on the following one time scheme. A session key is chosen, and is broken into shares using a Shamir $k+1$ out of $n+k$ secret sharing scheme. Each of the n users receives a share. To broadcast, the remaining k shares are transmitted, along with the data which is encrypted using the session key. A user now knows $k+1$ shares, so can reconstruct the session key and so can decrypt the data. Someone outside the scheme only knows the k public shares, and so cannot decrypt. Moreover, before the broadcast k or fewer users cannot construct another share of the scheme (as $k+1$ shares are required to do this in the Shamir scheme) and so we have a form of the k-IPP property. The Kurosawa and Desmedt scheme takes this idea and removes the 'one time only' restriction, by disguising the users' shares using a discrete log problem. A paper of Tzeng and Tzeng [54] uses similar ideas, to create a scheme where a traitor's keys can be revoked without the need for innocent users to update their keys.

Boneh and Franklin [12] point out that the original Kurosawa and Desmedt scheme is insecure if the pirate aims to build their own decoder box (as opposed to supplying a key to a standard decoder). The point being that even though the traitors cannot compute a share of a non-traitor, they can compute some information that is sufficient to decrypt and which cannot be traced back to the traitors. Boneh and Franklin then propose another scheme, also relying on the difficulty of discrete logs, which fixes this deficiency. Kurosawa and Yoshida [29] point out that the deficiency of the original Kurosawa and Desmedt scheme can be easily fixed, by using a $2k$ out of $n + 2k - 1$ scheme to share the secret.

We mention a paper of Mitsunari, Sakai and Kasahara [32], that uses the Weil pairing on elliptic curves to construct traitor tracing schemes. Dan Boneh [11] has found an attack on this scheme: he shows how two users can create a decryption device that cannot be traced.

Copyrighted functions, a notion due to Naccache, Shamir and Stern [33], can be used to construct traitor tracing schemes. In this setting, the provider distributes different implementations of a function to the users. This is done in such a way that any efficient implementation of the function created by a pirate can be traced back to one of the users. Naccache *et al.* show how this can be done in the case of 2 users with a function based on RSA. (Kiayias and Yung [27] construct copyrighted functions based on the discrete log problem.) Naccache *et al.* then use collusion secure codes to build systems for larger numbers of users. In the traitor tracing setting, the function should be the decryption process for the provider's encrypted data.

6 A Conclusion

This paper has collected together combinatorial results which are useful in the setting of copyright protection. But the point should be made that the work mentioned here is just a small part of a huge area. For example, a great deal of work has been done on broadcast encryption schemes and revocation schemes, vital parts of many systems that use the combinatorial constructions we have been considering. But even in this small corner of the field, there is a wealth of questions waiting to be answered.

Acknowledgements

I would like to thank Ki Hyoung Ko and Keith Martin, for kindly reading and commenting on some of the earlier versions of this manuscript.

References

[1] N. Alon, G. Cohen, M. Krivelevich and S. Litsyn, Generalized hashing and applications to digital fingerprinting, in *Proc. 2002 IEEE Internat. Symposium on Information Theory*, p. 436.

[2] N. Alon, E. Fischer and M. Szegedy, Parent-identifying codes, *J. Combin. Theory Ser. A* **95** (2001), 349–359.

[3] N. Alon and U. Stav, Some new results on parent-identifying codes, preprint, 2002.

[4] A. Barg, G.R. Blakley and G. Kabatiansky, Digital fingerprinting codes: Problems statements, constructions, identification of traitors, DIMACS Technical Report, 2001-52, available from `http://dimacs.rutgers.edu/TechnicalReports/`.

[5] A. Barg, G. Cohen, S. Encheva, G. Kabatiansky and G. Zémor, A hypergraph approach to the identifying parent property: the case of multiple parents, *SIAM J. Discrete Math.* **14** (2001), 423–431.

[6] A. Barg and G. Kabatiansky, A class of IPP codes with efficient identification, DIMACS Technical Report, 2002-36, available from `http://dimacs.rutgers.edu/TechnicalReports/`.

[7] O. Berkman, M. Parnas and J. Sgall, Efficient dynamic traitor tracing, *SIAM J. Computers* **30** (2001), 1802–1828.

[8] S.R. Blackburn, Combinatorics and Threshold Cryptography, in *Combinatorial Designs and their Applications* (eds. F.C. Holroyd, K.A.S. Quinn, C. Rowley & B.S. Webb), CRC Press, London (1999), pp. 49–70.

[9] S.R. Blackburn, Frameproof codes, *SIAM J. Discrete Math.*, in press.

[10] S.R. Blackburn, An upper bound on the size of a code with the k-identifiable parent property, *J. Combin. Theory Ser. A*, in press.

[11] D. Boneh, Personal communication, 14 October 2002.

[12] D. Boneh and M. Franklin, An efficient public key traitor tracing scheme, in *Advances in Cryptography — CRYPTO '99* (ed. M. Wiener), *Lecture Notes in Computer Science*, 1666, Springer, Berlin (1999), pp. 338–353.

[13] D. Boneh and J. Shaw, Collision-secure fingerprinting for digital data, *IEEE Trans. Inform. Theory* **44** (1998), 1897–1905.

[14] B. Chor, A. Fiat and M. Naor, Tracing traitors, in *Advances in Cryptology — CRYPTO '94* (ed. Y.G. Desmedt), *Lecture Notes in Computer Science*, 839, Springer, Berlin (1994), pp. 257–270.

[15] B. Chor, A. Fiat, M. Naor and B. Pinkas, Tracing traitors, *IEEE Trans. Inform. Theory* **46** (2000), 893–910.

[16] G.D. Cohen, S.B. Encheva and H.G. Schaathun, On separating codes, Ecole Nationale Supérieure des Télécommunications, Tech. Report, to appear.

[17] I.J. Cox, J. Kilian, F.T. Leighton and T. Shamoon, Secure spread spectrum watermarking for multimedia, *IEEE Trans. Image Process.* **6** (1997), 1673–1687.

[18] S. Encheva and G. Cohen, Some new p-ary two-secure frameproof codes, *Appl. Math. Lett.* **14** (2001), 177–182.

[19] P. Erdős, P. Frankl and Z. Füredi, Families of finite sets in which no set is covered by the union of r others, *Israel J. Math.* **51** (1985), 79–89.

[20] P. Erdős and L. Lovász, Problems and results on 3-chromatic hypergraphs and some related questions, in *Infinite and Finite Sets* (eds. A. Hajnal, R. Rado & V. Sós), North–Holland, Amsterdam (1975), pp. 609–627.

[21] A. Fiat and T. Tassa, Dynamic traitor tracing, in *Advances in Cryptology — CRYPTO '99* (ed. M. Weiner), *Lecture Notes in Computer Science*, 1666, Springer, Berlin (1999), pp. 354–371.

[22] A.D. Friedman, R.L. Graham and J.D. Ullman, Universal single transition time asynchronous state assignments, *IEEE Trans. Inform. Theory* **18** (1969), 541–547.

[23] A. Garcia and H. Stichtenoth, A tower of Artin–Schreier extensions of function fields attaining the Drinfeld–Vladut bound, *Invent. Math.* **121** (1995), 211–222.

[24] R.L. Graham and N.J. Sloane, Lower bounds for constant weight codes, *IEEE Trans. Inform. Theory* **26** (1980), 37–43.

[25] V. Guruswami and M. Sudan, Improved decoding of Reed–Solomon and algebraic-geometry codes, *IEEE Trans. Inform. Theory* **45** (1999), 1757–1767.

[26] H.D.L. Hollmann, J.H. van Lint, J.-P. Linnartz and L.M.G.M. Tolhuizen, On codes with the identifiable parent property, *J. Combin. Theory Ser. A* **82** (1998), 121–133.

[27] A. Kiayias and M. Yung, Traitor tracing with constant transmission rate, in *Advances in Cryptology — EUROCRYPT 2002* (ed. L.R. Knudsen), *Lecture Notes in Computer Science*, 2332, Springer, Berlin (2002), pp. 450–465.

[28] K. Kurosawa and Y. Desmedt, Optimum traitor tracing and asymmetric schemes, in *Advances in Cryptology — EUROCRYPT '98* (ed. K. Nyberg), *Lecture Notes in Computer Science*, 1403, Springer, Berlin (1998), pp. 145–157.

[29] K. Kurosawa and T. Yoshida, Linear code implies public-key traitor tracing, in *Public Key Cryptography (PKC 2002)* (ed. P. Paillier), Springer, Berlin (2002), pp. 172–187.

[30] J.H. van Lint, *Introduction to Coding Theory*, 3rd edition, Springer, New York (1999).

[31] K. Mehlhorn, *Data Structures and Algorithms*,1: Sorting and Searching Springer, Berlin (1984).

[32] S. Mitsunari, R. Sakai and M. Kasahara, A new traitor tracing, *IEICE Trans. Fundamentals* **E85-A** (2002), 481–484.

[33] D. Naccache, A. Shamir and J.P. Stern, How to copyright a function?, in *Public Key Cryptography (PKC '99)* (eds. H. Imai & Y. Zheng), *Lecture Notes in Computer Science*, 1560, Springer, Berlin (1999), pp. 188–196.

[34] M. Naor and B. Pinkas, Threshold traitor tracing, in *Advances in Cryptology — CRYPTO '98* (ed. H. Krawczyk), *Lecture Notes in Computer Science*, 1462, Springer, Berlin (1998), pp. 502–517.

[35] A. Nilli, Perfect hashing and probability, *Combin. Probab. Comput.* **3** (1994), 407–409.

[36] R. Safavi-Naini and Y. Wang, Sequential traitor tracing, in *Advances in Cryptology — CRYPTO 2000* (ed. M. Bellare), *Lecture Notes in Computer Science*, 1880, Springer, Berlin (2000), pp. 316–332. (Final version to appear in *IEEE Trans. Inform. Theory*.)

[37] R. Safavi-Naini and Y. Wang, New results on frameproof codes and traceability schemes, *IEEE Trans. Inform. Theory* **47** (2001), 3029–3034.

[38] Yu.L. Sagalovich, Separating systems, *Problems Inform. Transmission* **30** (1994), 105–123.

[39] P. Sarkar and D.R. Stinson, Frameproof and IPP codes, in *Progress in Cryptology — INDOCRYPT 2001* (eds. C. Pandu Rangan & C. Ding), *Lecture Notes in Computer Science*, 2247, Springer, Berlin (2001), pp. 117–126.

[40] F. Sebé and J. Domingo-Ferrer, Short 3-secure fingerprinting codes for copyright protection, in *Information Security and Privacy, ACISP 2002* (eds. L. Batten & J. Seberry), *Lecture Notes in Computer Science*, 2384, Springer, Berlin (2002), pp. 316–327.

[41] A. Shamir, How to share a secret, *Comm. ACM* **22** (1979), 612–613.

[42] A. Silverberg, J. Staddon and J. Walker, Efficient traitor tracing algorithms using list decoding, in *Advances in Cryptology — Asiacrypt 2001* (ed. C. Boyd), *Lecture Notes in Computer Science*, 2248, Springer, Berlin (2001), pp. 175–192.

[43] J. Spencer, Probabilistic methods, *Graphs Combin.* **1** (1985), 357–382.

[44] J.N. Staddon, D.R. Stinson and R. Wei, Combinatorial properties of frameproof and traceability codes, *IEEE Trans. Inform. Theory* **47** (2001), 1042–1049.

[45] D.R. Stinson, An explication of secret sharing schemes, *Des. Codes Cryptogr.* **2** (1992), 357–390.

[46] D.R. Stinson, *Cryptography, Theory and Practice*, 2nd edition, Chapman & Hall, CRC Press, Boca Raton (2002).

[47] D.R. Stinson, T. van Trung and R. Wei, Secure frameproof codes, key distribution patterns, group testing algorithms and related structures, *J. Statist. Plann. Inference* **86** (2000), 595–617.

[48] D.R. Stinson and R. Wei, Combinatorial properties and constructions of traceability schemes and frameproof codes, *SIAM J. Discrete Math.* **11** (1998), 41–53.

[49] D.R. Stinson and R. Wei, Key preassigned traceability schemes for broadcast encryption, in *Selected Areas in Cryptography (Proceedings SAC'98)* (eds. S. Tavares & H. Meijer), *Lecture Notes in Computer Science*, 1556, Springer, Berlin (1999), pp. 144–156.

[50] M.D. Swanson, M. Kobayashi and A.H. Tewfik, Multimedia data-embedding and watermarking technologies, *Proc. IEEE* **86** (1998), 1064–1087.

[51] V.D. Tô, R. Safavi-Naini and Y. Wang, A 2-secure code with efficient tracing algorithm, in *Progress in Cryptology - INDOCRYPT'02* (eds. A. Menezes & P. Sarkar), *Lecture Notes in Computer Science*, 2551, Springer, Berlin (2002), pp. 149–162.

[52] T. van Trung and S. Martirosyan, Constructions for efficient IPP codes, preprint, University of Essen, 2002.
Available from `http://www.exp-math.uni-essen.de/~trung/`.

[53] T. van Trung and S. Martirosyan, On a class of traceability codes, preprint, University of Essen, 2002.
Available from `http://www.exp-math.uni-essen.de/~trung/`.

[54] W.-G. Tzeng and Z.-J. Tzeng, A public-key traitor tracing scheme with revocation using dynamic shares, in *Public Key Cryptography (PKC 2001)* (ed. K. Kim), *Lecture Notes in Computer Science*, 1992, Springer, Berlin (2001), pp. 207–224.

[55] C. Xing, Asymptotic bounds on frameproof codes, *IEEE Trans. Inform. Theory*, in press.

[56] Y. Yemane, *Codes with the k-Identifiable Parent Property* Ph.D. thesis, University of London, 2002.

[57] F. Zane, Efficient watermark detection and collusion security, in *Financial Cryptography 2000* (ed. Y. Frankel), *Lecture Notes in Computer Science*, 1962, Springer, Berlin (2000), pp. 21–32.

Department of Mathematics
Royal Holloway, University of London
Egham, Surrey TW20 0EX
United Kingdom
s.blackburn@rhul.ac.uk

Matroids and Coxeter groups

A.V. Borovik

Abstract

The paper describes a few ways in which the concept of a Coxeter group (in its most ubiquitous manifestation, the symmetric group) emerges in the theory of ordinary matroids:

- Gale's maximality principle which leads to the Bruhat order on the symmetric group;
- Jordan–Hölder permutation which measures distance between two maximal chains in a semimodular lattice and which happens to be closely related to Tits' axioms for buildings;
- matroid polytopes and associated reflection groups;
- Gaussian elimination procedure, BN-pairs and their Weyl groups.

These observations suggest a very natural generalisation of matroids; the new objects are called Coxeter matroids and are related to other Coxeter groups in the same way as (classical) matroids are related to the symmetric group.

Introduction

Combinatorics studies structures on a finite set; many of the most interesting of these arise from elimination of continuous parameters in problems from other mathematical disciplines.

Matroid is a combinatorial concept which arises from the elimination of continuous parameters from one of the most fundamental notions of mathematics: that of linear dependence of vectors.

Indeed, let E be a finite set of vectors in a vector space $\mathbb{R}^n$. Vectors $\alpha_1, \ldots, \alpha_k$ are linearly dependent if there exist real numbers $c_1, \ldots, c_k$, not all of zero, such that $c_1\alpha_1 + \cdots + c_k\alpha_k = 0$. In this context, the coefficients $c_1, \ldots, c_k$ are continuous parameters; what properties of the set E remain after we decide never to mention them? The solution was suggested by Hassler Whitney in 1936 [30]. He noticed that the set of linearly independent subsets of E has some very distinctive properties. In particular, if $\mathcal{B}$ is the set of *maximal* linearly independent subsets of E, then, by a well known result from linear algebra, it satisfies the following *Exchange Property*:

> *For all* $A, B \in \mathcal{B}$ *and* $a \in A \setminus B$ *there exists* $b \in B \setminus A$, *such that* $(A \setminus \{a\}) \cup \{b\}$ *lies in* $\mathcal{B}$.

Whitney introduced the term *matroid* for a finite structure consisting of a set E with a distinguished collection $\mathcal{B}$ of subsets (called *bases*) satisfying the Exchange Property.

The aim of these notes is to look again at the fundamentals of the classical concept of matroid. Indeed, let us make a step further: the most fundamental structure on a finite set—even in the absence of any other structures—is provided by its symmetric group acting on it. The symmetric group already lurks between the lines of the Exchange Property in the form of transpositions (a, b) responsible for the exchange of elements. We take this observation as the starting point and outline an approach to the theory of matroids in terms of the symmetric groups. As we shall soon see, one of its advantages is that it exposes the hidden symmetries of matroids. Moreover, our approach opens up directions of further generalisation.

Indeed, the symmetric group Sym_n is the simplest example of a finite Coxeter group (or, equivalently, a finite reflection group). It can be interpreted geometrically as the group of symmetries of the regular $(n-1)$-dimensional simplex in $\mathbb{R}^n$ with the vertices

$$(1,0,\dots,0),(0,1,0,\dots,0),\dots,(0,\dots,0,1).$$

We can replace the symmetric group with the group of symmetries of another Platonic solid in $\mathbb{R}^n$, the n-cube $[-1,1]^n$. (This group is called the *hyperoctahedral* group.) Then we get a very natural generalisation of matroids, called *symplectic matroids*. We will usually refer to matroids as *ordinary matroids*, to distinguish them from the more general symplectic matroids and from even more general Coxeter matroids. Some special classes of symplectic matroids have been already studied under the names Δ-matroids, metroids [20], symmetric matroids [17], or 2-matroids [19]. Symplectic matroids are related to the geometry of vector spaces endowed with bilinear forms, although in a more intricate way than ordinary matroids to ordinary vector spaces.

Furthermore, Sym_n is naturally embedded in the group of symmetries of the n-cube because we can make Sym_n permute the coordinate axes without changing their orientation; this action obviously preserves the n-cube $[-1,1]^n$. Thus ordinary matroids can be also understood as symplectic matroids, the latter becoming the most natural generalisations of the former.

However, the scope of the paper does not allow us to discuss in much detail the rich and structured theory of symplectic matroids; the interested reader may wish to consult [12] or [6]. Instead, we prefer to concentrate on demonstrating how naturally the language of Coxeter groups arises in the classical matroid theory. In particular, it helps to view matroids as special cases of more general combinatorial structures, *Coxeter matroids*. The latter were introduced by Gelfand and Serganova [23]; they are related to finite Coxeter groups in the same way as classical matroids are to the symmetric group.

One of the important tools of the theory is geometric interpretation of matroids—ordinary, symplectic, Coxeter—as convex polytopes with certain symmetry properties; this interpretation is provided by the Gelfand–Serganova theorem. This leads to a surprisingly simple cryptomorphic (equivalent but not obviously so) definition of a Coxeter matroid.

Let Δ be a convex polytope. For every edge of Δ, take the hyperplane that cuts the edge in its midpoint and is perpendicular to the edge and imagine this hyperplane being a semitransparent mirror. Now mirrors multiply by reflecting in other mirrors, as in a kaleidoscope. If we end up with only finitely many mirrors, we call Δ a *Coxeter matroid polytope*, which, in view of the Gelfand–Serganova interpretation, is equivalent to a Coxeter matroid.

Essentially, Coxeter matroids are n-dimensional kaleidoscopes which generate only finitely many mirror images. Rarely does a mathematical theory come to a more intuitive re-interpretation of its basic concept.

In the final section we return to the basics and look at the underlying combinatorics of the Gaussian elimination procedure. This classical routine involves permutation of rows of a matrix. The rules these permutations obey are extremely simple; when axiomatised in group-theoretic terms, they become what are known as axioms for a BN-pair (or a *Tits system*) and very quickly lead to Coxeter groups (and Coxeter matroids) appearing on the scene.

Every BN-pair has an associated geometric object, called a *building*. In its most compact axiomatisation, it is a set with a 'distance' function which takes values in a Coxeter group W. Buildings provide a natural way to represent Coxeter matroids. Indeed, the classical representation of matroids in vector spaces turns out to be a special case of representation in buildings. For symmetric groups, a weaker notion of W-distance arises in semimodular lattices as the Jordan–Hölder permutation which measures combinatorial distance between two maximal chains. Every matroid turns out to be representable in an appropriate semimodular lattice, thus eliminating continuous parameters from the concept of representation of matroid.

We freely use in the text some basic concepts and facts related to matroids, they all can be found in the standard reference books [25, 28, 29]; see also the survey paper [26]. We also refer to some standard facts about root systems and Coxeter groups [24]. All proofs are omitted; the interested reader can find the detailed exposition of the theory in the forthcoming book [6]. Most unattributed theorems in the paper are either classical results or belong to A.V. Borovik, I.M. Gelfand and N. White.

1 Matroids and Flag Matroids

1.1 Maximality Property

The intimate connection between matroids and the symmetric group Sym_n can be seen most clearly in the Maximality Property, which is really just a reformulation of the well-known characterisation of matroids in terms of the Greedy Algorithm. It says, briefly, that for every linear ordering of the set of elements of the matroid, there is a unique maximal basis. But linear orderings of a finite set can be interpreted as its permutations. This brings the symmetric group into a pivotal role in matroid theory.

Amazingly, a definition of matroid in terms of orderings was anticipated by Boruvka [16] before the invention of matroids. It was given by Gale [21] as a solution of the problem of optimal assignment in applied combinatorics, and then later but independently introduced, in a wider context, by Gelfand and Serganova [23]. We shall see that it naturally leads to the Bruhat ordering on Sym_n viewed as a Coxeter group.

Let $\mathcal{P}_{n,k}$ be the collection of all k-element subsets in $[n]$ (where, as is usual, $[n] = \{1, \ldots, n\}$). For $A, B \in \mathcal{P}_{n,k}$, where

$$A = \{\, i_1, \ldots, i_k \,\},\ i_1 < i_2 < \cdots < i_k$$

and

$$B = \{\, j_1, \ldots, j_k \,\},\ j_1 < j_2 < \cdots < j_k,$$

we set

$$A \leqslant B \text{ if and only if } i_1 \leqslant j_1, \ldots, i_k \leqslant j_k.$$

Let $W = \mathrm{Sym}_n$ be the group of all permutations of $[n]$. Then we can associate an ordering of $\mathcal{P}_{n,k}$ with each $w \in W$ by putting

$$A \leqslant^w B \text{ if and only if } w^{-1}A \leqslant w^{-1}B.$$

We call $\leqslant^w$ the *Gale ordering* on $\mathcal{P}_{n,k}$ induced by w. For brevity, we write $i \leqslant^w j$ instead of $\{\, i \,\} \leqslant^w \{\, j \,\}$ and say that i precedes j in w. This term has a very natural interpretation: $i \leqslant^w j$ if and only if i precedes j in the bottom row of the standard two-rowed notation for permutations:

$$w = \begin{pmatrix} 1 & 2 & \ldots & n \\ i_1 & i_2 & \ldots & i_n \end{pmatrix}.$$

Thus, the permutation w can be interpreted as the reordering

$$i_1 <^w i_2 <^w \cdots <^w i_n$$

of the set $[n]$.

The starting point of our approach to matroids is the following theorem, which allows us to define matroids in terms of orderings.

Theorem 1.1 (Gale [21]) *Let $\mathcal{B} \subseteq \mathcal{P}_{n,k}$. Then $\mathcal{B}$ is (the collection of bases of) a matroid if and only if $\mathcal{B}$ satisfies the following* Maximality Property:

> *for every $w \in \mathrm{Sym}_n$ the collection $\mathcal{B}$ contains a unique member $A \in \mathcal{B}$ maximal in $\mathcal{B}$ with respect to $\leqslant^w$; that is, $B \leqslant^w A$ for all $B \in \mathcal{B}$.*

We call A the *w-maximal* basis in $\mathcal{B}$.

1.2 What does this definition mean in terms of Coxeter groups?

Probably it is time to open the cards and reinterpret Theorem 1.1 in terms of Coxeter groups.

Recall that a *Coxeter group* is a group W with a finite set of generators S, subject to the following relations: $s^2 = 1$ for all $s \in S$ and, for every pair $s, r \in S$, a relation $(sr)^{m_{sr}} = 1$ for some positive integer m_{sr} which depends on s and r. Note that $m_{sr} = m_{rs}$ and m_{sr} is not necessarily finite. The set S is called the *set of distinguished generators* for W. In this paper, we consider only finite Coxeter groups (although much of the theory works in the infinite case as well).

Let $w \in W$. An expression $w = s_1 \cdots s_l$ of the minimal possible length is called a *reduced expression* for w. The *Bruhat order* $\leqslant$ on W is defined as follows. Let $u, v \in W$. Then $u \leqslant v$ if and only u can be obtained by deleting some generators s_i from a reduced expression $v = s_1 \cdots s_k$ for v.

Let P be a *standard parabolic subgroup* of W, i.e. $P = \langle R \rangle$ for some subset $R \subset S$. It can be shown that every coset wP of P contains a maximal, with respect to the Bruhat ordering, element $\max wP$. We define the Bruhat order on the factor set $W^P = W/P$ by setting $uP \leqslant vP$ if $\max uP \leqslant \max wP$. It can be shown that this is equivalent to the condition that $u' \leqslant v'$ for some (or any) representatives $u' \in uP$ and $v' \in vP$.

For $w \in W$, we introduce the w-shifted order $\leqslant^w$ by setting $uP \leqslant^w vP$ if and only if $w^{-1} \cdot uP \leqslant w^{-1} \cdot vP$.

Now a *Coxeter matroid* for W and P is a subset $\mathcal{B} \subseteq W^P$ which satisfies the *Maximality Property*:

for any $w \in W$, there is a unique $A \in \mathcal{B}$ such that, for all $B \in \mathcal{B}$,

$$B \leqslant^w A.$$

The elements of a Coxeter matroid $\mathcal{B}$ are called *bases*.

It is well-known that the symmetric group $W = \mathrm{Sym}_n$ is a Coxeter group (called *Coxeter group of type A_{n-1}*) with transpositions

$$s_1 = (1,2), s_2 = (2,3), \ldots, s_{n-1} = (n-1, n)$$

as generators, and with relations

$$\left\{ \begin{array}{rcl} s_i^2 & = & 1, \quad i = 1, \ldots, n-1 \\ (s_i s_j)^2 & = & 1, \quad |i-j| \geqslant 2, \quad i, j = 1, \ldots, n-1 \\ (s_i s_{i+1})^3 & = & 1, \quad i = 1, \ldots, n-1 \end{array} \right. .$$

Now consider the maximal parabolic subgroup

$$P = \langle s_1, \ldots, s_{k-1}, s_{k+1}, \ldots, s_{n-1} \rangle.$$

Obviously, P is the stabilizer of the k-set $\{1, 2, \ldots, k\}$, and the collection $\mathcal{P}_{n,k}$ of k-subsets in $[n]$ can be identified with the factor set $W^P = W/P$. In this identification, it can be shown that the Gale order of $\mathcal{P}_{n,k}$ coincides with the Bruhat order on W^P, and matroids of rank k on $[n]$ are exactly Coxeter matroids for $W = \mathrm{Sym}_n$ and the maximal parabolic subgroup P.

1.3 Flag matroids

Theorem 1.1 translates the definition of matroid into the algebraic language of permutations and orderings. Before moving to arbitrary Coxeter groups, let us stay for a while with the symmetric group Sym_n.

Even there we already have a natural generalisation of matroids. Indeed, when we look at Coxeter matroids for Sym_n and a *non-maximal* parabolic subgroup, we come to the concept of *flag matroid* which happens to be cryptomorphically equivalent to that of *matroid quotient.*

Flag matroids can be easily defined in more elementary terms of Gale order and maximality property, and that is what we shall do now.

1.3.1 Flags. A *flag* F is a strictly increasing sequence

$$F^1 \subset F^2 \subset \cdots \subset F^m$$

of finite sets. Denote by k_i the cardinality of the set F^i; the m-tuple $(k_1, \dots, k_m)$ will be called the *rank* of F. We write $F = (F^1, \dots, F^m)$; the set F^i is called i-th *constituent*, or constituent of rank k_i, of the flag F. The collection of all flags of rank $(k_1, \dots, k_m)$ in $[n]$ is denoted $\mathcal{F}_n^{k_1 \cdots k_m}$.

For every reordering $w \in \mathrm{Sym}_n$ of the set $[n]$ we define the Gale ordering $\leqslant^w$ on $\mathcal{F}_n^{k_1 \cdots k_m}$ as follows. If $F = (F^1, \dots, F^m)$ and $G = (G^1, \dots, G^m)$ are two flags, we set $F \leqslant^w G$ if and only if $F^i \leqslant^w G^i$ for all $i = 1, \dots, m$. Thus the Gale ordering on subsets induces the Gale ordering on flags.

1.3.2 Flag matroids. A collection $\mathcal{F}$ of flags of rank $(k_1, \dots, k_m)$ is called a *flag matroid* if and only if $\mathcal{F}$ satisfies the *Maximality Property*:

> for every $w \in \mathrm{Sym}_n$ the collection $\mathcal{F}$ contains a unique element maximal in $\mathcal{F}$ with respect to the ordering $\leqslant^w$.

For any collection $\mathcal{F}$ of flags, we call the collection of ith constituents F_i for $F \in \mathcal{F}$ the *i-th constituent* of $\mathcal{F}$. It follows immediately from this definition of flag matroid that the i-th constituent of a flag matroid $\mathcal{F}$ is a matroid $\mathcal{B}_i$ of rank k_i, called the *ith constituent matroid* of $\mathcal{F}$.

Notice that every matroid is a flag matroid and that $\mathcal{P}_{n,k} = \mathcal{F}_n^k$.

1.3.3 Matroid quotients. Let M and M' be matroids on the same set $[n]$. One says that M' is a *quotient* of M if every circuit of M is a union of circuits of M'. (Recall that a *circuit* is a minimal subset not belonging to any basis of the matroid.) An equivalent statement found in some texts is that the identity map on $[n]$ is a *strong map* from M to M' (cf. [28]).

If $\{ M_1, M_2, \ldots, M_m \}$ is a collection of distinct matroids on $[n]$, we say that these matroids are *concordant* if for every $i \neq j$, either M_i is a quotient of M_j or vice-versa. Notice that this implies that the matroids have distinct ranks and that M_i is a quotient of M_j precisely when rank $M_i <$ rank M_j. A collection of flag matroids of various ranks is called *concordant* if the collection of all their (distinct) constituent matroids is concordant.

We can now state the main theorem characterizing flag matroids; it translates the concept of flag matroids into the more traditional language of matroid theory.

Theorem 1.2 *A collection $\mathcal{F}$ of flags of rank $(k_1, \ldots, k_m)$ is a flag matroid if and only if*

(1) *each constituent M_i of $\mathcal{F}$ is a matroid,*

(2) *the constituents $M_1, \ldots, M_m$ are concordant, and*

(3) *every flag*
$$B_1 \subset \cdots \subset B_m$$
such that B_i is a basis of M_i for $i = 1, \ldots, m$ belongs to $\mathcal{F}$.

Corollary 1.3 *A collection of distinct matroids is concordant if and only if, for every $w \in \mathrm{Sym}_n$, the $\leqslant^w$-maximal bases of each matroid form a flag.*

1.4 Matroid polytopes

Finite Coxeter groups are finite reflection groups; the underlying geometric structures are extremely important for the theory of Coxeter groups, and, as we shall soon see, also shape the theory of Coxeter matroids. The root system associated with the symmetric group (called the *root system of type A_{n-1}*) is probably the most important geometric object of matroid theory.

1.4.1 Roots and reflections. The root system Φ of type A_{n-1} can be introduced *ad hoc* as the system of vectors (called *roots*) in $V = \mathbb{R}^n$ of the form $\epsilon_i - \epsilon_j$, $i \neq j$, where $\epsilon_1, \ldots, \epsilon_n$ is the standard orthonormal basis of V.

The group $W = \mathrm{Sym}_n$ acts on V in the natural way by permuting n vectors $\epsilon_1, \ldots, \epsilon_n$, which, obviously, induces the action of W on Φ. This action preserves the standard scalar product associated with the orthonormal basis $\epsilon_1, \ldots, \epsilon_n$. Therefore W acts on V by orthogonal transformations.

The natural correspondence between transpositions $r = (ij)$ in W and pairs of opposite roots $\pm(\epsilon_i - \epsilon_j)$ has a deep meaning. In its action on V the transposition r acts as the *reflection* in the mirror of symmetry perpendicular to the root $\rho = \epsilon_i - \epsilon_j$. It means that r inverts ρ ($r\rho = -\rho$), and fixes every point in the hyperplane $x_i = x_j$ perpendicular to ρ. It can be easily seen that every reflection in W is a transposition.

1.5 Polytopes associated with flag matroids

If $A = \{ a_1, \ldots, a_k \}$ is a subset in $[n]$, we denote

$$\delta_A = \epsilon_{a_1} + \cdots + \epsilon_{a_k},$$

and, for a flag $F = (F_1, \ldots, F_m)$, we denote by δ_F the point $\delta_{F_1} + \cdots + \delta_{F_m}$. This defines the map

$$\delta : \mathcal{F}_n^{k_1 \cdots k_m} \longrightarrow \mathbb{R}^n.$$

With any set $\mathcal{F}$ of flags in $[n]$ of the given rank $(k_1, \ldots, k_m)$ we associate a polytope $\Delta_{\mathcal{F}}$, the convex hull of points in $\delta(\mathcal{F})$. Notice that all points in $\delta(\mathcal{F}_n^{k_1 \cdots k_m})$ are vertices of the convex hull of $\delta(\mathcal{F}_n^{k_1 \cdots k_m})$. Therefore the set $\delta(\mathcal{F})$ is exactly the set of vertices of $\Delta_{\mathcal{F}}$.

We say that a polytope Δ in the real vector space $V = \mathbb{R}^n$ is a *matroid polytope* if Δ is convex, its edges are all parallel to the roots in Φ, and there exists a point equidistant from all of its vertices.

Theorem 1.4 *Let $\mathcal{F}$ be a set of flags of the same rank on $[n]$. Then the following conditions are equivalent.*

(1) *$\mathcal{F}$ is a flag matroid.*

(2) *$\Delta_{\mathcal{F}}$ is a matroid polytope.*

We call $\Delta_{\mathcal{F}}$ the *canonical flag matroid polytope* of $\mathcal{F}$.

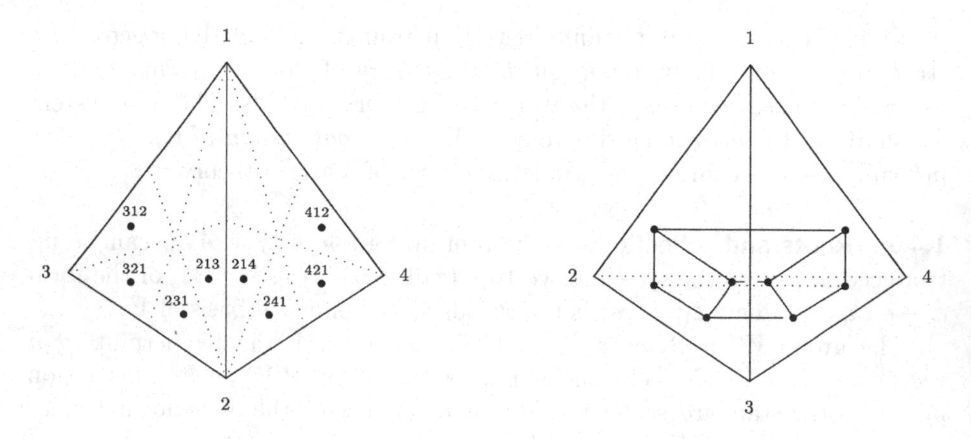

Figure 1: *A flag matroid polytope.*

Example 1.5 We can represent a flag matroid polytope in the projective space associated with $\mathbb{R}^n$ by labeling the vertices of an $(n-1)$-dimensional simplex with the elements of $[n]$. In this representation, the roots of Sym_n

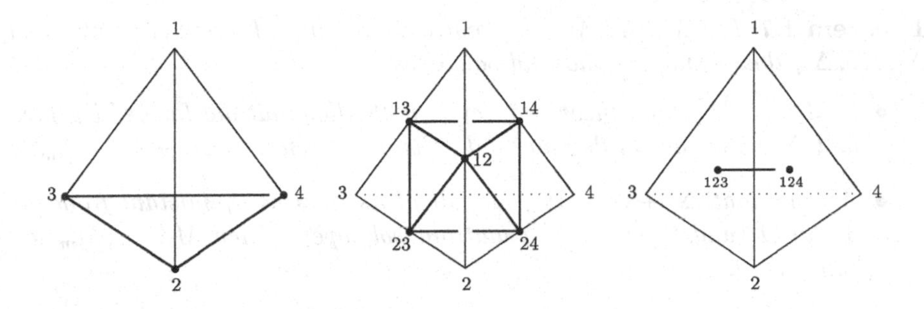

Figure 2: *Matroid polytopes of the constituent matroids of the flag matroid of Figure* 1.

are parallel to edges of the simplex. A k-element subset A corresponds to a $(k-1)$-dimensional face of the simplex, and we represent δ_A by the barycentre p_A of that face. For a flag F, we represent δ_F by the barycentre of the simplex which is the convex hull of all the p_A for A running over the constituents of F. For example, Figure 1 shows the polytope for the flag matroid on [4] with constituent matroids $\{2, 3, 4\}$, $\{12, 13, 14, 23, 24\}$ and $\{123, 124\}$, which are shown in Figure 2. We use the convention that, for example, 312 denotes the flag with constituents $\{3\}, \{1, 3\}, \{1, 2, 3\}$. The vertex 312 in Figure 1 is the barycentre of the simplex with the vertices 3, 13 (the barycentre of the edge of the tetrahedron connecting the vertices 1 and 3), and 123 (the barycentre of the facet of the tetrahedron with the vertices 1, 2 and 3).

The condition that the polytope is a matroid polytope is easily checked for the polytopes in Figures 1 and 2: one merely has to notice that each edge of the polytope is parallel to some edge of the tetrahedron. The existence of a point equidistant from all vertices of the polytope follows immediately from the construction of the polytope since the bases all have the same rank.

1.6 Minkowski sums

Now we explain how a flag matroid polytope is related to the polytopes of its constituents.

Let Δ_1 and Δ_2 be convex polytopes. Their *Minkowski sum* is defined as

$$\Delta_1 + \Delta_2 = \{\, x_1 + x_2 \mid x_i \in \Delta_i \,\}.$$

Let $\mathcal{F}_1$ and $\mathcal{F}_2$ be two concordant flag matroids. Then, by Theorem 1.2, the collection of all of their constituents forms a flag matroid, which we will denote $\mathcal{F}_1 \vee \mathcal{F}_2$. We will call $\mathcal{F}_1$ and $\mathcal{F}_2$ *rank disjoint* if no two constituents from them have the same rank.

Theorem 1.6 *Let the flag matroids $\mathcal{F}_1$ and $\mathcal{F}_2$ on $[n]$ be concordant and rank disjoint, Δ_1 and Δ_2 their canonical flag matroid polytopes, and Δ the canonical flag matroid polytope of $\mathcal{F} = \mathcal{F}_1 \vee \mathcal{F}_2$. Then $\Delta = \Delta_1 + \Delta_2$.*

Theorem 1.7 *Let $M_1, \ldots, M_m$ be matroids on $[n]$ of distinct ranks and $\Delta_1, \ldots, \Delta_m$ their canonical matroid polytopes.*

- *If $M_1, \ldots, M_m$ are concordant, let $\mathcal{F}$ is the flag matroid formed by them and Δ the canonical flag matroid polytope. Then $\Delta = \Delta_1 + \cdots + \Delta_m$.*
- *Assume that $\Delta_1 + \cdots + \Delta_m$ has all of its vertices equidistant from the origin O (and therefore is a matroid polytope). Then $M_1, \ldots, M_m$ are concordant.*

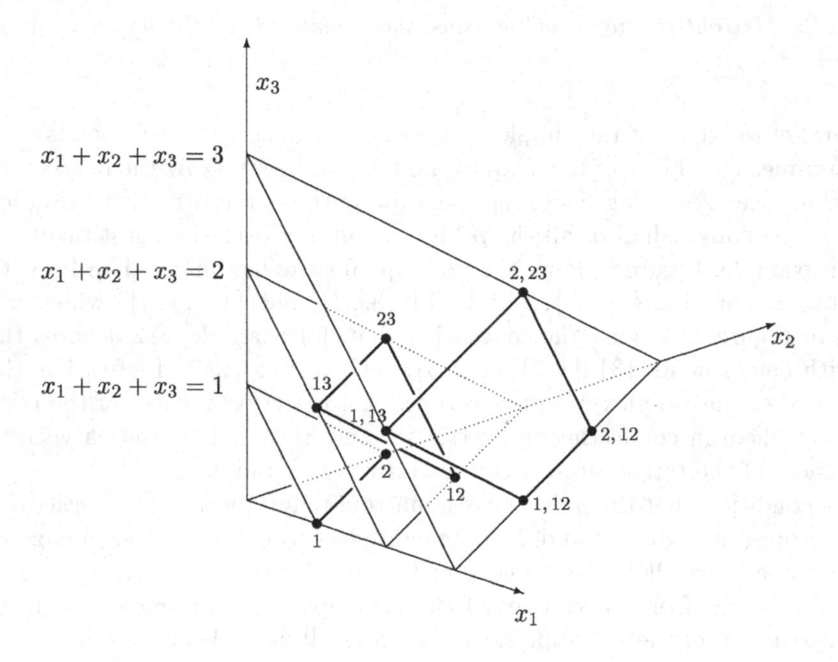

Figure 3: *The Minkowski sum of the canonical matroid polytopes for two concordant matroids* $\{1, 2\}$ and $\{12, 23, 13\}$ *on* $[3]$ *is the canonical polytope for the flag matroid* $\{(1, 12), (2, 12), (1, 13), (2, 23)\}$.

1.7 Flag matroids as Coxeter matroids

Let $W = \mathrm{Sym}_n$. A flag $F = (F^1, \ldots, F^n)$ of rank $(1, 2, \ldots, n)$ on $[n]$ is called *complete.* To every permutation w we assign the flag

$$w([1]) \subset w([2]) \subset \cdots \subset w([n]).$$

This correspondence is obviously W-equivariant in the following sense: if F is the flag corresponding to the permutation w, then, for all $u \in W$, the flag uF corresponds to uw.

The following description of the Bruhat ordering on Sym_n is classical. It relates the Bruhat order on Sym_n with the Gale order on $\mathcal{F}_n^{12\cdots n}$; both orderings are denoted by the same symbol $\leqslant$.

Theorem 1.8 *If u and v are two permutations in Sym_n corresponding to the flags F and G in $\mathcal{F}_n^{12\cdots n}$, then $u \leqslant v$ if and only if $F \leqslant G$. In particular, $u \leqslant v$ if and only if $u([k]) \leqslant v([k])$ for all $k = 1, \ldots, n$.*

If now P is the stabiliser in W of the flag $[k_1] \subset \cdots \subset [k_m]$ then one can easily see that

$$P = \langle \{s_1, \ldots, s_{n-1}\} \setminus \{s_{k_1}, \ldots, s_{k_m}\} \rangle$$

is a parabolic subgroup. The set $\mathcal{F}_n^{k_1 \cdots k_m}$ of all flags of rank $(k_1, \ldots, k_m)$ can be identified with the factor set $W^P = W/P$. An extension of Theorem 1.8 shows that the Bruhat order on W^P coincides with the Gale order on $\mathcal{F}_n^{k_1 \cdots k_m}$. Thus, flag matroids are exactly Coxeter matroids for the Coxeter group Sym_n.

A notion equivalent to Coxeter matroid is that of a *matroid map*

$$\mu : W \to W^P,$$

defined by the property that μ satisfies the *matroid inequality*

$$\mu(u) \leqslant^v \mu(v) \quad \text{for all } u, v \in W.$$

If $\mathcal{M} \subseteq W^P$ is a Coxeter matroid, the map

$$\begin{aligned} \mu : W &\to W^P \\ w &\mapsto \text{the } \leqslant^w\text{-maximal element of } \mathcal{M} \end{aligned}$$

is obviously a matroid map. Conversely, it is easy to see that the image $\mathcal{M} = \mu[W]$ of a matroid map $\mu : W \to W^P$ satisfies the Maximality Property and thus is a Coxeter matroid for W and P.

Henceforth we will often refer to a matroid map simply as a matroid. From context it should always be clear whether a matroid map or its image is meant.

Matroid maps provide a very efficient way to formulate and prove results related to flag matroids.

For example, let $\mathcal{F}_1$ and $\mathcal{F}_2$ be two flag matroids and $\mu_i : W \to W^{P_i}$ corresponding matroid maps. Then one can show that the concordancy condition is just the requirement that the intersection of the cosets $\mu_1(w) \cap \mu_2(w)$ is non-empty for all $w \in W$. Elementary group theory shows that in that case $\mu_1(w) \cap \mu_2(w)$ is a coset of the parabolic subgroup $P_1 \cap P_2$. Moreover, the matroid map for $\mathcal{F}_1 \vee \mathcal{F}_2$ is

$$\mu : w \mapsto \mu_1(w) \cap \mu_2(w).$$

If μ and ν are two concordant Coxeter matroids, we denote the Coxeter matroid $w \mapsto \mu(w) \cap \nu(w)$ by $\mu \vee \nu$.

Given a Coxeter matroid μ for the Coxeter group W and parabolic subgroup P, one can consider a matroid map $\mathsf{u}(\mu) : W \to W$ defined by

$$\mathsf{u}(\mu) = \max_{\leqslant^w} \mu(w).$$

It is a matroid map; we call it the *underlying flag matroid map* of μ and its image $\mathcal{F} = \mathsf{u}(\mu)[W]$ the *underlying (flag) matroid* for the Coxeter matroid $\mathcal{M} = \mu[W]$. The original matroid map $\mu : W \longrightarrow W^P$ can be found as

$$\mu(w) = \mathsf{u}(\mu)(w)P.$$

If $\mu : W \longrightarrow W^P$ is a matroid map and Q is a parabolic subgroup in W, then the map

$$\begin{array}{rcl} \mu^Q : W & \to & W^Q \\ w & \mapsto & \mathsf{u}(\mu)(w)Q \end{array}$$

is a Coxeter matroid for W and Q. If $\mu : W \longrightarrow W^P$, is a matroid map and $P_1, \dots, P_k$ are maximal parabolic subgroups in W such that $P = P_1 \cap \dots \cap P_k$ then the matroids $\mu^{P_1}, \dots, \mu^{P_k}$ are pairwise concordant and

$$\mu = \mu^{P_1} \vee \dots \vee \mu^{P_k}.$$

This generalises, to arbitrary Coxeter matroids, the way in which a flag matroid is built of constituent matroids.

Essentially, these simple observations mean that we can restrict our attention to matroid maps $\mu : W \to W$; they are called *W-matroids.*

For an (ordinary) matroid $\mathcal{B}$ of rank k on $[n]$, the underlying flag matroid is formed by all truncations and all Higgs' lifts of $\mathcal{B}$. Here, if $m \leqslant \operatorname{rank}(\mathcal{B})$, the *truncation* of $\mathcal{B}$ to rank m is the matroid whose bases are all independent sets of cardinality m (recall that an *independent* set is a subset of a basis). Every truncation of a matroid $\mathcal{B}$ is a quotient of $\mathcal{B}$, as is easily verified. Similarly, if $m > k$, the *Higgs' lift* of $\mathcal{B}$ to rank m is the matroid whose bases are all sets of cardinality m which contain a basis of $\mathcal{B}$.

Our experience shows that, instead of looking at individual matroids, it might be useful to work with maximal chains of matroids on $[n]$ with respect to the quotient relations, since these chains correspond to matroids maps on Sym_n. The set of all maximal chains of matroids has a natural structure of simplicial complex. The latter has many nice properties [7]; for example, it is shellable. The proofs in [7] show how useful is the language of Coxeter matroids even in the classical context of ordinary matroids.

1.8 From polytopes to flats

Adjacency of vertices in matroid polytopes encodes some crucial information about matroids. Here we give only two examples.

We say that two bases of a matroid are *adjacent* if they are related by a transposition, that is, by an elementary exchange. The following theorem is a refinement of the Gelfand–Serganova Theorem 1.4.

Theorem 1.9 (I.M. Gelfand *et al.* [22]) *Let Δ be the matroid polytope of a matroid $\mathcal{B}$. Two vertices δ_A and δ_B of Δ are adjacent (i.e. are incident to a common edge) if and only if the corresponding bases A and B of $\mathcal{B}$ are adjacent.*

Even more interesting is the case of flag matroids. As the following result shows, the adjacency of vertices in a certain flag matroid polytope associated with a matroid encodes the information about the flats of the matroid. Recall that *flats* of a matroid are combinatorial analogues of vector subspaces in a vector space and are defined as follows. Let $\mathcal{B}$ be a matroid on a set $[n]$; the *rank* $\operatorname{rk}(X)$ of a subset $X \subseteq [n]$ is the maximum of cardinalities of independent subsets of X. A *flat* of rank k is a maximal (with respect to inclusion) set of rank k.

Theorem 1.10 (A.V. Borovik, I.M. Gelfand, A. Vince and N. White [8]) *Let $\mathcal{B}$ be a matroid of rank k on $[n]$ and $\mathcal{F}$ the flag matroid formed by $\mathcal{B}$ and all its truncations. Thus, the vertices of $\Delta = \Delta_{\mathcal{F}}$ correspond to* ordered *bases $(a_1, \ldots, a_k)$ of $\mathcal{B}$.*

If $A = (a_1, \ldots, a_k)$ and $B = (b_1, \ldots, b_k)$ are two ordered bases of $\mathcal{B}$ which correspond to adjacent vertices of $\Delta_{\mathcal{F}}$ then either A and B are two reorderings of the same basis of $\mathcal{B}$, or the sequences of points $(a_1, \ldots, a_k)$ and $(b_1, \ldots, b_k)$ span the same flag of flats in the matroid $\mathcal{B}$, that is, flats spanned by a_1 and b_1, by a_1, a_2 and b_1, b_2, and so on, are pairwise equal.

It is well known that flats of a matroid form a semimodular lattice (see Section 2.1 for dicussion of lattices), and one can easily see that every maximal chain of flats is spanned by an ordered basis. Therefore every flat of the matroid $\mathcal{B}$ is represented by at least one vertex of $\Delta_{\mathcal{F}}$. Theorem 1.10 associates with $\Delta_{\mathcal{F}}$ another combinatorial object: the set $\mathcal{C}$ of all maximal chains of flats of $\mathcal{B}$. The next section will show that the latter is interesting on its own. For the purpose of this paper, the most important aspect of $\mathcal{C}$ is that its properties are controlled by the symmetric group Sym_k, $k = \operatorname{rank}(\mathcal{B})$—we shall see that in Section 2.

However, before moving to semimodular lattices, it would be useful to briefly review the concept of matroid representation.

1.9 Representable flag matroids

Let E be a *vector configuration*, that is, a family $E = \{ e_i \}_{i \in I}$ of vectors in a finite-dimensional vector space V. Notice that this allows for repeated elements in E: it might happen that $e_i = e_j$ although $i \neq j$. It is well known that the collection of all maximal subsets $J \subseteq I$ such that the vectors $e_j, j \in J$

are linearly independent is a matroid on I. Abusing notation, we shall identify the sets E and I (thus allowing repeated elements in E).

If E is a family of n vectors (possibly repeating) in a k-dimensional vector space V over the field K, then by expanding these vectors in terms of a basis of V, we can represent E as the set of columns of a $k \times n$ matrix A.

Consider now the action of elementary row operations on A. Obviously, they preserve the matroid on E. Thus the row-space U of A determines the equivalence class (with respect to elementary row operations) of representations of the matroid M.

It is important to remember, however, that in this definition we use some coordinate system of V, or, equivalently, a basis $b_1, \dots, b_n$ in V. Therefore a representation of a matroid of rank k on n letters is made of the following ingredients: a vector space of dimension n with a fixed basis and a subspace of dimension k.

This approach allows us to define the notion of representation of a flag matroid. Indeed, let $U_1 \subset U_2 \subset \cdots \subset U_m \subseteq K^n$ be a flag of subspaces of K^n of dimensions $k_1, k_2, \dots, k_m$, respectively. Fix some basis in K^n. Then each subspace U_i represents a matroid M_i of rank k_i on $[n]$.

Theorem 1.11 *The collection of matroids* $\{ M_1, M_2, \dots, M_m \}$ *represented by a flag of subspaces forms a flag matroid.*

A flag matroid arising from a flag of subspaces in this manner is called a *representable flag matroid.* In the next section we shall see that representation of a flag matroid is a combinatorial concept in the sense that the only structure needed is the semimodular lattice formed by subspaces of the vector space V. Interestingly, the Coxeter group structure on the symmetric group plays the crucial role in this construction.

2 Matroids and Semimodular Lattices

In this section, we continue the process of explaining the intimate connection between matroids and the symmetric group, by first switching to semimodular lattices and seeing how they are related to the symmetric group.

The crucial concept of this section is that of *Jordan–Hölder permutation.* Its meaning can best be seen in the context of group theory. Let G be a group. A subgroup $H \leqslant G$ is *subnormal* if there exist a chain of subgroups

$$H = H_0 \lhd H_1 \lhd \cdots \lhd H_l = G$$

in which every subgroup is normal in the next one. A composition series of G is a maximal chain

$$1 = G_0 \lhd G_1 \lhd \cdots \lhd G_n = G$$

of subnormal subgroups, while the factor groups G_i/G_{i-1} are called the composition factors (of the chain). The celebrated Jordan–Hölder Theorem states that if

$$1 = H_0 \lhd H_1 \lhd \cdots \lhd H_m = G$$

is another composition series of G then $n = m$ and there is a one-to-one correspondence between the composition factors of the two series such that the corresponding factors are isomorphic. It is natural to call the permutation π of $[n]$ such that $G_i/G_{i-1} \simeq H_{\pi(i)}/H_{\pi(i-1)}$ a Jordan–Hölder permutation. The analysis of the proof of the Jordan–Hölder Theorem shows that there is a combinatorial way to construct the 'canonical' Jordan–Hölder permutation. In fact, the partially ordered set of subnormal subgroups of a finite group is a semimodular lattice; the Jordan–Hölder permutation can be defined for any two maximal chains in a semimodular lattice and used as a measure of 'distance' between the maximal chains.

2.1 Semimodular lattices

We recall some standard definitions concerning partially ordered sets.

Let $\leqslant$ be a partial ordering of the set X. We set $x < y$ if $x \leqslant y$ and $x \neq y$. An element x *covers* element y if $y < x$ and if $y \leqslant z \leqslant x$ implies that either $z = y$ or $z = x$. If X has a minimum element 0, then an *atom* is an element which covers 0. A *chain* is a totally ordered subset $x_0 < x_1 < \cdots < x_k$, and the *length* of this chain is k. If X has the minimum element 0, then the *height* $h(x)$ is the maximum of lengths of chains $0 < x_1 < \cdots < x_k = x$ between 0 and x. The *height* of X is the maximum of heights of its elements.

A partially ordered set X satisfies the *Jordan–Hölder condition* if, given any two elements $x < y$ in X, all maximal chains between x and y have the same length.

A *lattice* is a partially ordered set L such that L contains, with any two elements x and y, a unique *least upper bound* $x \vee y$ and *greatest lower bound* $x \wedge y$. The elements $x \vee y$ and $x \wedge y$ are also called, respectively, the *join* and *meet* of x and y. If L has no infinite chains, it must have minimum element 0 and maximum element 1. In this paper, all lattices are assumed to have finite height. A lattice L is *semimodular* if and only if it satisfies the Jordan–Hölder condition and its height function satisfies the *semimodular inequality*

$$h(x) + h(y) \geqslant h(x \wedge y) + h(x \vee y),$$

for all $x, y \in L$.

2.2 Jordan–Hölder permutation

As we shall soon see, semimodular lattices are intimately linked to symmetric groups. For now, we will show that for any two maximal chains in a semimodular lattice, we can assign a 'distance' between them which is a permutation. This distance function provides the representation of flag matroids in semimodular lattices.

Let L be a semimodular lattice of height n and $\mathcal{C}$ the set of maximal chains in L. Let

$$c = \{\, c_0 < c_1 < \cdots < c_n \,\} \text{ and } d = \{\, d_0 < d_1 < \cdots < d_n \,\}$$

be two maximal chains in $\mathcal{C}$. We define a map

$$\pi(c,d) : [n] \longrightarrow [n]$$

by the following formula:

$$\pi(c,d)(i) = j \text{ if } \left\{ \begin{array}{ll} d_{i-1} \vee c_k = d_i \vee c_k & \text{for all } k \geqslant j \\ d_{i-1} \vee c_k < d_i \vee c_k & \text{for all } k < j. \end{array} \right.$$

Theorem 2.1 (Abels [1]) *The map $\pi(c,d)$ satifies*

$$\pi(c,d) \circ \pi(d,c) = e,$$

the identity permutation. In particular, $\pi(c,d)$ is a permutation of $[n]$.

Following Abels [1], we shall call the permutation $\pi(c,d)$ the *Jordan–Hölder permutation.*

2.3 Independent sets and bases

We say that a set of atoms $A = \{\, a_1, \ldots, a_k \,\}$ is *independent* if

$$a_{i_0} \wedge (a_{i_1} \vee \cdots \vee a_{i_m}) = 0$$

for any set of pairwise distinct atoms $a_{i_0}, a_{i_1}, \ldots, a_{i_m}$. A *basis* $B = \{\, b_1, \ldots, b_n \,\}$ in L is an independent set of atoms such that $b_1 \vee \cdots \vee b_n = 1$. Obviously,

$$0 < b_1 < b_1 \vee b_2 < \cdots < b_1 \vee \cdots \vee b_n$$

is a maximal chain in L; we say that the chain is *spanned* by the (ordered) basis $b_1, \ldots, b_n$.

A *geometric lattice* is a semimodular lattice of finite height in which every element is a join of atoms. It can be easily shown that a semimodular lattice of finite height is geometric if and only if every maximal chain is spanned by some ordered basis. Another well-known observation is that the lattice of flats of a matroid is geometric.

2.4 Representation of flag matroids

Let L be a semimodular lattice and $B = \{b_1, \ldots, b_n\}$ a basis in L. For a permutation $w \in \mathrm{Sym}_n$, we denote by $\alpha(w)$ the maximal chain

$$0 < b_{w(1)} < b_{w(1)} \vee b_{w(2)} < \cdots < b_{w(1)} \vee \cdots \vee b_{w(n)} = 1.$$

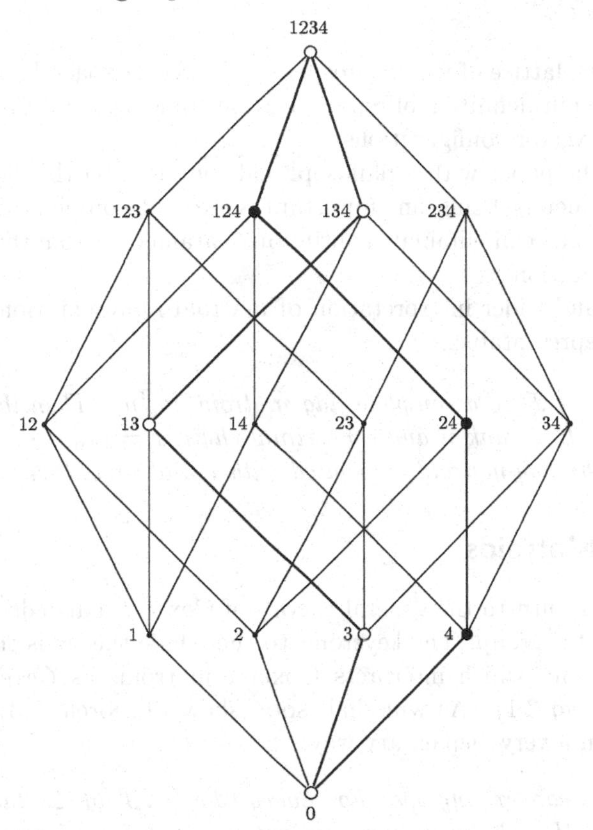

Figure 4: *The lattice of subsets of* $\{1,2,3,4\}$, *with the maximal chains* $a = \{0 < 3 < 13 < 134 < 1234\}$ *and* $b = \{0 < 4 < 24 < 124 < 1234\}$. *To compute* $\pi(a,b)$, *we proceed as follows. Observe that* $b_0 \vee a_3 = 134 = b_1 \vee a_3$, *whereas* $b_0 \vee a_2 = 13 < 134 = b_1 \vee a_2$, *hence* $\pi(a,b)(1) = 3$. *Similarly,* $\pi(a,b)(2) = 4$, $\pi(a,b)(3) = 2$, *and* $\pi(a,b)(4) = 1$; *hence* $\pi(a,b)$ *is the permutation* (1324).

Theorem 2.2 *In this notation, let* $x = \{\, 0 < x_1 < x_2 < \cdots < x_n = 1\,\}$ *be a maximal chain in* L. *The map*

$$\begin{aligned} \mu_x : \mathrm{Sym}_n &\longrightarrow \mathrm{Sym}_n \\ w &\mapsto w \cdot \pi(\alpha(w), x) \end{aligned}$$

is a matroid map.

Now the representation of a matroid can also be explained in terms of the Jordan–Hölder permutation on semimodular lattices. Indeed, if $y = \{y_1, \ldots, y_n\}$ is another flag in L with $y_k = x_k$ then the flag matroids on $[n]$ corresponding to the matroids maps μ_x and μ_y have the same rank k constituent M_k. Therefore we can say that the element x_k of height k in the semimodular lattice L of height n represents the matroid M_k of rank k on $[n]$.

When L is the lattice of vector subspaces of a vector space V of dimension n over a field K, our definition of representation coincides with the traditional one, in terms of vector configurations.

We started the paper with a philosophical comment on the elimination of continuous parameters being an important source of combinatorial objects. Now we see how one can eliminate continuous parameters from the concept of matroid representation.

In this new and wider interpretation of matroid representation, every matroid becomes representable.

Theorem 2.3 *Let $\mathcal{F}$ be a complete flag matroid on $[n]$. Then there exists a geometric lattice L of rank n and a maximal chain $x = \{x_0 < \cdots < x_n\}$ in L such that $\mathcal{F}$ is the flag matroid associated with x and some basis B of L.*

3 Coxeter Matroids

It is time to return to the general theory of Coxeter matroids for an arbitrary finite Coxeter group. The keystone to the whole theory is the Gelfand–Serganova Theorem, which interprets Coxeter matroids as *Coxeter matroid polytopes* (Theorem 3.1). As will shall soon show (Theorem 3.3), the latter can be defined in a very elementary way:

> *Let Δ be a convex polytope. For every edge $[\alpha, \beta]$ of Δ, take the hyperplane $H_{\alpha\beta}$ that cuts the segment $[\alpha, \beta]$ in its midpoint and is perpendicular to $[\alpha, \beta]$. Let W be the group generated by the reflections in all such hyperplanes. Then W is a finite group if and only if Δ is a Coxeter matroid polytope.*

The 'kaleidoscope' version of this definition, as given in the introduction, is a simple geometric observation: the system of mirrors $\{ H_{\alpha\beta} \}$ forms a kaleidoscope with finitely many reflections if and only if the group W is finite.

It is a classical result of the theory of reflection groups that W, being a finite reflection group (that is, a finite group generated by reflections), is a Coxeter group.

The converse is also true: every finite Coxeter group W has a finite dimensional representation by orthogonal transformations in which the images of the distinguished generators $s_1, \ldots, s_n$ are reflections. Therefore we can work in the standard setting for finite reflection groups: W is a finite reflection group acting in the Euclidean space V with a W-invariant scalar product $(\ ,\ )$, Φ is its root system, Σ is the mirror system associated with Φ (the set of hyperplanes perpendicular to roots in Φ).

Now let $\Pi = \{\rho_1, \ldots, \rho_n\}$ be the simple root system corresponding to the system of standard generators $s_1, \ldots, s_n$ of W. Let $J \neq \emptyset$ be a subset of $[n]$ and

$$P = W_J = \langle s_i \mid i \in J \rangle$$

the corresponding parabolic subgroup in W. We wish to represent the factor set $W^P = W/P$ by points in V. For that purpose, we have to find a point $\omega \in V$ such that its isotropy group $C_W(\omega) = \{w \in W \mid w\omega = \omega\}$ is P. The choice of ω is, of course, not unique; one of the possibilities is defined by the system of inequalities

$$\frac{(\omega, \rho_i)}{(\rho_i, \rho_i)} \begin{cases} < 0 & \text{for} \quad \rho_i \notin J, \\ = 0 & \text{for} \quad \rho_i \in J. \end{cases}$$

Notice that this is possible since $\{\rho_1, \ldots, \rho_n\}$ is a basis of V.

Therefore we can define a mapping

$$\delta : W^P \longrightarrow V$$

that sends wP to $w\omega$. We denote $\delta(A)$ by δ_A for all $A \in W^P$. The mapping δ identifies the factor set W^P with the orbit $W \cdot \omega$.

3.1 The Gelfand–Serganova Theorem

With any subset $\mathcal{M} \subseteq W^P$ we associate a polytope $\Delta_{\mathcal{M}}$, the convex hull of points in $\delta(\mathcal{M})$. Notice that, since the group W acts transitively on the set $W \cdot \omega$, all points in $W \cdot \omega$ are vertices of the convex hull of $W \cdot \omega$. Therefore the set $\delta(\mathcal{M})$ is exactly the set of vertices of $\Delta_{\mathcal{M}}$.

Theorem 3.1 (I.M. Gelfand and V.V. Serganova [23]) *Let W be a finite Coxeter group, Φ its root system with mirror system Σ, P a parabolic subgroup in W, $\mathcal{M}$ a subset in $\mathcal{W}^P$, and $\Delta = \Delta_{\mathcal{M}}$ the polytope associated with $\mathcal{M}$.*

Then the following conditions are equivalent.

(1) *$\mathcal{M}$ is a Coxeter matroid.*

(2) *Every edge of Δ is parallel to a root in Φ.*

(3) *Every edge of Δ is perpendicular to one of the mirrors in Σ.*

Theorem 3.1 can be restated in the following form.

Theorem 3.2 *A subset $\mathcal{M} \subseteq W^P$ is a Coxeter matroid if and only if, for any pair of adjacent vertices δ_A and δ_B of $\Delta_{\mathcal{M}}$, there is a reflection $s \in W$ such that $s\delta_A = \delta_B$ (and also $s\delta_B = \delta_A$, $sB = A$ and $sA = B$).*

3.2 Coxeter matroids and polytopes

Now we wish to offer a very elementary approach to Coxeter matroids.

Let Δ be a convex polytope in the real affine Euclidean space $\mathbb{A}^n$. For any two vertices α and β of Δ which are adjacent (i.e. connected by an edge) we can consider the reflection $s_{\alpha\beta}$ in the mirror of symmetry of the edge $[\alpha, \beta]$.

All these reflections generate a group $W(\Delta)$ of affine isometries of the space $\mathbb{A}^n$. We say that Δ is a *(Coxeter) matroid polytope* if the group $W(\Delta)$ is finite.

Examples of matroid polytopes are abundant. Obviously, Platonic solids (as well as most regular and semi-regular polytopes) are matroid polytopes. It immediately follows from Theorem 3.2 that polytopes associated with Coxeter matroids are Coxeter matroid polytopes. In particular, the matroid polytopes from ordinary matroid theory are Coxeter matroid polytopes, which justifies the use of this terminology.

If Δ is a matroid polytope, the group $W = W(\Delta)$ will be called the *exchange group* of Δ. Being a finite group, W fixes the barycentre of each of its (finite) orbits, so we can assume without loss of generality that W fixes the origin O of the vector space $\mathbb{R}^n$ and hence is a linear group. Moreover it is a finite reflection group and hence a Coxeter group. By definition of W all vertices of Δ belong to one W-orbit. Choose a vertex δ of Δ. Then the isotropy group P of δ is a standard parabolic subgroup of W, i.e. is generated by some elements r_i.

We find ourselves in the precise setting of our construction of polytopes associated with Coxeter matroids. Therefore the set of vertices of Δ can be identified with some subset $\mathcal{M}$ of the factor set W^P. Now the following result is an immediate corollary of Theorem 3.2.

Theorem 3.3 *If Δ is a matroid polytope, then $\mathcal{M}$ is a Coxeter matroid for $W(\Delta)$ and P.*

3.3 Exchange groups of Coxeter matroids

Now return to Theorem 3.1 and our construction of a matroid polytope Δ for a Coxeter matroid $\mathcal{M} \subseteq W^P$. We had some freedom in choosing the point ω. However, it can be shown that the combinatorial type of Δ and the exchange group $W(\Delta)$ of Δ do not depend on choice of the point ω. Notice, however, that the exchange group $W(\Delta)$ does not necessary coincide with the Coxeter group W we started with.

For example, consider a matroid of rank 2 on [4] given by the collection of its bases $\mathcal{B} = \{12, 14, 23, 34\}$. The corresponding Coxeter group $A_3 = \mathrm{Sym}_4$ acts, in its reflection representation, as the group of symmetries of the regular tetrahedron in $\mathbb{R}^3$. Using the representation of the polytope from Section 1, a basis $\{i, j\}$ in $\mathcal{B}$ is represented by the midpoint of the edge ij. Again up to affine transformation, the basis matroid polytope is just the convex hull of these vertices, as shown on Figure 5.

Obviously $W(\Delta_{\mathcal{B}}) \simeq \mathbb{Z}_2 \times \mathbb{Z}_2 \simeq \mathrm{Sym}_2 \times \mathrm{Sym}_2$ is the Klein four-group, though originally $\mathcal{B}$, being a matroid of rank 2 on [4], is a Coxeter matroid for Sym_4 and its parabolic subgroup $\mathrm{Sym}_2 \times \mathrm{Sym}_2$. It can be shown that, in general, the exchange group of a matroid on $[n]$ equals the direct product $\mathrm{Sym}_{n_1} \times \cdots \times \mathrm{Sym}_{n_d}$ for some n_i with $n_1 + \cdots + n_d \leqslant n$, where these direct product factors correspond to the non-trivial components of the matroid.

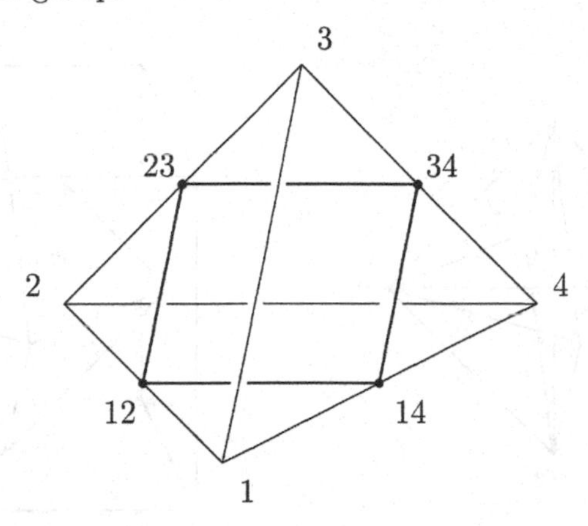

Figure 5: *The exchange group of this matroid polytope is the Klein four-group.*

It is significant that the concept of the exchange group of a matroid polytope sheds some light on interactions between different types of Coxeter matroids; we shall see more examples of this interaction in Section 4.

4 Symplectic Matroids

We have seen how matroids and semimodular lattices are intimately related to the symmetric group. Now we replace the symmetric group by another Coxeter group, namely, BC_n, the hyperoctahedral group. The resulting structures are called symplectic matroids, and they are in some sense rather general Coxeter matroids, as they include ordinary matroids as special cases. We are not attempting to develop or even outline the theory of symplectic matroids; the interested reader may find the detailed exposition of the theory in the book [6]. Instead, we only touch on relations between the ordinary and symplectic matroids because they provide further illustration of the fundamental principle that the true nature of a Coxeter matroid is determined by its exchange group.

4.1 Definition of symplectic matroids

4.1.1 Hyperoctahedral group and admissible permutations.

Define

$$[n]^* = \{1^*, 2^*, \ldots, n^*\}.$$

Define the map $* : [n] \to [n]^*$ by $i \mapsto i^*$ and the map $* : [n]^* \to [n]$ by $i^* \mapsto i$. In other words, we are defining $i^{**} = i$. Then $*$ is an involutive permutation of the set $J = [n] \sqcup [n]^*$ where $\sqcup$ is used to indicate union and remind us the sets are disjoint.

We say that a subset $K \subset J$ is *admissible* if and only if $K \cap K^* = \emptyset$.

Let W be the group of all permutations of the set J which commute with the involution *, i.e. a permutation w belongs to W if and only if $w(i^*) = w(i)^*$

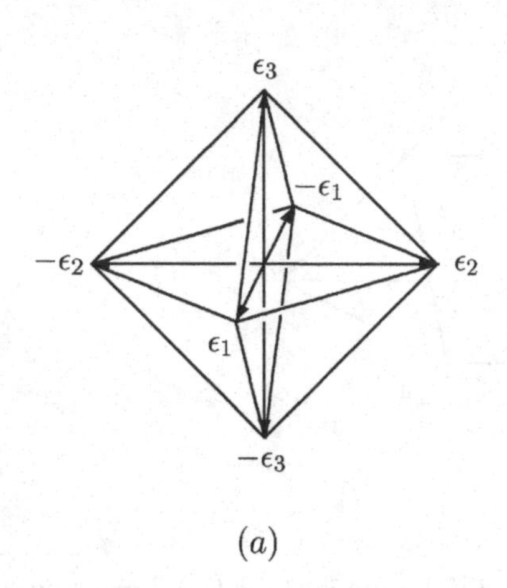

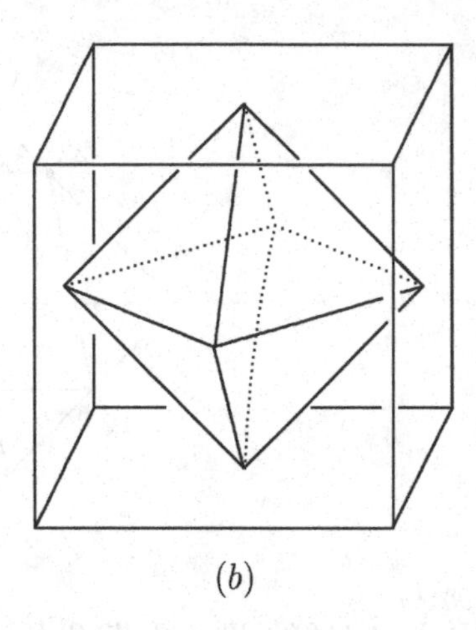

(*a*) (*b*)

Figure 6: *The hyperoctahedron ('octahedron' in dimension $n = 3$) or n-cross polytope is the convex hull of the points $\pm\epsilon_i$, $i = 1, \dots, n$ in $\mathbb{R}^n$ (picture* (a)*). Obviously the hyperoctahedron is the dual polytope to the unit cube (picture* (b)*).*

for all $i \in J$. We shall call permutations with this property *admissible*. The group W is known under the name of the *hyperoctahedral group* BC_n. It is easy to see that W is isomorphic to the group of symmetries of the n-cube $[-1, 1]^n$ in the n-dimensional real Euclidean space $\mathbb{R}^n$, where J can be thought of as the set of labels of the facets of the n-cube, with i being the opposite facet to i^*. Indeed, if $\epsilon_1, \epsilon_2, \dots, \epsilon_n$ is the standard orthonormal basis in $\mathbb{R}^n$, then W acts on $\mathbb{R}^n$ by the following orthogonal transformations: for $i \in [n]$ we set $\epsilon_{i^*} = -\epsilon_i$ and $w\epsilon_i = \epsilon_{w(i)}$. It can be easily seen that W is exactly the group of all orthogonal transformations of $\mathbb{R}^n$ preserving the unit cube $[-1, 1]^n$.

The name 'hyperoctahedral' for the group W is justified by the fact that the group of symmetries of the n-cube coincides with the group of symmetries of its dual polytope, whose vertices are the centers of the faces of the cube. The dual polytope for the n-cube is known under the name of *n-cross polytope* or *n-dimensional hyperoctahedron*, Figure 6.

4.1.2 Admissible orderings. We shall order the set J in the following way:

$$n > n - 1 > \cdots > 1 > 1^* > 2^* > \cdots > n^*.$$

Now if $w \in W$ then we define a new ordering $\leqslant^w$ of the set J by the rule

$$i \leqslant^w j \text{ if and only if } w^{-1}i \leqslant w^{-1}j.$$

Orderings of the form $\leqslant^w$, $w \in W$, are called *admissible* orderings of J. They can be characterised by the following property:

> *an ordering $\prec$ on J is admissible if and only if $\prec$ is a linear ordering and from $i \prec j$ it follows that $j^* \prec i^*$.*

Notice a natural one-to-one correspondences between admissible permutations of the set J; and admissible orderings of the set J. Indeed, for every admissible permutation $w \in W$ we have the admissible ordering $\leqslant^w$ of J. *Vice versa*, if $\prec$ is an admissible ordering of J, then the permutation

$$w = \begin{pmatrix} n^* & (n-1)^* & \dots & 1^* & 1 & \dots & n-1 & n \\ j_1 & j_2 & \dots & j_n & j_{n+1} & \dots & j_{2n-1} & j_{2n} \end{pmatrix}$$

where

$$j_1 \prec j_2 \prec \cdots \prec j_{2n-1} \prec j_{2n},$$

is admissible and the ordering $\prec$ coincides with $\leqslant^w$.

Denote by J_k the collection of all admissible k-subsets in J, for some $k \leqslant n$. If $\prec$ is an arbitrary ordering on J, it induces the partial ordering (which we denote by the same symbol $\prec$) on J_k: if $A, B \in J_k$ and

$$A = \{a_1 \prec a_2 \prec \cdots \prec a_k\} \text{ and } B = \{b_1 \prec b_2 \prec \cdots \prec b_k\},$$

we set $A \prec B$ if

$$a_1 \prec b_1, a_2 \prec b_2, \dots, a_k \prec b_k.$$

This is the same idea we used in Section 1 to induce a partial order on subsets from an order on individual elements, so we refer to this as the Gale order.

4.1.3 Symplectic matroids. Now let $\mathcal{B} \subseteq J_k$ be a collection of admissible k-element subsets of the set J. We say that the triple $M = (J, {}^*, \mathcal{B})$ is a *symplectic matroid* if it satisfies the following *Maximality Property*:

> *for every admissible order $\prec$ on J, the collection $\mathcal{B}$ contains a unique maximal element, i.e. a subset $A \in \mathcal{B}$ such that $B \prec A$ (in the order on J_k induced from $\prec$), for all $B \in \mathcal{B}$.*

The collection $\mathcal{B}$ is called the *collection of bases* of the symplectic matroid M, its elements are called *bases* of M, and the cardinality k of the bases is the *rank* of M. A *Lagrangian* matroid is a symplectic matroid of rank n, the maximum possible.

4.2 Root systems of type C_n and the Gelfand–Serganova Theorem

In order to study symplectic matroid polytopes, we need an *ad hoc* description of root system Φ of type C_n. It consists of the vectors $2\epsilon_j$, $j \in J$ (called *long roots*), together with the vectors $\epsilon_{j_1} - \epsilon_{j_2}$, where $j_1, j_2 \in J$, $j_1 \neq j_2$ or j_2^* (called *short roots*). Written in the standard basis $\epsilon_1, \epsilon_2, \dots, \epsilon_n$, the roots take the form $\pm 2\epsilon_i$, $i = 1, 2, \dots, n$, or $\pm\epsilon_i \pm \epsilon_j$, $i, j = 1, 2, \dots, n$, $i \neq j$.

Recall that if r is a non-zero vector in $\mathbb{R}^n$ then the *reflection* σ_r in the hyperplane perpendicular to r is the linear transformation of $\mathbb{R}^n$ determined by

$$\sigma_r(x) = x - \frac{2(x,r)}{(r,r)} r, \text{ for } x \in \mathbb{R}^n,$$

where (,) is the standard scalar product in $\mathbb{R}^n$.

It is easy to see that when r is one of the long roots $\pm 2\epsilon_i$, $i \in [n]$, then σ_r is the linear transformation corresponding to the element $s_r = (i, i^*)$ of W in its canonical representation. Analogously, if $r = \epsilon_i - \epsilon_j$, $i, j \in J$, is a short root, then the reflection σ_r corresponds to the admissible permutation $s_r = (i, j)(i^*, j^*)$.

If we now choose in Φ a *simple system of roots*

$$\Pi = \{ \epsilon_2 - \epsilon_1, \ldots, \epsilon_n - \epsilon_{n-1}, 2\epsilon_n \}$$

then the corresponding reflections

$$s_1 = (1, 2)(1^*, 2^*), \ldots, s_{n-1} = (n-1, n)((n-1)^*, n^*), s_n = (n, n^*)$$

are distinguished generators of W as a Coxeter group. If

$$P_k = \langle s_1, \ldots, s_{k-1}, s_{k+1}, \ldots, s_n \rangle$$

is the maximal parabolic subgroup generated by all s_i with the exception of s_k then, obviously, P_k is the stabiliser in W of the admissible subset $[k]$. Therefore the collection J_k of all admissible k-subsets can be identified with the factor space W/P_k and, as the reader already expects, the symplectic matroids on $[n] \sqcup [n]^*$ are exactly Coxeter matroids for the hyperoctahedral group W and maximal parabolic subgroup P_k.

The special case of the Gelfand–Serganova Theorem for symplectic matroids is formulated in the most natural way.

For an admissible set $A \in J_k$ define the point $\delta_A \in \mathbb{R}^n$ as

$$\delta_A = \epsilon_{i_1} + \epsilon_{i_2} + \cdots + \epsilon_{i_k} \text{ where } A = \{i_1, i_2, \ldots, i_k\}.$$

Theorem 4.1 *Let $\mathcal{B} \subseteq J_k$ be a collection of admissible k-sets in J. Let Δ be the convex hull of the points δ_A with $A \in \mathcal{B}$.*

Then δ_A are vertices of Δ for all $A \in \mathcal{B}$. Moreover, $\mathcal{B}$ is the collection of bases of a symplectic matroid on J if and only if all edges (i.e. 1-dimensional faces) of Δ are parallel to roots in Φ.

4.3 Representable symplectic matroids

Symplectic matroids arise naturally from symplectic geometry, in much the same way that ordinary matroids arise from projective geometry.

4.3.1 Isotropic subspaces. We begin with a *standard symplectic space*, which is a vector space V over K with a basis

$$E = \{e_1, e_2, \ldots, e_n, e_{1^*}, e_{2^*}, \ldots, e_{n^*}\}$$

and which is endowed with an anti-symmetric bilinear form $(.,.)$ such that $(e_i, e_j) = 0$ for all $i, j \in J, i \neq j^*$, and $(e_i, e_{i^*}) = 1 = -(e_{i^*}, e_i)$ for $i \in [n]$. An *isotropic subspace* of V is a subspace U such that $(u, v) = 0$ for all $u, v \in U$. Let $\dim U = k$; one can easily see that $k \leqslant n$. Now choose a basis $\{u_1, u_2, \ldots, u_k\}$ of U, and expand each of these vectors in terms of the basis E:

$$u_i = \sum_{j=1}^{n} a_{i,j} e_j + \sum_{j=1}^{n} b_{i,j} e_{j^*}.$$

This represents the isotropic subspace U as the row-space of a $k \times 2n$ matrix (A, B), $A = (a_{i,j}), B = (b_{i,j})$, with the columns indexed by J, specifically, the columns of A by $[n]$ and those of B by $[n]^*$.

It is easy to see that a subspace U of the standard symplectic space V is isotropic if and only if U is represented by a matrix (A, B) with AB^{t} symmetric. This property is preserved under elementary row operation on (A, B).

4.3.2 Symplectic matroids from isotropic subspaces. Now, given a $k \times 2n$ matrix $C = (A, B)$ with columns indexed by J, let us define a collection $\mathcal{B} \subseteq J_k$ by saying $X \in \mathcal{B}$ if X is an admissible k-set and the $k \times k$ minor formed by taking the columns of C indexed by elements of X is non-zero.

Theorem 4.2 *If U is isotropic, then $\mathcal{B}$ is the collection of bases of a symplectic matroid.*

A symplectic matroid $\mathcal{B}$ which arises from a matrix (A, B), with AB^{t} symmetric, is called a *representable symplectic matroid*, and (A, B) (with its columns indexed by J) is a *representation* or *coordinatisation* of it (over the field K). We sometimes refer to this type of representation of a symplectic matroid as a *C_n-representation* to distinguish it from some other types of representations. In particular, the notation C_n here is used to distinguish C_n from B_n as an algebraic group; see [4] for B_n-representations and [27] for D_n-representations.

Notice, in particular, that if A is an $(n \times n)$ symmetric matrix and Id is the $(n \times n)$ identity matrix, then the subspace represented by (A, Id) is isotropic. An admissible subset $K \subset J$ is a basis of the associated symplectic matroid if and only if the diagonal minor of A formed by rows and columns indexed by $K \cap [n]$ is non-degenerate (if $K = [n]^*$, we assume that the 'empty' diagonal minor has determinant 1). Of course, $K = (K \cap [n]) \cup ([n] \setminus K)^*$. Therefore the symplectic matroid associated with (A, Id) encodes the combinatorial properties of the set of non-degenerate diagonal minors of the symmetric matrix A; cf. Bouchet [18] who formulated this observation in terms of *Δ-matroids*.

Example 4.3 Consider the symplectic matroid represented by the following matrix:

$$(A, B) = \begin{pmatrix} 1 & 1 & 1 & 0 & 0 & 0 \\ 0 & 0 & 0 & 1 & 1 & -2 \end{pmatrix}.$$

(columns labelled $1\ 2\ 3\ 1^*\ 2^*\ 3^*$)

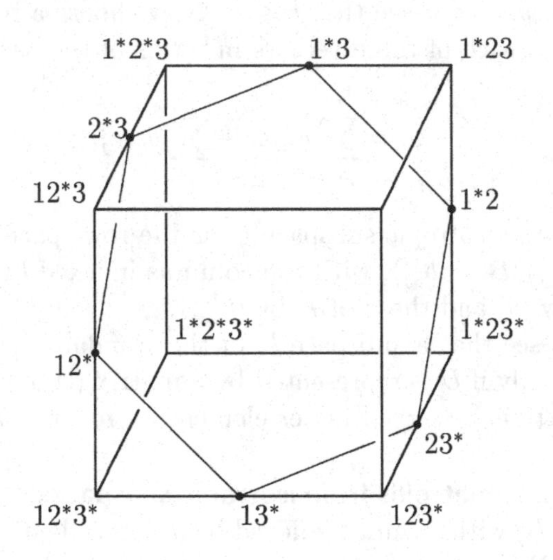

Figure 7: *Matroid polytope of a symplectic matroid.*

First we note that $AB^{\mathrm{t}} = 0$ is symmetric. The bases of the symplectic matroid are

$$\mathcal{B} = \{12^*, 13^*, 1^*2, 1^*3, 23^*, 2^*3\}.$$

Figure 7 shows a useful way of representing the corresponding matroid polytope Δ. Label the facets of the n-cube by the symbols in J, with i, i^* labeling opposite facets. In the Figure, label 1 is given to the front facet, 1^* to the back facet, 2 to the right facet, etc. In this case, each basis is a 2-element admissible set, which is therefore represented by the barycentre of the edge which is the intersection of the two facets involved. The six bases give the vertices of the polytope shown, which is a regular hexagon. Notice that the edges of Δ do satisfy the Gelfand–Serganova Theorem, in that they are all parallel to the roots of BC_n, which geometrically are parallel to edges of the cube or to diagonals of two-dimensional faces of the n-cube, regardless of what n is. Moreover, the exchange group $W(\Delta)$ is the symmetry group of the regular hexagon and is isomorphic to Sym_3. Therefore the polytope is the matroid polytope of a flag matroid on $[n]$ (actually of the *uniform* flag matroid on $[n]$ consisting of all complete flags). Theorem 4.4 below provides more examples of symplectic matroids which, because of the nature of their exchange groups, happen to be (ordinary) flag matroids.

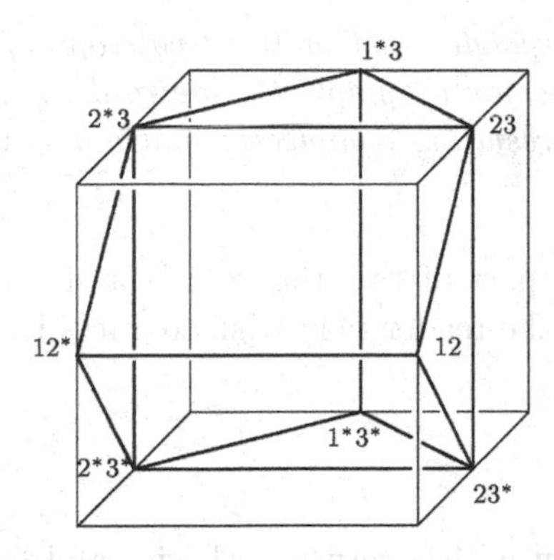

Figure 8: *It can be checked that this matroid polytope corresponds to a non-representable symplectic matroid.*

4.4 Homogeneous symplectic matroids

A collection $\mathcal{B} \subseteq J_k$ is said to be *m-homogeneous* if $|B \cap [n]| = m$ for every $B \in \mathcal{B}$. In other words, all members of $\mathcal{B}$ have the same number of unstarred elements, and consequently also the same number of starred elements. We shall soon see that a homogeneous symplectic matroid is equivalent to a flag of two ordinary matroids.

If A is an admissible set in $[n] \sqcup [n]^*$, denote by flag(A) a flag of two subsets built according to the following procedure. Denote $A_0 = A \cap [n]$ (the set of the non-starred elements in A), $A_1 = A \cap [n]^*$ (the set of the starred elements in A) and take $A_1^* \subseteq [n]$ (the set of the starred elements with stars stripped off). Since A is admissible, $A \cap A_1^* = \emptyset$. Now flag(A) is the pair $(A_0, [n] \setminus A_1^*)$.

Theorem 4.4 *An m-homogeneous collection $\mathcal{B}$ of subsets of $[n] \sqcup [n]^*$ of cardinality $m + l$ is a symplectic matroid if and only if* flag($\mathcal{B}$) *is the set of bases of a flag matroid on $[n]$ of rank $(m, n - l)$. A collection of flags $\mathcal{F} \subseteq \mathcal{F}_{k,l}$ is a flag matroid if and only if* $\text{flag}^{-1}(\mathcal{F})$ *is a k-homogeneous symplectic matroid of rank $k + n - l$ on $[n] \sqcup [n]^*$.*

4.5 Independent-set matroid polytope

Another example of interactions between ordinary and symplectic matroids is provided by the *independent-set matroid polytope.*

If we are given an ordinary matroid of rank k on $[n]$, then each of its independent sets (including the empty set) may be encoded as a vector of 0s and 1s in $\mathbb{R}^n$. We shall call the convex hull of these vectors the *independent-set matroid polytope.*

Theorem 4.5 *The independent-set matroid polytope of an ordinary matroid M is a matroid polytope for a symplectic matroid. Moreover, if M is representable over $\mathbb{R}$, the resulting symplectic matroid is C_n-representable over $\mathbb{R}$.*

The representability part of this theorem is a slightly disguised classical result of linear algebra, the reader may wish to guess it.

5 Buildings

We begin this section with a return to basics and look at the underlying combinatorics of the Gaussian elimination procedure. This classical routine involves permutation of rows and columns of a matrix. The rules these permutations obey are extremely simple; when axiomatised in group-theoretic terms, they become what are known as axioms for a BN-pair (or a *Tits system*) and very quickly lead to Coxeter groups appearing on the scene.

This algebraic approach is combined with the geometric ideas already used in Section 2. For every Coxeter group W, an analogue of the Jordan–Hölder permutation can be introduced, leading to the concept of *buildings*, the geometric objects introduced by Tits as generalisations of projective spaces. Buildings provide a natural way to represent Coxeter matroids. Indeed, the classical representation of matroids turns out to be a special case of representation in buildings.

5.1 Gaussian decomposition

We start our study with some comments on the classical mathematical procedure, Gaussian elimination. We work over an arbitrary field K and consider only non-degenerate $m \times m$ matrices g. They form the *general linear group* $\mathrm{GL}_m(K)$. For the purpose of our exposition we restrict the Gaussian Elimination to certain elementary row and column transformations: we allow ourselves to subtract a scalar multiple of a row (column) from some later row (column). As soon as we have a nonzero matrix element g_{ij} in the matrix g, we use these transformations to kill all non-zero elements in the ith row to the right of g_{ij} and in the jth column below g_{ij}. It is easy to see that the transformations of rows amount to multiplication of g by a lower unitriangular (i.e., triangular with all diagonal entries equal to 1) matrix t^- from the left, and the transformations of columns amount to multiplication of g by an upper unitriangular matrix t^+ from the right, and that we can proceed in this way until we convert g into a *monomial* matrix n, that is, a matrix which contains exactly one non-zero element in each row and column. Hence, in matrix notation,

$$t^- g t^+ = n.$$

Set $u^- = (t^-)^{-1}$ and $u^+ = (t^+)^{-1}$, then g is represented in the form

$$g = u^- n u^+, \tag{5.1}$$

where the matrices u^-, u^+, n belong to the lower unitriangular group U^-, upper unitriangular group U^+, and monomial group N, respectively. We call Equation 5.1 a *Gaussian decomposition* of g.

However, let us work in a slightly more general setting and call any decomposition

$$g = b^- n b^+, \tag{5.2}$$

a *generalised Gaussian decomposition* if the matrices b^-, b^+, n belong to the lower triangular group B^-, upper triangular group B^+, and monomial group N, respectively. Notice that $B^- \cap N = B^+ \cap N$ is the group of diagonal matrices and is normal in N. A minute's thought about how the Gaussian Elimination works convinces us that although the matrices b^-, b^+, n in Equation 5.2 for g are not uniquely defined by g, the image w of n in the factor group $W = N/(B^- \cap N) = N/(B^+ \cap N)$ is uniquely determined by the original matrix g. It is easy to see that W is isomorphic to the symmetric group Sym_m. Notice that w can be thought of as a coset of $B^- \cap N$ in N, so that the expressions $B^- w$, wB^+ make sense as products of cosets in $G = \mathrm{GL}_m(K)$.

In this notation, we have the (generalised) *Gaussian decomposition* of G:

$$G = \bigsqcup_{w \in W} B^- w B^+.$$

What is the underlying combinatorics of Gaussian decomposition?

Let s be a permutation matrix such that multiplication $g \mapsto sg$ by s from the left amounts to swapping two adjacent rows in g, then the permutation w' in the Gaussian decomposition for sg either coincides with w, or equals sw. This can be immediately seen from comparing the effects of elementary row and column transformations on the matrices g and sg. Hence,

$$sg \in B^- w B^+ \cup B^- sw B^+,$$

and, since this argument applies to an arbitrary element $g \in B^- w B^+$,

$$sB^- w B^+ \subseteq B^- w B^+ \cup B^- sw B^+. \tag{5.3}$$

This is a nice formula, but we can make it much more symmetric by converting it to a formula which contains only the lower triangular subgroup $B = B^-$. For that purpose, take the permutation matrix w_0 which corresponds to the permutation

$$\begin{pmatrix} 1 & 2 & \cdots & m-1 & m \\ m & m-1 & \cdots & 2 & 1 \end{pmatrix}.$$

Obviously, $w_0^2 = 1$. One can easily check that $w_0 B^+ w_0 = B^-$. Now we can multiply Equation 5.3 by w_0 from the right:

$$sB^- wB^+ w_0 \subseteq B^- wB^+ w_0 \cup B^- swB^+ w_0$$

and rewrite it as

$$sB^- ww_0 \cdot w_0 B^+ w_0 \subseteq B^- ww_0 \cdot w_0 B^+ w_0 \cup B^- sww_0 \cdot w_0 B^+ w_0,$$

which becomes, after abbreviating B^- as B,

$$sBww_0 B \subseteq Bww_0 B \cup Bsww_0 B.$$

After renaming an arbitrary element $ww_0 \in W$ as w, one gets

$$sBwB \subseteq BwB \cup BswB.$$

Finally, we rewrite the formula in terms of double cosets:

$$BsB \cdot BwB \subseteq BwB \cup BswB.$$

5.2 BN-pairs

The computation above motivates the definition of a BN-pair.

5.2.1 Definition of a BN-pair. Let G be a group. We say that a quadruple (B, N, W, S) is a *BN-pair* in G if the following axioms are satisfied.

BN0. B and N are subgroups in G which generate G, $B \cap N \lhd N$, $W = N/(B \cap N)$, S is a generating set in W.

BN1. For all $w \in W$ and $s \in S$,

$$BsB \cdot BwB \subseteq BwB \cup BswB.$$

BN2. For all $s \in S$,

$$s^{-1} Bs \not\subseteq B.$$

The group W is called the *Weyl group* of the BN-pair, elements in S are *standard generators* of W.

Notice that the axioms BN0–BN1 are satisfied in $G = \mathrm{GL}_n(K)$, if we take for B and N the lower triangular and monomial subgroups and for S the set of transpositions

$$S = \{ (12), (23), \dots, (n-1, n) \}$$

in $W = Sym_n$, where these transpositions are represented by matrices in the usual way. The validity of axiom BN2 in $\mathrm{GL}_n(K)$ in this setting is obvious. Hence the group $\mathrm{GL}_n(K)$ provides an example of a group with a BN-pair.

It can be shown that group with a BN-pair decomposes as

$$G = \bigsqcup_{w \in W} BwB;$$

this decomposition is called *Bruhat decomposition*.

5.2.2 Coxeter groups. Coxeter groups (and Coxeter matroids) appear on the scene by virtue of the following classical theorem of Tits.

Theorem 5.1 *In a group with a BN-pair, the Weyl group $W = N/B \cap N$ is a Coxeter group, with S being the set of distinguished generators.*

In particular, this means that every $s \in S$ has order 2 in W.

5.3 W-metric

As we know from Section 2, the Jordan–Hölder permutation measures the 'distance' between two maximal chains in a semimodular lattice. Description of combinatorial properties of BN-pairs requires introduction of similar 'metric' with values in a Coxeter group (the Weyl group of the BN-pair).

5.3.1 W-metrics. Let W be a Coxeter group with a system of standard generators S.

We say that a map $\pi : X \times X \longrightarrow W$ is a *W-metric* on the set X if the following axioms are satisfied.

D1. $\pi(x, y) = 1$ if and only if $x = y$.

D2. $\pi(x, y) = \pi(y, x)^{-1}$.

D3. If $\pi(x, y) = w$ and $\pi(y, z) = s$ for some standard generator $s \in S$, then

$$\pi(x, z) = w \text{ or } ws.$$

A canonical example of a W-metric is provided by the group W itself, with the map

$$\begin{array}{rcl} \pi : W \times W & \longrightarrow & W \\ (u, v) & \mapsto & u^{-1}v. \end{array}$$

The term *isometry*, in relation to a W-metric, has the obvious meaning.

5.3.2 Chain complex of a semimodular lattice admits a W-metric. Let L be a semimodular lattice of height n and C the set of maximal chains in L (it is called the *chain complex* of lattice L).

Denote the standard generators of Sym_n as

$$s_1 = (12), s_2 = (23), \ldots, s_{n-1} = (n-1, n).$$

Theorem 5.2 (Abels [2]) *Let L be a semimodular lattice of height n and C the set of maximal chains in L. Then the Jordan–Hölder permutation (see Section* 2.2*)*

$$\pi : C \times C \longrightarrow \mathrm{Sym}_n$$

satisfies the axioms D1–D3 *of Sym_n-metric.*

5.4 Buildings

5.4.1 Definition of buildings. Let W be a Coxeter group. We define a *building* of type W to be a set Δ with a W-metric $\pi : \Delta \times \Delta \to W$ which satisfies the additional property

D4. Any two elements of Δ belong to an image of some isometry

$$\alpha : W \to \Delta.$$

Elements of Δ are called *chambers*. Images of isometries $\alpha : W \to \Delta$ are called *apartments*.

Obviously the Coxeter group W is a building of type W.

5.4.2 Buildings of projective spaces. A semimodular lattice L of finite height is *modular* if it satisfies the *modular equality*

$$h(x) + h(y) = h(x \wedge y) + h(x \vee y),$$

for all $x, y \in L$.

A modular lattice is a *projective space lattice* if, for every pair of elements $a < b$ with $h(b) - h(a) = 2$, there are at least three distinct elements x in the interval $a < x < b$. It is easy to see that a projective space lattice is a geometric lattice.

Theorem 5.3 (Tits) *Let L be a projective space lattice of finite height n and C the set of maximal chains in L. The Jordan–Hölder permutation*

$$\pi : C \times C \longrightarrow \mathrm{Sym}_n$$

makes C a building of type $\mathrm{Sym}_n \cong A_{n-1}$.

5.4.3 Building associated with a BN-pair. BN-pairs provide the main source of buildings:

Theorem 5.4 (Tits) *Let G be a group with a BN-pair B, N and the standard set of generators $S \subset W = N/B \cap N$. The set $\Delta = G/B$ of left cosets with respect to B is a building with the W-metric*

$$\pi(gB, hB) = w \textit{ if and only if } g^{-1}h \in BwB.$$

The natural group action of G by left multiplication on G/B preserves the W-metric π.

5.5 Representing Coxeter matroids in buildings

In Section 2, we saw that any maximal chain in a semimodular lattice with specified basis produced a matroid map. We have the analogous statement for an arbitrary chamber in a building with specified apartment.

Theorem 5.5 *Let Δ be a building of type W and A an apartment in Δ which is an image of isometry $\alpha : W \to \Delta$, and x a chamber in Δ. Let*

$$\mu_x : W \to W$$

be defined by

$$\mu_x(w) = w \cdot \pi(\alpha(w), x)$$

for all $w \in W$. Then μ_x is a W-matroid map.

Thus every chamber determines a matroid map, and we say the resulting W-matroid is *represented in the building* Δ. Results of Section 2 show that every flag matroid (for $A_{n-1} = \mathrm{Sym}_n$) represented by a flag of subspaces of a vector space is always represented in a building as well. Indeed, it is easily seen that the matroid map determined by a maximal chain in the modular lattice L of subspaces of a vector space of dimension n is the same as that determined by the same chain when regarded as a chamber in the building (of type A_{n-1}) determined by L, provided the apartment specified in the building is that corresponding to the specified basis of L. Although we seen proven that every ordinary flag matroid may be so represented in some semimodular lattice, not all may be so represented in a building. Indeed, it can be shown that modular lattices are the only semimodular lattices whose chain complexes are buildings, and that furthermore, these are all of the buildings of type A_{n-1}. It follows that representability of an ordinary matroid or flag matroid in a building reduces to representability in a projective space, which for matroids of rank at least 4, is equivalent to classical representability over a division ring.

5.6 Schubert cells

If the building Δ arises from a BN-pair (B, N), Coxeter matroids represented in the building can be directly defined in terms of the BN-pair.

Let G be a group with a BN-pair (B, N). The image in G/B of a double coset $wBw^{-1}gB$ with respect to a pair of subgroups wBw^{-1} and B is called a *Schubert cell.*

It can be shown that G/B is the disjoint union of Schubert cells with representatives $u \in W$:

$$G/B = \bigsqcup_{u \in W} wBw^{-1}uB/B.$$

In this context, for $x \in B/G$ denote by $\mu_x(w)$ a (unique) element $u \in W$ such that $x \in wBw^{-1}uB/B$.

Then $\mu_x : W \to W$ is a W-matroid map; it coincides with the map constructed in Theorem 5.5. Notice again that flag matroids on $[n]$ represented in the n-dimensional vector space K^n are exactly the flag matroids represented in the BN-pair of the group $\mathrm{GL}_n(K)$.

Final comments

The principal aim of the present survey was to show how the concept of a Coxeter group arises, again and again, in the study of matroids.

However, we so far avoided the most important source of motivation for our theory: many interesting examples of Coxeter matroids (and, in particular, all examples of ordinary and symplectic matroids in this paper which are represented by a matrix of some kind) come from torus orbits on flag varieties of semisimple algebraic groups.

Theorem 5.6 *Let G be a semisimple algebraic group over $\mathbb{C}$, P a parabolic subgroup in G and $H < P$ a maximal torus. Let x be a point on the flag variety G/P and $X = \overline{Hx}$ the closure of the torus orbit of x. Let $\mathfrak{h}$ be the Lie algebra of H and Δ be the convex polytope in the weight vector space $\mathfrak{h}^*_{\mathbb{R}}$ canonically associated with X by means of a moment map* [23]. *Then Δ is a Coxeter matroid polytope; its exchange group $W(\Delta)$ is a subgroup of the Weyl group $W = N_G(H)/H$ of G.*

This theorem explains why different appearances of Coxeter groups in matroid theory are so closely intertwined. Indeed, a closer examination of the theorem reveals that it involves all the ingredients of our theory: convex polytopes, reflection groups, BN-pairs (because if B is a Borel subgroup containing H then the subgroups B and $N_G(H)$ form a BN-pair, while the parabolic subgroup P has the form BRB for some parabolic subgroup R of W). A combinatorial description of the closure $\overline{Hx}$ of the torus orbit necessitates the use of the Bruhat order because the latter describes the adjacency of Schubert cells: a Schubert cell BuB belongs to the closure of the cell BvB, $u, v \in W$, if and only if $u \leqslant v$.

For that reason, our secondary aim was to outline a synthetic approach to matroid theory which makes use of all of the above concepts. It is very elementary, and for good reason: the geometrically intuitive 'kaleidoscope' definition of Coxeter matroids has happened to be the one actually used in Theorem 5.6. We hope that, even at a very elementary level, our approach sheds new light at the classical chapters of matroids theory and opens up a new area of research, symplectic matroids. The latter, because of the natural embeddings of root systems $A_{n-1} < D_n < \mathrm{BC}_n$, cover all infinite series of Coxeter matroids and appear to be the most natural generalisations of classical matroids.

References

[1] H. Abels, The gallery distance of flags, *Order* **8** (1991), 77–92.

[2] H. Abels, The geometry of the chamber system of a semimodular lattice, *Order* **8** (1991), 143–158.

[3] R.F. Booth, A.V. Borovik, I.M. Gelfand and D.A. Stone, Lagrangian matroids and cohomology, *Ann. Comb.* **4** (2000), 171–182.

[4] R.F. Booth, A.V. Borovik and N. White, Lagrangian matroids: representations of type B_n, arXiv.org, math.CO/0209217, 2002.

[5] A.V. Borovik and I.M. Gelfand, WP-matroids and thin Schubert cells on Tits systems, *Adv. Math.* **103** (1994), 162–179.

[6] A.V. Borovik, I.M. Gelfand and N. White, *Coxeter Matroids*, Birkhäuser, Boston (2003).

[7] A.V. Borovik, I.M. Gelfand and D.A. Stone, On the topology of the combinatorial flag varieties, *Discrete Comput. Geom.* **27** (2002), 195–214.

[8] A.V. Borovik, I.M. Gelfand, A. Vince and N. White, The lattice of flats and its underlying flag matroid polytope, *Ann. Comb.* **1** (1997), 17–26.

[9] A.V. Borovik, I.M. Gelfand and N.L. White, Boundaries of Coxeter matroids, *Adv. Math.* **120** (1996), 258–264.

[10] A.V. Borovik, I.M. Gelfand and N. White, Coxeter matroid polytopes, *Ann. Comb.* **1** (1997), 123–134.

[11] A. V. Borovik, I. M. Gelfand and N. White, On exchange properties of Coxeter matroids and oriented matroids, *Discrete Math.* **179** (1998), 59–72.

[12] A.V. Borovik, I.M. Gelfand and N. White, Symplectic matroids, *J. Algebraic Combin.* **8** (1998), 235–252.

[13] A.V. Borovik, I.M. Gelfand and N. White, Representations of matroids in semimodular lattices, *European J. Combin.* **22** (2001), 789–799.

[14] A.V. Borovik, I.M. Gelfand and N. White, Combinatorial flag varieties, *J. Combin. Theory Ser. A* **91** (2000), 111–136.

[15] A.V. Borovik and A. Vince, An adjacency criterion for Coxeter matroids, *J. Algebraic Combin.* **9** (1999), 271–280.

[16] O. Boruvka, O jistém problému minimálním, *Prace Mor. Přírodově Spol. v Brně* (*Acta Societ. Scient. Natur. Moravicae*) **3** (1926), 37–58.

[17] A. Bouchet, Greedy algorithm and symmetric matroids, *Math. Program.* **38** (1987), 147–159.

[18] A. Bouchet, Representability of Δ-matroids, in *Proc. 6th Hungarian Colloquium of Combinatorics (July 1987) Colloquia Mathematica Societas Janos Bolyai*, (1987), pp. 167–182.

[19] A. Bouchet, Multimatroids I. Coverings by independent sets, *SIAM J. Discrete Math.* **10** (1997), 626–646.

[20] A. Bouchet, A. Dress, and T. Havel, Δ-matroids and metroids, *Adv. Math.* **91** (1992), 136–142.

[21] D. Gale, Optimal assignments in an ordered set: an application of matroid theory, *J. Combin. Theory* **4** (1968), 1073–1082.

[22] I.M. Gelfand, M. Goresky, R.D. MacPherson and V.V. Serganova, Combinatorial geometries, convex polyhedra, and Schubert cells, *Adv. Math.* **63** (1987), 301–316.

[23] I.M. Gelfand and V.V. Serganova, Combinatorial geometries and torus strata on homogeneous compact manifolds, *Russian Math. Surveys* **42** (1987), 133–168. (See also I.M. Gelfand, *Collected Papers*, Vol. III, Springer-Verlag, New York (1989) pp. 926–958.)

[24] J.E. Humphreys, *Reflection Groups and Coxeter Groups*, Cambridge University Press, Cambridge (1976).

[25] J.G. Oxley, *Matroid Theory*, Oxford University Press, Oxford (1992).

[26] J.G. Oxley, On the interplay between graphs and matroids, in *Surveys in Combinatorics 2001* (ed. J.W.P. Hirschfeld), *London Math. Soc. Lecture Notes*, 288, Cambridge University Press, Cambridge (2001), pp. 199–239.

[27] A. Vince and N. White, Orthogonal Matroids, *J. Algebraic Combin.* **13** (2001), 295–315.

[28] D.J.A. Welsh, *Matroid Theory*, Academic Press, London (1976).

[29] N. White, ed., *Theory of Matroids*, Cambridge University Press, Cambridge (1986).

[30] H. Whitney, On the abstract properties of linear dependence, *Amer. J. Math.* **57** (1935), 509–533.

Department of Mathematics
UMIST
PO Box 88
Manchester M60 1QD
United Kingdom
borovik@umist.ac.uk

Defining sets in combinatorics: a survey

Diane Donovan, E. S. Mahmoodian,
Colin Ramsay and Anne Penfold Street

Abstract

In a given class of combinatorial structures there may be many distinct objects with the same parameters. Two questions arise naturally.

- Given two such objects, where and how do they differ?
- How much of an individual object is needed to identify it uniquely?

These questions are obviously related, the first leading to the concept of a *trade*, and the second to that of a *defining set*. This survey deals with defining sets in block designs, graphs and some related structures. The corresponding trades in each structure are also discussed briefly.

1 Introduction

We start with a simple example. A *graph* $G = (V, E)$ consists of a finite set V of elements called *vertices*, and a set E of unordered pairs of vertices, called *edges*. The *complete graph* on v vertices, K_v, is a graph in which all pairs of distinct vertices constitute edges, so that any graph on v or fewer vertices may be considered as a subgraph of K_v.

If $v = 2n$, then a *one-factor* of K_v is a set of n unordered pairs which between them contain each element of V precisely once. A *defining set* of a one-factor is a subset of its edges which uniquely identifies it. More generally, a *perfect matching* in a graph G on $2n$ vertices is a set of n edges incident with each vertex of V. In this context, a defining set has also been termed a *forcing set*. Figure 1 provides examples of one-factors in K_2, K_4 and K_6. K_2 has only one edge so its defining set is empty. In K_4 either edge in a one-factor forms a defining set, and in K_6 any two edges in a one-factor form a defining set.

A set of $n-1$ one-factors which between them contain every edge precisely once is called a *one-factorization* of K_v, and every complete graph on an even number of vertices has a one-factorization. A set of one-factors which uniquely identifies a one-factorization is termed a defining set for a one-factorization.

The difference between any two one-factors or any two one-factorizations provides us with examples of trades within the corresponding structures.

The complete graph on four vertices, K_4, has three one-factors, say,

$$F_1 = 01,\ 23;\ F_2 = 02,\ 13;\ F_3 = 03,\ 12$$

and these form its only one-factorization. So for $v = 4$, there is a unique one-factorization of K_v, the defining set is empty and no trade exists. But for $v = 6$, the situation becomes more interesting. K_6 has 15 one-factors, shown in Figure 2; each one-factor belongs to two of the six possible one-factorizations.

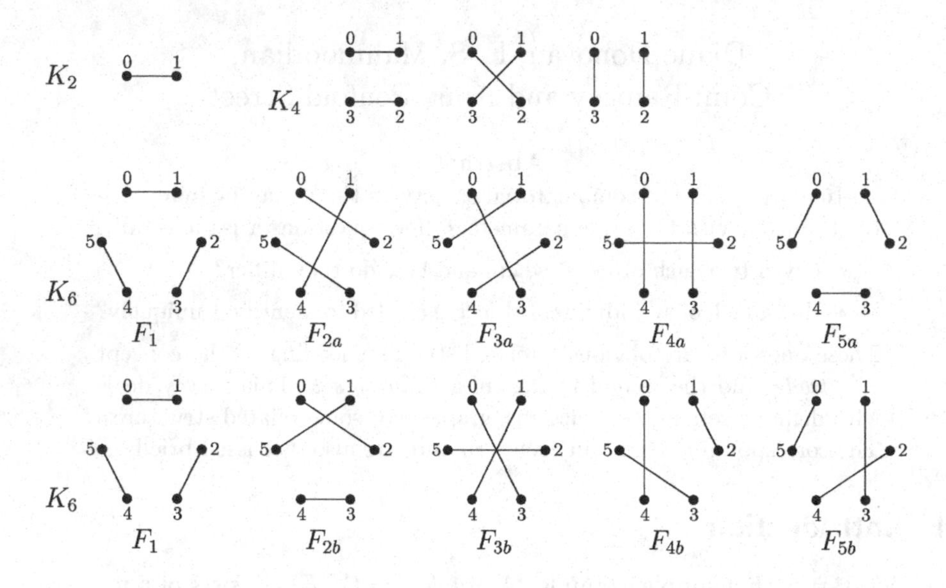

Figure 1: One-factors and one-factorizations in K_2, K_4 and K_6

F_1 is arbitrarily chosen to be the one-factor containing the edge 01, and one-factors with an edge in common with F_1 are left unlabelled. Thus F_1 is disjoint from two of the three one-factors that contain the edge 02, namely, F_{2a} and F_{2b}. The two one-factors F_1 and F_{2a} form a defining set of the one-factorization, since it can only be completed in one way, with F_{3a}, F_{4a} and F_{5a}. Again F_1 and F_{2b} can be completed to a one-factorization only with F_{3b}, F_{4b} and F_{5b}. Similarly, any pair of edge-disjoint one-factors forms a defining set of a unique one-factorization of K_6. Since any two one-factorizations with a common one-factor differ in four mutually edge-disjoint one-factors, these four one-factors form a *trade.* Thus for example $F_{2a}, F_{3a}, F_{4a}, F_{5a}$ cover precisely the same collection of edges as $F_{2b}, F_{3b}, F_{4b}, F_{5b}$.

Not only do any two disjoint one-factors of K_6 define a one-factorization, but also partial one-factors may form a defining set. For instance, if we specify that one of the one-factors must contain the edges 01 and 23, that another must contain 02, and a third must contain 15, then we have in fact specified F_1, F_{2a}, and F_{3a}, and hence the one-factorization. This is analogous to a *pointwise* defining set in a block design, where partial blocks may suffice to define the whole design. This idea is discussed further in Section 3.1.

F_1	01	23	45		01	24	35		01	25	34
	02	13	45	F_{2a}	02	14	35	F_{2b}	02	15	34
	03	12	45	F_{3b}	03	14	25	F_{3a}	03	15	24
F_{4b}	04	12	35	F_{4a}	04	13	25		04	15	23
F_{5a}	05	12	34	F_{5b}	05	13	24		05	14	23

Figure 2: Choosing one-factors for a one-factorization of K_6

In this survey, we generalise these ideas to other combinatorial structures, notably to block designs and to graphs, first discussing briefly the relationships between these other structures and latin squares, insofar as they have been applied in finding defining sets. We make no attempt to discuss critical sets in latin squares in any detail, as they have been surveyed previously [84] and a further survey is forthcoming [83].

Fuchs [50] in 1958 considered defining sets for Cayley tables of groups. Later work by Dénes [37], Frisch [49], Drápal [42, 43], Drápal and Kepka [41] and Donovan, Oates-Williams and Praeger [40] has led to results on defining sets and trades, in latin squares derived from both groups and closely related quasigroups.

The earliest generalization of these ideas to arbitrary latin squares seems to be that of Nelder [114] who asked in 1977 how many, and which, entries of a latin square uniquely determine the remaining entries. He called such a set *critical*.

A similar question was asked by Curtis [29] in 1984 for the Steiner system $S(5, 8, 24)$. He showed that one could choose eight of its 759 blocks so that they *defined* (his usage) the rest. This was the first result on defining sets in designs, and he applied it to check whether an element of the symmetric group S_{24} was contained in the Mathieu group, M_{24}, that is, the automorphism group of the design. In 1990 K. Gray considered the idea of defining sets of designs in a general context, proving several basic results [68, 67, 66].

Many puzzles make incidental use of defining sets, either implicitly or explicitly. For example, Ramsay [119] discusses them in relation to a particular polyomino puzzle, and the logic puzzle given in [27] is, in effect, a defining set of a particular integer partition.

Although these ideas were originally of theoretical interest, some applications have now been proposed. These include secret-sharing schemes based on defining sets of designs (Seberry and Street [131]) and tests of fluid intelligence requiring completion of partial latin squares (Birney and Halford [13]). But in graph theory, the subject seems to have developed in reverse order, with the need for practical applications motivating the theoretical research.

A *Kekule structure* is a polycondensed aromatic hydrocarbon (PAH) and the graph corresponding to the arrangement of the carbon atoms in such a compound contains a number of polygons, mainly hexagons, which share common edges; see Figure 3. Since each carbon atom has four electrons available for bonding, most of the carbon atoms are linked to one hydrogen atom each. Klein and Randić [89] in 1987 raised two questions in organic chemistry: first, when can the double bonds between the carbon atoms in a PAH be placed appropriately throughout the compound; secondly, how many such bonds will determine the rest? Further discussion of this idea followed in 1991 [78].

Framed in graph theoretic terms, this is equivalent to asking, first, when can a perfect matching be found in such a graph and, secondly, how many (and which) edges of the matching will determine the rest? Note that rings of

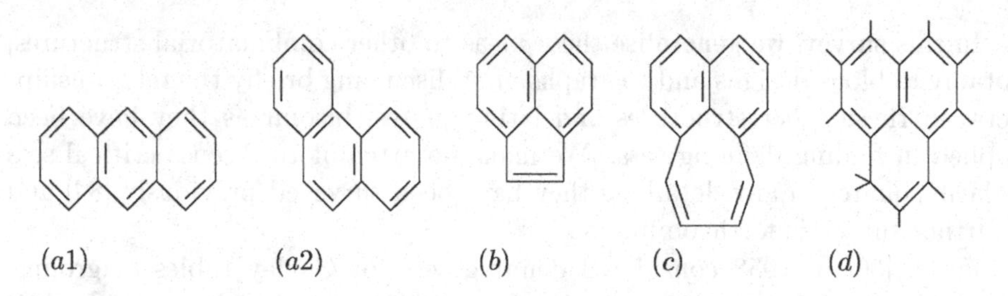

Figure 3: Examples of Kekule structures

five or seven carbon atoms are possible; see Figure 3(b) and (c). Also the lack of a perfect matching in the underlying graph does not preclude the existence of the corresponding compound, since two hydrogen atoms may be linked to a carbon atom which adjoins no double bond, as shown in Figure 3(d).

In 1993 Harary [77] introduced 'forcing concepts' as one of three new directions in graph theory. He was concerned with the smallest number of elements (vertices or edges) of a graph whose selection for inclusion in a specified set leaves no options in the choice of the remaining elements of the set. There is a natural relationship between critical sets in latin squares and defining sets of graph colourings; see for instance Keedwell [84], Mahmoodian [100, 101].

In the next section, we state the essential background on latin squares, graphs, block designs and trades. Section 3 deals with defining sets of block designs and Section 4 with defining sets of special features of graphs. In Section 5 we review these ideas, noting other structures to which they are relevant, and pointing out open problems in the area.

2 Background

For general background on block designs, latin squares and certain aspects of graph theory, see [26]; for an overall introduction to graph theory, see [140].

2.1 Complete and partial latin squares

A *partial latin square of order* n is an $n \times n$ array based on some set of n symbols, where each row and each column contains each symbol at most once. We usually take the set of symbols to be $N = \{0, 1, \ldots, n-1\}$ and use the convention that the rows and columns are numbered from 0 to $n-1$. Equivalently, a partial latin square $P = [p_{ij}]$ can be written as a set of at most n^2 ordered triples, namely the set of triples $(i, j; p_{ij})$. We use the notation P^k to denote the partial latin square based on the set of symbols $N + nk = \{nk, nk+1, \ldots, nk+n-1\}$ for appropriate positive integers k.

In the special case where each symbol of the set N occurs *precisely* once in each row and each column, the partial latin square is in fact a *latin square*, which we usually denote by L.

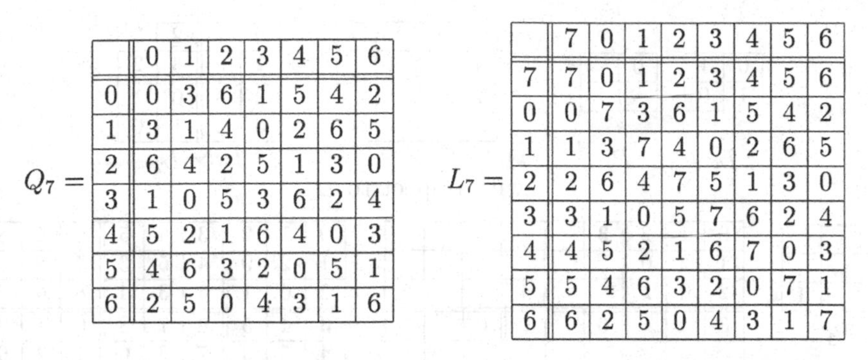

$Q_7 =$

	0	1	2	3	4	5	6
0	0	3	6	1	5	4	2
1	3	1	4	0	2	6	5
2	6	4	2	5	1	3	0
3	1	0	5	3	6	2	4
4	5	2	1	6	4	0	3
5	4	6	3	2	0	5	1
6	2	5	0	4	3	1	6

$L_7 =$

	7	0	1	2	3	4	5	6
7	7	0	1	2	3	4	5	6
0	0	7	3	6	1	5	4	2
1	1	3	7	4	0	2	6	5
2	2	6	4	7	5	1	3	0
3	3	1	0	5	7	6	2	4
4	4	5	2	1	6	7	0	3
5	5	4	6	3	2	0	7	1
6	6	2	5	0	4	3	1	7

Figure 4: Quasigroup Q_7 of order 7 and loop L_7 of order 8

A *quasigroup* is a set V closed under a binary operation $\circ$ such that in the equation $x \circ y = z$ the choice of any two of the three elements x, y and z uniquely determines the third. A *loop* is a quasigroup which has an identity element, e, with respect to the binary operation $\circ$. Removing the headline and sideline of the operation table of a quasigroup (in particular a loop) gives a latin square. Figure 4 illustrates this for a quasigroup Q_7 of order seven and a loop L_7 of order eight with $e = 7$.

A partial latin square is one of six associated *conjugates*, namely, P itself, and the partial latin squares: $P^T = \{(j, i; k) : (i, j; k) \in P\}$; $\{(k, j; i) : (i, j; k) \in P\}$; $\{(i, k; j) : (i, j; k) \in P\}$; $\{(k, i; j) : (i, j; k) \in P\}$; $\{(j, k; i) : (i, j; k) \in P\}$.

Let P be a partial latin square and assume the cell (i, j) of P is empty; that is, for all $k \in N$, we have $(i, j; k) \notin P$. The adjunction to P of a triple $(i, j; k)$ is said to be *forced* if either:
(1) for every $h \neq k$, there exists z such that $(i, z; h) \in P$ or $(z, j; h) \in P$; or
(2) a conjugate of P satisfies (1).

Let L be a latin square and P a partial latin square, both of order n. Then P is *uniquely completable* to L if L is the only latin square of order n containing P. (Note that P need not be contained in any latin square.) That is, P is a defining set for L. If P is minimal with respect to this property, then P is a *critical set* for L. In other words, if every partial latin square properly contained in P has more than one completion to a latin square of order n, then P is a critical set for L. Following Keedwell's usage [84], a *minimal* critical set for a *particular* latin square is one of smallest cardinality, and a *smallest* critical set of order n (denoted by $\mathrm{scs}(n)$) is one of the smallest cardinality which can exist in *any* latin square of order n. Analogously the *largest* critical set in any latin square of order n is denoted by $\mathrm{lcs}(n)$. The terminology here differs from that for block designs; see Section 2.3.2.

If $P \subseteq L$ is uniquely completable to L, and if $M \subseteq L$ is a latin subsquare of L, then $P \cap M$ is uniquely completable to M.

Further, P has a *strong unique completion* to L if $P \subseteq L$ and there exists a collection of $m = n^2 - |P|$ partial latin squares $P_1, \ldots, P_m$ such that each

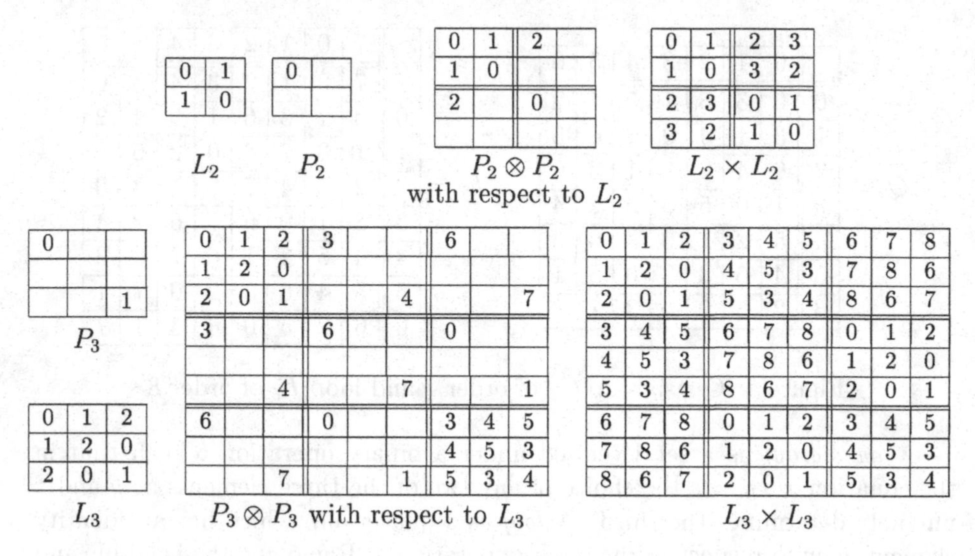

Figure 5: Two examples of the completable product of partial latin squares

triple $t \in P_{i+1} \setminus P_i$ is forced in P_i for $0 \leqslant i \leqslant m-1$, and P_0 is taken to be P.

Any k rows of a latin square, where $k \leqslant n$, form a *latin rectangle*, that is, a $k \times n$ array such that each row contains each symbol precisely once, and each column contains each symbol at most once. Any $k \times n$ latin rectangle can be extended to an $n \times n$ latin square; see for example Ryser [128].

Let L and M be latin squares of orders n and m respectively. Let $P \subseteq L$ and $Q \subseteq M$ be partial latin squares of orders n and m. The *completable product* of P and Q, with respect to L and M is the set

$$\begin{aligned} P \otimes Q &= \{(um+x, vm+y; wm+z) : ((u,v;w) \in P) \wedge ((x,y;z) \in M)\} \\ &\cup \{(um+x, vm+y; wm+z) : ((u,v;w) \in L \setminus P) \wedge ((x,y;z) \in Q)\}. \end{aligned}$$

That is, the partial latin square $P \otimes Q$ of order nm is obtained by replacing each cell of P containing the entry k, $0 \leqslant k \leqslant n-1$, by an $m \times m$ array containing M^k and each empty cell of P by an $m \times m$ array containing Q^k, where k belongs to the corresponding cell of $L \setminus P$. This definition includes the usual definition for the *direct product* of two latin squares, and so

$$L \times M = \{(um+x, vm+y; wm+z) : ((u,v;w) \in L) \wedge ((x,y;z) \in M)\}.$$

Figure 5 shows two examples of the completable product of a partial latin square with itself: first, where the partial latin square is P_2, of order 2; then where the partial latin square is P_3, of order 3.

If Q is a partial latin square with unique completion then the partial latin squares $P_2 \otimes Q$ and $P_3 \otimes Q$ have unique completion. This is a special case of results by Gower [55] and Bedford and Whitehouse [9] who showed that, with certain additional restrictions on the uniquely completable partial latin square P, $P \otimes Q$ will be uniquely completable.

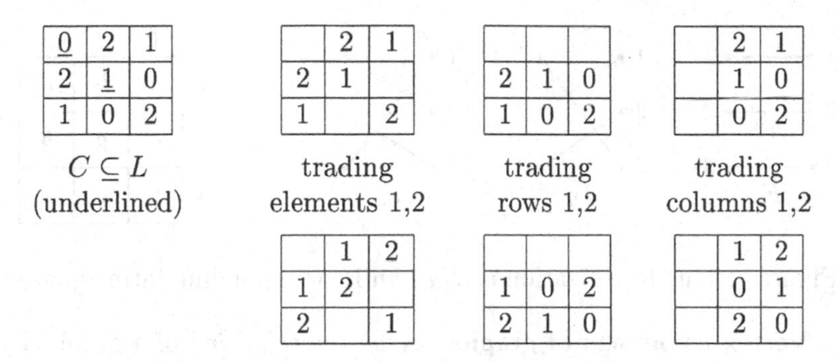

Figure 6: Critical set C in L, and latin trades intersecting C in (1,1;1)

A partial latin square, P, forms a *latin trade* if there exists a corresponding partial latin square, P', such that for all i, j, cell (i, j) of P contains $P_{ij} \in N$ if and only if cell (i, j) of P' contains $P'_{ij} \in N$ where $P_{ij} \neq P'_{i,j}$, and the symbol $x \in N$ occurs in row (column) i of P if and only if x occurs in row (column) i of P'. There is a strong connection between critical sets and latin trades.

Theorem 2.1 *Let L be a latin square of order n and $C \subseteq L$. The partial latin square C is a critical set if and only if:*
(1) for all latin trades $I \subseteq L$, $I \cap C \neq \emptyset$;
(2) for all $x \in C$, there exists a latin trade $I \subseteq L$ such that $I \cap C = \{x\}$.

A latin square L of order 3, with a critical set $C = \{(0, 0; 0), (1, 1; 1)\}$, is shown in Figure 6. L contains three latin trades which intersect C in only the element $(1, 1; 1)$. These three latin trades are those shown level with L in Figure 6, with their disjoint trade mates below them. The trades are between elements 1 and 2, between rows 1 and 2, and between columns 1 and 2 respectively. Latin trades also exist which intersect C in only the element $(0, 0; 0)$.

2.2 Graphs and latin squares

A graph G is said to be *bipartite* if its vertex set $V(G) = X \cup Y$, where X and Y are disjoint sets, and if each of its edges xy joins a vertex $x \in X$ to a vertex $y \in Y$. G is a *complete* bipartite graph if every vertex in X is joined to every vertex in Y. If $|X| = m$ and $|Y| = n$, then the complete bipartite graph is written $K_{m,n}$. For $K_{m,n}$ to have a one-factor it is necessary that $m = n$.

A one-factorization of $K_{n,n}$ is equivalent to a latin square $L = [\ell_{ij}]$ of order n, since we can define the vertex sets to be $X = \{x_1, x_2, \ldots, x_n\}$ and $Y = \{y_1, y_2, \ldots, y_n\}$, and the one-factor F_k to consist of the pairs $\{x_i, y_j\}$ where $\ell_{ij} = k$; see Figure 7 for an example. Latin squares of order n exist for all positive integers n, so $K_{n,n}$ has a one-factorization for all n.

As we see in the next section, latin squares are also associated with graph colourings. There is also a natural correspondence between critical sets for latin squares and defining sets for vertex colourings of graphs.

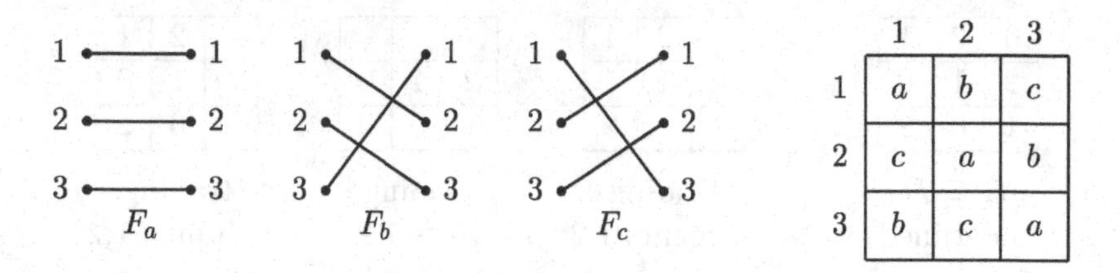

	1	2	3
1	a	b	c
2	c	a	b
3	b	c	a

Figure 7: One-factorization of $K_{3,3}$ and corresponding latin square

2.2.1 Vertex colourings of graphs A *vertex colouring* of a graph G is an allocation of colours to its vertices so that no two adjacent vertices receive the same colour; if n colours are used, we refer to this as an n-colouring. The smallest number of colours possible for such a vertex colouring is called the *chromatic number* of G and denoted by $\chi(G)$. If G is bipartite with at least one edge, then $\chi(G) = 2$.

Suppose that a graph G has chromatic number $\chi(G)$, and that a set $S \subseteq V(G)$ has a specified colouring in $\chi(G)$ or fewer colours. If this colouring can be extended to a unique colouring of all of $V(G)$ in $\chi(G)$ colours, then S is said to be a *defining set* of the colouring. The *defining number*, $d(G, \chi)$, of G is the cardinality of its smallest defining set. For example, for a connected bipartite graph, $d(G, \chi) = 1$; for a bipartite graph with n components, $d(G, \chi) = n$. A *minimal* defining set is one which does not properly contain a defining set.

Figure 8 shows minimal defining sets of several graphs with $\chi(G) = 3$. The triangle with a tail has a minimal defining set of size two, shown in Figure 8(a), and another of size three, shown in Figure 8(b). This second set is minimal, since uncolouring either the vertex of degree three or that of degree one gives a colouring with two quite different completions, and uncolouring the coloured vertex of degree two allows two isomorphic completions. Whether isomorphic colourings should be considered equal is a question related to ideas discussed later in connection with class and member defining sets for block designs. However West [140] states in Remark 1.1.16 that 'When we draw a graph, its vertices are named by their physical locations, even if we give them no other

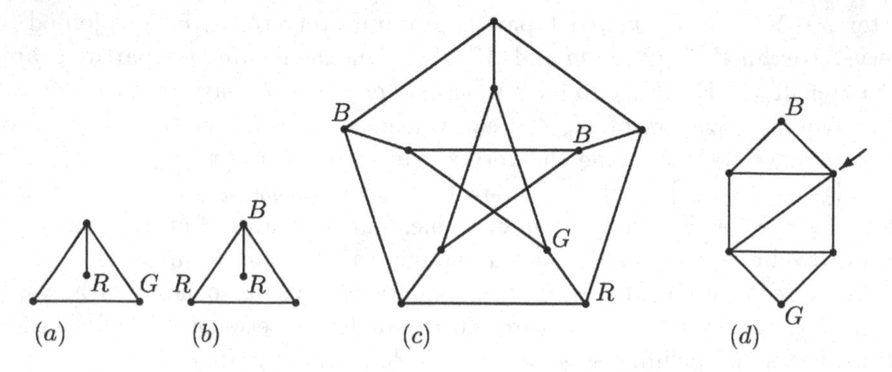

Figure 8: Minimal defining sets for some vertex colourings where $\chi(G) = 3$

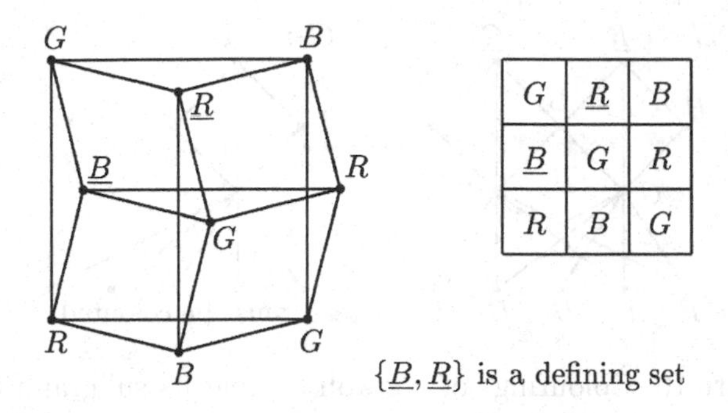

Figure 9: The Cartesian product of K_3 with itself

names'. Following this convention the defining set in Figure 8(b) is minimal.

Figure 8(c) shows a minimal defining set of size four for a colouring of the Petersen graph. In each of Figure 8(a), (b) and (c), the colouring of vertices outside each defining set is forced at each step, that is, the defining set is *strong*. (Compare this idea with that of strong critical sets for latin squares; see Section 2.1.) The defining set of the graph shown in Figure 8(d) does not have this pleasant property; at first glance, the vertex indicated by the arrow might be coloured red or green, but if we attempt to colour it red, we are unable to complete the colouring. However colouring it green gives a unique vertex colouring. We have similar problems if we start colouring at any other vertex outside the defining set.

Defining sets of graph colourings raise a spectrum question, namely: if $\mathcal{P}$ is a family of graphs with $\chi(G) = k$ and $|V(G)| = n$, what are the possible values of $d(G, \chi)$ for $G \in \mathcal{P}$?

The *Cartesian product* G of two graphs G_1 and G_2, written $G_1 \times G_2$, has vertex set $V(G) = V(G_1) \times V(G_2)$, and two vertices (u_1, u_2) and (v_1, v_2) of G are adjacent if and only if *either* $u_1 = v_1$ and $u_2v_2 \in E(G_2)$ *or* $u_2 = v_2$ and $u_1v_1 \in E(G_1)$. Note that $G_1 \times G_2$ and $G_2 \times G_1$ are isomorphic. In particular, the *n-cube* Q_n is defined as

$$Q_n = \begin{cases} K_2 & \text{if } n = 1, \\ Q_{n-1} \times K_2 & \text{if } n > 1. \end{cases}$$

The Cartesian product $K_m \times K_n$ of the complete graphs K_m and K_n has mn vertices and contains m 'horizontal' copies of K_n, say $K_n^{(1)}, K_n^{(2)}, \ldots, K_n^{(m)}$, ordered from top to bottom, and similarly n 'vertical' copies of K_m, say $K_m^{(1)}, K_m^{(2)}, \ldots, K_m^{(n)}$, ordered from left to right. A horizontal copy $K_n^{(i)}$ and a vertical copy $K_m^{(j)}$ have exactly one vertex in common. Figure 9 shows a small example. Obviously for the chromatic number we have $\chi(K_m \times K_n) = n$ if $m \leqslant n$. So an n-colouring of $K_m \times K_n$ is equivalent to an $m \times n$ latin rectangle.

Critical sets in latin squares and rectangles are related to defining sets of graph colourings. For instance, colouring two vertices, B and R, as shown in

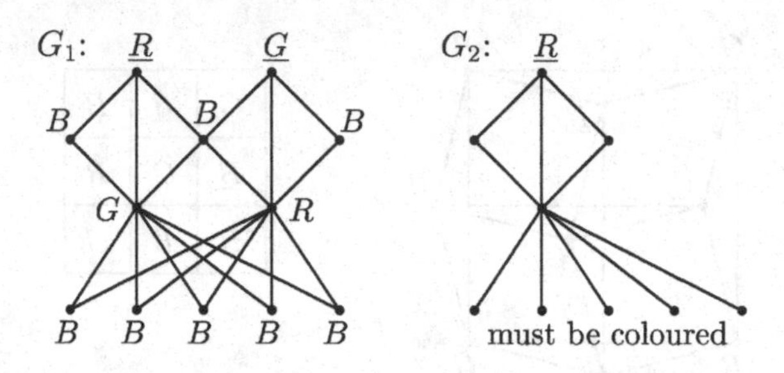

Figure 10: Colourings of a graph G_1 and its subgraph G_2

Figure 9, defines both a unique vertex colouring for the whole of $K_3 \times K_3$, and a unique latin square of order three.

With regard to substructures, the behaviour of coloured graphs is more complicated than that of latin squares. For example, colouring two vertices, $\underline{R}$ and $\underline{G}$, in the graph G_1 of Figure 10, defines the vertex colouring of the whole graph. However in its subgraph G_2, which contains only one of these coloured vertices, namely $\underline{R}$, the colouring of G_2 is not determined. Further, at least five more of the vertices of G_2 need to be coloured to force the rest uniquely, so that the defining set of the subgraph is bigger than that of the graph.

Mahmoodian, Naserasr and Zaker [103] give a lower bound on $d(G,\chi)$ for an arbitrary graph G with at least one edge. Let S be a defining set of size $d(G,\chi)$ which extends to a $\chi(G)$ colouring of G. Let r_i and v_i be the numbers of vertices coloured i in these colourings of S and G respectively. In a graph, just as in a latin square or a block design, every defining set must intersect every trade. Thus if G_{ij} is the induced subgraph of G over the vertices coloured i or j, then S must contain at least one vertex from each component of G_{ij}, so the number of components of G_{ij} is at most $r_i + r_j$. Now every connected graph with n vertices has at least $n-1$ edges, so every graph with n vertices and c components has at least $n-c$ edges. Since $|V(G_{ij})| = v_i + v_j$, and the number of components of G_{ij} is at most $r_i + r_j$, we have $|E(G_{ij})| \geqslant (v_i + v_j) - (r_i + r_j)$. If we sum all these inequalities, for all pairs i, j, then each vertex of G appears $\chi(G) - 1$ times on the righthand side as a summand with a positive sign, and each vertex of S appears $\chi(G) - 1$ times with a negative sign, giving $|E(G)| \geqslant (\chi(G) - 1)(|V(G)| - d(G,\chi))$. From this follows a general result.

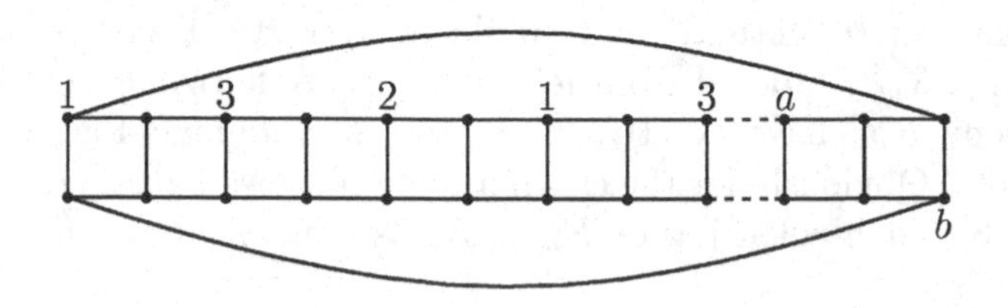

Figure 11: $G = K_2 \times C_{2n+1}$; labels indicate colours

Theorem 2.2 *For any graph G,*

$$d(G,\chi) \geqslant |V(G)| - \frac{|E(G)|}{\chi(G)-1}.$$

For $G = K_2 \times C_{2n+1}$, Theorem 2.2 implies that $d(G,\chi) \geqslant n + (1/2)$. By the construction indicated in Figure 11 we have, in fact, that $d(G,\chi) = n+1$. Another general result in [103] is the following.

Theorem 2.3 *For any graph G and any integer $n \geqslant \chi(G)$, we have*

$$d(G \times K_n, \chi) \geqslant |V(G)|(n-1) - 2|E(G)|.$$

2.3 Incomplete block designs

2.3.1 The basics A *block design* $D = (V, \mathcal{B})$ is a pair consisting of a set V of v points and a set $\mathcal{B}$ of b k-subsets (blocks) chosen from V in such a way that every element of V occurs in exactly r blocks. If every t-subset of elements of V belongs to exactly λ_t blocks, the design is called a *t-design* or, more fully, a t-(v,k,λ_t) design. Often the parameter λ_t is simply written λ. A block containing the elements $x_1, \ldots, x_k$ of V is written as $B = \{x_1 \cdots x_k\}$ or possibly just as $x_1 \cdots x_k$.

A design with $\lambda_t = 1$ is known as a *Steiner system*, often written $S(t,k,v)$. Steiner systems with $k=3$ or $k=4$ are known as Steiner *triple* or *quadruple* systems and written $STS(v)$ or $SQS(v)$ respectively. A design is called *incomplete* if $k < v$ and *simple* if no block is repeated. A Steiner system is necessarily simple. Standard counting arguments (see [26]) lead to the following.

Lemma 2.4 *For a t-(v,k,λ) design D,*

$$vr = bk \quad \textit{and} \quad \lambda\binom{v}{t} = b\binom{k}{t}.$$

In particular if $t=2$ then $\lambda(v-1) = r(k-1)$. Further, $v \leqslant b$, $k \leqslant r$ and D is an s-(v,k,λ_s) design for $0 \leqslant s \leqslant t$. The values of λ_s are given by

$$\lambda_s = \lambda\binom{v-s}{t-s} \Big/ \binom{k-s}{t-s}.$$

Note that $\lambda_0 = b$, $\lambda_1 = r$ and $\lambda_t = \lambda$.

Let S_v denote the symmetric group of permutations on the v-set V. For a block $B = \{x_1 \cdots x_k\}$ of a t-(v,k,λ) design $(V,\mathcal{B})$, and for $\rho \in S_v$, we define $\rho B = \{\rho x_1 \cdots \rho x_k\}$ and $\rho\mathcal{B} = \{\rho B : B \in \mathcal{B}\}$. We say designs $D_1 = (V,\mathcal{B}_1)$ and $D_2 = (V,\mathcal{B}_2)$ are *distinct* if $\mathcal{B}_1 \neq \mathcal{B}_2$. They are *isomorphic* if there exists $\rho \in S_v$ such that $\mathcal{B}_2 = \rho\mathcal{B}_1$, and *non-isomorphic* if no such permutation exists.

Given a design D, if $\rho \in S_v$ is such that $\rho\mathcal{B} = \mathcal{B}$, then ρ is called an *automorphism* of D, and the set of all automorphisms of D is a subgroup of S_v, denoted by $\mathrm{Aut}(D)$. If $\mathrm{Aut}(D)$ contains a cycle of length v, then D is said to be *cyclic*. If D is a t-(v,k,λ) design, then the class of designs isomorphic to D has size $|S_v|/|\mathrm{Aut}(D)| = v!/|\mathrm{Aut}(D)|$.

If $V_7 = \{0,1,\ldots,6\}$ and $\mathcal{B}_7 = \{013, 124, 235, 346, 450, 561, 602\}$, for example, then $\mathcal{F} = (V_7, \mathcal{B}_7)$ is an $STS(7)$, that is, a 2-$(7,3,1)$ design with $k = r = 3$ and $b = v = 7$. Again, if $V_9 = \{0,1,\ldots,8\}$ and $\mathcal{B}_9 = \{012, 345, 678\} \cup \{036, 147, 258\} \cup \{048, 156, 237\} \cup \{057, 138, 246\}$, then $\mathcal{A} = (V_9, \mathcal{B}_9)$ is an $STS(9)$ or 2-$(9,3,1)$ design with $r = 4$ and $b = 12$. Each of these is unique up to isomorphism. Note that $\mathcal{A}$ is *resolvable*, that is, its block set can be partitioned into r sets of v/k blocks each, such that each set contains each element of V precisely once; this partition is indicated in the list of blocks given above.

A *symmetric* design is one with $v = b$, and consequently $r = k$. It is called *trivial* if $k = 1$ or $v - 1$. It is *linked,* that is, any two of its blocks have λ elements in common. Its *residual* design with respect to a block, say B_v, is a 2-$(v-k, k-\lambda, \lambda)$ design based on the set $V \setminus B_v$ with the $v-1$ blocks $B_1 \setminus B_v, \ldots, B_{v-1} \setminus B_v$, such that each element appears in k of these blocks. Further, any two blocks of the residual have at most λ elements in common. A design with precisely two block-intersection sizes is said to be *quasi-symmetric.*

For example, the $STS(7)$ is symmetric and hence linked, but the $STS(9)$ is quasi-symmetric. The residual design of $\mathcal{F}$ is a 2-$(4,2,1)$ design; if it is taken with respect to the block 602, it is based on the set $\{1,3,4,5\}$ and has blocks $13, 14, 35, 34, 45, 15$, with intersection sizes 0 and 1.

The *order* of a symmetric design is taken to be $n = k - \lambda$ and the parameter v must satisfy $4n - 1 \leqslant v \leqslant n^2 + n + 1$; see [5, 136] for different approaches. The designs in which v meets these bounds are the *Hadamard* designs with parameters 2-$(4n-1, 2n-1, n-1)$ and the *projective planes* with parameters 2-$(n^2+n+1, n+1, 1)$. The $STS(7)$ is the only non-trivial design that meets both bounds.

A Hadamard design can always be extended by complementation to a 3-design. For instance, starting from the 2-$(7,3,1)$ design $\mathcal{F}$ we adjoin the new element 7 to each block of $\mathcal{B}_7$ and add the new blocks which are the complements in $V \cup \{7\}$ of these extended blocks, giving the 3-$(8,4,1)$ design with additional blocks 2456, 3560, 4601, 5012, 6123, 0234, 1345.

If A is a collection of blocks, then $A^x = \{B \setminus \{x\} : x \in B \text{ and } B \in A\}$ and $\overline{A^x} = \{B : x \notin B \text{ and } B \in A\}$. If x is a new element not in any block of A, then we write xA for the set of blocks $\{\{x\} \cup B : B \in A\}$.

We say that x has *multiplicity* r_x in A if x is contained in r_x blocks of A; that is, $r_x = |A^x|$. Suppose that $D = (V, \mathcal{B})$ is a t-(v,k,λ) design, and let x be any point in V. The *restriction* of D on x is the $(t-1)$-$(v-1, k-1, \lambda)$ design $(V \setminus \{x\}, \mathcal{B}^x)$. Suppose that x is a new point not in V. Then for some set of blocks A, it may be possible to *extend* D to a $(t+1)$-$(v+1, k+1, \lambda)$ design $(V \cup \{x\}, x\mathcal{B} \cup A)$, called an *extension* of D. If A is the complement of $x\mathcal{B}$ (with respect to $V \cup \{x\}$), then the process of extension is called *extension by complementation.* The set of blocks of an extended design formed by complementation is necessarily self-complementary.

The set of all subspaces of the $(n+1)$-dimensional vector space over the field $GF[q]$ is called the *projective geometry of dimension n over $GF[q]$*, denoted by $PG(n,q)$. The points and lines of $PG(2,q)$ form a *projective plane* of order q, that is, a 2-$(q^2+q+1, q+1, 1)$ design, one of the extreme cases of symmetric designs. Such a design is cyclic and linked, with index of linkage one, and exists for all prime powers q.

The residual design of $PG(n,q)$ with respect to a hyperplane is the *affine geometry of dimension n over $GF[q]$*, denoted by $AG(n,q)$. The points and lines of $AG(2,q)$ form an *affine plane* of order q, that is, a 2-$(q^2,q,1)$ design. This design is resolvable and quasi-symmetric, with block intersections of sizes 0 and 1. An affine plane can be extended to a 3-$(q^2+1, q+1, 1)$ design, that is, *an inversive plane* of order n.

For example, the points and lines of $PG(2,3)$ form a 2-$(13,4,1)$ design which can be developed from the starter block 0139 (mod 13). Its residual design with respect to any block is a 2-$(9,3,1)$ design, that is, an $STS(9)$ or affine plane $AG(2,3)$, isomorphic to $\mathcal{A}$. This can be extended to the inversive plane or 3-$(10,4,1)$ design, which also happens to be the unique $SQS(10)$.

The other extreme case of the symmetric design is the *Hadamard design*, a 2-$(4n-1, 2n-1, n)$ design. The classical example of such a design is that developed from the quadratic residues (mod $4n-1$) when $4n-1$ is a prime power. Thus the $STS(7)$ is also a Hadamard design, as is the 2-$(11,5,2)$ developed from the starter block of quadratic residues 13459 (mod 11).

2.3.2 Trades and defining sets Let T_1 and T_2 be collections of m k-subsets of elements of the v-set V. If each t-subset of elements of V occurs in the k-subsets belonging to T_1 with precisely the same multiplicity that it occurs in the k-subsets belonging to T_2, then T_1 and T_2 are said to be *mutually t-balanced.* If T_1 and T_2 are disjoint (that is, they have no k-subset in common) then they form a (v,k,t) *trade* of *volume m*, often denoted by $T=(T_1,T_2)$. The volume of T is denoted by $m(T)$, and a trade of zero volume is said to be *void.* T_1 and T_2 are known as *trade mates.* Sometimes the trade T is named by the single collection T_1, a risky practice since trade mates need not be isomorphic.

For instance, consider the sets $T_1=\{13,14,25,26\}$, $T_2=\{12,12,34,56\}$ and $T_3=\{12,15,23,46\}$. No two of these are isomorphic. (T_1,T_2) and (T_1,T_3) are $(6,2,1)$ trades but T_2 and T_3 are not disjoint and hence not trade mates. Given T_2, then its mate T_1 is forced, up to a permutation of $\{3,4,5,6\}$.

The *foundation* of the trade, $f(T)$, is the set of elements of V that occur in T_1 (and of course in T_2). A trade is *minimal* if it does not properly contain a trade. If the k-subsets of T_1 are in fact blocks of a t-(v,k,λ_t) design D, then D is said to *contain* the trade.

For example, $\mathcal{F}$ contains seven trades of volume 4, that is, *Pasch* trades. Each set of the four blocks of $\mathcal{F}$ that omit one point of V forms a Pasch configuration. The collection of blocks $T_1=\{124,156,235,346\}$, which omit the element 0, trades with $T_2=\{125,146,324,356\}$ to give a distinct design with the same parameters, that is, another $STS(7)$.

Early work on trades was due to Graham, Li and Li [57] and Graver and Jurkat [59] among others. Trades have been used in constructing designs with particular properties, such as avoiding certain blocks or ensuring that certain blocks are repeated; these properties may be relevant to their use in experimental work. More recent material on this topic includes that of Mahmoodian and Soltankhah [106] and Ajoodani-Namini and Khosrovshahi [86].

A set of blocks which is a subset of a unique t-(v, k, λ_t) design D is a *defining set* of that design, denoted by $d(t\text{-}(v, k, \lambda_t))$ or $d(D)$. A defining set is: *minimal*, denoted by $d_m(D)$, if it does not properly contain a defining set of D; *smallest*, denoted by $d_s(D)$, if no defining set of D has fewer blocks; *largest minimal*, denoted by $d_\ell(D)$, if no minimal defining set of D has more blocks. Note the difference from the latin square usage. The proportion of the blocks of a design which form a smallest defining set is denoted by $\mu(D)$, that is, $\mu(D) = |d_s(D)|/b$.

For example, in $\mathcal{F}$, any set of three blocks with no common element is a smallest defining set. Up to isomorphism, this is the only minimal defining set. But in $\mathcal{A}$, as well as the smallest defining set with four blocks, such as $\{012, 345, 036, 147\}$, there is also a minimal defining set with five blocks, such as $\{012, 036, 156, 057, 138\}$.

We define the *spectrum* of minimal defining sets of a design D by $spec(D) = \{|M| : M \text{ is a minimal defining set of } D\}$. For example $spec(STS(7)) = \{3\}$ and $spec(STS(9)) = \{4, 5\}$. Call h a *hole* in $spec(D)$ if there is no minimal defining set of D with cardinality h, but there are minimal defining sets of D with cardinalities both larger and smaller than h. If $spec(D)$ does not contain a hole, then it is said to be *continuous*.

Theorem 2.5 ([68]) *Let D be a t-(v, k, λ_t) design, containing distinct trades $T_1, T_2, \dots, T_n$, and let $S \subseteq \mathcal{B}$. Then S is a defining set of D if and only if $S \cap T_i \neq \emptyset$ for each $i = 1, \dots, n$.*

For instance, in $\mathcal{F}$ the defining set $S = \{013, 124, 235\}$ intersects all seven of the trades of volume four. If $T_{1,i}$ denotes the set of four blocks which do not contain the element i, we have $S \cap T_{1,0} = \{124, 235\}$, $S \cap T_{1,1} = \{235\}$, $S \cap T_{1,2} = \{013\}$ and so on.

Theorem 2.6 ([68]) *Suppose that S is a defining set of a t-(v, k, λ_t) design D. If $\rho \in Aut(D)$, then $\rho(S)$ is also a defining set of D and $Aut(S) \leqslant Aut(D)$.*

This result often rules out a potential defining set of blocks, R say, if the order of $Aut(R)$ does not divide the order of $Aut(D)$. Again in $\mathcal{F}$ the set $R = \{013, 450, 602\}$ is not a defining set, since the permutation $(13) \in Aut(R)$, but $(13) \notin Aut(\mathcal{F})$.

Theorem 2.7 ([68]) *Let $D = (V, \mathcal{B})$ be a t-(v, k, λ_t) design, with subdesign $E = (W, \mathcal{C})$ and $S \subseteq \mathcal{B}$. If S is a defining set of D then $S \cap \mathcal{C}$ is a defining set of E.*

Even if S is a smallest defining set of D, there is no reason to think that

$S \cap \mathcal{C}$ is a minimal (let alone a smallest) defining set of E. For example, the $STS(15)$ based on $V = \{0, \ldots, 9, a, \ldots, e\}$ and developed (mod 15) from the starter blocks $A = 014$, $B = 028$ and $C = 05a$ has 15 2-$(7,3,1)$ subdesigns and smallest defining sets of 16 blocks each. One such smallest defining set contains all 15 blocks developed from A, together with any one of those developed from B. Such a defining set intersects 12 of the 15 2-$(7,3,1)$ subdesigns in three blocks each, and the remaining three subdesigns in four blocks each.

If the set of blocks of a t-(v, k, λ) design D can be partitioned into subsets in such a way that each subset forms the set of blocks of a t-(v, k, λ_i) design, where $\sum_i \lambda_i = \lambda$ and $0 < \lambda_i < \lambda$, then D is called *decomposable*. If no such partition exists, the design is *indecomposable*.

For instance, let $V_7 = \{0, 1, \ldots, 6\}$ as before, let $\mathcal{B}_1$ be the set of 14 blocks generated cyclically from the starter blocks $\{013, 023\}$ (mod 7), let $\mathcal{B}_2 = \{012, 034, 056, 135, 146, 236, 245\}$ and let $\mathcal{C}$ be the set of 21 blocks generated cyclically from the starter blocks $\{012, 024, 036\}$ (mod 7). If each design is based on V_7, then the blocks of $\mathcal{B}_2$ are those of a 2-$(7,3,1)$ design, $\mathcal{B}_1 \cup \mathcal{B}_2$ of a decomposable 2-$(7,3,3)$ design, and $\mathcal{C}$ of an indecomposable 2-$(7,3,3)$ design.

Theorem 2.8 ([67]) *Let $D = (V, \mathcal{B}_1 \cup \mathcal{B}_2)$ be a decomposable t-$(v, k, \lambda_1 + \lambda_2)$ design where $D_1 = (V, \mathcal{B}_1)$ and $D_2 = (V, \mathcal{B}_2)$ are t-(v, k, λ_1) and t-(v, k, λ_2) designs respectively. If the designs D, D_1 and D_2 have smallest defining sets of sizes s, s_1 and s_2 respectively, then $s \geqslant s_1 + s_2$.*

This bound is not in general tight. For instance, the 36 non-isomorphic 2-$(9,3,2)$ designs include nine which are decomposable. Six of these have smallest defining sets of eight blocks, but three need nine blocks [85].

For given t, v and k, the number of indecomposable designs is finite (see Engel [44]) but in general this number is not known. However, when all the indecomposable designs are known, all t-(v, k, λ) designs can be constructed (see Grüttmüller [75]).

A group of permutations on a set V of elements is said to be *single-transposition-free* (STF) if no single transposition (ij) is an element of the group. If a t-(v, k, λ_t) design based on V has an STF automorphism group, then it is said to be an STF design. The set of STF designs includes all Steiner designs and all symmetric designs.

Theorem 2.9 ([66]) *Any defining set of an* STF *design based on a v-set has at least $v - 1$ elements contained in its blocks.*

The last results we give here show some simple properties of trades. Note that Lemma 2.11 follows from Lemma 2.10(1).

Lemma 2.10 ([80], Hwang [82]) *Suppose that $T = (T_1, T_2)$ is a non-void (v, k, t) trade. Then:*
(1) *T is a (v, k, s) trade for all $0 < s < t$;*
(2) *$m(T) \geqslant 2^t$;*
(3) *$f(T) \geqslant k + t + 1$.*

Lemma 2.11 ([82]) *Suppose that $T = (T_1, T_2)$ is a (v, k, t) trade. Then:*
(1) $T^x = (T_1^x, T_2^x)$ *is a* $(v-1, k-1, t-1)$ *trade of volume* r_x*;*
(2) $\overline{T^x} = (\overline{T_1^x}, \overline{T_2^x})$ *is a* $(v-1, k, t-1)$ *trade of volume* $m(T) - r_x$*;*
(3) $xT^x = (xT_1^x, xT_2^x)$ *is a* $(v, k, t-1)$ *trade of volume* r_x*.*

2.4 Latin squares and partial Steiner triple systems

A *partial Steiner triple system* of order v is a pair $\mathcal{S} = (V, \mathcal{B})$ where V is a set of v elements, and $\mathcal{B}$ is a collection of subsets of V, each of size three, such that each pair of elements of V occurs in at most one of these subsets. If each pair of elements of V occurs in precisely one of these subsets, we have an $STS(v)$. Its blocks are often called *triples*. An $STS(v)$ can be thought of as a decomposition of K_v into edge-disjoint triangles, generalising the idea of one-factorization.

Steiner triple systems can be used to construct both quasigroups and loops, with special properties. A *Steiner quasigroup* or *squag* is based on the set V with quasigroup operation defined as follows: (1) for all $v \in V$, $v \circ v = v$; (2) for all $u, v \in V$, $u \neq v$, $v \circ u = z = u \circ v$, if and only if uvz is a block of the Steiner triple system.

Similarly a *Steiner loop* or *sloop* is based on the set $V \cup \{e\}$ with loop operation defined as follows: (1) for all $v \in V \cup \{e\}$, $v \circ v = e$; (2) for all $v \in V$, $v \circ e = v = e \circ v$; (3) for all $u, v \in V$, $u \neq v$, $v \circ u = z = u \circ v$, if and only if uvz is a block of the Steiner triple system.

For example, the squag and sloop associated with $\mathcal{F}$ were given in Figure 4. The latin squares defined by the squag and sloop associated with $\mathcal{A}$ are given in Figure 12. Note that Q_9, representing the squag, is a latin square of order 9 but that L_9, representing the sloop, is a latin square of order 10.

This process can be reversed, to construct an $STS(v)$ from a squag or sloop of order v. For example, let $(Q, \circ)$ be a squag, that is, a quasigroup satisfying the identities: (1) for all $q \in Q$, $q \circ q = q$; (2) for all $p, q \in Q$, $p \circ q = q \circ p$; (3) for all $p, q \in Q$, $p \circ (p \circ q) = q$. Defining $\mathcal{B}$ as $\{\{p, q, p \circ q\}$: for all $p, q \in Q, p \neq q\}$

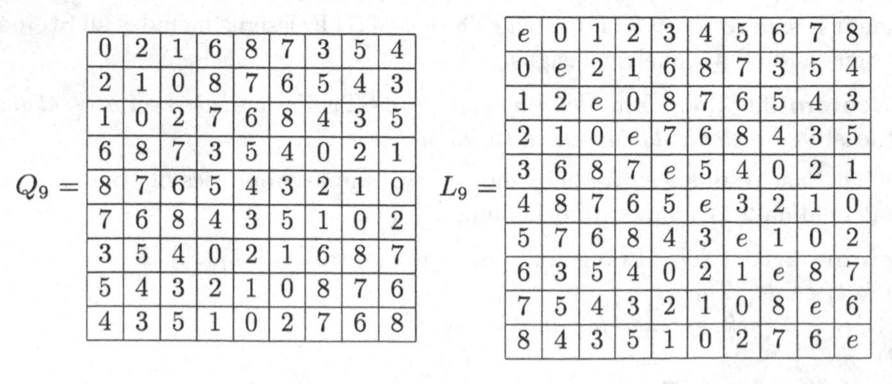

$Q_9 =$

0	2	1	6	8	7	3	5	4
2	1	0	8	7	6	5	4	3
1	0	2	7	6	8	4	3	5
6	8	7	3	5	4	0	2	1
8	7	6	5	4	3	2	1	0
7	6	8	4	3	5	1	0	2
3	5	4	0	2	1	6	8	7
5	4	3	2	1	0	8	7	6
4	3	5	1	0	2	7	6	8

$L_9 =$

e	0	1	2	3	4	5	6	7	8
0	e	2	1	6	8	7	3	5	4
1	2	e	0	8	7	6	5	4	3
2	1	0	e	7	6	8	4	3	5
3	6	8	7	e	5	4	0	2	1
4	8	7	6	5	e	3	2	1	0
5	7	6	8	4	3	e	1	0	2
6	3	5	4	0	2	1	e	8	7
7	5	4	3	2	1	0	8	e	6
8	4	3	5	1	0	2	7	6	e

Figure 12: Latin squares from a quasigroup of order 9 and a loop of order 10

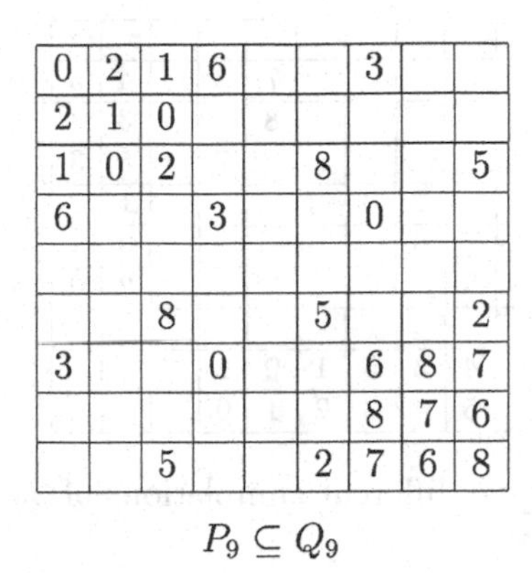

0	2	1	6			3		
2	1	0						
1	0	2			8			5
6			3			0		
		8			5			2
3			0			6	8	7
						8	7	6
		5			2	7	6	8

$P_9 \subseteq Q_9$

e	0	1	2	3	4	5	6	7	8
0	e	2	1	6			3		
1	2	e	0			6	5		
2	1	0	e		6		4		
3	6			e	5	4	0		
4			6	5	e	3	2		
5		6		4	3	e	1		
6	3	5	4	0	2	1	e	8	7
7							8	e	6
8							7	6	e

$J_9 \subseteq L_9$

Figure 13: Partial squag $P_9 \subseteq Q_9$ and partial sloop $J_9 \subseteq L_9$

gives a Steiner triple system $(Q, \mathcal{B})$.

Again suppose that $(Q\cup\{e\}, \circ)$ is a sloop, that is, a quasigroup with identity element e satisfying the following: (1) for all $p \in Q \cup \{e\}$, $p \circ e = p$; (2) for all $p, q \in Q\cup\{e\}$, $p\circ q = q \circ p$; (3) for all $p, q \in Q\cup\{e\}$, $p\circ(p\circ q) = q$. Defining $\mathcal{C}$ as $\{\{p, q, p \circ q\}$: for all $p, q \in Q, p \neq q\}$ gives a Steiner triple system $(Q, \mathcal{C})$. Thus there are one-to-one correspondences between Steiner triple systems, squags and sloops; see Ganter and Werner [52].

If a set, D, of blocks of an $STS(v)$ is used to construct a partial latin square of either type (squag or sloop), and if this partial latin square has unique completion, then D is a defining set of the $STS(v)$. Figure 13 shows a partial squag P_9 with unique completion to the squag Q_9. Unfortunately the partial latin square corresponding to a defining set of an $STS(v)$ need not be uniquely completable. For an example of this, see the appendix of [39].

The partial squag P_9 completes uniquely to Q_9, provided we know that the missing element of V is in fact 4. It corresponds to the set of blocks in $\mathcal{A}$, namely, $\{012, 678, 036, 258\}$ which again, given that the missing element is 4, completes uniquely to $\mathcal{A}$, illustrating the fact that a uniquely completable partial squag corresponds to a defining set of the corresponding Steiner triple system. These four blocks form a minimal defining set of the $STS(9)$ $\mathcal{A}$, and the partial squag is minimal with respect to being uniquely completable, and is thus a critical set of Q_9. From a larger minimal defining set of $\mathcal{A}$, say $\{012, 345, 036, 048, 138\}$ with five blocks, we obtain a corresponding partial squag which is again a critical set of Q_9, but a larger one.

Figure 13 also shows the partial sloop J_9, contained in L_9; J_9 has two completions. In $\mathcal{A}$, the set of six blocks $\{012, 345, 678, 036, 156, 246\}$ corresponds to J_9, and can be completed with *either* the remaining blocks of $\mathcal{A}$, namely, 147, 258, 048, 237, 057, 138, *or* with six different blocks, namely, 148, 257, 047, 238, 058, 137, to give a different $STS(9)$. These two sets, of six blocks each, form a trade in $\mathcal{A}$ and Figure 14 shows the corresponding partial sloops T_9, contained in L_9, and its trade mate T'_9.

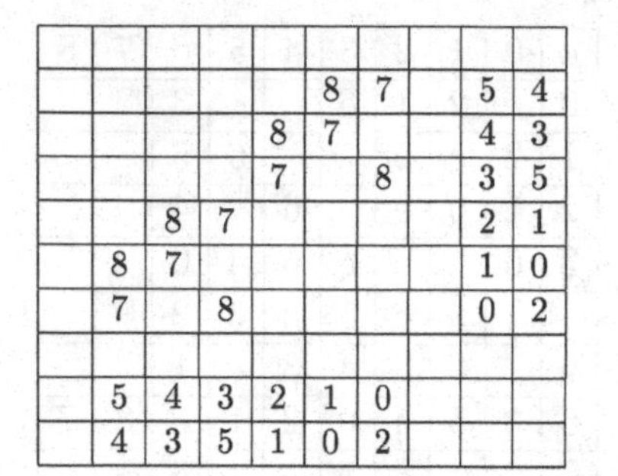

					8	7		5	4
				8	7			4	3
				7		8		3	5
		8	7					2	1
	8	7						1	0
	7		8					0	2
	5	4	3	2	1	0			
	4	3	5	1	0	2			

					7	8		4	5
				7	8			3	4
				8		7		5	3
		7	8					1	2
	7	8						0	1
	8		7					2	0
	4	3	5	1	0	2			
	5	4	3	2	1	0			

Figure 14: T_9 and its trade mate T'_9 give different completions of J_9

3 Defining sets of block designs

3.1 Pointwise defining sets

Our discussion so far has concentrated on defining sets which consist of whole blocks of a design. Now we allow the possibility of partial blocks as well; see Delaney, B. Gray, K. Gray, Maenhaut, Sharry and Street [34].

If D is a t-(v, k, λ_t) design with blocks $\mathcal{B} = \{B_i : i = 1, \dots, b\}$, then a *partial design* S of D is a set of subsets of some of those blocks. So $S = \{S_i : \emptyset \neq S_i \subseteq B_i; i \in I \subseteq \{1, \dots, b\}\}$. If $|S_i| = k$ for every $i \in I$ then S is said to be a *blockwise* partial design. Otherwise it is a *pointwise* partial design. The *cardinality* of a partial design is defined to be: for a blockwise partial design, the number of blocks it contains, $|I|$; for a pointwise partial design, the total number of points in all blocks, $\Sigma|S_i|$. Note that a blockwise partial design may also be considered as a pointwise partial design so that, for instance, a set of three blocks of a 2-(7, 3, 1) design has cardinality three as a blockwise partial design and cardinality nine as a pointwise partial design.

Now a defining set of the t-design D is a partial design S contained in D but in no other design with the same parameters as D. S is said to be a *blockwise defining set* if full blocks *only* are being considered, or a *pointwise defining set* if partial blocks are permitted. Note also that if S is a partial design of a design D, and S is not a defining set, then neither is any partial design contained in S.

As before a defining set of D is said to be *minimal* if it contains no proper partial design which is also a defining set of D; a *smallest* defining set of D is any such set with smallest cardinality of all defining sets of D and is necessarily minimal. This general definition may be applied to either blockwise or pointwise defining sets with the appropriate interpretation of "partial design" and "cardinality".

Example 3.1 In the design $F_1 = \{124, 235, 346, 457, 561, 672, 713\}$, no two blocks form a defining set so $S = \{124, 235, 346\}$ is a smallest (hence minimal) blockwise defining set. $S' = \{124, 235, 34, 47\}$ is easily shown to be a pointwise defining set, though not a smallest one since it has cardinality 10, whereas

S, considered as a pointwise defining set, has cardinality 9. However, S' is a minimal pointwise defining set, since removing any point from any of its blocks leaves a partial design contained in at least two 2-(7, 3, 1) designs.

Much of the earlier discussion for blockwise defining sets carries over for pointwise defining sets, but the concept of *significance* is new for pointwise defining sets. Given a design D with blocks $\mathcal{B} = \{B_i : i = 1, \ldots, b\}$ and a partial design $\mathcal{P}$ satisfying $\mathcal{P} = \{P_i : i = 1, \ldots, p; \emptyset \neq P_i \subseteq B_i; p \leqslant b\}$, we say that a partial block $P \in \mathcal{P}$ with $P \subseteq B_j$ for some $j \leqslant p$ is *significant in* $\mathcal{P}$ if $\mathcal{P}$ has fewer completions than $\mathcal{P} \setminus P$ does; that is, if retaining P in the collection leaves $\mathcal{P}$ closer to being a defining set. Similarly, a partial block $P \subseteq B_j$, $1 \leqslant j \leqslant b$, such that $P \notin \mathcal{P}$, is *significant to* $\mathcal{P}$ if adjoining P to $\mathcal{P}$ gives a collection which has fewer completions than $\mathcal{P}$.

The following examples illustrate the concept of significance with respect to a 2-(7, 3, 1) design. If $\mathcal{P} = \{123, 145\}$, the block 167 is not significant to $\mathcal{P}$; it is forced by the blocks of $\mathcal{P}$. However the block 246 is significant to $\mathcal{P}$; $\mathcal{P}$ has two completions but $\mathcal{P} \cup \{246\}$ is a defining set. If $\mathcal{P} = \{123, 14\}$, the partial block 26 is not significant to $\mathcal{P}$; we already know it must occur, though in neither block of $\mathcal{P}$. If $\mathcal{P} = \{123, 14\}$, the block 16 is significant to $\mathcal{P}$; adjoining it to $\mathcal{P}$ forbids completions containing the block 146.

All discussion of trades so far has been specific to full blocks. Now we consider how these concepts relate to finding pointwise defining sets. A *configuration* of p points (or *p-configuration*) of a t-design D is a collection of partial blocks of D consisting of a total of p points from specified blocks. These partial blocks may be of different sizes.

A configuration differs from a pointwise partial design in that we know which full block of D contains each partial block of a configuration. Any s-set for $1 \leqslant s \leqslant t$ occurs in λ_s blocks in D; in a configuration we know which of those blocks is referred to. Essentially, a partial design is an unordered collection of subsets of the blocks of the design, and a configuration is an ordered collection of subsets of the blocks.

If a design D is given by a list of blocks, say $\mathcal{B} = \{123, 145, 167, 246, 257, 347, 356\}$, then a configuration P is written as a list of the partial blocks in the same order, including $\emptyset$ for any empty partial blocks. For example, $\{\emptyset, 145, 16, 26, \emptyset, \emptyset, \emptyset\}$ is a 7-configuration of D.

If $P = \{P_i : i = 1, ..., b; P_i \subseteq B_i\}$ is a configuration of a design D with blocks $\mathcal{B} = \{B_i : i = 1, ..., b\}$, then its *complement* $\overline{P}$ is given by $\overline{P} = \{\overline{P_i} : \overline{P_i} = B_i \backslash P_i; i = 1, ..., b\}$; if any P_i are empty then $\overline{P}$ will include the corresponding full blocks of D.

A *trade core* C of design D is a configuration in D such that $\overline{C}$ is not a pointwise defining set. C is a *minimal trade core* if removing any element from any block in C gives C' where C' is not a trade core, that is, $\overline{C'}$ is a pointwise defining set. This leads to the following result.

Lemma 3.2 (1) *If C is a trade core of design D, and T is the set of blocks of D corresponding to the non-empty partial blocks of C, then T is a trade.*

	$\mathcal{B}$	C_1	$\overline{C_1}$	C_2	$\overline{C_2}$
1	123	∅	123	∅	123
2	145	∅	145	∅	145
3	167	∅	167	6	17
4	246	6	24	∅	246
5	257	7	25	7	25
6	347	7	34	7	34
7	356	6	35	6	35

Table 1: Example of trade cores and their complements

(2) *If a p-configuration P of design D is a pointwise defining set, then it intersects every minimal trade core of D.*

It is important to consider trade cores as configurations rather than partial designs. It is possible that the partial blocks of a trade core could be subsets of other blocks of the design, but can only be traded when considered as part of the blocks specified by the configuration. That is, it may be possible to fit the partial blocks of the trade core into other blocks of the design and for these same partial blocks configured in that way not to form a trade core. For instance, in Table 1, C_1 is a trade core since $\overline{C_1}$ can be completed in two ways ($\mathcal{B}$ as given, or with 6 and 7 swapped in blocks $4, \ldots, 7$). C_2 is not a trade core since $\overline{C_2}$ is a pointwise defining set (because blocks 1, 2 and 4 are a defining set contained within $\overline{C_2}$).

Lemma 3.3 *If P is a p-configuration of a design $D = (V, \mathcal{B})$ such that P contains at least one point of every trade core, then P is a pointwise defining set.*

All this information is of little use unless we have some way of finding trade cores. Note that a trade T may have many trade cores.

A minimal trade core can be found by matching the blocks of $T_1 \subseteq \mathcal{B}$ to the blocks of T_2, where T_1 is a minimal trade with trade mate T_2, and removing from each block of T_1 all the points common to it and its matched block of T_2. Only points common to T_1 and T_2 can be removed, otherwise the resulting configuration will not be a trade core. To see this, let T_1 be a trade of volume v with trade mate T_2. Let C_1 and C_2 be configurations corresponding to T_1 and T_2 respectively. That is, for $1 \leqslant i \leqslant r$, label the blocks of T_1 and T_2 as B_{1i} and B_{2i} respectively, making $C_1 = \{B_{11}, B_{12}, \ldots, B_{1r}, \emptyset, \ldots, \emptyset\}$ and $C_2 = \{B_{21}, B_{22}, \ldots, B_{2r}, \emptyset, \ldots, \emptyset\}$. Let x be a point such that $x \in B_{1,i}$ but $x \notin B_{2,i}$; note that, since T_1 and T_2 are disjoint, there may be values of i for which no such x exists. Let C_1^x be C_1 with x removed from $B_{1,i}$. Then $x \in \overline{C_1^x}$ is a point distinguishing between the two trades which give different possible ways of completing the design. Since T_1 is a minimal trade, there is no smaller portion of it which could be replaced by different k-sets. So $\overline{T_1^x}$ is a defining set, and T_1^x is not a trade core.

Depending on the ordering of blocks in T_2, that is, on how they are matched

T_1	order 1	C_1	order 2	C_2
246	247	6	346	2
257	256	7	357	2
347	346	7	247	3
356	357	6	256	3

Table 2: Different trade cores for different orders of the blocks

to blocks of T_1, different trade cores will be produced.

Let D be the design given in Table 1, with blocks $\mathcal{B}$. Let T_1 consist of its blocks 4, 5, 6 and 7. They form a trade with the blocks $T_2 = \{247, 256, 346, 357\}$. Changing the order of the blocks of T_2 and removing the intersecting points from the blocks of T_1 gives rise to different trade cores, as shown in Table 2.

In fact, it follows from the Marriage Theorem that we can always match the blocks of a trade so that matched blocks intersect in at least t points. Such a configuration gives an ordering of the trades which results in a smaller trade core than other orderings. This is useful since smaller trade cores give stricter testing conditions for pointwise defining sets.

To find every possible minimal trade core, we try every possible ordering of the k-sets in T_2, that is, the k-sets not in our given design, which trade with the blocks of the design, in T_1. Since the collections T_1 and T_2 contain all the same t-sets, we order the k-sets of T_2 by matching them to blocks of T_1 which have a common t-set, and find all such orderings by a recursive (depth first) search.

In order to find a smallest pointwise defining set, we first note that any smallest pointwise defining set can be found by removing points from other pointwise defining sets, and ultimately from a blockwise defining set. It is possible that a set which is not significant at one stage of the calculation may become significant at a later stage, so caution is needed before deleting such sets. Algorithms have been based essentially on the analogues for pointwise defining sets of Theorems 2.5, 2.6 and 2.9, and on Lemmas 3.2 and 3.3.

As an example of results obtained, let $V = \{0, 1, \ldots, 9, a\}$ and let D be the cyclic 2-(11, 5, 2) design developed from the block 13459, the set of quadratic residues modulo 11. Up to isomorphism, D has precisely two blockwise defining sets [72]; we choose as representatives of these classes $S_1 = \{13459, 2456a, 35670, 479a0, 12690\}$ and $S_2 = \{13459, 2456a, 03567, 14678, 25789\}$. $P_1 = \{459, 245a, 3670, 479a0, 1690\}$ is a pointwise defining set contained in S_1 and $P_2 = \{3459, 456a, 3567, 1678, 2789\}$ in S_2.

In summary, the smallest pointwise defining sets within five blocks of this design consist of 20 points; eight non-isomorphic smallest defining sets were found in S_1 and 26 in S_2. Of these 34 non-isomorphic pointwise defining sets, one (P_2) comprises five 4-sets, 21 (including P_1) comprise one 3-set, three 4-sets and one 5-set, and 12 comprise two 3-sets, one 4-set and two 5-sets. Thus the points of these sets spread fairly evenly over the five blocks. Partial designs with six or more blocks have not been searched completely, though preliminary exploration suggests that it is unlikely that pointwise defining sets exist with fewer points spread over more blocks. The results on the 2-(11, 5, 2) design illustrate the following.

Theorem 3.4 *Let D be a t-(v,k,λ) design with a smallest pointwise defining set S and a minimal blockwise defining set M, such that each block of S is contained in a block of M. Then each partial block of S contains at least $t+1$ points.*

Note that, for instance, in Example 3.1, the minimal pointwise defining set S' of F_1 is not contained in any minimal blockwise defining set, nor do its partial blocks contain at least $t+1=3$ points. There is no analogue of Theorem 3.4 for minimal pointwise defining sets as far as we know.

3.2 Class and member defining sets

Again we broaden our definition of defining sets, but in a different way. We consider only blockwise defining sets so, given a t-(v,k,λ) design $D=(V,\mathcal{B})$, the set of blocks $d(D)\subseteq\mathcal{B}$ is a defining set if for all t-(v,k,λ) designs $D_0=(V,\mathcal{B}_0)$ distinct from D, the collection of blocks $d(D)$ is not a subset of $\mathcal{B}_0$. But now we seek a collection of k-subsets which uniquely determines a design within an isomorphism class, or which uniquely determines the isomorphism class. Trades which distinguish between two members of an isomorphism class are said to be *member trades* or *m-trades* and trades which distinguish between two designs from different isomorphism classes are said to be *class trades* or *c-trades*.

To explore these ideas let $\mathcal{D}=\{D_0,\dots,D_{\delta-1}\}$ represent a partition of the set of all t-(v,k,λ) designs into δ isomorphism classes and assume $D\in D_i$. A collection of k-subsets, denoted $c(D)$, is a *class defining set* if, whenever there exist distinct t-(v,k,λ) designs $D=(V,\mathcal{B})$ and $D'=(V,\mathcal{B}')$ such that $c(D)\subseteq\mathcal{B}\cap\mathcal{B}'$, then $D'\in D_i$. That is, if a t-(v,k,λ) design D' is a completion of $c(D)$, then D' is a member of the same isomorphism class as D. Let $c_s(D)$ denote the size of the smallest such class defining set. A collection of k-subsets, denoted $m(D)$, is a *member defining set* if, whenever there exist distinct t-(v,k,λ) designs $D=(V,\mathcal{B})$ and $D'=(V,\mathcal{B})'$ such that $m(D)\subseteq\mathcal{B}\cap\mathcal{B}'$, then D and D' are not isomorphic. We let $m_s(D)$ denote the size of the smallest such member defining set.

Many combinatorial objects, including designs, are unique up to isomorphism. They can be regarded as having class defining sets of size zero. Examples of this include the one-factorizations given in the introduction and the Fano plane $\mathcal{F}$.

Note that S is a defining set of D if and only if S is both a class and a member defining set of D. Clearly $c(D)$ intersects all c-trades in D and $m(D)$ intersects all m-trades in D. Consequently,

$$|m_s(D)|,\ |c_s(D)|\leqslant|d_s(D)|\leqslant|m_s(D)|+|c_s(D)|. \tag{3.1}$$

	$\mathcal{N}_1$	$\mathcal{N}_2$	$\mathcal{N}_3$	$\mathcal{N}_4$	$\mathcal{N}_5$	$\mathcal{N}_6$	$\mathcal{N}_7$
$\mathcal{N}_1$	1619	5300	23760	34560	2556	9540	36480
$\mathcal{N}_2$	1664	522	54	1248	2124	630	2868
$\mathcal{N}_3$	1632	598	1362	264	2316	606	1176
$\mathcal{N}_4$	1466	749	876	1475	2352	624	357
$\mathcal{N}_5$	1184	3200	8480	12000	3326	10	10560
$\mathcal{N}_6$	1568	616	1584	1936	2234	149	1600
$\mathcal{N}_7$	1518	807	1107	234	2358	639	1048

Table 3: The number of minimal m-trades and minimal c-trades

Ramsay [123] studied t-designs with parameter sets 2-$(8,4,3)$, 2-$(10,4,2)$, 2-$(9,4,3)$, 3-$(10,5,3)$ and 2-$(10,5,4)$, and enumerated the m-trades and c-trades associated with these designs. For each of these designs, his results extended to identifying the size of the smallest defining set, the smallest class defining set and the smallest member defining set. We include below the results for the 3-$(10,5,3)$ designs, which highlight some interesting features.

There are a maximum of seven non-isomorphic 3-$(10,5,3)$ designs and hence seven isomorphism classes, labelled $\mathcal{N}_i$, for $i = 1, \ldots, 7$. In Table 3 the entry in cell (x,x) is the number of minimal m-trades in class x, and the entry in cell (x,y), where $x \neq y$, is the number of minimal c-trades from designs in class x to designs in class y. In Table 4 the total number of minimal c-trades and minimal trades contained in different isomorphism classes are compared. Table 5 displays the size of the smallest defining sets of each type.

It is interesting to note that the total number of minimal trades can be less than the sum of the number of c-trades. This phenomenon is explained by the fact that a trade may have a number of distinct trade mates, and these trade mates need not be isomorphic. Trading blocks may result in a number of non-isomorphic t-(v,k,λ) designs. This variation in the number of trade mates also explains why the results on c-trades are not symmetric. For the isomorphism classes $\mathcal{N}_2$ and $\mathcal{N}_3$, we note that the number of minimal c-trades equals the number of minimal trades. For isomorphism class $\mathcal{N}_6$ the number of minimal c-trades is greater than the total number of minimal trades, which seems curious until we recognise that a c-trade, T, is minimal only if all trades contained in T distinguish between the same isomorphism classes as T does.

class	c-trades	trades
$\mathcal{N}_1$	24056	25271
$\mathcal{N}_2$	380	380
$\mathcal{N}_3$	306	306
$\mathcal{N}_4$	352	370
$\mathcal{N}_5$	858	951
$\mathcal{N}_6$	1562	759
$\mathcal{N}_7$	402	412

Table 4: Total number of minimal c-trades and minimal trades

class	$m_s(D)$	$c_s(D)$	$d_s(D)$
$\mathcal{N}_1$	5	4	5
$\mathcal{N}_2$	5	8	8
$\mathcal{N}_3$	6	8	8
$\mathcal{N}_4$	6	8	8
$\mathcal{N}_5$	5	4	6
$\mathcal{N}_6$	5	4	6
$\mathcal{N}_7$	6	8	8

Table 5: Size of smallest defining sets

The results in Table 5 show that in the inequality (3.1) there are cases where equality holds and where it does not. For instance, $m_s(\mathcal{N}_1) = d_s(\mathcal{N}_1)$ but $m_s(\mathcal{N}_i) < d_s(\mathcal{N}_i)$ for $i = 2, \dots, 7$. Again, $c_s(\mathcal{N}_i) < d_s(\mathcal{N}_i)$ for $i = 1, 5$ or 6, but $c_s(\mathcal{N}_i) = d_s(\mathcal{N}_i)$ for $i = 2$, 3, 4 or 7. Finally for each of the 3-(10, 5, 3) designs, $d_s(\mathcal{N}_i) < m_s(\mathcal{N}_i) + c_s(\mathcal{N}_i)$. We also note that $m_s(\mathcal{N}_i) < c_s(\mathcal{N}_i)$ for $i = 2$, 3, 4 or 7, and $m_s(\mathcal{N}_i) > c_s(\mathcal{N}_i)$ for $i = 1$, 5 or 6.

For examples of the remaining cases we have to go beyond the 3-(10, 5, 3) designs. For the full design D (or multiples thereof) we have $|Aut(D)| = v!$, so that $m_s(D) = 0$ and hence $d_s(D) = m_s(D) + c_s(D)$. For the 2-(4, 3, 2) design C, we have $m_s(C) = c_s(C) = d_s(C) = 0$. For the 2-(10, 4, 2) design H_3, with automorphism group of order 24, we have $m_s(H_3) = c_s(H_3) = 5$; see [73, 72, 123]. For other results on member and class defining sets see also [65].

3.3 Families of designs and the sizes of their smallest defining sets

We now return to the blockwise defining sets discussed earlier. A number of authors have focused on the question of how varying the parameters of a family of t-(v, k, λ) designs affects the sizes of the smallest defining sets. For instance, let P_n be the 2-$(2^{n+1} - 1, 3, 1)$ design obtained from the points and lines of $PG(n, 2)$ and let μ_n be the fraction of its blocks in a smallest defining set. B. Gray [61] observed that $\{\mu_n\}_{n=2}^{\infty}$ is a non-decreasing sequence, and so $\mu_n \to l$ as $n \to \infty$, for some limiting value l. Moran [110] showed that $\mu_3 = 16/35$, so $l \geqslant 16/35$, but the exact value of l is unknown. This is an example of the problem of determining the limit (if it exists) of the size of a smallest defining set as v increases.

B Gray and Ramsay [65] found the limiting value of the fraction of blocks in a smallest *member* defining set for Steiner designs. More precisely, they proved the following.

Theorem 3.5 *Let m_D be the size of a smallest member defining set of a Steiner design D. Fix $k > t > 1$, let $m_v = \max\{m_D : D$ is a t-$(v, k, 1)$ design$\}$, and let b_v be the number of blocks in a t-$(v, k, 1)$ design. Then $m_v/b_v \to 0$ as $v \to \infty$.*

Now, if S is a defining set of a design D then S is both a class and a member defining set of D. So, for Steiner designs, the problem of finding the asymptotic value for the fraction of blocks in a smallest defining set reduces to the problem of finding the fraction of blocks in a smallest class defining set. This result applies to the designs P_n derived from $PG(n, 2)$, as follows.

Corollary 3.6 *Let c_n and μ_n be the fraction of blocks in smallest class defining sets and smallest defining sets respectively of P_n. Then*

$$\lim_{n\to\infty} c_n = \lim_{n\to\infty} \mu_n = l, \text{ for some } l \geqslant \frac{16}{35}.$$

The method of Theorem 3.5 can be applied to other combinatorial structures, for example the problem of finding critical sets of latin squares. A number of authors have studied the sizes of critical sets in latin squares; see, for example, Keedwell [84]. It is conjectured that the number of entries in a smallest *uniquely completable* (UC) set of a latin square of order n is at least $n^2/4$ if n is even, and $(n^2 - 1)/4$ if n is odd.

An *isotopism class* of a latin square L of order n consists of all the latin squares obtained from L by permuting rows, columns or entries of L. So an isotopism class contains at most $(n!)^3$ members. Define in the natural way a member UC set and a class UC set of a latin square.

Theorem 3.7 *For a latin square L of order n, let m_L be the size of a smallest member UC set. Let $m_n = \max\{m_L : L \text{ is a latin square of order } n\}$. Then $m_n/n^2 \to 0$ as $n \to \infty$.*

So, asymptotically, the unsolved problem of determining the size of a smallest UC set is equivalent to determining the size of a smallest class UC set.

Next we consider some families of designs in which the block size k varies. B. Gray, Hamilton and O'Keefe [63] showed that if $D_q = PG(2, q)$, then $\mu_q \leqslant 1/2 + \epsilon_q$, where $\epsilon_q \to 0$ as $n \to \infty$. In contrast, B. Gray [61] showed that if D_n is the symmetric design obtained from the points and hyperplanes of $PG(n, 2)$, then $\mu_n \to 1$ as $n \to \infty$.

If q is an odd prime power and D_q is the Hadamard design cyclically generated from the quadratic residues of $GF(q)$, then Seberry [130] together with Sarvate [129] and Kunkle [91] has conjectured that $\mu_q \leqslant 1/2$. This, and a related conjecture concerning the value of μ_q for residuals of these designs, have been verified for 15 small primes.

Little is known in general about the effect on the sizes of defining sets in a t-design of varying t, though empirically ([138, 139]) the higher the value of t, the smaller the fraction of the blocks needed to define a design. A classic example of this fact is the result by Curtis [29] that only eight of the 759 blocks are needed to define the unique 5-$(24, 8, 1)$ Matthieu design. K. Gray [68] showed that the unique $SQS(8)$ has a smallest defining set of three blocks, whereas when this design is considered as one of the four 2-$(8, 4, 3)$ designs, its smallest defining set has six blocks. Similarly Greenhill and Street [72] showed that the unique $SQS(10)$ has a smallest defining set of four blocks, whereas when this design is considered as a 2-$(10, 4, 4)$ design (of which there are more than 1.7×10^6, see Mathon and Rosa [108]), its smallest defining set has 16 blocks.

There are two main results here. The first concerns the sizes of a smallest defining set of a design D considered as a t-design and as an s-design respectively, where $0 < s < t$. We apply earlier results, namely, Lemmas 2.4, 2.10 and 2.11 and Theorem 2.5. If S is a defining set of D considered as an s-design, then S is also a defining set of D considered as a t-design, showing that $s_t \leqslant s_s$. But equality is not in fact possible, which follows from one further lemma.

Lemma 3.8 (1) *The set of blocks of a (v, k, t) trade can be partitioned into at least 2^{t-s} non-void (v, k, s) trades.*
(2) *For any minimal defining set M of a t-design D, there is a trade in D which intersects M in just one block.*

This gives us the bound $s_s \geqslant s_t + 2^{t-s} - 1$. As results for the $SQS(8)$ and the $SQS(10)$ show, this bound is certainly not tight. From Lemma 3.8(2) it also follows that the smallest defining set of a t-$(v, k, 1)$ design considered as a 1-design contains all but one of its blocks.

Finally we consider results in [65] on how the sizes of smallest defining sets vary for designs in some families of designs in which λ varies. First, by a counting argument, and then by an application of Lemma 2.10(3) we find the following bounds.

Lemma 3.9 *The size s of a smallest defining set of a t-$(v, t+1, \lambda_t)$ design, $D = (V, \mathcal{B})$, with b blocks satisfies $s \leqslant b - r$.*

Lemma 3.10 *Let $D = (V, \mathcal{B})$ be a t-$(k+t+1, k, \lambda_t)$ design. For $x \in V$, let D^x be the $(t-1)$-$(k+t, k-1, \lambda_t)$ design $(V \setminus \{x\}, \mathcal{B}^x)$ (so D^x is the restriction of D on x). Suppose that s and s^x are the sizes of smallest defining sets of D and D^x respectively. Then $s \leqslant s^x \leqslant r$.*

For example, the size of a smallest defining set of a 4-$(11, 5, 1)$ design is five. From the previous lemma, the size of a smallest defining set of a 5-$(12, 6, 1)$ design is at most five. It is, in fact, straightforward to see that this size must equal five; see [72].

Suppose $k = t+1$ and $v = k+t+1$, so that $r = b/2$. Lemmas 3.9 and 3.10 each imply that $s \leqslant r$, but combining the approaches of these two lemmas can improve the bound further.

Lemma 3.11 *The size s of a smallest defining set of a t-$(2t+2, t+1, \lambda_t)$ design, $D = (V, \mathcal{B})$, satisfies $s \leqslant r - \lambda_2$.*

Combining some new upper bounds on the size of smallest defining sets with information on the decomposability of the designs leads to solutions for the sizes of smallest defining sets of the 2-$(6, 3, \lambda)$ designs and to good bounds on the sizes for the 2-$(7, 3, \lambda)$ and 3-$(8, 4, \lambda)$ designs. First, if D is the unique 2-$(6, 3, 2)$ design, then its smallest defining set has size three; see [67]. Results cited in Gronau [74] imply that any 2-$(6, 3, \lambda)$ design is necessarily decomposable into copies of D. Since $b = 10\lambda/2$, $r = 5\lambda/2$ and $\lambda_2 = \lambda$, Theorem 2.8 and Lemma 3.11 imply the following.

Theorem 3.12 *Let $\lambda \geqslant 2$ be even, J be a 2-$(6, 3, \lambda)$ design, s_λ be the size of a smallest defining set of J, and μ_λ be the proportion of blocks in a smallest* *defining set of J. Then* $s_\lambda = 3\lambda/2$ *and* $\mu_\lambda = 3/10$.

Grüttmüller notes in [75] that Langdev [93] has shown that there are exactly two indecomposable 2-$(7,3,\lambda)$ designs, namely the 2-$(7,3,1)$ design $(V_7, \mathcal{B}_2)$ and the 2-$(7,3,3)$ design $(V_7, \mathcal{C})$, both introduced in Section 2.3.2. The number of blocks in the smallest defining sets of these designs are 3 and 7 respectively; see [67]. An argument similar to that applied to the 2-$(6,3,\lambda)$ design yields a similar result.

Theorem 3.13 *If D is a* 2-$(7,3,\lambda)$ *design, then its smallest defining set has at most* $16\lambda/5$ *blocks.*

Corollary 3.14 *For any* 2-$(7,3,\lambda)$ *design J,* $1/3 \leqslant \mu \leqslant 16/35$. *If J can be decomposed into Fano planes, then* $3/7 \leqslant \mu \leqslant 16/35$.

The 3-$(8,4,\lambda)$ designs are self-complementary, and can all be obtained by extending (uniquely) the 2-$(7,3,\lambda)$ designs by complementation; see Khosrovshahi and Vatan [88]. The extension of an indecomposable 2-$(7,3,\lambda)$ design is an indecomposable 3-$(8,4,\lambda)$ design, while any restriction of an indecomposable 3-$(8,4,\lambda)$ design is an indecomposable 2-$(7,3,\lambda)$ design. So there are precisely two indecomposable 3-$(8,4,\lambda)$ designs, one with $\lambda = 1$, one with $\lambda = 3$. Let μ be the proportion of blocks in a smallest defining set of a 3-$(8,4,\lambda)$ design. Since extension by complementation doubles the number of blocks, the following corollary of Lemma 3.10 and Theorem 3.13 is immediate.

Corollary 3.15 *For any* 3-$(8,4,\lambda)$ *design J,* $1/6 \leqslant \mu \leqslant 8/35$. *If J can be decomposed into* 3-$(8,4,1)$ *designs, then* $3/14 \leqslant \mu \leqslant 8/35$.

Note that if $\lambda \leqslant 4$, then Corollaries 3.14 and 3.15 are sufficient to determine the sizes of smallest defining sets for those designs which can be decomposed into Fano planes or into 3-$(8,4,1)$ designs.

3.4 A few more results on Steiner systems

General bounds for the sizes of minimal defining sets in Steiner systems are given by K. Gray [68] (lower bound) and B. Gray, Mathon, Moran and Street [64] (upper bound), as follows.

Theorem 3.16 *If D is an* $S(t,k,v)$ *and* $x \in spec(D)$, *then*

$$2\left(\frac{v-1}{k+1}\right) \leqslant x \leqslant \binom{v}{t}\Big/\binom{k}{t} - 2\binom{v}{t-1}\Big/\left(\binom{k}{t-1}+1\right).$$

These bounds are interesting but they are not tight. For instance, if D is an $STS(15)$ and $x \in spec(D)$, then Theorem 3.16 tells us that $7 \leqslant x \leqslant 27$. There are precisely 80 nonisomorphic $STS(15)$ (see Mathon and Rosa [108]) and Ramsay [120] has determined the sizes of their smallest defining sets as 11, 12, 13, 14 or 16 in every case, so the lower bound is too low. Among these triple systems, the only one for which we know the size of the largest minimal defining set is P_3, that is, the $STS(15)$ based on points and lines of $PG(3,2)$ and in this case the size of the set is 22 (see [55, 64]) so the upper bound is too high. We now improve these bounds for the particular case of the Steiner triple systems P_n, associated with the projective geometries $PG(n, 2)$. Note that in this notation, $\mathcal{F} = P_2$.

First we show that the largest minimal defining set of P_3 has 22 blocks, starting from the construction of Gower [55].

Theorem 3.17 *Consider the Steiner triple system P_n of order $2^{n+1} - 1$ which is isomorphic to the point-line design of $PG(n, 2)$ for $n \geqslant 2$. Let $\mathcal{H}$ be a set of $n + 1$ hyperplanes of $PG(n, 2)$ for $n \geqslant 2$, such that no point of $PG(n, 2)$ is common to all these hyperplanes. The set of h_n blocks of P_n which correspond to the set of lines in the hyperplanes of the set $\mathcal{H}$ is a minimal defining set for the design, where*

$$h_n = \left(\frac{2}{3}\right) 4^n - \left(\frac{1}{2}\right) 3^n - 2^n + \left(\frac{5}{6}\right).$$

This yields a minimal defining set of 22 blocks for P_3. That no minimal defining set of P_3 contains more than 22 blocks follows by the combination of a counting argument and a backtrack search [64]. Moran [110] used the structure of P_3 to obtain two smallest defining sets of 16 blocks each, and the spectrum was proved continuous by the construction of one minimal defining set of each order from 17 to 21 [64].

From the structure of the projective geometry $PG(n, 2)$, we find that $95 \in spec(P_4)$ and that $403 \in spec(P_5)$. Since P_{n+1} contains $2^{n+2} - 1$ subsystems P_n and since each block of P_{n+1} is contained in exactly $2^n - 1$ subsystems P_n, Moran's argument generalises, giving a better bound for this particular case.

Theorem 3.18 *Let m_n be the size of $d_s(P_n)$, a smallest defining set of P_n. Then*

$$m_{n+1} \geqslant \left\lceil \frac{2^{n+2} - 1}{2^n - 1} m_n \right\rceil.$$

Since $m_3 = 16$, we obtain $m_4 \geqslant 71$ and $m_5 \geqslant 299$ from Theorem 3.18, whereas the general bound of Theorem 3.16 gives only $m_4 \geqslant 15$ and $m_5 \geqslant 31$. The limiting value of the lower bound on μ from Theorem 3.18 is $0.4594 \cdots$; see also Corollary 3.6. From Gower's construction we find the following bound.

Theorem 3.19 *Let ℓ_n be the size of $d_\ell(P_n)$, a largest minimal defining set of P_n. Then*

$$\ell_n \geqslant \left(\frac{2}{3}\right) 4^n - \left(\frac{1}{2}\right) 3^n - 2^n + \left(\frac{5}{6}\right).$$

Again we obtain $\ell_4 \geqslant 115$ and $\ell_5 \geqslant 530$, from Theorem 3.18. The general bound of Theorem 3.16 gives $\ell_4 \leqslant 139$ and $\ell_5 \leqslant 619$.

Ramsay [122] pointed out that the lower bound of Theorem 3.16 implies that a Steiner triple system $STS(v)$ must have $\mu \geqslant 3/v$. But $\lim_{v\to\infty} 3/v = 0$. However he also remarked [122] that any t-design containing m mutually disjoint trades must have at least m blocks in its smallest defining set, and for all $v \equiv 1, 9 \pmod{25}$, $v \geqslant 25$, there exists an $STS(v)$ decomposable into

Pasch configurations by results of Adams, Billington and Rodger [2]. Hence for all such v, there exists an $STS(v)$ with $\mu \geqslant 1/4$.

Grannell, Griggs and Wallace [58] extended these ideas by observing that the mutually disjoint trades in question did not need to be isomorphic, and that the sizes of the smallest defining sets of the trades in the design were also relevant. By constructing appropriate $STS(v)$, they showed that for all $v \equiv 1, 3 \pmod 6$, $v > 7$, there exists an $STS(v)$ with $\mu \geqslant 1/4$. They also proved other similar results, in particular the following.

Theorem 3.20 *Let $STS(u)$ and $STS(v)$ be Steiner triple systems with smallest defining sets of sizes s_u and s_v respectively. Let $\sigma_{uv} = us_v + vs_u + \max\{u(u-1)s_v, v(v-1)s_u\}$. Then the Steiner triple system $STS(uv)$ formed as the product of the systems $STS(v)$ and $STS(v)$ has $\mu \geqslant 6\sigma_{uv}/uv(uv-1)$.*

Now let $\overline{\mu_v}$ be the maximum value of μ for all $STS(v)$, and let $M = \sup\{\overline{\mu_v} : v \text{ is admissible}\}$. Then Theorem 3.20, together with the fact that $\mu \geqslant 299/651$ for the Steiner triple system P_5, leads to two corollaries.

Corollary 3.21 $\overline{\mu_v} \to M$ *as* $v \to \infty$.

Corollary 3.22 *For every $\epsilon > 0$, there exists U_0 such that for all $u > U_0$, $\overline{\mu_u} > \frac{299}{651} - \epsilon$.*

3.5 Steiner triple systems and latin squares

The completable product of two sloops (two squags) is also a sloop (squag); see Section 2.1 and [52]. This observation can be used to prove that if Q is a partial sloop which has strong unique completion to a sloop M, then the partial Steiner triple system corresponding to Q is a defining set, as is the partial Steiner triple system corresponding to the completable product $P_2 \otimes Q$. Similarly, if Q is a partial squag which has strong unique completion to a squag M, then the partial Steiner triple system corresponding to Q is a defining set, as is the partial Steiner triple system corresponding to the completable product $P_3 \otimes Q$. Since this technique can be used to identify defining sets, it seems reasonable to ask whether such a technique can be used to identify minimal defining sets in the direct product.

Gower [55, 56] and Donovan, Khodkar and Street [38, 39] have shown that if a suitable restriction is placed on the partial Steiner triple system associated with Q, then the completable product gives rise to a minimal defining set.

To understand this work, we must introduce a construction for critical sets in latin squares developed by Stinson and van Rees [137]. Let L_2 and P_2 be the partial latin squares given in Figure 5. Let M be a latin square of order m, with the property that for some $i, j, k, i', j', k' \in N$ we have $I \subseteq M$, where

$$I = \{(i, j; k), (i, j'; k'), (i', j; k'), (i', j'; k)\}.$$

Note that the four cells of I define a subsquare corresponding to the intersection of rows i and i' with columns j and j'. (This is like the intersection of rows 1 and 2 with columns 3 and 4 in Figure 15.) It is easy to see that I gives rise

		$\underline{w}$	y	$\underline{v}$	x
		y	w	x	v
$\underline{w}$	y			$\underline{u}$	z
y	w			z	u
$\underline{v}$	x	$\underline{u}$	z		
x	v	z	u		

Figure 15: Partial latin square corresponding to $\mathcal{T}_0$

to a number of isomorphic subsquares in $L_2 \times M$. In particular $L_2 \times M$ will contain the subsquares:

$$J = \{(i,j;k),(i,j+m;k+m),(i+m,j;k+m),(i+m,j+m;k)\};$$
$$J^* = \{(i,j;k),(i,j'+m;k'+m),(i'+m,j;k'+m),(i'+m,j'+m;k)\};$$
$$J_{0,1} = \{(i,j+m;k+m),(i,j'+m;k'+m),(i',j+m;k'+m),(i',j'+m;k+m)\};$$
$$J_{1,0} = \{(i+m,j;k+m),(i+m,j';k'+m),(i'+m,j;k'+m),(i'+m,j';k+m)\};$$
$$J_{1,1} = \{(i+m,j+m;k),(i+m,j'+m;k'),(i'+m,j+m;k'),(i'+m,j'+m;k)\}.$$

It is shown [137] that, if Q is a critical set in a latin square M, with the property that for each entry $(i,j;k) \in Q$ there exists a set of cells

$$I = \{(i,j;k),(i,j';k'),(i',j;k'),(i',j';k)\} \subseteq M$$

such that $I \cap Q = \{(i,j;k)\}$, then the completable product $P_2 \otimes Q$ is a critical set in $L_2 \times M$.

Now assume that $\mathcal{B}$ is a minimal defining set in a Steiner triple system $\mathcal{D}$ with the property that for every block $B = \{uvw\}$ of $\mathcal{B}$ there exists a Steiner trade $\mathcal{T}_0 = \{uvw, uxy, zvy, zxw\}$ in $\mathcal{D}$ such that $\mathcal{T}_0 \cap \mathcal{B} = \{uvw\}$. Let $S_\mathcal{B}$, $S_\mathcal{D}$ and $S_{\mathcal{T}_0}$ be the partial sloops corresponding to $\mathcal{B}$, $\mathcal{D}$ and $\mathcal{T}_0$ respectively. The partial sloop $S_{\mathcal{T}_0}$ is displayed in Figure 15; note the ordering u, z, v, x, w, y has been used to label rows and columns. Observe that this partial latin square $S_{\mathcal{T}_0}$ can be partitioned into six subsquares isomorphic to I (given above). Further it is easy to see that each of these subsquares intersects $S_\mathcal{B}$ in a single entry; such entries are underlined.

It follows from results in [137] that the partial sloop $P_2 \otimes S_\mathcal{B}$ is a critical set in $L_2 \times S_\mathcal{D}$. Let $\mathcal{B}^*$ and $\mathcal{D}^*$ represent the partial Steiner triple systems corresponding to $P_2 \otimes S_\mathcal{B}$ and $L_2 \otimes S_\mathcal{D}$ respectively. Further, the subsquares of $S_{\mathcal{T}_0}$ can be used to identify subsquares in $L_2 \times S_\mathcal{D}$ which, when recombined, show that for each block in $B^* \in \mathcal{B}^*$ there exists a Steiner trade $\mathcal{T}_0^*$ isomorphic to $\mathcal{T}_0$ such that $\mathcal{T}_0^* \cap \mathcal{B}^* = B^*$. Consequently, if $S_\mathcal{B}$ has a unique completion, then $\mathcal{B}^*$ is a minimal defining set in $\mathcal{D}^*$.

In [55] and [38] these techniques have been used to show the existence of minimal defining sets of volume t in Steiner triple systems of order $2^n - 1$ with

$$t \in \left\{\alpha - \frac{3^n}{6}, \alpha - \frac{37.3^n}{162}, \alpha - \frac{29.3^n}{162}\right\},$$

		$\underline{t}$	v	x	$\underline{s}$	u	w
		v	x	t	w	s	u
$\underline{t}$	v				$\underline{r}$	z	
v	x					r	z
x	t				z		r
$\underline{s}$	w	$\underline{r}$	z				
u	s	z	r				
w	u		z	r			

Figure 16: Partial latin square corresponding to $\mathcal{T}_1$

where

$$\alpha = \frac{4^n - 3.2^n + 5}{6}.$$

These minimal defining sets exist in Steiner triple systems corresponding to the projective geometry $PG(n, 2)$.

In a similar fashion the construction in [137] has been adapted in [56] and [38, 39] to give minimal defining sets in Steiner triple systems of order 3^n for all positive integers $n \geqslant 1$.

So assume $\mathcal{B}$ is a minimal defining set in a Steiner triple system $\mathcal{D}$ with the property that for every block $B = \{rst\}$ of $\mathcal{B}$ there exists a Steiner trade $\mathcal{T}_1 = \{rst, ruv, rwx, zsv, zux, zwt\}$ in $\mathcal{D}$ such that $\mathcal{T}_1 \cap \mathcal{B} = \{rst\}$. Let $S_\mathcal{B}$, $S_\mathcal{D}$ and $S_{\mathcal{T}_1}$ be the partial squags corresponding to $\mathcal{B}$, $\mathcal{D}$ and $\mathcal{T}_1$ respectively. The partial squag $S_{\mathcal{T}_1}$ is displayed in Figure 16; note the ordering r, z, s, u, w, t, v, x, has been used to label rows and columns. Observe that this partial latin square $S_{\mathcal{T}_1}$ can be partitioned into six subarrays which are not isomorphic, but each has a conjugate which is isomorphic to the other subarrays. Further it is easy to see that each of these subarrays intersects $S_\mathcal{B}$ in a single entry; such entries are underlined.

It can be shown that the partial squag $P_3 \otimes S_\mathcal{B}$ is a critical set in $M_3 \times S_\mathcal{D}$. Let $\mathcal{B}^*$ and $\mathcal{D}^*$ represent the partial Steiner triple systems corresponding to $P_3 \otimes S_\mathcal{B}$ and $M_3 \times S_\mathcal{D}$ respectively. Further, the subarrays of $S_{\mathcal{T}_1}$ can be used to identify subarrays in $M_3 \times S_\mathcal{D}$ which when recombined show that for each block in $B^* \in \mathcal{B}^*$ there exists a Steiner trade $\mathcal{T}_1^*$, isomorphic to $\mathcal{T}_1$ such that $\mathcal{T}_1^* \cap \mathcal{B}^* = B^*$. So once again, if $S_\mathcal{B}$ has a unique completion, then $\mathcal{B}^*$ is a minimal defining set in $\mathcal{D}^*$.

In [55] and [39] these techniques have been used to prove the existence of minimal defining sets of volume t in Steiner triple systems of order 3^n with

$$t \in \left\{\beta - \frac{7^n}{6}, \beta - \frac{7^{n-2}}{6} - 7^{n-1}, \beta - \frac{7^{n-3}}{6} - 52.7^{n-3}\right\},$$

where

$$\beta = \frac{9^n - 3^n + 1}{6}.$$

Minimal defining sets of volume t and order 3^n were shown to exist in Steiner triple systems corresponding to the affine geometry $AG(n, 3)$. In addition minimal defining sets of volume $\beta - (7^n/6)$ were found in non-affine Hall triple systems of order 3^n; see Beneteau [11] for the properties of these systems.

3.6 Algorithms for finding defining sets

Questions regarding defining sets can, in theory, be answered simply by considering all possibilities. Unfortunately, the search-space grows exponentially with the parameters, so this method is practical only for small cases, even with the assistance of a computer. It is known that deciding whether a partial STS$(2v + 1)$ can be embedded in an STS(w) for any $w \leqslant 4v + 1$ is NP-complete [23], and similar results have been obtained for Latin squares [24, 25]. So, in general, we do not expect efficient algorithms to exist. However, appropriate heuristics can extend the reach of exhaustive techniques, and many of the results we have on defining sets were obtained by computer-based methods.

A key tool for investigating the defining set problem is an algorithm to *complete* a partial design; that is, given the parameters of a t-(v, k, λ) design, a v-set V, and a collection $\mathcal{P}$ of k-subsets of V, find all the t-(v, k, λ) designs, based on V, which contain $\mathcal{P}$. Such designs are called *completions* of $\mathcal{P}$, and $\mathcal{P}$ may have zero, one, or more than one completion.

To prove computationally that $\mathcal{P}$ is a defining set, a deterministic and exhaustive technique is necessary. (To prove that $\mathcal{P}$ is *not* a defining set or that it can be embedded in (that is, completed to) *some* design, random or non-exhaustive techniques may suffice.) A back-track search through a tree of partial designs is straightforward to programme, and is the usual technique (see, for example, [54, 90]). Recall Lemma 2.4 and the definition of the indices λ_s. An effective heuristic to improve the efficiency of a back-track completion search is to keep track of the number of s-subsets of V in the current partial design, for one or more $1 \leqslant s \leqslant t$, and to maintain a list of the k-subsets which can be added to the current partial; this information can then be used to prune the search tree. However, it is not clear how best to organise this information, and there is still scope for further improvements in this area [95]. Of course, if all the designs with the given parameters have some particular property, this can be used to guide the search; for example, if all the designs are simple or are linked [33, 91], or if the designs are Steiner [120].

Algorithm A: Given an efficient completion routine, it is straightforward to find the size s of a smallest defining set of a design D. Starting at some lower bound for s, simply test all subsets of D of successively larger sizes until a defining set is found, and then stop. If all smallest defining sets are required, then defer stopping until all subsets of size s have been considered. If all minimal defining sets are required, then all subsets of D which are not supersets of a previously found minimal defining set need to be tested.

Algorithm A was first described and implemented by Greenhill [71], in the context of simple STF designs. Delaney [33] reimplemented Algorithm A for general designs, and also produced versions for pointwise defining sets and for defining sets of large and overlarge sets (see [132, 133] for material on large and overlarge sets).

Obviously, the completion properties of a set of blocks are invariant under isomorphisms, so we need only check a representative of each isomorphism class using our completion routine. To put sets of blocks into classes, the graph isomorphism programme `nauty` [109] is commonly used. This programme generates the automorphism group of a set of blocks (considered as a particular bipartite graph) and can provide a canonic version of the set of blocks (which allows isomorphism testing). The relative speeds of the completion routine and `nauty` for a particular design determine whether or not it is faster to classify the subsets via `nauty` and then run the completion routine on a class representative, or whether it is more efficient to check every subset via the completion routine and then classify, if this is required, the (minimal) defining sets using `nauty`. The implementations of Algorithm A described in [33, 71] adopt the former approach, while [120] describes a case where the latter approach was more efficient.

If any theoretical results are available which yield necessary conditions on a defining set, then these can be used to reject some of the subsets of D, and avoid the need to run `nauty` or the completion routine. For example, Theorem 2.9 tells us that only subsets of D which cover at least $v-1$ points of V need be considered if D is STF. Information on the automorphism groups of the designs and the partials can often be used to reject some classes of partials (recall Theorem 2.6).

Algorithm A is capable of providing full information on all the defining sets of a design, and is easy to optimise for a particular parameter set. It has been used extensively to investigate the smaller designs, and many of the results quoted in Tables 6 and 7 depend on it.

One necessary condition on a defining set of D is that it must intersect all trades contained in D (recall Theorem 2.5). Implementations of Algorithm A commonly make use of a list of trades in D for this purpose, and can generate further trades by considering the completions of partial designs which are not defining sets or by applying the automorphism group of D to known trades in D. Given a list of trades in D, the requirement that any defining set intersects all the trades in the list can be reformulated as an integer linear programme. If such a programme has an optimal solution in m blocks, then m is a lower bound on $|d_s(D)|$.

Algorithm B: Find some trades in D and put these in a list $\mathcal{T}$. Now perform steps (1) and (2) repeatedly until S has only one completion.

(1) Form the integer programme corresponding to $\mathcal{T}$, find an optimal solution, and form the set of blocks S corresponding to this solution.

(2) If S does not complete uniquely, then completions not equal to D define

additional trades in D, and these are added to $\mathcal{T}$.

The running time for step (1) can be improved by *minimising* $\mathcal{T}$. That is, if $T_a \in \mathcal{T}$ and $T_b \in \mathcal{T}$, and if $T_a \subsetneq T_b$, then T_b can be removed from $\mathcal{T}$. Also, as for Algorithm A, additional trades can be generated in step (2) by applying the automorphism group of D to existing trades.

The integer linear programme technique was first explicitly used by Khodkar [85], who found smallest defining sets for the 36 2-(9, 3, 2) designs. Algorithm B was formalised by Adams, Khodkar and Ramsay [3], who used it to find smallest defining sets for the 104 STS(19) with $|\mathrm{Aut}(D)| \geqslant 9$.

Note that Algorithm B yields only a single smallest defining set, and it requires the use of a utility for solving integer linear programmes (CPLEX [28] and *opbdp* [6] have been used for this). Which of Algorithms A and B is 'better' for any particular design D is not clear; Algorithm A gives much more information, while Algorithm B is more efficient (particularly for larger designs).

Algorithms A and B require a completion routine. We can eliminate this requirement by extending the ideas behind Algorithm B. Instead of iteratively generating more and more trades, via our completion routine, until we 'converge' on a defining set, we can find defining sets directly by considering *all* the trades in a design.

Algorithm C: Let $\mathcal{D} = \{D_1, \ldots, D_n\}$ be a transversal of the t-(v, k, λ) designs; that is, any t-(v, k, λ) design D is isomorphic to precisely one of the D_i. For each $D_i \in \mathcal{D}$, all the trades in D_i can be generated by applying all the permutations of the point set to all the designs in $\mathcal{D}$ and 'comparing' D_i and the permuted designs. This list of trades yields an integer linear programme, any optimal solution of which is a smallest defining set.

Since the list of trades in a design is complete, Algorithm C is easily extended to yield all smallest or minimal defining sets. Also, member and class defining sets for each D_i can be found by considering only permutations of D_i or $\mathcal{D} \setminus D_i$ respectively. Algorithm C was described by Ramsay [121], who used it to find smallest defining sets, including member and class, for five parameter sets of designs.

Algorithm C has high complexity, since it must process all $v!$ permutations, and it generates large integer programmes. Information on the automorphism groups of the D_i can eliminate the need to consider a particular permutation for a particular design, while, as in Algorithm B, the lists of trades can be minimised to improve the running time. In general, Algorithm C can only be run in its entirety when a transversal is available and v is not too large. However, even when this is not the case, an incomplete run is a useful method of generating a list of trades in a design. For example, a member defining set can be found for a design D even if no other designs of the same parameters are known.

Note that when working with trades in a non-simple design, care must be taken to consider all copies of a trade. For example, recall the STS(7) $\mathcal{F}$ and the Pasch trade $T_1 = \{124, 156, 235, 346\}$ which it contains. A 2-(7, 3, 2) design, say $\mathcal{F}^2$, is easily constructed by taking two copies of $\mathcal{F}$. Now each of the four blocks of T_1 occurs twice in $\mathcal{F}^2$, which thus contains sixteen copies of T_1. Any defining set of $\mathcal{F}^2$ must intersect all of these copies; in fact, both copies of one of the blocks are necessary (and sufficient). These ideas were developed into the concept of *discriminating sets* in [123]. This allowed the trade enumeration technique discussed in Algorithm C to be efficiently applied to many non-simple designs, yielding results on smallest defining sets, including member and class, for the 2-(9, 3, 2) designs and many 2-(6, 3, λ), 2-(7, 3, λ) and 3-(8, 4, λ) designs.

3.7 Defining sets of some small designs

In Tables 6 and 7, designs are listed in the order used in the Handbook of Combinatorial Designs [26]. The notation s^m indicates that, among the designs with these parameters, those belonging to m of the isomorphism classes have smallest defining sets of s blocks each. If $m = 1$, it is omitted.

These tables refer only to smallest defining sets, and do not include any minimal defining sets of larger size. They also do not include information on: smallest defining sets of complementary designs; designs with empty defining sets; or member or class defining sets.

4 Defining sets of special features of graphs

4.1 Graph colourings

Most papers on defining sets of graph colourings deal with finding $d(G, \chi)$ for some special classes of graphs; see Section 2.2.1 for notation. The next theorem [103] is a step towards finding the value of $d(K_n \times K_n, \chi)$, equivalent to scs(n), the size of the smallest critical set for latin squares of order n.

Theorem 4.1 *If $n \geqslant m^2$, then $d(K_m \times K_n, \chi) = m(n - m)$.*

The proof of Theorem 4.1 uses Theorem 2.3 and gives a construction for a defining set of a graph colouring of size $m(n - m)$. The defining numbers of several graphs are determined in [103, 99], where C_n denotes the cycle with n vertices and n edges, and P_n the path with n vertices and $n - 1$ edges.

Theorem 4.2 *We have:*

(1) $d(C_m \times K_3, \chi) = \lfloor m/2 \rfloor + 1$;

(2) $m \leqslant d(C_m \times K_4, \chi) \leqslant m + 1$;

(3) $d(C_m \times K_5, \chi) = \begin{cases} 2m & \text{for } m \text{ even,} \\ 2m \text{ or } 2m+1 & \text{for } m \text{ odd;} \end{cases}$

(4) *for* $n \geqslant 6$, $d(C_m \times K_n, \chi) = m(n - 3)$;

(5) *for* $n \geqslant 6$ *we have* $d(P_m \times K_n, \chi) = m(n - 3) + 2$.

t	v	b	r	k	λ	s	references
2	7	7	3	3	1	3	[68]
2	9	12	4	3	1	4	[67]
2	13	13	4	4	1	6	[67, 72]
2	6	10	5	3	2	3	[67]
2	16	20	5	4	1	7	[67, 72]
2	21	21	5	5	1	8	[67, 72]
2	11	11	5	5	2	5	[67, 72]
2	13	26	6	3	1	9, 8	[71]
2	7	14	6	3	2	6^4	[68]
2	10	15	6	4	2	8, 6, 5	[62, 72]
2	25	30	6	5	1	10	[62]
2	31	31	6	6	1	11	[62]
2	16	16	6	6	2	9, 7^2	[60, 62, 72]
2	15	35	7	3	1	16, 14^3, 13^4, 12^{20}, 11^{52}	[110, 120]
2	8	14	7	4	3	6^4	[68]
2	15	15	7	7	3	9, 8, 7^3	[61, 69, 70]
2	9	24	8	3	2	10^5, 9^{21}, 8^{10}	[85]
2	9	18	8	4	3	8^9, 6^2	[111]
2	19	57	9	3	1	23^2, 22^{14}, 21^{68}, 20^{17}, 19^2, 18 (note 4)	[3]
2	7	21	9	3	3	9^9, 7	[68]
2	10	18	9	5	4	8, 7^{18}, 6^2	[87, 121]
2	19	19	9	9	4	8^6	[112]
2	23	23	11	11	5	8 (note 3)	[112]
2	7	28	12	3	4	12^{34}, 10	[123]
2	10	30	12	4	4	16 (note 1)	[72]
2	27	27	13	13	6	$\leqslant 11$ (note 3)	[112]
2	31	31	15	15	7	$\leqslant 10$ (note 3), 24 (note 5)	[61, 112]
2	63	63	31	31	15	52–55 (note 5)	[61]
2	6	5λ	$5\lambda/2$	3	λ	$3\lambda/2$ (note 2)	[65]
2	7	7λ	3λ	3	λ	$7\lambda/3$–$16\lambda/5$	[65]

1) the unique 3-(10, 4, 1) design, regarded as a 2-design
2) λ even
3) the design from quadratic residues
4) the 104 designs D with $|\mathrm{Aut}(D)| \geqslant 9$
5) the point/hyperplane design from $PG(n, 2)$

Table 6: Smallest defining sets for some small 2-designs

t	v	b	r	k	λ	s	references
3	8	14	7	4	1	3	[68]
3	12	22	11	6	2	5	[112]
3	10	30	12	4	1	4	[72]
3	8	28	14	4	2	6^4	[68]
3	16	30	15	8	3	9, 8, 7^3	[112]
3	10	36	18	5	3	8^4, 6^2, 5	[111, 121]
3	20	38	19	10	4	8^3	[112]
3	17	68	20	5	1	7	[123]
3	8	42	21	4	3	9^9, 7	[68]
3	22	77	21	6	1	8	[110]
3	24	46	23	12	5	8 (note 1)	[112]
3	28	54	27	14	6	$\leqslant$11 (note 1)	[112]
3	8	56	28	4	4	12^{30}, 10	[123]
3	32	62	31	16	7	$\leqslant$10 (note 1)	[112]
3	8	14λ	7λ	4	λ	$7\lambda/3$–$16\lambda/5$	[65]
4	11	66	30	5	1	5	[72]
4	23	253	77	7	1	8	[110]
5	12	132	66	6	1	5	[72]
5	24	759	253	8	1	8	[29, 110]

1) extension of the 2-design from quadratic residues

Table 7: Smallest defining sets for some small t-designs, $t \geqslant 3$

4.1.1 Defining numbers of regular graphs Suppose that $\mathcal{P}$ is the family of graphs with $\chi(G) = k$ and $|V(G)| = n$. It is natural to study the possible values of $d(G, \chi)$ for $G \in \mathcal{P}$ and easy to see that they make up the set $\{k - 1, k, \ldots, n - 1\}$.

Now we ask the same question for families of graphs having some specified structure, for example, the case where $\mathcal{P}$ is the family of r-regular graphs with $\chi(G) = k$ and $|V(G)| = n$. For $k = 2$ it is easy to see that the answer is the set $\{1, 2, \ldots, \lfloor \frac{n}{2r} \rfloor\}$, but for $k \geqslant 3$ the solution is no longer easy. Let $d(n, r, \chi = k)$ be the smallest value of $d(G, \chi)$ for all r-regular graphs with $\chi(G) = k$ and $|V(G)| = n$. For regular graphs we must first find the value of $d(n, r, \chi = k)$ for different values of k, r and n. This is studied in a series of papers [102, 105, 135, 134]. We need the following definition for our discussion.

A set S of size s, with an assignment of colours in a graph G, is called a *strong defining set* if there exists an ordering $\{v_1, v_2, \ldots, v_{n-s}\}$ of the vertices of the induced subgraph $\langle G - S \rangle$ such that, in the induced list of colours in each of the subgraphs $\langle G - S \rangle$, $\langle G - (S \cup \{v_1\}) \rangle$, $\langle G - (S \cup \{v_1, v_2\}) \rangle$, ..., $\langle G - (S \cup \{v_1, v_2, \ldots, v_{n-s}\}) \rangle$, there exists at least one vertex whose list of colours has cardinality 1. For example in Figure 8(c), the defining set in the Petersen graph is strong, but that in the graph of Figure 8(d) is not.

By a well-known theorem of Brooks [140], in any r-regular k-chromatic graph other than a complete graph or an odd cycle, we have $r \geqslant k$. A useful tool in the proof of Theorem 4.4 is the following Lemma; see [102].

Lemma 4.3 *Any defining set of a k-regular k-chromatic graph is strong.*

In [102], the following two results are proved for the cases where $r = k$ and where $k = 3$, respectively.

Theorem 4.4 *For $k = 3, 4$ and 5 we have*

$$d(n, k, \chi = k) = \left\lceil \frac{k-2}{2(k-1)}n + \frac{2 + (k-2)(k-3)}{2(k-1)} \right\rceil$$

with one exception, namely, $d(10, 5, \chi = 5) = 6$. In general,

$$d(n, k, \chi = k) \geqslant \left\lceil \frac{k-2}{2(k-1)}n + \frac{2 + (k-2)(k-3)}{2(k-1)} \right\rceil. \qquad (4.1)$$

Theorem 4.5 *For each n and each $r \geqslant 4$ we have $d(n, r, \chi = 3) = 2$.*

The results of Theorems 4.4 and 4.5 lead to two further natural questions [102], subsequently settled by Mahmoodian and Soltankhah [105] and Soltankhah, Omoomi and Mahmoodian [135] respectively. The first of these results is that it is *not* true that for every k there exists $n_0(k)$ such that for all $n \geqslant n_0(k)$ equality holds in (4.1). The second result is that there *do* exist $n_0(k)$ and $r_0(k)$ such that for all $n \geqslant n_0(k)$ and $r \geqslant r_0(k)$ we have $d(n, r, \chi = k) = k - 1$. For more information on these results, see [134, 115].

4.2 Total colouring

Behzad, Mahmoodian, and Soltankhah [10] have studied the defining sets of total colouring of graphs.

The minimum number of different colours required to colour the *elements* (the vertices and the edges) of a graph G so that no two associated elements have the same colour is called the *total chromatic number* of G, denoted by $\chi''(G)$. In a given graph G, a set of elements D, $D \subset V \cup E$, with an assignment of colours is said to be a *defining set of a total colouring*, or simply, a total defining set, if there exists a unique extension of the colours of D to a $\chi''(G)$-colouring of the elements of G. A total defining set with minimum cardinality is called a *minimum total defining set*, denoted by $S''(G)$, and its cardinality is denoted by $d(G, \chi'')$. For the path P_n and the cycle C_{3n} this parameter is $d(P_n, \chi'') = d(C_{3n}, \chi'') = 2$.

Behzad, Mahmoodian, and Soltankhah [10] have studied the defining sets of total colouring of graphs and determined $d(G, \chi'')$ for some families of graphs as follows.

Theorem 4.6 *For cycles we have:*

(1) $d(C_n, \chi'') = 2$ *if* $n \equiv 0 \pmod 3$;
(2) $d(C_n, \chi'') = \lceil \frac{2n}{3} \rceil$ *if* $n \equiv 2 \pmod 3$;
(3) $d(C_n, \chi'') = \lceil \frac{2n}{3} \rceil + 1$ *if* $n \equiv 1 \pmod 3$.

Theorem 4.7 *Every total defining set for the graph K_{2n} is a total defining set for the graph K_{2n+1} as well.*

Theorem 4.8 *If $n > 6$, then for the wheel W_n, we have $d(W_n, \chi'') = 3n - 4$.*

4.3 Generalizations for $k > \chi(G)$

Morrill and Pritikin [113] have generalized the concept of defining sets of vertex colouring to any $k \geqslant \chi(G)$. They proved a more general theorem which leads to Theorem 2.2 as a corollary.

A *k-colouring* of a graph G is an assignment of k different colours to the vertices of G in such a way that no two adjacent vertices receive the same colour. In a given graph G, a set of vertices S with an assignment of colours is called a *defining set of the k-colouring* if there exists a unique extension of the colours of S to a k-colouring of the vertices of G. A defining set with minimum cardinality is called a *minimum defining set* and its cardinality is the *defining number*, denoted by $d(G, k)$.

In [98] this generalization is applied to analogous concepts in latin squares, raising such challenging problems as evaluating $d(K_n \times K_n, k)$ for $n \leqslant k \leqslant 2n - 1$. The case $k = 2n - 1$ has been completely answered in [98] as follows.

Theorem 4.9

$$d(K_n \times K_n, 2n-1) = \begin{cases} n^2 - n & \text{for } n \text{ even,} \\ n^2 - n + 1 & \text{for } n \text{ odd.} \end{cases}$$

The case $k = n$ is equivalent to finding $\mathrm{scs}(n)$ for latin squares. Consequently, Mahmoodian [100] and independently Bate and van Rees [7] have conjectured that $d(K_n \times K_n, n) = \lfloor \frac{n^2}{4} \rfloor$. The best general lower bound, due to Horak, Aldred and Fleischner [81], is $(4n - 8)/3$. For back circulant latin squares, Cavenagh [15] has the interesting lower bound $O(n^{4/3}/2)$ for large n.

4.3.1 Other related concepts

We mention here a few of the other papers related to the concept of defining sets in graph colourings. In [141] Zaker discusses the "greedy defining sets of graphs". Daneshgar introduces a related concept called the "fixing number of graphs" in a series of papers, some with other researchers, giving numerous results on this subject. See [30, 31, 32].

Another related subject introduced in [97] deals with the so-called uniquely 2-list colourable graphs. Let G be a graph with n vertices such that, for each vertex i, there is a list S_i of at least two colours available for use at that vertex. Then G has property $M(2)$, if either there is no proper colouring for G from these lists, or there are at least two proper colourings for G. A graph which is not $M(2)$ is called a *uniquely 2-list colourable graph.* For example, the graph $K_4 - e$ shown in Figure 17 is uniquely 2-list colourable graph.

The main result of [97], quoted below, has played a central role in the proof of many results in this area. For more results on the uniquely 2-list colourable graphs, see [51, 53].

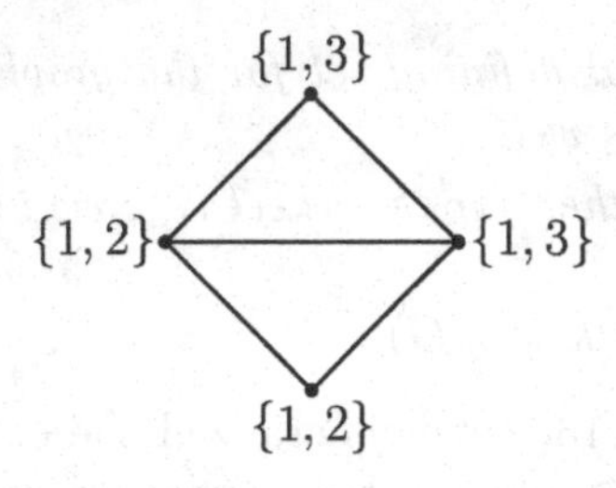

Figure 17: A uniquely 2-list colourable graph

Theorem 4.10 *A graph has property $M(2)$ if and only if each of its blocks is either a cycle, a complete graph, or a complete bipartite graph.*

A graph G is called *uniquely vertex colourable* if any colouring of its vertices with $\chi(G)$ colours gives rise to a unique partition of its vertices. If G is such a graph, clearly its smallest defining sets are of size $\chi(G) - 1$. Hajiabolhassan, Mehrabadi, Tusserkani and Zaker [76] gave the following characterization of uniquely vertex colourable graphs.

Theorem 4.11 ([76]) *If for a colouring of a graph G, all minimal defining sets are of size $\chi(G) - 1$, then G is a uniquely vertex colourable graph.*

4.4 Forcing sets in graphs

The idea of a defining set in a graph need not be restricted to colourings and has been extended to other properties exhibited by subsets of either the vertex set or the edge set of a graph. For instance, we may seek to identify a subset of edges which uniquely determines a perfect matching for a given graph, or we may seek a subset of vertices which uniquely determines a convex hull or a dominating set. In the literature defining sets which pertain to perfect matchings, strong or unilateral orientations, dominating sets, geodetic sets or convex hulls have been termed forcing sets; for consistency with the literature we adhere to this terminology.

Let $G = (V(G), E(G))$ be a connected graph on a finite set of vertices. Let $S \subseteq V(G)$ or $S \subseteq E(G)$, where S exhibits property P with respect to graph G. A subset $T \subseteq S$ is said to be a *forcing set* for S if, whenever $S' \subseteq V(G)$ or $S' \subseteq E(G)$ such that $T \subseteq S'$, and S' exhibits property P in G, then $S' = S$. The *forcing number* for set S with respect to property P is denoted $f(S,P)$ and is the minimum cardinality among the forcing sets of S. A forcing set of S of cardinality $f(S,P)$ is said to be a minimum forcing set with respect to property P. The *forcing number* for graph G is denoted $f(G,P)$ and is the smallest forcing number among the minimum forcing sets with respect to property P.

4.4.1 Matchings in graphs Let $G = (V(G), E(G))$ be a graph that admits a perfect matching M (that is, $M \subseteq E(G)$ such that if $v \in V(G)$ then v is incident with precisely one edge of M). As stated above a *forcing set* $T \subseteq M$ is

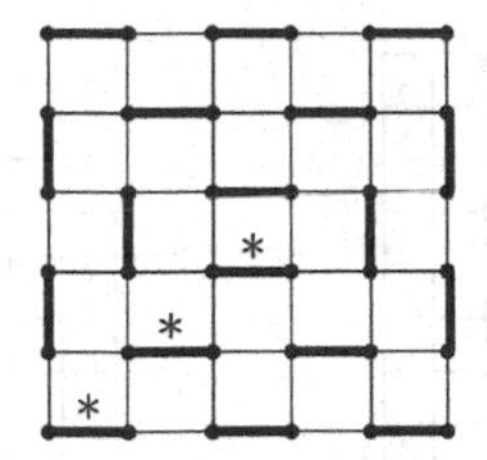

Figure 18: $f(P_6 \times P_6, M) = 3$

a subset of edges which is contained in no other perfect matching. The *forcing number* of a perfect matching M is denoted by $f(G, M)$, and is the smallest cardinality of any forcing set of M. (Note that this notation deviates slightly from the general format given above, but we keep to it to remain consistent with the literature.) The graph in Figure 18 is the grid $P_6 \times P_6$. The thick edges in the graph indicate a perfect matching M. The set of three edges indicated by '*' form a forcing set for the given matching. It can easily be shown that this is a forcing set of smallest cardinality, so $f(P_6 \times P_6, M) = 3$.

Let G be a graph, M a perfect matching in G and C a cycle in G. Denote the edge set of C by $E(C)$. If $M \cap E(C)$ is a perfect matching for C, then $M \cap E(C)$ is said to be an *alternating cycle* in M. The following useful proposition is the analogue for perfect matchings of Theorem 2.5; as noted in [117], it is easy to verify.

Lemma 4.12 *Let G be a graph and M a perfect matching in G. A subset T of M is a forcing set for M if and only if it contains at least one edge from each alternating cycle in M.*

In the introduction we mentioned that perfect matchings arise naturally in the study of Kekule structures, where the associated theory ([124, 89, 78]) involves a study of forcing sets of graphs similar to PAH, sometimes called 'polyhexes'. In [117], Pachter and Kim extend this concept to other graphs such as $2n \times 2n$ square grids and the hypercube Q_n. Their basic results are as follows.

Theorem 4.13 *For any perfect matching M and any positive integer n we have*

$$n \leqslant f(P_{2n} \times P_{2n}, M) \leqslant n^2.$$

Pachter and Kim [117] also show that both bounds given in Theorem 4.13 are sharp, and they raise some interesting conjectures on forcing numbers in a broader class of graphs. One such conjecture [116] is as follows.

Conjecture 4.14 *For every perfect matching M in Q_n, $f(Q_n, M) \geqslant 2^{n-2}$.*

Riddle [125] proved Conjecture 4.14 for any even number n and went on to prove the following.

Theorem 4.15 *The forcing number of a perfect matching on a $2n \times 2m$ torus $(C_{2n} \times C_{2m})$, with $m \geqslant n$, is at least $2n$. Furthermore, this bound is sharp.*

Another relevant conjecture made by Riddle [125] is the following.

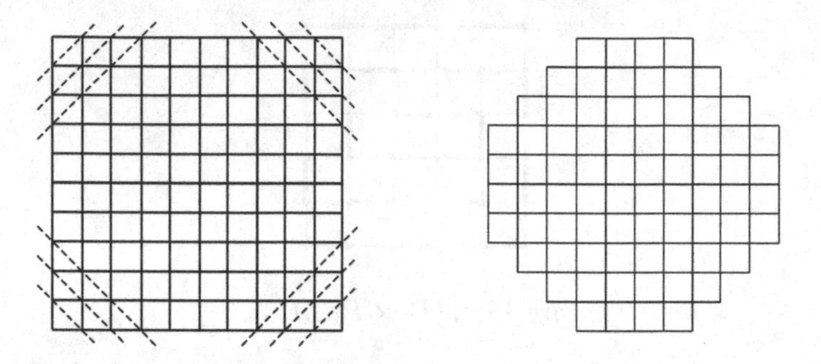

Figure 19: The $(5,3)$ stop sign obtained from the 10×10 square grid

Conjecture 4.16 *For every perfect matching M in the torus $(C_{2n} \times C_{2m})$, we have $f(C_{2n} \times C_{2m}, M) \leqslant mn$.*

An (n,k) *stop sign*, $k \leqslant n-1$, is a graph obtained from the $2n \times 2n$ square grid by deleting all of the squares lying along the k diagonals closest to each of the four corners. A $(5,3)$ stop sign is given in Figure 19. Lam and Pachter [92] found the forcing number of stop signs and went on to prove the following theorem.

Theorem 4.17 *For any perfect matching M in an (n,k) stop sign P we have*

$$n \leqslant f(P,M) \leqslant \begin{cases} \left(n - \frac{k}{2}\right)^2 & \text{for } k \text{ even,} \\ \left(n - \frac{k-1}{2}\right)\left(n - \frac{k+1}{2}\right) & \text{for } k \text{ odd.} \end{cases}$$

Recently Adams, Mahdian and Mahmoodian [1] discussed the problem of the *spectrum* for forcing numbers of perfect matchings, where the spectrum of G, $spec(G)$, is defined to be the set of k for which there is a perfect matching M of G with $f(G,M) = k$. It is natural to ask which finite subsets of the natural numbers are the spectra for forcing numbers of perfect matching for some graph.

For Q_n, one can easily find a perfect matching with a forcing number equal to 2^{n-2}. Indeed we have $spec(Q_n) = \{2^{n-2}\}$, for $n = 2$, 3 and 4. In [1] it is shown that $spec(Q_5) = \{8, 9\}$. The perfect matching M in Q_5, for which $f(Q_5, M) = 9$, was the first known example of a perfect matching in any Q_n with forcing number other than 2^{n-2}. These results enabled the authors to develop general constructions and to prove the following.

Theorem 4.18 *For each positive integer $n \geqslant 5$ and each $r \in \{2^{n-2}, 2^{n-2} + 1, \ldots, 2^{n-2} + 2^{n-5}\}$, there exists a perfect matching in Q_n with forcing number equal to r.*

In [4] the spectrum of possible forcing numbers for perfect matchings in the grids $P_m \times P_n$ was studied, as well as spectra for other specific classes of graphs. For certain classes of planar graphs, it was proved that there is no hole in the spectrum. More specifically the authors generalized Theorem 4.13 as follows.

Theorem 4.19 *For each positive integer n and each $r \in \{n, \dots, n^2\}$, there exists a perfect matching in $P_{2n} \times P_{2n}$ with forcing number equal to r.*

The computational complexity of finding the forcing number of a perfect matching for graphs was discussed in [1]. The following problem was studied and a proof for its NP-completeness given.

SMALLEST FORCING SET PROBLEM
INSTANCE: A graph G, a perfect matching M in G, and an integer k.
QUESTION: Is there any subset S of at most k edges of M, such that S is a forcing set for M?

Theorem 4.20 ([1]) SMALLEST FORCING SET PROBLEM *is NP-complete for bipartite graphs with maximum degree* 3.

The authors left an open question, subsequently answered in [4].

SMALLEST FORCING NUMBER OF GRAPH
INSTANCE: A graph G and an integer k.
QUESTION: Is there a perfect matching in G with forcing number at most k?

Theorem 4.21 SMALLEST FORCING NUMBER OF GRAPH *is NP-complete for bipartite graphs with maximum degree* 4.

We observe that there exists a polynomial time algorithm [117] for the problem of finding a smallest forcing set for a given perfect matching in a bipartite planar graph. This leads to several open questions.

4.5 Orientations in graphs

An *orientation* of a graph G is a digraph D, with the same vertex set, whose underlying graph is G. A *partial orientation* of an undirected graph G is a subset of the edges of an orientation of G.

A graph has a *strong orientation* if its edges can be oriented in such a way that it is strongly connected; that is, for every two vertices u and v there is a directed path from u to v and a directed path from v to u. A graph has a *unilateral orientation* if its edges can be oriented in such a way that for every two vertices u and v there is either a path from u to v or a path from v to u.

In [17], Chartrand, Harary, Schultz and Wall studied the concept of forcing sets for orientations of graphs. A partial orientation F is called a *strong orientation forcing set* (*unilateral orientation forcing set*) for a strong (unilateral) orientation D of a graph G, if D is the only strong (unilateral) orientation of G which contains F. The *strong orientation forcing number* of a strong orientation D of G is the minimal cardinality of a subset S of $E(G)$ for which there exists an assignment of directions to the edges of S which has a unique extension to the strong orientation D. The *unilateral orientation forcing number* of a unilateral orientation D of G is the minimal cardinality of a subset S of $E(G)$ for which there exists an assignment of directions to the edges of S which has a unique extension to the unilateral orientation D.

Graph G	$f(G, unilateral)$
$K_{2,n}$, where $n \geqslant 4$	n
$K_{2,3}$	4
$K_{3,3}$	7

Table 8: Some smallest unilateral orientation forcing numbers

Given a graph G we are interested in the *spectrum* of the strong and unilateral orientation forcing numbers. The *spectrum of strong orientation forcing numbers* is the set of all integers k for which G has an orientation with strong orientation forcing number k. The strong orientation *forcing number* of G, denoted $f(G, strong)$, is the cardinality of the smallest partial strong orientation which forms a forcing set for some strong orientation of G. The *spectrum of unilateral orientation forcing numbers* is the set of all integers k for which G has an orientation with forced unilateral orientation number k. The unilateral orientation *forcing number* of G, denoted $f(G, unilateral)$, is the cardinality of the smallest partial unilateral orientation which forms a forcing set for some unilateral orientation of G.

Robbins [126] showed that a graph has a strong orientation if and only if it is 2-edge-connected (connected and bridgeless). This result was used in [17] to obtain bounds on the size of the smallest strong orientation forcing number.

Theorem 4.22 *If G is a 2-edge-connected graph with n vertices and m edges, then $f(G, strong) = m - n + 1$.*

As a corollary to this result, it was proved in [17] that a 2-edge-connected graph G has smallest forcing number equal to one if and only if G is a cycle. Pascovici [118] obtained a similar result for unilateral orientation forcing numbers and proved the following theorem.

Theorem 4.23 *If G is a 1-edge-connected graph with n vertices and m edges, then $f(G, unilateral) = m - n + 2$.*

In [17] it was shown that for every positive integer k, there is a graph with smallest unilateral forcing number equal to k. In addition, $f(G, unilateral)$ was determined in certain cases and the following results for $f(G, unilateral) = 1$ and $f(G, unilateral) = 2$ were obtained.

Theorem 4.24 *Let G be a graph that has a unilateral orientation. Then:*
(1) $f(G, unilateral) = 1$ if and only if $G = P_n$ for some integer $n \geqslant 2$, and
(2) $f(G, unilateral) = 2$ if and only if G is a unicyclic graph that is not a cycle (and such that every bridge of G lies on a common path).

Pascovici [118] generalized these results for the $f(G, unilateral) = 3$ case and determined the unilateral orientation forcing numbers given in Table 8.

For a given graph G let $F(G, strong)$ denote the largest integer k for which G has a strong orientation with forcing number k. Farzad, Mahdian, Mahmoodian, Seberi and Sadri [46] have characterized graphs with strong orientation for which $F(G, strong)$ is as large as possible.

Theorem 4.25 *For a graph G with m edges we have $F(G, strong) = m$ if and only if G is a 4-edge-connected graph.*

Table 9 summarises results presented in [46] on the spectrum of strong orientation forcing numbers for certain graphs.

4.6 Dominating sets in graphs

A vertex of a graph G *dominates* itself and its neighbours. A set S of vertices of G is a *dominating set* if each vertex of G is dominated by some vertex of S. The *domination number*, $\gamma(G)$, of G is the minimum cardinality among the dominating sets of G. A *minimum dominating set* is one of cardinality $\gamma(G)$. A subset T of a minimum dominating set S is a *forcing subset for S* if S is the unique minimum dominating set containing T. The *forcing domination number* of S is denoted $f(S, \gamma)$, and is the minimum cardinality among the forcing subsets of S. The *forcing domination number* of G is denoted $f(G, \gamma)$, and is the minimum cardinality among the minmum dominating sets of G.

To illustrate the above definitions consider the Petersen graph, P, as given in Figure 20. Vertex v_1 dominates vertices v_2, v_5 and u_1, vertex v_4 dominates vertices v_3, v_5 and u_4, and vertex u_5 dominates vertices u_2, u_3 and v_5. Therefore the set of vertices $S_P = \{v_1, v_4, u_5\}$ is a dominating set for P. Each vertex of the Petersen graph has degree three, and so each vertex dominates exactly four vertices. Since the Petersen graph has ten distinct vertices, the smallest dominating set must contain at least three vertices, implying $\gamma(P) = 3$. Consider the subset $T_P = \{v_1, v_4\}$ of S_P. All vertices except u_2, u_3 and u_5 are dominated by T_P. To extend T_P to a minimum dominating set, we must adjoin at least one of the vertices u_2, u_3 or u_5 to T_P. To obtain a dominating set of cardinality three, u_5 must be adjoined to T_P. Thus T_P is a forcing set for S_P. Repeating this argument for the sets $\{u_5, v_4\}$ and $\{v_1, u_5\}$ confirms that $f(S_P, \gamma) = 2$.

Now consider any vertex of P. Without loss of generality we may choose v_1. Vertex v_1 dominates three vertices distinct from itself and when these vertices and associated edges are removed from P the remaining vertices and edges constitute a cycle of length six. Thus there are three distinct ways of choosing a dominating set for the remaining vertices, and so v_1 does not force

Graph G	Spectrum
K_3	1
K_4	3
K_n, where $n \geqslant 5$	$\binom{n}{2} - n + 1, \ldots, \binom{n}{2}$
$K_{2,2}$	1
$K_{2,n}$, where $n \geqslant 3$	$n-1, n$
$K_{3,3}$	$4, 6$
$K_{3,n}$, where $n \geqslant 4$	$2n-2, 2n-1, 2n$

Table 9: Spectra of strong orientation forcing numbers

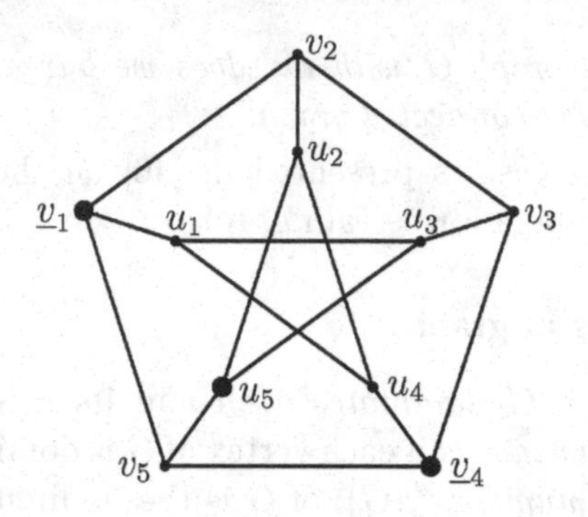

Figure 20: The Petersen graph with a forcing dominating set of size 2

a dominating set. Hence any forcing set for a dominating set containing vertex v_1 must contain at least two distinct vertices, implying $f(G, \gamma) = 2$.

Chartrand, Gavlas, Vandell and Harary in [16] studied the forcing domination number of a graph. For every graph G, $f(G, \gamma) \leqslant \gamma(G)$. In [16] it is shown that for integers a and b with $0 \leqslant a \leqslant b$, there exists a graph G such that $f(G, \gamma) = a$ and $\gamma(G) = b$. By viewing a graph G as a Cayley graph, the algebraic aspects of minimum dominating sets in G and forcing subsets are explored in [16]. The authors of [16] also obtain results on the forcing domination number of the Cartesian product G of k copies of the cycle C_{2k+1} and for specific graphs such as multipartite graphs, paths, cycles, ladders and prisms. These results are summarised in Table 10.

4.7 Geodetics and convex hulls in graphs

Let G be a connected graph and $u, v \in V(G)$. A u-v *geodesic* in G is a shortest path from u to v. The set $H(u, v)$ consists of all vertices lying on some u-v geodesic. For $S \subseteq V(G)$,

$$H(S) = \bigcup_{u,v \in S} H(u, v).$$

G	$\gamma(G)$	$f(G,\gamma)$	G	$\gamma(G)$	$f(G,\gamma)$
$K_{1,n}$	1	0	C_{3k}	k	1
P	3	2	C_{3k+1}	$k+1$	2
K_n	1	1	C_{3k+2}	$k+1$	2
$K_{1,p_2,\dots,p_k}$	1	0	$P_{2k-1} \times K_2$	k	1
$K_{1,1,p_3,\dots,p_k}$	1	1	$P_{2k} \times K_2$	$k+1$	2
$K_{p_1,p_2,\dots,p_k}$	2	2	$C_{4k} \times K_2$	$2k$	1
P_{3k}	k	0	$C_{4k+1} \times K_2$	$2k+1$	2
P_{3k+1}	$k+1$	2	$C_{4k+2} \times K_2$	$2k+2$	3
P_{3k+2}	$k+1$	1	$C_{4k+3} \times K_2$	$2k+2$	2

Table 10: Some forcing domination numbers

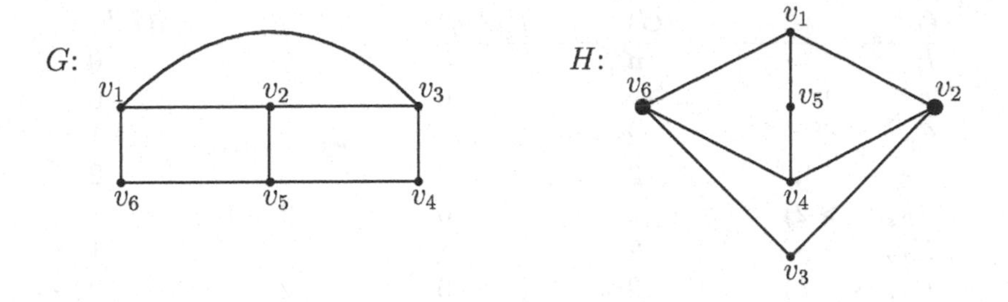

Figure 21: The graphs G and H

If $H(S) = V(G)$, then the set S is said to be a *geodetic set*. Let $S \subseteq V(G)$ be a geodetic set of minimum cardinality. In this case we say S is a *minimum geodetic set* and the *geodetic number* of G, denoted $g(G)$, equals $|S|$. A subset T of a minimum geodetic set S is called a *forcing subset* for S if S is the unique minimum geodetic set containing T. The minimum cardinality among all such forcing subsets of S is denoted $f(S, g)$. The smallest forcing geodetic number among all minimum geodetic sets is the *forcing geodetic number* of G, denoted $f(G, g)$.

A set $C \subseteq V(G)$ is *convex* if $H(C) = C$. The smallest convex set containing C is said to be the *convex hull* of C in G. The *hull number*, $h(G)$, of the graph G is the smallest cardinality of a set C whose convex hull is the entire graph G (see [45, 18]). In this case C is called a *minimum hull set* of G. A *forcing subset* of a minimum hull set C of G is a subset D of C, with the property that C is the unique minimum hull set containing D. The *forcing hull number* of C is denoted $f(C, h)$, and is the minimum cardinality of the forcing subsets for C. The *forcing hull number* of G is denoted $f(G, h)$ and is the smallest forcing hull number among all minimum hull sets of G.

To illustrate the above definitions consider the graph G given in Figure 21. Note that for every pair of vertices $u, v \in V(G)$ there exists a path of length 1 or 2 from u to v. Let u and v be two vertices with u-v geodesic of length 2. For each such pair there exists a third vertex $w \in V(G)$ such that the shortest path from u to v which is incident with w has length 3. Consequently $g(G) \geqslant 3$. There are three non-isomorphic minimum geodetic sets of cardinality three in G, namely $S_1 = \{v_2, v_4, v_6\}$, $S_2 = \{v_1, v_3, v_5\}$ and $S_3 = \{v_1, v_4, v_5\}$. Hence $g(G) = 3$. The vertex v_2 belongs to precisely one minimum geodetic set, and so $\{v_2\}$ is a forcing subset for S_1, implying $f(S_1, g) = 1$. All remaining vertices of G belong to more than one minimum geodetic set, giving $f(S_i, g) \geqslant 2$, $i = 2, 3$. Thus we deduce that $f(G, g) = 1$.

The convex hull of a set $C \subseteq V(G)$ may be found by evaluating the sequence $\{H^k(C) : k \geqslant 0\}$, where $H^0(C) = C$ and $H^k(C) = H(H^{k-1}(C))$, $k > 0$. Once again consider the graph G given in Figure 21. Take the set of vertices $\{v_2, v_5\}$. Then $H^0(\{v_2, v_5\}) = \{v_2, v_5\}$ and $H^1(\{v_2, v_5\}) = \{v_2, v_5\}$, and so $\{v_2, v_5\}$ is a convex hull, but not a minimum hull set. It can be shown that no minimum hull set of G has cardinality 1 or 2, implying $3 \leqslant h(G) \leqslant 6$. Alternatively, take the

G	$g(G)$	$f(G,g)$	$h(G)$	$f(G,h)$
K_n	n	0	n	0
T (tree*)	ω	0	ω	0
C_{2k}	2	1	2	1
C_{2k+1}	3	2	3	2
$K_{2,s}$ $(s > 2)$	2	0	2	1
$K_{2,2}$	2	1	2	1
$K_{3,s}$ $(s > 3)$	3	0	2	2
$K_{3,3}$	3	1	2	2
$K_{4,s}$ $(s \geqslant 4)$	4	3	2	2
$K_{5,s}$ $(s \geqslant 5)$	4	4	2	2

* ω represents the number of pendant vertices

Table 11: Some forcing geodetic and hull numbers

set of vertices $X = \{v_1, v_2, v_4\}$. Then $H^0(X) = X$, $H^1(X) = \{v_1, v_2, v_3, v_4, v_5\}$, and $H^2(X) = V(G)$. Thus $H(X) = V(G)$, implying X is a minimum hull set in G and $h(G) = 3$. Similarly, consider the set $Y = \{v_3, v_5, v_6\}$. $H^0(Y) = Y$, $H^1(Y) = \{v_1, v_2, v_3, v_4, v_5, v_6\}$, or $H^1(Y) = V(G)$. Thus Y is a minimum hull set in G. Since these two minimum hull sets are disjoint, $f(G, h) \geqslant 1$. However, for each $z \in \{v_1, v_2, \ldots, v_6\}$, there exists distinct minimum hull sets C_1 and C_2 such that $z \in C_1 \cap C_2$, and so $f(G, h) \geqslant 2$. In addition, it can be seen that $\{v_1, v_4, v_5\}$ and $\{v_2, v_4, v_6\}$ are both minimum hull sets. Consequently, $\{v_1, v_4\}$ and $\{v_2, v_4\}$ are not forcing sets for X. In contrast the only minimum hull set containing the vertices v_1 and v_2 is X, implying $\{v_1, v_2\}$ is a forcing set for X. Consequently $f(X, h) = f(G, h) = 2$. Also note that $\{v_5, v_6\}$ is a forcing set for Y, giving $f(Y, h) = 2$.

Alternatively, take the graph H presented in Figure 21. The sets $\{v_1, v_3\}$ and $\{v_2, v_6\}$ are examples of minimum hull sets. Since there are other examples of minumum hull sets containing the vertices v_1 or v_3, then $f(\{v_1, v_3\}, h) = 2$. In contrast, the only minimum hull set containing the vertices v_2 or v_6 is the set $\{v_2, v_6\}$. Hence $f(\{v_2, v_6\}, h) = 1$ and so $f(H, h) = 1$.

Chartrand and Zhang [19, 20] studied the forcing geodetic numbers and forcing hull number of graphs. They proved the following general result, and documented further results for specific graphs. These results are summarised in Theorem 4.26 and Table 11.

Theorem 4.26 *For all integers a and b, with $0 \leqslant a \leqslant b$:*
(1) a connected graph G such that $f(G, g) = a$ and $g(G) = b$ exists if and only if $(a, b) \notin \{(1, 1), (2, 2)\}$;
(2) there exists a graph G such that $f(G, h) = a$ and $h(G) = b$.

5 Future directions and open problems

Topics discussed here belong to a general class of questions, all essentially asking how much of a particular structure is enough to specify it uniquely. Many problems relating to defining sets remain open; we mention a few of them.

A: Combinatorial structures whose defining sets have not yet been studied in general include the following:
(1) group divisible designs [104];
(2) large and overlarge sets of designs [94, 35, 36];
(3) directed designs, such as Mendelsohn and transitive triple systems: see [104, 107, 26, 12];
(4) G-designs [14], including cycle systems [96];
(5) F-squares [8, 47, 48], Youden squares [127] and Room squares [22, 21].

B: Open problems relating to defining sets in graphs include the following:
(1) finding possible values of the defining number, $d(G, \chi)$, of the vertex colouring of a graph G, perhaps a graph with special properties;
(2) finding the relation between the values of this defining number for a graph and for its subgraphs;
(3) answering analogous questions for edge and total colourings of graphs;
(4) generalizing Theorem 4.10 for uniquely k-list colourable graphs;
(5) answering a challenging problem related to Theorem 4.9, namely, evaluating $d(K_n \times K_n, k)$ for $n \leq k \leq 2n - 1$;
(6) deciding in particular whether the conjecture made independently in [100] and in [8] is correct, that is, whether $d(K_n \times K_n, n) = \lfloor \frac{n^2}{4} \rfloor$;
(7) finding possible values of forcing numbers of perfect matchings for graphs;
(8) finding results like those in Tables 8, 9, 10 and 11, for other special graphs.

C: Open problems relating to defining sets in designs include the following.

Gower showed that, if a partial latin square (squag or sloop) is constructed from a set D of blocks of an $STS(v)$, and if that partial latin square has unique completion, then the set D is a defining set of the $STS(v)$; see Figure 13 for example. The converse of this theorem is not true in general. Under what conditions *might* it be true?

Generalizing Gower's technique [55, 56] has led to new minimal defining sets in Steiner triple systems related to geometries, and in non-affine Hall triple systems; see Section 3.5. Can her approach be generalized to deal with additional classes of Steiner triple systems?

Bounds on the possible sizes of the minimal defining set of a t-design are still weak, and the question of whether the spectrum of such minimal defining sets of a design may have a hole is still undecided; see Section 2.3.2 and [64]. In particular, these questions have not been settled even for the size of the smallest defining set of the $STS(2^{n+1}-1)$ isomorphic to the point-line design of $PG(n, 2)$, nor for the $STS(3^n)$ isomorphic to the point-line design of $AG(n, 3)$.

Work of K. Gray [68, 66] laid the foundations for the study of the relationships between the smallest defining sets of a design, of its extensions and of its residuals. Greenhill [71] built on Gray's results in developing her original algorithm for finding defining sets of designs. Subsequently Gray's results were used extensively by Moran [111, 112, 110] in determining smallest defining sets of several families of designs. For example, starting from known smallest defining sets of the 2-$(21, 5, 1)$ and 5-$(24, 8, 1)$ designs, Moran filled in the information about the smallest defining sets of 3-$(22, 6, 1)$ and 4-$(23, 7, 1)$ designs, thus giving a new proof of the result of Curtis [29]. It seems unlikely that these ideas have been fully used yet.

Hartman and Yehudai [79] discussed the construction of Steiner designs by greedy algorithms, and the starting of the algorithm with a non-empty partial design, called the *hint*. The *greediness* is defined to be the minimal size of a hint which, in our context, is a form of defining set. This gives a concise method of describing a design, while avoiding the computational complexity of completing it. Properties of hints present interesting questions.

Acknowledgements

Two of the authors (ESM, APS) thank the Department of Mathematics, The University of Queensland and the Department of Discrete and Statistical Sciences, Auburn University, respectively, for hospitality during part of the time that this paper was being written. This project was supported by Australian Research Grants DP0210416, DP0344078, A49801961, A49801415 and A49937047.

References

[1] P. Adams, M. Mahdian and E.S. Mahmoodian, On the forced matching number of graphs, *Discrete Math.*, in press.

[2] Peter Adams, Elizabeth J. Billington and C.A. Rodger, Pasch decompositions of lambda-fold triple systems, *J. Combin. Math. Combin. Comput.* **15** (1994), 53–63.

[3] Peter Adams, Abdollah Khodkar and Colin Ramsay, Smallest defining sets of some $STS(19)$, *J. Combin. Math. Combin. Comput.* **38** (2001), 225–230.

[4] P. Afshani, H. Hatami and E.S. Mahmoodian, On the spectrum of the forced matching number of graphs, *Australas. J. Combin.*, in press.

[5] Ian Anderson, *Combinatorial Designs and Tournaments*, Clarendon Press, Oxford (1997).

[6] Peter Barth, A Davis-Putman based enumeration algorithm for linear pseudo-boolean optimisation, Technical Report MPI-I-95-2-003, Max-Planck-Institut für Informatik, 1995.

[7] J.A. Bate and G.H.J. van Rees, The size of the smallest strong critical set in a Latin square, *Ars Combin.* **53** (1999), 73–83.

[8] J.A. Bate and G.H.J. van Rees, A note on critical sets, *Australas. J. Combin.* **25** (2002), 299–302.

[9] David Bedford and David Whitehouse, Products of uniquely completable partial latin squares, *Util. Math.* **58** (2000), 195–201.

[10] M. Behzad, E.S. Mahmoodian and N. Soltankhah, Defining sets in total coloring of graphs, in *Proceedings of the 28th Annual Iranian Mathematics Conference, Part 1 (Tabriz, 1997)*, Tabriz University, Tabriz (1997), pp. 521–526.

[11] Lucien Beneteau, Hall triple systems, in Colbourn and Dinitz [26], pp. 377–380.

[12] Frank E. Bennett and Alireza Mahmoodi, Directed designs, in Colbourn and Dinitz [26], pp. 317–321.

[13] Damian P. Birney and Graeme S. Halford, The psychometric properties of relational complexity in adult cognition, in *XVI British Psychological Society Cognitive Section Conference (poster session)*, University of Essex (2000).

[14] Darryn E. Bryant and Barbara M. Maenhaut, Defining sets of G-designs, *Australas. J. Combin.* **17** (1998), 257–266.

[15] Nicholas J. Cavenagh, A lower bound for the size of a critical set in the back circulant latin square, submitted for publication.

[16] Gary Chartrand, Heather Gavlas, Robert C. Vandell and Frank Harary, The forcing domination number of a graph, *J. Combin. Math. Combin. Comput.* **25** (1997), 161–174.

[17] Gary Chartrand, Frank Harary, Michelle Schultz and Curtiss E. Wall, Forced orientation numbers of a graph, *Congr. Numer.* **100** (1994), 183–191.

[18] Gary Chartrand, Frank Harary and Ping Zhang, On the hull number of a graph, *Ars Combin.* **57** (2000), 129–138.

[19] Gary Chartrand and Ping Zhang, The forcing geodetic number of a graph, *Discuss. Math. Graph Theory* **19** (1999), 45–58.

[20] Gary Chartrand and Ping Zhang, The forcing hull number of a graph, *J. Combin. Math. Combin. Comput.* **36** (2001), 81–94.

[21] Ghulam Chaudhry and Jennifer Seberry, On the influence of entries in critical sets of Room squares, *Bull. Inst. Combin. Appl.* **28** (2000), 67–74.

[22] Ghulam R. Chaudhry and Jennifer Seberry, Minimal critical set of a Room square of order 7, *Bull. Inst. Combin. Appl.* **20** (1997), 90.

[23] Charles J. Colbourn, Embedding partial Steiner triple systems is NP-complete, *J. Combin. Theory Ser. A* **35** (1983), 100–105.

[24] C.J. Colbourn, The complexity of completing partial Latin squares, *Discrete Appl. Math.* **8** (1984), 25–30.

[25] C.J. Colbourn, M.J. Colbourn and D.R. Stinson, The computational complexity of recognizing critical sets, in *Graph Theory (Singapore, 1983)* (eds. K.M. Koh and H.P. Yap), *Lecture Notes in Mathematics*, 1073, Springer-Verlag, Berlin (1984), pp. 248–253.

[26] C.J. Colbourn and J.H. Dinitz, eds., *CRC Handbook of Combinatorial Designs*, CRC Publishing Co., Boca Raton, Florida (1996). See also `http://www.emba.uvm.edu/~dinitz/hcd.html`.

[27] Combinatorial Mathematics Society of Australasia, Recent conference reports, *E-newsletter* **10** (October 2002). Available electronically via `http://www.maths.uq.edu.au/~ejb/cmsa-newsletters.html`.

[28] CPLEX callable library, CPLEX Optimization Inc., 1989–1995.

[29] R.T. Curtis, Eight octads suffice, *J. Combin. Theory Ser. A* **36** (1984), 116–123.

[30] A. Daneshgar, Forcing structures and uniquely colorable graphs, in *Proceedings of the 28th Annual Iranian Mathematics Conference, Part 1 (Tabriz, 1997)*, Tabriz University, Tabriz (1997), pp. 121–128.

[31] Amir Daneshgar, On r-type-constructions and Δ-colour-critical graphs, *J. Combin. Math. Combin. Comput.* **29** (1999), 183–206.

[32] Amir Daneshgar and Reza Naserasr, On small uniquely vertex-colourable graphs and Xu's conjecture, *Discrete Math.* **223** (2000), 93–108.

[33] Catherine M. Delaney, Computational aspects of defining sets of combinatorial designs, Master's thesis, University of Queensland (1995).

[34] Cathy Delaney, Brenton D. Gray, Ken Gray, Barbara M. Maenhaut, Martin J. Sharry and Anne Penfold Street, Pointwise defining sets and trade cores, *Australas. J. Combin.* **16** (1997), 51–76.

[35] Cathy Delaney, Adelle M. Howse, Julie L. Lawrence, Monica Mangold, Martin J. Sharry and Anne Penfold Street, Defining sets of large sets of designs, talk presented at 23rd ACCMCC, Brisbane, 1998.

[36] Cathy Delaney, Adelle M. Howse, Julie L. Lawrence and Anne Penfold Street, Defining sets of overlarge sets of designs, talk presented at 23rd ACCMCC, Brisbane, 1998.

[37] József Dénes, On a problem of L. Fuchs, *Acta Sci. Math. (Szeged)* **23** (1962), 237–241.

[38] Diane Donovan, Abdollah Khodkar and Anne Penfold Street, On doubling and tripling constructions for defining sets in Steiner triple systems, *Graphs Combin.*, in press.

[39] Diane Donovan, Abdollah Khodkar and Anne Penfold Street, Affine geometries over $GF[3]$: their minimal defining sets, in *Designs 2002: Further Computational and Constructive Design Theory* (ed. W.D. Wallis), Kluwer Acad. Publ., Dordrecht (2002), pp. 103–131.

[40] Diane Donovan, Sheila Oates-Williams and Cheryl E. Praeger, On the distance between distinct group Latin squares, *J. Combin. Des.* **5** (1997), 235–248.

[41] A. Drápal and T. Kepka, Group modifications of some partial groupoids, *Ann. Discrete Math.* **18** (1983), 319–332.

[42] Aleš Drápal, On quasigroups rich in associative triples, *Discrete Math.* **44** (1983), 251–265.

[43] Aleš Drápal, How far apart can the group multiplication tables be?, *European J. Combin.* **13** (1992), 335–343.

[44] Konrad Engel, The number of indecomposable designs is finite, *Ars Combin.* **17** (1984), 33–34.

[45] Martin G. Everett and Stephen B. Seidman, The hull number of a graph, *Discrete Math.* **57** (1985), 217–223.

[46] B. Farzad, M. Mahdian, E.S. Mahmoodian, A. Saberi and B. Sadri, Forced orientation of graphs, submitted for publication.

[47] L.F. Fitina and Jennifer Seberry, On the spectrum of an F-square, *Australas. J. Combin.* **22** (2000), 81–89.

[48] L.F. Fitina, Jennifer Seberry and Dinesh Sarvate, On F-squares and their critical sets, *Australas. J. Combin.* **19** (1999), 209–230.

[49] Sophie Frisch, On the minimal distance between group tables, *Acta Sci. Math. (Szeged)* **63** (1997), 341–351.

[50] L. Fuchs, *Abelian Groups*, Hungarian Academy of Sciences, Budapest (1958).

[51] Y.G. Ganjali, M. Ghebleh, H. Hajiabolhassan, M. Mirzazadeh and B.S. Sadjad, Uniquely 2-list colorable graphs, *Discrete Appl. Math.* **119** (2002), 217–225.

[52] Bernhard Ganter and Heinrich Werner, Co-ordinatizing Steiner systems, *Ann. Discrete Math.* **7** (1980), 3–24.

[53] M. Ghebleh and E.S. Mahmoodian, On uniquely list colorable graphs, *Ars Combin.* **59** (2001), 307–318.

[54] P.B. Gibbons, Computational methods in design theory, in Colbourn and Dinitz [26], pp. 718–753.

[55] Rebecca A.H. Gower, Minimal defining sets in a family of Steiner triple systems, *Australas. J. Combin.* **8** (1993), 55–73.

[56] Rebecca A.H. Gower, Defining sets for the Steiner triple systems from affine spaces, *J. Combin. Des.* **5** (1997), 155–175.

[57] R.L. Graham, S-Y.R. Li and W-C.W. Li, On the structure of t-designs, *SIAM J. Algebraic Discrete Methods* **1** (1980), 8–14.

[58] M.J. Grannell, T.S. Griggs and J. Wallace, The smallest defining set of a Steiner triple system, *Util. Math.* **55** (1999), 113–121.

[59] J.E. Graver and W.B. Jurkat, The module structure of integral designs, *J. Combin. Theory Ser. A* **15** (1973), 75–90.

[60] Brenton D. Gray, The maximum number of trades of volume four in a symmetric design, *Util. Math.* **52** (1997), 193–203.

[61] Brenton D. Gray, Smallest defining sets of designs associated with $PG(d, 2)$, *Australas. J. Combin.* **16** (1997), 87–98.

[62] Brenton D. Gray, Defining sets of projective planes and biplanes and their residuals, *J. Combin. Math. Combin. Comput.* **30** (1999), 171–193.

[63] Brenton D. Gray, Nicholas Hamilton and Christine M. O'Keefe, On the size of a smallest defining set of $PG(2, q)$, *Bull. Inst. Combin. Appl.* **21** (1997), 91–94.

[64] Brenton D. Gray, Rudolf Mathon, Tony Moran and Anne Penfold Street, The spectrum of minimal defining sets of some Steiner systems, *Discrete Math.* **261** (2003), 277–284.

[65] Brenton D. Gray and Colin Ramsay, Some results on defining sets of t-designs, *Bull. Austral. Math. Soc.* **59** (1999), 203–215.

[66] Ken Gray, Defining sets of single-transposition-free designs, *Util. Math.* **38** (1990), 97–103.

[67] Ken Gray, Further results on smallest defining sets of well-known designs, *Australas. J. Combin.* **1** (1990), 91–100.

[68] Ken Gray, On the minimum number of blocks defining a design, *Bull. Austral. Math. Soc.* **41** (1990), 97–112.

[69] Ken Gray and Anne Penfold Street, Smallest defining sets of the five non-isomorphic 2-$(15,7,3)$ designs, *Bull. Inst. Combin. Appl.* **9** (1993), 96–102.

[70] Ken Gray and Anne Penfold Street, The smallest defining set of the 2-$(15,7,3)$ design associated with $PG(3,2)$: a theoretical approach, *Bull. Inst. Combin. Appl.* **11** (1994), 77–83.

[71] Catherine S. Greenhill, An algorithm for finding smallest defining sets of t-designs, *J. Combin. Math. Combin. Comput.* **14** (1993), 39–60.

[72] Catherine S. Greenhill and Anne Penfold Street, Smallest defining sets of some small t-designs and relations to the Petersen graph, *Util. Math.* **48** (1995), 5–31.

[73] Catherine Suzanne Greenhill, An algorithm for finding smallest defining sets of t-designs, Master's thesis, University of Queensland (1993).

[74] Hans-Dietrich O.F. Gronau, A survey of results on the number of t-(v,k,λ) designs, *Ann. Discrete Math.* **26** (1985), 209–220.

[75] Martin Grüttmüller, On the number of indecomposable block designs, *Australas. J. Combin.* **14** (1996), 181–186.

[76] H. Hajiabolhassan, M.L. Mehrabadi, R. Tusserkani and M. Zaker, A characterization of uniquely vertex colorable graphs using minimal defining sets, *Discrete Math.* **199** (1999), 233–236.

[77] Frank Harary, Three new directions in graph theory, in *Proceedings of the First Estonian Conference on Graphs and Applications (Tartu-Kääriku, 1991)*, Tartu University, Tartu (1993), pp. 15–19.

[78] Frank Harary, Douglas J. Klein and Tomislav P. Živković, Graphical properties of polyhexes: perfect matching vector and forcing, *J. Math. Chem.* **6** (1991), 295–306.

[79] Alan Hartman and Zvi Yehudai, Greedesigns, *Ars Combin.* **29**(C) (1990), 69–76.

[80] A.S. Hedayat, The theory of trade-off for t-designs, in *Coding Theory and Design Theory, Part II: Design Theory* (ed. D.K. Ray-Chaudhuri), IMA Vol. Math. Appl., 21, Springer, New York (1990) pp. 101–126.

[81] P. Horak, R.E.L. Aldred and H. Fleischner, Completing Latin squares: critical sets, *J. Combin. Des.* **10** (2002), 419–432.

[82] H.L. Hwang, On the structure of (v, k, t) trades, *J. Statist. Plann. Inference* **13** (1986), 179–191.

[83] A.D. Keedwell, Critical sets in latin squares and related matters: an update, preprint.

[84] A.D. Keedwell, Critical sets for Latin squares, graphs and block designs: a survey, *Congr. Numer.* textbf113 (1996), 231–245.

[85] A. Khodkar, Smallest defining sets for the 36 non-isomorphic twofold triple systems of order nine, *J. Combin. Math. Combin. Comput.* **17** (1995), 209–215.

[86] G.B. Khosrovshahi and S. Ajoodani-Namini, A new basis for trades, *SIAM J. Discrete Math.* **3** (1990), 364–372.

[87] G.B. Khosrovshahi and H.R. Maimani, Smallest defining sets for 2-$(10, 5, 4)$ designs, *Australas. J. Combin.* **15** (1997), 31–35.

[88] G.B. Khosrovshahi and F. Vatan, On 3-$(8, 4, \lambda)$ designs, *Ars Combin.* **32** (1991), 115–120.

[89] D.J. Klein and M. Randić, Innate degree of freedom of a graph, *J. Comput. Chem.* **8** (1987), 516–521.

[90] Donald L. Kreher and Douglas R. Stinson, *Combinatorial Algorithms: generation, enumeration and search*, CRC Press, Boca Raton, Florida (1999).

[91] Thomas Kunkle and Jennifer Seberry, A few more small defining sets for $SBIBD(4t-1, 2t-1, t-1)$, *Bull. Inst. Combin. Appl.* **12** (1994), 61–64.

[92] Fumei Lam and Lior Pachter, Forcing numbers for stop signs, preprint.

[93] I. Langdev, Nondecomposable block designs with $v = 7$ and $\lambda = 3$ (Bulgarian), in *Mathematics and mathematical education (Albena, 1989)* (ed. G. Gerov), Bulgarian Academy of Sciences, Sofia (1989), pp. 386–389.

[94] Julie L. Lawrence, Algorithms for combinatorics, Master's thesis, University of Queensland, in preparation.

[95] Julie L. Lawrence, Computational techniques for finding smallest defining sets of some discrete structures, talk presented at 23rd ACCMCC, Brisbane, 1998.

[96] Barbara M. Maenhaut, Substructures of cycle systems with applications to access schemes, Ph.D. thesis, University of Queensland (1999).

[97] M. Mahdian and E.S. Mahmoodian, A characterization of uniquely 2-list colorable graphs, *Ars Combin.* **51** (1999), 295–305.

[98] M. Mahdian and E.S. Mahmoodian, The roots of an IMO97 problem, *Bull. Inst. Combin. Appl.* **28** (2000), 48–54.

[99] M. Mahdian, E.S. Mahmoodian, R. Naserasr and F. Harary, On defining sets of vertex colorings of the cartesian product of a cycle with a complete graph, in *Combinatorics, Graph Theory and Algorithms* (eds. Y. Alavi, D.R. Lick and A. Schwenk), New Issues Press, Kalamazoo (1999), pp. 461–467.

[100] E.S. Mahmoodian, Some problems in graph colorings, in *Proceedings of the 26th Annual Iranian Mathematics Conference, Vol. 2 (Kerman, 1995)* (eds. S. Javadpour and M. Radjabalipour), Iranian Mathematical Society, University of Kerman, Kerman (1995), pp. 215–218.

[101] E.S. Mahmoodian, Defining sets and uniqueness in graph colorings: a survey, *J. Statist. Plann. Inference* **73** (1998), 85–89.

[102] E.S. Mahmoodian and E. Mendelsohn, On defining numbers of vertex colouring of regular graphs, *Discrete Math.* **197/198** (1999), 543–554.

[103] E.S. Mahmoodian, Reza Naserasr and Manouchehr Zaker, Defining sets in vertex colorings of graphs and Latin rectangles, *Discrete Math.* **167/168** (1997), 451–460.

[104] E.S. Mahmoodian and M.A. Shokrollahi, Open problems at the combinatorics workshop of AIMC25 (Tehran, 1994), in *Combinatorics Advances* (eds. C.J. Colbourn and E.S. Mahmoodian), Kluwer Acad. Publ., Dordrecht (1995), pp. 321–324.

[105] E.S. Mahmoodian and N. Soltankhah, On defining numbers of k-regular k-chromatic graphs, submitted for publication.

[106] E.S. Mahmoodian and Nasrin Soltankhah, On the existence of (v,k,t) trades, *Australas. J. Combin.* **6** (1992), 279–291.

[107] E.S. Mahmoodian, Nasrin Soltankhah and Anne Penfold Street, On defining sets of directed designs, *Australas. J. Combin.* **19** (1999), 179–190.

[108] Rudolf Mathon and Alexander Rosa, 2-(v, k, λ) designs of small order, in Colbourn and Dinitz [26], pp. 3–41.

[109] Brendan D. McKay, *nauty* user's guide (version 1.5), Technical Report TR-CS-90-02, Department of Computer Science, Australian National University, 1990.

[110] Anthony Terence Moran, Block designs and their defining sets, Ph.D. thesis, University of Queensland (1997).

[111] Tony Moran, Smallest defining sets for 2-$(9, 4, 3)$ and 3-$(10, 5, 3)$ designs, *Australas. J. Combin.* **10** (1994), 265–288. Corrigendum, *ibid.* **14** (1996), 311–312.

[112] Tony Moran, Defining sets for 2-$(19, 9, 4)$ designs and a class of Hadamard designs, *Util. Math.* **55** (1999), 161–187.

[113] T. Morrill and D. Pritikin, Defining sets and list-defining sets in graphs, preprint, 1997.

[114] J. Nelder, Critical sets in Latin squares, *Newsletter (CSIRO Division of Mathematics and Statistics)* **38** (1977).

[115] B. Omoomi, New bound for defining numbers of regular graphs, preprint.

[116] L. Pachter, Domino tiling, gene recognition, and mice, Ph.D. thesis, MIT (1999).

[117] Lior Pachter and Peter Kim, Forcing matchings on square grids, *Discrete Math.* **190** (1998), 287–294.

[118] Dana Pascovici, On the forced unilateral orientation number of a graph, *Discrete Math.* **187** (1998), 171–183.

[119] Colin Ramsay, The IQ-Block: an interesting polyomino puzzle, *J. Recreational Math.*, in press.

[120] Colin Ramsay, An algorithm for completing partials, with an application to the smallest defining sets of the $STS(15)$, *Util. Math.* **52** (1997), 205–221.

[121] Colin Ramsay, An algorithm for enumerating trades in designs, with an application to defining sets, *J. Combin. Math. Combin. Comput.* **24** (1997), 3–31.

[122] Colin Ramsay, On a family of $STS(v)$ whose smallest defining sets contain at least $b/4$ blocks, *Bull. Inst. Combin. Appl.* **20** (1997), 91–94.

[123] Colin Ramsay, Trades and defining sets: theoretical and computational results, Ph.D. thesis, University of Queensland (1998).

[124] M. Randić and D.J. Klein, Kekule valence structures revisited, in *Innate degrees of freedom of π-electron couplings in mathematical and computational concepts in chemistry* (ed. N. Trinajstic), Wiley, New York (1985).

[125] M.E. Riddle, The minimum forcing number for the torus and hypercube, *Discrete Math.* **245** (2002), 283–292.

[126] H.E. Robbins, A theorem on graphs, with an application to a problem in traffic control, *Amer. Math. Monthly* **46** (1939), 281–283.

[127] Kenneth Russell and Jennifer Seberry, The minimum critical set of a class of Youden squares, talk presented at 23rd ACCMCC, Brisbane, 1998.

[128] Herbert John Ryser, *Combinatorial Mathematics*, number 14 in Carus Math. Monogr., Mathematical Association of America Inc. (1963).

[129] Dinesh Sarvate and Jennifer Seberry, A note on small defining sets for some $SBIBD(4t-1, 2t-1, t-1)$, *Bull. Inst. Combin. Appl.* **10** (1994), 26–32.

[130] Jennifer Seberry, On small defining sets for some $SBIBD(4t-1, 2t-1, t-1)$, *Bull. Inst. Combin. Appl.* **4** (1992), 58–62. Corrigendum, *ibid.* **6** (1992), 62.

[131] Jennifer Seberry and Anne Penfold Street, Strongbox secured secret sharing schemes, *Util. Math.* **57** (2000), 147–163.

[132] Martin J. Sharry and Anne Penfold Street, Another look at large sets of Steiner triple systems, in *Computational and Constructive Design Theory* (ed. W.D. Wallis), Kluwer Acad. Publ., Dordrecht (1996), pp. 255–335.

[133] Martin James Sharry, Large and overlarge sets of block designs, Ph.D. thesis, University of Queensland (1992).

[134] N. Soltankhah and B. Omoomi, Defining numbers of some regular graphs, preprint.

[135] N. Soltankhah, B. Omoomi and E.S. Mahmoodian, On the defining number of r-regular k-chromatic graphs, submitted for publication.

[136] R.G. Stanton and Anne Penfold Street, Three results on symmetric balanced incomplete block designs, *Bull. Inst. Combin. Appl.* **1** (1991), 93–96.

[137] D.R. Stinson and G.H.J. van Rees, Some large critical sets, *Congr. Numer.* **34** (1982), 441–456.

[138] Anne Penfold Street, Defining sets for block designs: an update, in *Combinatorics Advances* (eds. C.J. Colbourn and E.S. Mahmoodian), Kluwer Acad. Publ., Dordrecht (1995), pp. 307–320.

[139] Anne Penfold Street, Trades and defining sets, in Colbourn and Dinitz [26], pp. 474–478.

[140] Douglas B. West, *Introduction to Graph Theory*, Prentice-Hall (2nd edition, 2001).

[141] Manouchehr Zaker, Greedy defining sets of graphs, *Australas. J. Combin.* **23** (2001), 231–235.

Centre for Discrete Mathematics and Computing
The University of Queensland
Brisbane
Queensland 4072, Australia
dmd@maths.uq.edu.au
cram@itee.uq.edu.au
aps@maths.uq.edu.au

Department of Mathematical Sciences
Sharif University of Technology
P.O. Box 11365 - 9415
Tehran, I.R. Iran
emahmood@sina.sharif.edu

Finite projective planes with a large abelian group

Dina Ghinelli and Dieter Jungnickel

Abstract

Let Π be a finite projective plane of order n, and let G be a large abelian (or, more generally, quasiregular) collineation group of Π; to be specific, we assume $|G| > (n^2 + n + 1)/2$. Such planes have been classified into eight cases by Dembowski and Piper in 1967. We survey the present state of knowledge about the existence and structure of such planes. We also discuss some geometric applications, in particular to the construction of arcs and ovals. Technically, a recurrent theme will be the amazing strength of the approach using various types of difference sets and the machinery of integral group rings.

1 Introduction

A *projective plane* is a geometry consisting of points and lines such that any two distinct lines meet in exactly one point, any two distinct points are on exactly one common line, and there are four points no three of which are collinear. Well-known examples are provided by the *desarguesian* (or *classical*) projective planes $\mathrm{PG}(2, q)$, q a prime power, which can be defined over the finite field $\mathrm{GF}(q)$ as follows: points and lines are the 1- and 2-dimensional subspaces of the vector space $\mathrm{GF}(q)^3$, and a point is incident with a line if and only if it is a subset of the line. A standard reference for arbitrary projective planes is Hughes and Piper [104]; for the desarguesian case, see Hirschfeld [89]. For finite geometries in general, the reader is referred to Dembowski [46] and Beth, Jungnickel and Lenz [23].

In the case of a finite projective plane, it can be shown that the number of points is $n^2 + n + 1$ for some $n \geqslant 2$ which is called the *order* of the plane. Also, each line has $n + 1$ points, each point is on $n + 1$ lines, and there are $n^2 + n + 1$ lines. In particular, the plane $\mathrm{PG}(2, q)$ has order q. Thus projective planes of order n exist for all prime powers n, but no example is known for any other value of n. This motivates the following longstanding conjecture.

Conjecture 1.1 (The prime power conjecture – PPC)
A projective plane of order n exists if and only if n is a prime power.

The nonexistence of a projective plane of order n is known for all $n \equiv 1$ or 2 (mod 4) which are not the sum of two squares (Bruck–Ryser theorem [36]), for $n = 10$ (see Lam, Thiel and Swiercz [131]) and for no other values of n. For proper prime powers $q = p^c$, one knows many projective planes besides the desarguesian ones. This is not the case for prime orders p which has led to the following conjecture.

Conjecture 1.2 *Let Π be a projective plane of order p, where p is a prime. Then Π is isomorphic to* $\mathrm{PG}(2,p)$.

Trying to prove the preceding conjectures in general seems hopeless with the present methods of mathematics; thus it is natural to add extra assumptions, for instance, the existence of a nice collineation group. We recall that a *collineation* of a projective plane Π is a permutation of the points which maps lines to lines. The set of all collineations of Π forms a group Aut Π under composition. Any subgroup of Aut Π is called a *collineation group* of Π. For the time being, we mention just two famous results in this direction; the seminal one is the celebrated Ostrom-Wagner theorem proved in 1959, see [148].

Theorem 1.3 (Ostrom-Wagner theorem) *Let Π be a projective plane of order n admitting a doubly transitive collineation group G. Then Π is desarguesian, and $PSL(3,n)$ is a subgroup of G.*

If we weaken the hypothesis of the Ostrom-Wagner theorem a little and only require a flag-transitive group (where a *flag* is an incident point-line pair), the result should be much the same as before; but unfortunately – in spite of a lot of effort by many authors – this case is not yet settled. What we know at present is summarised in the following result; the notation dev D used there will be explained later when we discuss difference sets.

Theorem 1.4 *Let Π be a projective plane of order n admitting a flag-transitive collineation group G. Then*

(i) *Π is desarguesian, and $PSL(3,n)$ is a subgroup of G; or*

(ii) *G acts regularly on flags and is a Frobenius group. Also, n^2+n+1 is a prime p. If $n \neq 2$ or 8, then Π is a non-classical plane, n is divisible by 8 but is not a power of 2, and $\Pi \cong \operatorname{dev} D$, where D consists of the n-th powers in $\mathbb{Z}_p^*$.*

It is conjectured that only the desarguesian planes of orders $n = 2$ and 8 actually occur in case (ii). We refer the reader to Fink [61], Kantor [124] and Feit [59] – to mention just a few papers concerning Theorem 1.4 – and to the recent survey by Koen Thas [171].

In the present survey paper, Π will be a projective plane of order n with a *quasiregular* collineation group G, that is, a group inducing a regular action on each orbit: each group element fixes either none or all elements in the orbit. This condition is satisfied in particular when G is abelian, for it is easy to prove that all permutation representations of a group G are quasiregular if and only if every subgroup of G is normal. We will mainly concentrate on the abelian case, as much less is known in the general case.

If a quasiregular group G is *"large"* in the sense that

$$|G| > \frac{1}{2}(n^2 + n + 1),$$

then there are unique faithful point and/or line orbits ([46], p.181). Let us also recall that the number of point orbits of a collineation group agrees with the number of line orbits by the *orbit theorem*, see [104, Theorem 13.4]. Dembowski and Piper classified planes of this type into the following eight cases.

Theorem 1.5 (Dembowski-Piper theorem) *Let G be a collineation group acting quasiregularly on the points and lines of a projective plane of order n, and assume $|G| > \frac{1}{2}(n^2 + n + 1)$. Let t denote the number of point orbits and F denote the incidence structure consisting of the fixed points and fixed lines. Then one of the following holds.*

(a) $|G| = n^2 + n + 1$, $t = 1$, $F = \emptyset$. *Here G is transitive.*

(b) $|G| = n^2$, $t = 3$, *F is a flag, that is, an incident point-line pair (∞, L_∞).*

(c) $|G| = n^2$, $t = n+2$, *F is either a line and all its points or, dually, a point together with all its lines.*

(d) $|G| = n^2 - 1$, $t = 3$, *F is an antiflag – a non-incident point line pair (∞, L_∞).*

(e) $|G| = n^2 - \sqrt{n}$, $t = 2$, $F = \emptyset$. *In this case one of the point orbits is precisely the set of points of a Baer subplane Π_0 of Π – a subplane of order $\sqrt{n}$.*

(f) $|G| = n^2 - n$, $t = 5$, *F consists of two points, the line joining them and another line through one of the two points.*

(g) $|G| = n^2 - 2n + 1$, $t = 7$, *F consists of the vertices and sides of a triangle.*

(h) $|G| = (n^2 - \sqrt{n} + 1)^2$, $t = 2\sqrt{n} + 1$, $F = \emptyset$. *In this case there are $t - 1$ disjoint subplanes of order $\sqrt{n} - 1$ whose point sets constitute $t - 1$ orbits, each of length $n - \sqrt{n} + 1$.*

Theorem 1.5 is actually a consequence of a more general result on tactical decompositions not necessarily deriving from the orbit decomposition of a collineation group, see Dembowski [45] and Dembowski and Piper [48]. For the proof in cases (b), (c) see [45]; for the remaining cases, see [48]. Some of these results are also contained in two papers of Hughes [99, 100], with more specialised hypotheses. If the assumption $|G| > \frac{1}{2}(n^2 + n + 1)$ is replaced by the weaker condition that there is only one point (or line) orbit on which G

is faithful, then the only other possibilities besides (a)–(h) occur either for $n \leqslant 4$ or for F consisting of $k \geqslant 2$ collinear points and the line joining them (or the dual), see Piper [153]. We also note that Ganley and Piper showed in Theorems 2 and 3 of [66] that a quasiregular group G without any faithful point orbit has to be a p-group and that n is a power of the same prime p; in this case the orbit structure of G is also determined.

Planes of type (c) are translation planes or their duals. Here a *translation plane* and a *dual translation plane* are projective planes of order n with a collineation group G of order n^2 such that every element of G fixes all points on a suitable line, and all lines through a suitable point, respectively. By a classical result of André (see [2] or [138, Theorem 1.7]), in the case of translation planes and dual translation planes the collineation group G is always an elementary abelian p-group and hence the PPC holds. We shall not treat case (c) here, as there is a vast literature on such planes, see Kallaher [120, 121], Lüneburg [138], or Biliotti, Jha and Johnson [24]. Hence we confine ourselves to a few remarks. Let us first mention that there is a close connection to Galois geometry, since Π can also be described via a spread, see André [2] and Bruck and Bose [35]. A general classification of (dual) translation planes is clearly impossible, due to the explosive growth of the number of isomorphism types; for instance, there are just 8 isomorphism types of translation planes of order 16, see [50]; 7 isomorphism types of translation planes of order 27, see [49]; but already 1347 types of order 49, see [145] and [38]. We finally note that in the special case of flag-transitive affine planes, which by a result of Wagner [175] are necessarily translation planes, there are important classification results due mainly to Baker and Ebert, see [14, 15, 16, 17, 55].

The purpose of the present paper is to survey most known results on the remaining cases (a), (b) and (d)–(h) and to summarise some applications to geometry, in particular to arcs and ovals. Throughout the paper, G will be a quasiregular group of order $|G| > \frac{1}{2}(n^2+n+1)$ acting on a projective plane Π of order n; in most cases, G is actually assumed to be abelian. As we shall see, there is a close connection to certain types of difference sets which allows one to apply algebraic methods. The fundamental technical tools are group rings, characters, and algebraic number theory. These connections will be explained in the next section.

2 Preliminaries

A k-subset D of an additively written group G of order $v = mc$ is called a *relative difference set* with parameters (m, c, k, λ) (for short, an (m, c, k, λ)-RDS) provided that the list of differences $(d - d' : d, d' \in D, d \neq d')$ covers every element in $G \setminus N$ exactly λ times, and the elements in $N \setminus \{0\}$ not at all, where N is a specified subgroup of G of order c, usually called the *forbidden subgroup*. A relative difference set D is called *cyclic* or *abelian* if G has the respective property.

Counting differences in two ways, we get the fundamental equation

$$k(k-1) = \lambda c(m-1). \tag{2.1}$$

Thus only three parameters are essential. We also have the obvious inequality $k \leqslant m$, since otherwise at least one coset of N would contain more than one element from D which would lead to a difference in N.

If $c = 1$, the relative difference set becomes a *difference set* in the usual sense. Here every element different from the identity of the group G has exactly λ representations as a difference of elements in D. In this case, we simply speak of a (v, k, λ)-difference set.

Relative difference sets first appear in the work of Bose [29], although he was only concerned with a special case and did not use this term which was introduced by Butson [37]. The first systematic investigations are in Elliott and Butson [57] and Lam [130]. For further results and references see the excellent survey on relative difference sets by Pott [162] and the recent updates [117] and [118] of the survey on difference sets of the second author [113]. A comprehensive introduction to difference sets can be found in Chapter VI of Beth, Jungnickel and Lenz [23]; see also Baumert [22] and Lander [133] for earlier – now in many aspects obsolete – systematic treatments.

The main motivation for studying relative difference sets is provided by the fact that the existence of an (m, c, k, λ)-RDS in G relative to a subgroup N is equivalent to the existence of a *divisible semisymmetric design* with the same parameters (briefly: a divisible (m, c, k, λ)-SSD) admitting G as a regular automorphism group (so that G is sharply 1-transitive on both points and blocks). Here a *divisible semisymmetric design* is an incidence structure with $v = mc$ points and blocks, and with a partition of the point set into m *point classes* of equal size c such that two distinct points in the same class are not joined whereas points in distinct point classes are joined by exactly λ blocks. For $c = 1$, this reduces to a *symmetric* (v, k, λ)*-design*, and for $\lambda = 1$ to a *divisible semiplane* (or *elliptic semiplane*). The group G is called a *Singer group* for the SSD, since Singer proved in his seminal paper [168] that the symmetric design of points and hyperplanes of a projective geometry admits a regular cyclic group. The following basic result is due to the second author [106].

Theorem 2.1 *Assume the existence of an* (m, c, k, λ)-RDS D *in a group* G *relative to a subgroup* N. *Then the incidence structure*

$$\operatorname{dev} D = (G, \mathbf{B}, \in) \qquad \textit{with} \quad \mathbf{B} = \{D + g \colon g \in G\}$$

is a divisible (m, c, k, λ)-SSD *admitting* G *as a Singer group, where* N *acts as the stabiliser of the point class of* 0. *Moreover, any* SSD *with a Singer group* G *may be represented in this way. Finally,* N *is a normal subgroup of* G *if and only if it stabilises each point class of* $\operatorname{dev} D$.

The reader is referred to Ghinelli [71, 72, 75, 76], Hughes [103] and Jungnickel [106] for more details on semisymmetric designs and relative difference sets. Unless stated otherwise, we shall from now on assume N to be a normal subgroup, which is the case relevant to the study of quasiregular groups. In general, not much is known when this assumption is dropped. A few examples without normality may be found in [106].

Four of the eight cases in the Dembowski-Piper theorem are connected to relative difference sets with $\lambda = 1$, as shown by Ganley and Spence [68]. If we are in one of the cases (a), (b), (d), and (e) of Theorem 1.5, then the faithful point and line orbit of G form a divisible semiplane, and if p and L are a point and line in these orbits, respectively, then $D = \{g \in G : p^g \in L\}$ is an $(m, c, k, 1)$-RDS; of course, D depends on the choice of the *base point* p and the *base line* L. More precisely, we have the following by [68, Lemma 2.2].

Type (a) Here D is an ordinary $(n^2+n+1, n+1, 1)$-difference set; one usually calls D a *planar difference set* of order n.

Type (b) Here D has parameters $(n, n, n, 1)$, and the forbidden subgroup N is the pointwise stabiliser of the fixed line L_∞.

Type (d) Here D has parameters $(n+1, n-1, n, 1)$, and the forbidden subgroup N is the pointwise stabiliser of the fixed line L_∞; one usually calls D an *affine difference set* of order n.

Type (e) Here D has parameters $(n+\sqrt{n}+1, n^2-\sqrt{n}, n, 1)$, and the forbidden subgroup N is the stabiliser of the Baer subplane Π_0.

In each case, the group G acts as a Singer group on the *development* dev D of D. Moreover, Π may be uniquely reconstructed from dev D, when $n > 1$, see Ganley and Spence [68, Theorem 3.1]. That an RDS of type (e) extends in a unique manner is due to Piper [154].

With two exceptions – namely, an example of Baker [13] with parameters $n = 3, m = 15$ and $k = 7$, and an example of Mathon [144] with parameters $n = 3, m = 45$ and $k = 12$ – which do not admit a Singer group, all known divisible semiplanes Δ arise by omitting the empty set or a *Baer subset* from a projective plane Π, see [46]. Thus $\Delta = \Pi \setminus S$ where S is one of the following:

(i) S is the empty set;

(ii) (∞, L_∞) is a flag, and S consists of a line L_∞ with all its points, and a point ∞ with all the lines through it;

(iii) (∞, L_∞) is an antiflag, and S consists of a line L_∞ with all its points, and a point ∞ with all the lines through it;

(iv) S is a Baer subplane of Π.

The resulting structure Δ is a divisible semiplane with parameters $(m, c, k, 1)$. More precisely,

(i) this case just yields $\Delta = \Pi$ again: the projective plane itself is a divisible semiplane with parameters $(n^2 + n + 1, 1, n + 1, 1)$;

(ii) we obtain a *symmetric net* of order n – a divisible semiplane with parameters $(n, n, n, 1)$.

(iii) In case (iii) we obtain a *biaffine plane* of order n – a divisible semiplane with parameters $(n + 1, n - 1, n, 1)$.

(iv) in this case Δ is a *Baer semiplane* of order n, a divisible semiplane with parameters $(n + \sqrt{n} + 1, n - \sqrt{n}, n, 1)$.

In particular, any divisible semiplane of one of these four types which admits a Singer group G corresponds to an RDS and – as we assume the forbidden subgroup N to be normal – to a quasiregular group of one of the types (a), (b), (d) and (e).

Let us briefly discuss the known examples of relative difference sets for divisible semiplanes. As the two sporadic semiplanes mentioned above do not admit such a representation, we only meet examples corresponding to semiplanes arising from projective planes. Whether or not this is necessarily so is an interesting problem, and probably a difficult one.

(i) *Projective planes.* The desarguesian plane PG$(2, q)$ admits a cyclic Singer group; indeed, this classical result of Singer [168] has provided the reason for present day terminology. This case is discussed in detail in Section 3.

(ii) *Symmetric nets.* The desarguesian plane PG$(2, q)$ yields a symmetric net admitting an abelian Singer group. But there are also many non-classical examples in this case: each projective plane over a finite semifield gives rise to a symmetric net Δ admitting a Singer group. Moreover, any planar function of degree n leads to such a net; this yields one further class of examples, namely the Coulter-Matthews planes [44]. We will give the relevant definitions and discuss these examples in detail in Section 4.

(iii) *Biaffine planes.* The desarguesian plane PG$(2, n)$ yields the biaffine plane BAG$(2, n)$ admitting a cyclic Singer group, see [29]. This case is discussed in detail in Section 5.

(iv) *Baer semiplanes.* The only known abelian example is given by PG$(2, 4) \setminus$ PG$(2, 2)$, but there is also a non-abelian example associated with the Hughes plane of order 9; see Section 6.

According to a result of Lam [130] all cyclic relative difference sets with $\lambda = 1$ and $k \leqslant 50$ are of the types discussed above.

One of the most important tools in investigating (relative) difference sets is the use of the *integral group ring* $\mathbb{Z}G$ of G. One advantage of this approach is that subsets and even lists (or submultisets of G) can be represented just as elements of the group ring; more important is the fact that it enables us to use algebraic techniques. Indeed, the amazing strength of the approach using various types of difference sets and the machinery of integral group rings will be a recurrent theme which we want to stress in this survey.

In order to work with group rings, we have to write G multiplicatively. Then

$$\mathbb{Z}G = \{\sum_{g\in G} a_g g : a_g \in \mathbb{Z}\}$$

is the free $\mathbb{Z}$-module with G as basis, equipped with the multiplication

$$(\sum_{g\in G} a_g g)\cdot(\sum_{h\in G} b_h h) = \sum_{g,h\in G} a_g b_h gh.$$

We will use the following conventions. For $X = \sum a_g g \in \mathbb{Z}G$, we write $|X| = \sum a_g$ and $[X]_g = a_g$ (the coefficient of g in X). For $r \in \mathbb{Z}$ we write r for the group ring element $r1$, and for $S \subseteq G$ we write, by abuse of notation, S instead of $\sum_{g\in S} g$. If $\alpha\colon G \to H$ is any mapping of G into a group H, we extend α to a linear mapping from the group ring $\mathbb{Z}G$ into the group ring $\mathbb{Z}H$:

$$X^\alpha = \sum a_g g^\alpha \quad \text{for} \quad X = \sum a_g g.$$

In the special case $G = H$ and $\alpha\colon g \mapsto g^t$ for some $t \in \mathbb{Z}$, we write $X^{(t)}$ instead of X^α, so that $X^{(t)} = \sum a_g g^t$. In particular, $S^{(-1)} = \sum_{g\in S} g^{-1}$. Sometimes we will also need the group ring RG for a commutative ring R other than $\mathbb{Z}$, in particular for $R = \mathbb{Z}_p$, the field of residues modulo a prime p. The same definitions and notation apply in this more general setting.

Using the conventions just introduced, we immediately obtain the following simple but fundamental lemma.

Lemma 2.2 *Let G be a multiplicative group of order mc, let N be a subgroup of G of order c (not necessarily normal in G), and let $D \in \mathbb{Z}G$. Then D is an (m, c, k, λ)-RDS in G relative to N if and only if the following equation holds in $\mathbb{Z}G$:*

$$DD^{(-1)} = k + \lambda(G - N). \tag{2.2}$$

We shall also need another trivial observation which shows how to compute intersection sizes using the group ring $\mathbb{Z}G$.

Lemma 2.3 *Let A and B be subsets of a finite group G. Then*

$$|A \cap Bg| = [AB^{(-1)}]_g.$$

Later we will also require several variants of relative difference sets in order to study the types (f), (g) and (h). These will be defined in the relevant sections. Once a difference set condition is translated into a group ring equation, such as (2.2), these objects can be studied in a purely algebraic setting. We shall see quite a few examples of this approach throughout the rest of this survey. We will confine ourselves to elementary – by which we mean character-free – methods which can already yield surprisingly strong results. Of course there are important methods which are more advanced. One may apply complex characters to (2.2) to obtain a characterisation of relative difference sets in terms of character sums, which gives a powerful tool for nonexistence results. The values $\chi(D)$ are in fact algebraic integers in $\mathbb{Q}(\zeta)$, where ζ is a primitive v^*-th root of unity and where v^* denotes the exponent of the underlying abelian group G. This makes it possible to apply the machinery of algebraic number theory to the study of hypothetical (relative) difference sets. We refer the reader to Pott [161], Beth, Jungnickel and Lenz [23] and Schmidt [165] for examples of this approach which goes back to the seminal paper of Turyn [172].

3 Type (a): Planar difference sets

In this section Π will denote a plane of type (a), a projective plane with an abelian Singer group. As we have seen, Π may be represented as the development of an abelian planar difference set D in G. The classical case is given by the seminal result of Singer [168].

Theorem 3.1 (Singer's theorem) *The classical projective plane* $\mathrm{PG}(2,q)$ *admits a cyclic Singer group and therefore can be represented by a cyclic* $(q^2+q+1, q+1, 1)$*-difference set.*

The planes $\mathrm{PG}(2,q)$ are the only known projective planes admitting a difference set representation. Both the scarcity of known examples and the strong nonexistence results to be discussed below have led to some famous unsolved conjectures on planar difference sets.

Conjecture 3.2 *Any finite projective plane admitting a Singer group is desarguesian.*

In 1960, Bruck [33] proved that all cyclic planes of order n or n^2 with $n \leqslant 9$ are indeed desarguesian. The following remarkable result was obtained by Ott [149] in 1975 for the cyclic case, and generalised to the abelian case by Ho [94] in 1998. It is the best general result providing evidence for Conjecture 3.2.

Theorem 3.3 (Ott-Ho theorem) *A finite projective plane* Π *is desarguesian if and only if it admits two distinct abelian Singer groups* $G_1 \neq G_2$. *In other words, if there exists an abelian Singer group which is not normal in the full collineation group* $\mathrm{Aut}\,\Pi$, *then* Π *is desarguesian.*

Ott's proof of the cyclic case of Theorem 3.3 first shows that the collineation group generated by G_1 and G_2 acts 2-transitively on the points, and then applies the Ostrom-Wagner theorem. A new and shorter proof of Ott's theorem was given in 1994 by Müller [146], by applying a theorem of Schur on finite permutation groups, the classification of finite simple groups, and some ideas of Kantor [123]. Ho's generalisation to the abelian case uses entirely different methods; it is comparatively elementary, but far from simple.

Since Conjecture 3.2 does not seem to be accessible to present day methods, even if it is probably true, work has concentrated on the weaker prime power conjecture. The numerical evidence for the validity of the PPC is quite large. Many existence tests have been devised, as we shall see below. Already in 1951, Evansand Mann [58] verified the PPC in the cyclic case for $n \leqslant 1,600$. This has been extended to abelian groups in general up to $n \leqslant 2,000,000$, see Gordon [79]. We finally mention a third conjecture.

Conjecture 3.4 *Any abelian planar difference set is cyclic.*

We remark that the existence of an abelian planar difference set of order $n \equiv 1 \pmod 3$ implies that of a non-abelian example associated with the same projective plane; this is due to Bruck [32], see also [23, Theorem VI.7.1]. For $PG(2,q)$, all Singer groups – including the non-abelian ones – are known explicitly from a result of Ellers and Karzel [56], see also Dembowski [46, 1.4.17]. In particular, Conjecture 3.4 holds for the classical planes and would therefore follow from the validity of Conjecture 3.2; see Dembowski [46, 1.4.19].

The representation of a projective plane with a Singer group by a planar difference set D exhibits a regular group of automorphisms; but, of course, the plane may have more automorphisms besides. For instance, the full automorphism group of a desarguesian projective plane is transitive on quadrangles, and hence, in particular, 2-transitive on the set of points. It is often possible to find some of these additional automorphisms in terms of the difference set representation using the notion of multipliers. In the abelian case, a *multiplier* of D can be defined as an automorphism α of the Singer group G which induces an automorphism of $\Delta = \operatorname{dev} D$. If G is cyclic, then every multiplier is a *numerical multiplier*, having the form $\alpha : x \mapsto tx$ for some integer t coprime to $|G|$. Then the condition for t to be a multiplier is $tD = D + g$ for some $g \in G$.

The importance of the concept of multipliers lies in the fact that, very often, just the parameters of a hypothetical abelian difference set D force the existence of numerical multipliers which then may be used to help with either the construction of D or a nonexistence proof. These fundamental ideas are due to Hall [82] who considered them in the special case of cyclic planar difference sets. His result (and proof) admits a rather straightforward generalisation due to Chowla and Ryser [42].

Theorem 3.5 (First multiplier theorem) *Let D be an abelian (v,k,λ)-difference set, and let p be a prime dividing $n = k - \lambda$ but not v. If $p > \lambda$, then p is a multiplier of D.*

Corollary 3.6 *Let D be a planar difference set of order n. Then any divisor t of n is a multiplier of D.*

Hall's original proof – which is the proof still used in some text books, for example [1, 83, 104] – covered several pages, was quite technical, and was certainly not very illuminating. It took more than thirty years before Eric Lander [132] gave a much more transparent, conceptual proof in his 1980 Ph.D. thesis. Further simplifications of his proof are due to Pott [155, 158] and van Lint and Wilson [137]. We shall provide such a proof, as it will give a good first impression of the power of the group ring approach, and also for its sheer beauty and elegance. There are two main ingredients, a purely combinatorial lemma and an algebraic argument. The combinatorial part is the following characterisation of the blocks of a symmetric design which can be proved by a simple counting argument; see, for example, [23, Lemma VI.4.2].

Lemma 3.7 *Let S be a k-subset of a symmetric (v,k,λ)-design and assume that S meets each block in at least λ points. Then S is itself a block.*

Using Lemma 3.7, the first multiplier theorem follows if we can show that

$$|pD \cap (D+g)| \geqslant \lambda \qquad \text{for all } g \in G. \tag{3.1}$$

Now the hypothesis $p > \lambda$ allows us to deduce this condition from the following congruence:

$$|pD \cap (D+g)| \equiv \lambda \pmod{p} \qquad \text{for all } g \in G. \tag{3.2}$$

All known proofs require algebraic arguments to prove (3.2); more precisely, they all make use of group rings. So we shall switch to multiplicative notation for G and use the integral group ring $\mathbb{Z}G$. By Lemma 2.2, the difference set property of D translates into the following group ring equation:

$$DD^{(-1)} = (k-\lambda) + \lambda G, \tag{3.3}$$

and assertion (3.2) now reads

$$|D^{(p)} \cap Dg| \equiv \lambda \pmod{p} \qquad \text{for all } g \in G. \tag{3.4}$$

Lemma 2.3 suggests that (3.4) should be proved by evaluating the group ring element $D^{(p)}D^{(-1)}$ modulo p. In order to reduce all the coefficients modulo p, we now work in the group algebra $\mathbb{Z}_pG$ over the field $\mathbb{Z}_p$ of residues modulo p, heavily using the hypothesis that p divides $n = k - \lambda$. We will apply (3.3) and the standard fact

$$D^p \equiv D^{(p)} \pmod{p} \qquad \text{for } D \in \mathbb{Z}_pG, \tag{3.5}$$

which follows from the multinomial theorem, see [23, Lemma VI.3.7]. Now we compute in $\mathbb{Z}_pG$ as follows:

$$\begin{aligned} D^{(p)}D^{(-1)} &= D^pD^{(-1)} = D^{p-1}(DD^{(-1)}) \\ &= D^{p-1}(\lambda G) = \lambda k^{p-1}G = \lambda^p G = \lambda G, \end{aligned}$$

which already gives the desired congruence (3.4).

There are further multiplier theorems for the general case $\lambda > 1$ which may be viewed as an attempt to get rid of the assumption $p > \lambda$ in Theorem 3.5 which is widely believed to be redundant (*multiplier conjecture*); the interested reader is referred to [23, Chapter VI]. Corollary 3.6 is all that is required for the planar case. To apply this result, we need another fact; again, we only give a simple version sufficient for the planar case and refer to [23] for more general results.

Lemma 3.8 *Let D be an abelian (v, k, λ)-difference set, where k is prime to v. Then there is an element b of G such that $D + b$ is fixed by every multiplier.*

Proof Write $D = \{d_1, \ldots, d_k\}$. Note that the map $x \mapsto kx$ is an automorphism of G, since G is abelian and k and v are coprime. Thus there is exactly one $b \in G$ satisfying $d_1 + \ldots + d_k + kb = 0$. For any multiplier α we have $(D + b)^\alpha = D + c$ for a suitable element c. Then

$$\begin{aligned} 0 &= (d_1 + \ldots + d_k + kb)^\alpha = ((d_1 + b) + \ldots + (d_k + b))^\alpha \\ &= (d_1 + c) + \ldots + (d_k + c) = d_1 + \ldots + d_k + kc, \end{aligned}$$

and therefore $kc = kb$, hence $c = b$. □

Corollary 3.9 *Let M be a group of multipliers of an abelian (v, k, λ)-difference set in G, where k is prime to v, and assume without loss of generality that D is fixed under M. Then D is a union of orbits of M on G, and hence k is a sum of some of the orbit sizes.*

In view of of Lemma 3.8 and Corollary 3.9, it is interesting to ask how large the multiplier group of a difference set can be. Ho [91] proved that the multiplier group M of a cyclic planar difference set D has cardinality at most k and that D necessarily contains a residue coprime to $v = n^2 + n + 1$, unless D belongs to the projective plane of order 4; in the exceptional case, $|M| = 6$, $v = 21$ and $D = \pm\{3, 6, 7, 12, 14\}$. See also Ho and Pott [95] for related results.

Example 3.10 Multipliers can be used to simplify the construction of planar difference sets by trial and error. If $q = p^a$ is a prime power, we know by Singer's Theorem 3.1 that there exists a cyclic planar difference set D of order q. By Corollaries 3.6 and 3.9, we may assume that D is fixed by the multiplier p, so $pD = D$. Thus D is the union of orbits of the automorphism $x \mapsto px$ of G. Let us give a few examples.

- For $n = 4$, we have already given an example above which could have been found as follows. There is a difference set $D \subset \mathbb{Z}_{21}$ with $D = 2D$ which therefore consists of orbits under $M = \{1, 2, 4, 8, 16, 11\}$. Because $k = 5$, the only candidates are the "small" orbits, namely $\{0\}$, $\{7, 14\}, \{3, 6, 12\}$ and $\{9, 18, 15\}$. Thus D has to be the union of $\{7, 14\}$ with one of the 3-element orbits. In fact, both possible choices D_1 and D_2 work. This is not surprising, since $D_2 = -D_1$. Thus we have also proved that D is uniquely determined up to equivalence. Here difference sets D and E in some abelian group G are called *equivalent* if $E = tD+g$ for some $t \in \mathbb{Z}$ and some $g \in G$.

- For $n = 2$, there is a difference set D with $D = 2D$. We may assume $1 \in D$, hence also $2, 4 \in D$, which gives $D = \{1, 2, 4\}$. Similarly, we easily find the difference sets $\{0, 1, 3, 9\} \subset \mathbb{Z}_{13}$ of order 3, $\{1, 5, 25, 11, 24, 27\} \subset \mathbb{Z}_{31}$ of order 5 and $\{1, 2, 4, 8, 16, 32, 64, 55, 37\} \subset \mathbb{Z}_{73}$ of order 8.

- For PG(2, 7), $D = 7D$ has to be the union of sets xM with $M = \{1, 7, 49\}$. If x is prime to 19, then $|xM| = 3$, and $19M = \{19\}$. Hence D is of the form $\{0, 19\} \cup xM \cup yM$. If one tries $x = 1$, then a lot of values for y do not work, but $y = 23$ does, and we get $D = \{0, 1, 7, 19, 23, 44, 47, 49\}$. The reader is asked to check that this is indeed a difference set.

Let us give a further simple example for the application of Lemma 3.8 and show the nonexistence of a planar abelian difference set D of order n, where n is any multiple of 6. Otherwise, we may assume that D is fixed by the multipliers 2 and 3. Choose $d \neq 0$ in D; then we also have $2d, 3d \in D$. Because $\lambda = 1$ and $3d - 2d = 2d - d$, we obtain the contradiction $d = 0$. Reasoning in a similar way, we obtain the following general results due to Hall [82] in the cyclic case and to Lander [133] and – for three of the cases – Jungnickel and Pott [115] in the abelian case.

Lemma 3.11 *Let D be a planar difference set of order n in an abelian group G. If t_1, t_2, t_3 and t_4 are numerical multipliers satisfying*

$$t_1 - t_2 \equiv t_3 - t_4 \pmod{\exp G},$$

then $\exp G$ *divides the least common multiple of* $t_1 - t_2$ *and* $t_1 - t_3$.

Theorem 3.12 *Let D be a planar abelian difference set of order n. Then n is not divisible by any of the numbers* 6, 10, 14, 15, 21, 22, 26, 33, 34, 35, 38, 39, 46, 51, 55, 57, 58, 62 *and* 65.

The reader is referred to [23, Theorem VI.7.19] for the proofs. A general discussion of the application of Lemma 3.11 using algebraic number theory is given by Gordon [80] who managed to extend Theorem 3.12 to many further numbers $\leqslant 1000$.

Corollary 3.6 shows that all divisors of n are multipliers in the planar case; however, not all multipliers have to divide n. For instance, $11 \equiv 2^5 \pmod{21}$ is a multiplier of the cyclic $(21, 5, 1)$-difference set for the projective plane of order 4. A multiplier t not dividing n is called an *extraneous multiplier*; this notion was introduced by Mann [141]. It is interesting to note that the small primes $p \leqslant 5$ cannot be extraneous multipliers of a planar difference set, by results of Pott [157], see also Ghinelli [74], and Wei, Gao and Xiang [176]. The reader is also referred to [23, §VI.4] for proofs and further results for abelian difference sets in general. It is not known whether 7 can be an extraneous multiplier.

Perhaps the most important existence criterion for difference sets is provided by the so-called *Mann test* which goes back to Mann [142]. We will just give a basic version of the rather technical result now known which incorporates improvements due to Jungnickel and Pott [115], and to Arasu et al. [6], and also applies to some non-abelian cases. The reader is referred to [23, §VI.6] for the statement and proof of the general result. The proof given there uses the group ring approach combined with ideas from algebraic coding theory which were introduced by Lander [133] and Pott [156]. A purely computational proof within the group algebra is also possible, see Jungnickel and Pott [115].

Theorem 3.13 (The Mann test) *Let D be a (v, k, λ)-difference set in an abelian group G of order v and let u^* denote the exponent of the quotient $H = G/U$, where U is a subgroup of order s and index $u \neq 1$ of G. If p is a prime not dividing u^*, and $tp^f \equiv -1 \pmod{u^*}$ for some suitable non-negative integer f and some multiplier t of D, then p does not divide the square-free part of $n = k - \lambda$.*

In the case of planar difference sets, the Mann test yields some powerful existence criteria for non-square orders n. Fortunately, this special case allows a simple geometric proof independent of Theorem 3.13; this fact seems not to have been noted before. For this purpose, we require the following result on planar abelian difference sets with a multiplier of order 2 taken from Blokhuis, Brouwer and Wilbrink [25].

Theorem 3.14 *Let D be a planar difference set in an abelian group G. If D admits a multiplier t of order 2, then n is a perfect square, say $n = m^2$, and necessarily $t = m^3$.*

Proof Define subgroups A and B of G as follows:

$$A = \{g \in G : tg = -g\} \quad \text{and} \quad B = \{g \in G : tg = g\}.$$

The mappings α and β defined by $g^\alpha = (g - tg)/2$ and $g^\beta = (g + tg)/2$ are homomorphisms from G to A and B, respectively. Then $A \cap B = \{0\}$ and $g = g^\alpha + g^\beta$ for each $g \in G$; thus $G = A \oplus B$. By assumption, t induces an involutory collineation τ of the projective plane $\Pi = \text{dev } D$. Thus τ is either an elation (with $n+1$ fixed points); a homology (with $n+2$ fixed points); or a Baer involution (with m^2+m+1 fixed points, where $n = m^2$), see Hughes and Piper [104]. Since the order of B divides that of G, the last case must occur, and B is the point set of a Baer subplane Π_0 of Π. Thus $n = m^2$ is a square, B has order m^2+m+1, and A has order m^2-m+1. As $\gcd(m^2-m+1, m^2+m+1) = 1$, G has unique subgroups with these two orders. So any multiplier of order 2 leads to the same representation $G = A \oplus B$ and acts on A and B in the same way that t does. The result now follows, since D admits the multiplier m and hence also the multiplier m^3 of order 2, by Corollary 3.6. □

We can now easily prove the planar version of the Mann test.

Theorem 3.15 *Let D be a planar abelian difference set of order n in G. Then either n is a square or every multiplier of D has odd order modulo $u = \exp G$. In particular, let p and q be prime divisors of n and of $v = n^2+n+1$, respectively. Then each of the following conditions implies that n is a square:*

(a) *D has a multiplier which has even order modulo q;*

(b) *p is a quadratic non-residue modulo q;*

(c) *$n \equiv 4$ or $6 \pmod 8$;*

(d) *$n \equiv 1$ or $2 \pmod 8$ and $p \equiv 3 \pmod 4$;*

(e) *$n \equiv m$ or $m^2 \pmod{m^2+m+1}$ and p has even order $\pmod{m^2+m+1}$.*

Proof Let s be a multiplier which has even order modulo u, say $2h$. Then $t = s^h$ is a multiplier of order 2, and the general assertion is an immmediate consequence of Theorem 3.14. The particular criteria now follow without much difficulty, using some elementary number theory; see [23, Theorem VI.7.8] for details. □

Another nonexistence criterion for abelian projective planes was obtained by Ghinelli [73] using representation theory, Galois theory and the p-adic integers. As Lander [134] noted, this could have been deduced from the Mann test, though the connection is by no means obvious.

Theorem 3.14 also implies that the search for a possible counterexample to the PPC can be restricted to non-square orders. Returning to the proof of Theorem 3.14, B is obviously a Singer group for the Baer subplane Π_0. In fact one may even assume that Π_0 is represented by the planar difference set $D \cap B$. This establishes the following result due to Ostrom [147] in the cyclic case and to Jungnickel and Vedder [119] in general.

Corollary 3.16 *Assume the existence of a planar abelian difference set of order $n = m^2$ in G. Then there also exists a planar difference set of order m in some subgroup H of G.*

In the cyclic case, Ho [92] proved the following result that somewhat strengthens Corollary 3.16. Assume the existence of a planar cyclic difference set D of order $n = m^s$ in G. Then there exists a planar difference set D' of order m in the unique subgroup H of G of order $m^2 + m + 1$ if and only if s is not a multiple of 3. In this case, every multiplier of D is also a multiplier of D'. We refer the reader to [23, Chapter VI] for further results and references concerning the structure of the multiplier group and related topics.

The nonexistence results discussed up to now only apply to square orders $n = m^2$ if we can also exclude the value m. The following criterion which specifically applies to square orders was obtained by Arasu [3].

Theorem 3.17 *Let D be a cyclic planar difference set of order n, where either $n = m^{2s}$ for some positive integer s which is not a multiple of 3 or $n = m^{2^r}$ for some positive integer r, and let t be a multiplier of D. Assume the existence of a prime q dividing $m^2 - m + 1$ such that t has odd order modulo q. Then t has odd order modulo $m^4 + m^2 + 1$.*

For instance, assume $m \equiv 10$ or $12 \pmod{14}$. Then there does not exist a cyclic planar difference set of order $n = m^{2s}$, where s is any positive integer which is not divisible by 3; also, there is no abelian planar difference set of order $n = m^{2^r}$ for any positive integer r. The reader may derive this from Theorem 3.17 as an exercise.

Finally, there is one further result, applicable in the planar case, namely Wilbrink's theorem [177]. We shall adapt the proof of Jungnickel [111] for a generalisation of this result, see also [23, Theorem VI.6.10]. This proof merely requires the integral group ring and a few computations modulo p and p^2, respectively.

Theorem 3.18 (Wilbrink's theorem) *Let D be a planar difference set of order n in an abelian group G (written multiplicatively), let p be a prime dividing n such that p^2 does not divide n and assume, without loss of generality, that D is fixed under the multiplier p. Then the following identity holds in $\mathbb{Z}_p G$:*

$$D^{p-1} + \left(D^{(-1)}\right)^{p-1} = 1 + G. \tag{3.6}$$

Proof By Corollary 3.6, p is a multiplier for D, and by Lemma 3.8, we may indeed assume that D is fixed under p; thus, in group ring notation, $D^{(p)} = D$. By (3.5), we may write $D^p - D = pA$ for some $A \in \mathbb{Z}G$ and hence $(D^{(-1)})^p - D^{(-1)} = pA^{(-1)}$. This implies the following congruence modulo p^2:

$$0 \equiv (pA)(pA^{(-1)}) \equiv \left(DD^{(-1)}\right)^p + DD^{(-1)}\left[1 - D^{p-1} - \left(D^{(-1)}\right)^{p-1}\right] \pmod{p^2}.$$

By applying (3.3) and $DG = D^{(-1)}G = (n+1)G$, we obtain

$$n\left[D^{p-1} + \left(D^{(-1)}\right)^{p-1}\right] \equiv (n+G)^p + (n+G) - 2(n+1)^{p-1}G \pmod{p^2}. \tag{3.7}$$

Using $p|n$, it is easy to check that

$$(n+G)^p \equiv G^p \equiv (n+1)^{p-1}G \equiv (1-n)G \pmod{p^2}.$$

Therefore (3.7) reduces to

$$n\left[D^{p-1} + \left(D^{(-1)}\right)^{p-1}\right] \equiv -(1-n)G + (n+G) \equiv n + nG \pmod{p^2}. \tag{3.8}$$

The assertion follows, as $n = mp$ for some integer m not divisible by p. □

The following application of Theorem 3.18 is due to Wilbrink [177] and Jungnickel and Vedder [119]. It can also be obtained using the existence criterion of Ghinelli [73].

Corollary 3.19 *Let D be a planar abelian difference set of even order n. Then $n = 2$, $n = 4$, or n is a multiple of 8.*

Proof If n is not a multiple of 4, Theorem 3.18 gives the identity

$$D + D^{(-1)} = 1 + G \qquad \text{in } \mathbb{Z}_2G. \tag{3.9}$$

The right-hand side of (3.9) contains $n^2 + n$ group elements with coefficient 1, whereas the left-hand side contains at most $2(n+1)$ such elements. Hence $2(n+1) \geqslant n^2 + n$ which implies $n = 2$. Now assume $n \equiv 4 \pmod 8$. Then n is a perfect square by Theorem 3.15, say $n = m^2$. By Corollary 3.16, there also exists a planar abelian difference set of order m. As m is even and not a multiple of 4, we conclude $m = 2$, by the previous case. □

A similar, though a little more involved argument can be used in the case $p = 3$ to prove the following result of Wilbrink [177]. Nobody has yet succeeded in using Theorem 3.18 for $p \geqslant 5$.

Corollary 3.20 *Let D be a planar abelian difference set of order n divisible by 3. Then either $n = 3$ or n is a multiple of 9.*

Note that the PPC for planar difference sets admits the following somewhat clumsy reformulation.

Conjecture 3.21 (Reformulation of the PPC) *Assume the existence of a planar difference set of order n. If $p^i|n$ for some prime p but $p^{i+1} \nmid n$, where i is a positive integer, then $n = p^i$.*

By Corollaries 3.19 and 3.20, this holds for the abelian case provided that $p^i \in \{2, 3, 4\}$. These three cases are still the only known general results towards the validity of the PPC for planes of type (a). For instance, it does not seem possible yet to exclude such a specific series of values as $n = p(p+2)$ (even if both p and $p+2$ are primes), a case considered by Ho [90]. Clearly much remains to be done in this area.

4 Type (b): Semiregular relative difference sets and planar functions

In this section Π will denote a plane of type (b), that is a projective plane of order n with an abelian (or, more generally, a quasiregular) automorphism group G of order n^2 fixing a flag (∞, L_∞) and with three point (and line) orbits. By omitting from Π the line L_∞ with all its points, and the point ∞ with all the lines through it, we obtain a symmetric net Δ. Thus the n^2 points of Δ split into n point classes of n points each given by the lines through ∞, and a line class consists of the lines through a point on L_∞. As Δ is the structure consisting of the faithful point and line orbit, the group G is a Singer group for Δ. Therefore Δ – and hence Π – may be represented by a relative difference set D with parameters $(n, n, n, 1)$, as explained in Section 2.

We note that planes of type (b) are (∞, L_∞)-*transitive* for the flag (∞, L_∞). Thus for any two points $p, p' \notin L_\infty$ on a line through ∞ there is a collineation φ fixing L_∞ pointwise such that $\varphi(p) = p'$. In other words, the forbidden subgroup N is always a group of (∞, L_∞)-*elations*.

The classical example is given by discarding one parallel class from AG$(2, q)$. More generally, semifields and planar functions also give rise to planes of type (b). We shall now discuss these two notions, beginning with semifields.

Loosely speaking, a proper semifield may be thought of as a (not necessarily commutative) field with non-associative multiplication. To be precise, a finite *semifield* is a finite set S on which two operations, addition and multiplication $(\cdot)$, are defined with the following properties.

(S1) (S,+) is an abelian group with identity 0.

(S2) $a \cdot (b + c) = a \cdot b + a \cdot c$ and $(a + b) \cdot c = a \cdot c + b \cdot c$ for all a, b, $c \in S$.

(S3) There is an element $1 \neq 0$ with $1 \cdot a = a = a \cdot 1$ for all $a \in S$.

(S4) If $a \cdot b = 0$, then $a = 0$ or $b = 0$.

The *middle nucleus* of S is defined by

$$\{x \in S \mid (a \cdot x) \cdot b = a \cdot (x \cdot b) \text{ for all } a, b \in S\}.$$

It is easily seen that the middle nucleus is a field, and hence any semifield can be viewed as a left or right vector space over its middle nucleus. Whereas this notion is very important in studying semifields, we will not really need it here and have included it for the sake of completeness. There are many constructions for semifields. We just mention here that a *proper semifield* – a semifield which is not a field – of order $q = p^r$ exists if and only if $r \geqslant 3$ for $p \neq 2$ and $r \geqslant 4$ for $p = 2$, see [46, p. 244]. For a detailed discussion of semifields, we refer the reader to Dembowski [46] and Hughes and Piper [104], where the term "division ring" is used instead. Commutative semifields which are two-dimensional over their middle nucleus GF(q) always arise from a construction of Cohen and Ganley [43]. They are of particular importance because of their intimate connections to geometric objects such as spreads and spread sets, q-clans, translation planes, eggs, translation generalised quadrangles and ovoids of generalised quadrangles, see Ball and Brown [18] and Ball and Lavrauw [19].

In order to construct a $(q, q, q, 1)$-RDS associated with the projective semifield plane Π of order $q = p^r$ determined by S, we need an explicit description of the corresponding affine semifield plane Σ, see Hughes and Piper [104]. The points of Σ are the pairs (x, y) with $x, y \in S$, and the lines are all point sets

$$[m, k] = \{(x, y) : mx + y = k\} \quad \text{with } m, k \in S$$

and all

$$[k] = \{(k, y) : y \in S\} \quad \text{with } k \in S.$$

The divisible semiplane Δ is obtained by deleting the parallel class of lines $[k]$ which corresponds to the special point ∞ on the line L_∞ of Π. Then the q^2 bijections

$$\alpha_{ab} : (x, y) \mapsto (x + a, y + ax + b) \quad \text{with } a, b \in S \tag{4.1}$$

are collineations of Σ. Indeed,

$$\alpha_{ab} : [m, k] \mapsto [m - a, k + ma + b - a^2] \quad \text{and} \quad [k] \mapsto [k + a].$$

These collineations form a group G acting regularly on the points and lines of Δ, with product $\alpha_{ab}\alpha_{a'b'} = \alpha_{a+a',b+b'+a'a}$. To simplify notation, we identify α_{ab} with the ordered pair (a, b) and consider G to be defined on $S \times S$ by

$$(a, b) * (a', b') = (a + a', b + b' + a'a). \tag{4.2}$$

Note that G is abelian if and only if S is a commutative semifield; we now assume this to be the case. Using (4.2), it is easy to show by induction on m that

$$(a, b)^m = \left(ma, mb + \frac{m(m-1)}{2}a^2\right) \quad \text{in } (G, *). \tag{4.3}$$

Thus (a, b) has order p for all $(a, b) \neq (0, 0)$ if p is odd; and for $p = 2$, the element (a, b) has order 2 whenever $a = 0$ and $b \neq 0$ and order 4 for $a \neq 0$. Finally, we write down a corresponding $(q, q, q, 1)$-RDS D explicitly. To do so, we choose the point (0,0) as base point and the line $\{(x, x) : x \in R\} = [-1, 0]$ as base line. By (4.1), the unique element of G mapping (0,0) to (a, b) is α_{ab}; hence

$$D = \{(x, x) : x \in S\} \subset G. \tag{4.4}$$

Thus we have the following theorem essentially due to Hughes [99]; see also Dembowski [46] and Jungnickel [106].

Theorem 4.1 *Let S be a semifield of order $q = p^r$. Then the set $S \times S$ together with the operation (4.2) is a group G which acts as a quasiregular group of type (b) on the semifield plane associated with S, and a corresponding $(q, q, q, 1)$-RDS is given by (4.4). Moreover, G is abelian if and only if S is commutative; in this case, G is elementary abelian if p is odd, and a direct product of cyclic groups of order 4 if $p = 2$.*

As far as we are aware, the only known examples of planes of even order with an abelian collineation group of type (b) are those defined over a commutative semifield of even order. It is an interesting (but probably rather difficult) problem to decide if there are any other examples. In the odd order case, the situation is different, as we shall soon see.

Next, we explain the notion of a planar function as introduced by Dembowski and Ostrom [47]. Let H and K be additively written, but not necessarily abelian, groups of order n. A *planar function of order n* is a mapping $f : H \to K$ such that for every $h \in H \setminus \{0\}$ the induced mapping $f_h : x \mapsto f(h + x) - f(x)$ is a bijection. Every planar function gives rise to a projective plane; this is due to [47]. We mention in passing that planar functions on cyclic groups have important applications in information theory and the communication sciences, see [129].

If a planar function from H to K exists, then $G = H \oplus K$ is a group of type (b) and $D = \{(x, f(x)) : x \in H\}$ is easily seen to be an $(n, n, n, 1)$-RDS in $G = H \oplus K$ relative to $N = \{0\} \oplus K$. Conversely, every *splitting* $(n, n, n, 1)$-RDS – that is, every RDS for which the forbidden subgroup N is a direct factor of the underlying group G – is of this type; see Kumar [129] and Pott [162]. In particular, in view of Theorem 4.1, any commutative semifield plane of odd order can be described by a planar function; see Dembowski [46, p. 245] for a more detailed discussion and some explicit examples. In [85], Hiramine characterised the planar functions over $(\mathrm{GF}(q), +)$ corresponding to semifield planes.

The reader should note that the explicit description of D in equation (4.4) does not immediately give a corresponding planar function. To obtain one such function, we note that the forbidden subgroup of G – in the setting of Theorem 4.1 – is the group $N = \{(0, y) : y \in S\}$, and that the complement of N in G is the group

$$H = \{(x, \frac{1}{2}x^2) : x \in S\},$$

as is easily seen using equation (4.2). In terms of this decomposition of G, a typical element of D splits as

$$(x, x) = (x, \frac{1}{2}x^2) * (0, x - \frac{1}{2}x^2).$$

By identifying the element $(x, \frac{1}{2}x^2) \in H$ with $x \in S$, we see that the mapping

$$h : x \mapsto x - \frac{1}{2}x^2 \qquad \text{for } x \in S \tag{4.5}$$

is a planar function over $(S,+) \cong (\mathrm{GF}(q),+)$, where the multiplication in (4.5) is that of the semifield. The reader may easily check this assertion. We remark that Dembowski and Ostrom [47] give the planar function $x \mapsto x^2$ instead.

It is easy to see that the order of a planar function has to be odd, see [47]. All known examples of planar functions occur in the situation where $H = K$ is the additive group of a finite field $\mathrm{GF}(q)$ of order $q = p^e$, p odd. By the Lagrange interpolation theorem, every function $f : \mathrm{GF}(q) \to \mathrm{GF}(q)$ may be obtained as the evaluation map of some polynomial. Accordingly, $f \in \mathrm{GF}(q)[X]$ is said to be a *planar polynomial* if the polynomial $\Delta_{f,a}$ defined as

$$\Delta_{f,a}(X) = f(X+a) - f(X)$$

is a *permutation polynomial* over $\mathrm{GF}(q)$ for each $a \in \mathrm{GF}(q)^* = \mathrm{GF}(q) \setminus \{0\}$. In other words, $\Delta_{f,a}$ induces a permutation of $\mathrm{GF}(q)$. Any polynomial $f \in \mathrm{GF}(q)[X]$ may be reduced modulo $X^q - X$ to a polynomial of degree less than q which induces on $\mathrm{GF}(q)$ the same function as f; this is called the *reduced form* of f. The planes generated by a planar polynomial f and by $f+L$, where L is an additive polynomial, are isomorphic, as are those generated by $f(L)$ or $L(f)$ if L is a permutation polynomial; see Coulter and Matthews [44]. In their seminal paper on planar functions [47] Dembowski and Ostrom described a class of polynomials which sometimes give rise to planar functions. Such a polynomial, in reduced form, is given by

$$f(X) = \sum_{i,j=0}^{e-1} a_{ij} X^{p^i+p^j}. \tag{4.6}$$

Following Coulter and Matthews [44], we call any polynomial of this type a *Dembowski-Ostrom polynomial.* If the coefficients a_{ij} satisfy the condition

$$\sum_{i,j=0}^{e-1} a_{ij}(x^{p^i}y^{p^j} + x^{p^j}y^{p^i}) = 0 \quad \text{if and only if } x = 0 \text{ or } y = 0, \tag{4.7}$$

then f is a planar polynomial. Dembowski and Ostrom noted that any planar polynomial of the form (4.6) produces a translation plane. According to [44], all known planar Dembowsi-Ostrom polynomials are *equivalent* – in the sense that they define isomorphic planes – to one of the following types.

(i) $f(X) = X^2$, which gives the desarguesian plane over $\mathrm{GF}(q)$, q odd.

(ii) $f(X) = X^{p^\alpha+1}$ (already considered in [47]), which is planar over $\mathbb{F}_{p^e}$, p odd, if and only if $e/(\alpha, e)$ is odd.

(iii) $f(X) = X^{10} + X^6 - X^2$, which is planar over $\mathrm{GF}(3^e)$ if and only if $e = 2$ or e is odd.

There is just one further class of planar functions known. Coulter and Matthews [44] proved that the polynomial

$$f(x) = X^{(3^\alpha+1)/2} \tag{4.8}$$

is planar over $GF(3^e)$ provided that α is odd and $(\alpha, e) = 1$. The corresponding projective planes are not translation planes but of Lenz-Barlotti type[1] II.1 and are called the *Coulter-Matthews planes*. It is an interesting open problem whether or not there are similar constructions in the case of a characteristic other than 3.

In the case $q = p$ a prime, the classification of planar functions was settled in 1989 and 1990 by Gluck [78], Hiramine [84], and Rónyai and Szőnyi [164], who independently showed that any planar polynomial over a prime field must reduce to a quadratic. This established Conjecture 1.2 for projective planes with a group of type (b) and, more generally, the conjecture of Kallaher [120, p. 145] that any affine plane of prime order with a point-transitive collineation group must be desarguesian; see [84] for details.

Planar functions and abelian groups of type (b) have been investigated for more than three decades using geometric and algebraic methods, but until recently the restrictions obtained were comparatively weak. The first conclusive result appeared in 1976 when Ganley [64] proved the PPC for the even order case. Another result worth mentioning came in 1996, when Ma [139] proved that n can not be the product of two distinct primes. Further work in the area was done by Kumar [129], Hiramine [85, 86], Fung, Siu and Ma [63], Ma and Pott [140] and Leung, Ma and Tan [136]. Finally, Blokhuis, Jungnickel and Schmidt [27] proved the validity of the PPC for planes of type (b) when n is odd. In the remainder of this section, we will sketch a proof for the PPC, beginning with Ganley's result.

Theorem 4.2 *Let D be a relative difference set with parameters $(n, n, n, 1)$ in an abelian group G, where n is even. Then n is a power of 2, say $n = 2^b$, G is the direct product of b cyclic groups of order 4, and the forbidden subgroup N is isomorphic to $\mathbb{Z}_2^b$.*

Ganley's proof of Theorem 4.2 proceeded via coordinatising the corresponding projective plane by a certain type of cartesian group and studying the absolute points of suitable polarities. This rests on results of Dembowski and Ostrom [47]. A simpler, and considerably shorter, proof, which uses D to (implicitly) produce an oval of the associated projective plane, was given in 1987 by the second author [108]. In particular, he established the following lemma.

Lemma 4.3 *Let D be a relative difference set with parameters $(n, n, n, 1)$ in an abelian group G relative to N, where n is even, and assume $0 \in D$. Then:*

[1] A brief discussion of the famous Lenz-Barlotti classification of projective planes with respect to their central collineations will be given in Section 8.

(i) *D is a system of coset representatives for N;*

(ii) *N contains all involutions of G;*

(iii) $2D = N$.

We will sketch a proof for this result at the beginning of Subsection 10.4. It is then easy to finish the proof of Theorem 4.2. First one shows that $2N = 0$. By way of contradiction, assume $2b \neq 0$ for some $b \in N$, and choose $d \in D$ such that $2b = 2d$. If $d \notin N$, then $2(d-b) = 0$ contradicts (ii). Thus $d \in N$; but then $2x \neq 0$ for all $x \in D \setminus \{d\}$ by (i) and (ii), and thus $2d = 0$ by (iii), which is absurd. Hence $2N = 0$. Now $2G = N$, hence $4G = 0$, and so G is a direct product of factors isomorphic to $\mathbb{Z}_4$.

We now state the results of Blokhuis, Jungnickel and Schmidt [27].

Theorem 4.4 *Let D be a relative difference set with parameters $(n,n,n,1)$ in an abelian group G of order n^2, where n is odd. Then n is a prime power, say $n = p^b$, and G has rank $\geqslant b+1$.*

Corollary 4.5 *If there is a planar function of order n between abelian groups, then n is a prime power.*

The proof of Theorem 4.4 uses the RDS-equation (2.2) which here reads

$$DD^{(-1)} = n + G - N \quad \text{in } \mathbb{Z}G \tag{4.9}$$

and some simple auxiliary results. The first of these concerns the *p-rank* $r_p(G)$ of a finite abelian group G, namely the number of groups whose order is a power of p in a decomposition of G into cyclic groups of prime power order. The lemma in question gives this rank in terms of the coefficient of 1 in a suitable group ring element, and is straightforward to check.

Lemma 4.6 *Let G be a finite abelian group, let N be a subgroup of G, and let p be a prime. Then*

$$[G^{(p)}]_1 = p^{r_p(G)} \quad \text{and} \quad [G^{(p)}N]_1 = p^{r_p(G/N)}|N|. \tag{4.10}$$

The heart of the proof is the following lemma which uses a trick first introduced in [26].

Lemma 4.7 *Let G be an abelian group and $D \in \mathbb{Z}G$ with $|D| = k$, and assume*

$$DD^{(-1)} = k + X \quad \text{and} \quad DX = aG$$

for some integer a and some $X \in \mathbb{Z}G$. Furthermore, let p be an odd prime dividing k. Then

$$(p-1)k^2 \leqslant k[X + X^{(p)}]_1 + [XX^{(p)}]_1 \tag{4.11}$$

with equality if and only if the only coefficients in $D^{(-1)}D^{(p)}$ are 0 and p.

Proof Write $A := D^{(-1)}D^{(p)} = \sum a_g g$. Then $\sum a_g = k^2$. Since G is abelian, we have $D^{(p)} = D^p$ in $\mathbb{Z}_pG$. As k is divisible by p and $p \geqslant 3$, we get

$$A = (k+X)D^{p-1} = XD^{p-1} = aGD^{p-2} = akGD^{p-3} = 0$$

in $\mathbb{Z}_pG$. Hence all a_g are divisible by p and thus

$$\sum a_g^2 \geqslant p\sum a_g = pk^2$$

with equality if and only if $a_g \in \{0, p\}$ for all g. On the other hand,

$$AA^{(-1)} = (k+X)(k+X^{(p)}) = k^2 + k(X+X^{(p)}) + XX^{(p)}$$

and thus

$$\sum a_g^2 = [AA^{(-1)}]_1 = k^2 + k[X+X^{(p)}]_1 + [XX^{(p)}]_1.$$

This proves the lemma. □

As D is a system of coset representatives for N, we have $DN = G$, and in view of (4.9), we may apply Lemma 4.7 with $X = G - N$ and $k = n$. Together with Lemma 4.6, we obtain the following result.

Lemma 4.8 *Let D be an $(n, n, n, 1)$-RDS in an abelian group G relative to N, and let $p \geqslant 3$ be a prime divisor of n. Then*

$$(p-2)n \leqslant p^{r_p(G)} - p^{r_p(N)} - p^{r_p(G/N)}. \tag{4.12}$$

Now we are ready to prove Theorem 4.4. Recall that G has order n^2. If n is not a prime power, then there is a prime divisor p of n such that the Sylow p-subgroup S of G has order less than n. But then $p^{r_p(G)} \leqslant |S| < n$ contradicting (4.12). Thus n is a prime power, say $n = p^b$. Another application of (4.12) shows $p^{r_p(G)} > n$, and hence G must have rank at least $b+1$.

Though the results of [27] establish the PPC in case (b), they still fall short of what one might hope for. It is entirely reasonable to conjecture that G has rank $2b$ – and thus is the additive group of a finite vector space – in the situation of Theorem 4.4. Similarly, one may conjecture that planar functions of order $q = p^r$ only exist over the group $(\mathrm{GF}(q), +)$. For $r = 1$, this is true by the results of Gluck [78], Hiramine [84], and Rónyai and Szőnyi [164] already mentioned; and for $r = 2$ it holds by Ma and Pott [140]; all other cases remain open.

5 Type (d): Affine difference sets

In this section Π will denote a plane of type (d), that is, a projective plane with an abelian (or, more generally, quasiregular) automorphism group G of order $n^2 - 1$ fixing an antiflag (∞, L_∞), and with three point (and line) orbits. By omitting from Π the line L_∞ with all its points, and the point ∞ with all the lines through it, we obtain a biaffine plane Δ. Thus the $n^2 - 1$ points of Δ split into $n + 1$ point classes of $n - 1$ points each, given by the lines through ∞, and a line class consists of the lines through a point on L_∞. Note that Δ is the structure consisting of the faithful point and line orbits, hence the group G is a Singer group for Δ. Therefore Δ may be represented by an affine difference set D of order n, as explained in Section 2.

We note that planes of type (d) are (∞, L_∞)-*transitive* for the antiflag (∞, L_∞). So, for any two points $p, p' \neq \infty$ and $p, p' \notin L_\infty$ on a line through ∞ there is a collineation φ fixing ∞ and L_∞ pointwise such that $\varphi(p) = p'$. In other words, the forbidden subgroup N is always a group of (∞, L_∞)-*homologies.*

A fundamental result of Bose [29] provides the classical example, namely $\mathrm{AG}(2, q)$ *punctured* at its origin $(0, 0)$: that is, $\Delta = \mathrm{AG}(2, q) \setminus \{(0, 0)\}$ with the $n + 1$ lines through $(0, 0)$ deleted. To simplify notation we shall denote this example by $\mathrm{BAG}(2, q)$. We note that the special choice of the origin does not affect the result of puncturing, as $\mathrm{AG}(2, q)$ has a point-transitive group.

Theorem 5.1 $\mathrm{BAG}(2, q)$ *has a cyclic Singer group. Thus there exists an affine difference set of order* q *in* $\mathbb{Z}_{q^2-1}$ *for every prime power* q.

The proof of Theorem 5.1 is analogous to the standard proof of Singer's [168] theorem. This result of Bose [29] was the starting point for the investigation of *cyclic affine planes* – those affine planes of order n which admit a cyclic group of order $n^2 - 1$. They have been studied extensively, beginning with the work of Hoffman [96] who already stated the PPC for cyclic affine difference sets. The two papers of Bose [29] and of Hoffman [96] started the theory of affine difference sets in much the same way as the work of Singer [168], followed by that of Hall [82], started the theory of planar difference sets. Interestingly, the results – and to a considerable extent also the methods – for both the affine and the planar case of the PPC are quite parallel.

Some non-abelian affine difference sets describing desarguesian planes of odd order were given by Ganley and Spence [68], but no non-abelian examples seem to be known for even order planes. As mentioned in Section 3, abelian but non-cyclic planar difference sets would have to describe non-desarguesian planes. We do not know whether an analogous result holds for affine difference sets.

Arasu [5], extending a result of Hoffmann, obtained subaffine difference sets of certain families of affine difference sets. As a consequence he proved that if there exists a cyclic affine plane of order m^{2r+1}, where r is any nonnegative integer, then there exists one of order m. Thus any nonexistence result for a cyclic affine plane of order m extends to planes of order m^s, where s is any odd (positive) integer. In certain instances, these reduction theorems also hold for abelian affine planes.

As in the planar case, a stronger version of the PPC conjectures that any affine plane associated with an affine difference set is necessarily desarguesian. In [99] Hughes considered relative difference sets with parameters $(n+1, n-1, n, 1)$ – which he called partial difference sets – and mentioned that the only examples known arise from desarguesian planes. It is therefore conjectured that this has to be the case. However, the evidence for this conjecture is not at all extensive; in particular, no affine analogue of the Ott-Ho theorem is yet known. A systematic study of affine difference sets was given by Jungnickel [112] who concentrated on nonexistence results, giving some evidence for the validity of PPC. To a large extent, we shall follow his account.

As with ordinary difference sets, multipliers are a central tool in the theory of affine difference sets. Again, a *multiplier* for D is an automorphism α of G inducing a collineation of $\Delta = \operatorname{dev} D$. A multiplier of the special form $\alpha : x \mapsto tx$ for some integer t with $(t, n^2-1) = 1$ is called a *numerical multiplier*. By abuse of langage, t itself is also said to be a multiplier.

Hoffman [96] proved the affine analogue of Hall's multiplier theorem for planar difference sets: every prime divisor p of n is a multiplier of every cyclic affine difference set of order n. This result remains true for abelian affine difference sets, as a special case of the multiplier theorem of Elliott and Butson [57] for relative difference sets. A shorter and more transparent proof was given by the second author in [112]. This proof is a special case of Pott's proof [158] for the multiplier theorem of [57]. We now present an even simpler proof, similar to that for the first multiplier theorem.

Theorem 5.2 (Affine multiplier theorem) *Let D be an abelian affine difference set of order n. Then every prime divisor p of n is a multiplier of D.*

Proof We use the group ring setting, writing the underlying group G multiplicatively throughout. As D is an $(n+1, n-1, n, 1)$-RDS, it meets all but one of the cosets of N exactly once. We may assume, without loss of generality, that $D \cap N = \emptyset$ in what follows. It will suffice to prove that

$$|D^{(p)} \cap Dg| \geqslant 1 \qquad \text{for all } g \notin N, \tag{5.1}$$

for then the set L, obtained from $D^{(p)}$ by adjoining the point of the projective extension Π of $\operatorname{dev} D$ corresponding to the parallel class of D, meets every line of Π and hence is itself a line of Π, by Lemma 3.7. As in the proof of Theorem 3.5, (5.1) follows from the congruence

$$|D^{(p)} \cap Dg| \equiv 1 \pmod{p} \qquad \text{for all } g \notin N \tag{5.2}$$

which we prove by evaluating the group ring element $D^{(p)}D^{(-1)}$ modulo p. Using the RDS-equation $DD^{(-1)} = n+G-N$ and the congruence (3.5) together with $DG = nG$ and $DN = G - N$, we compute in $\mathbb{Z}_pG$ as follows, remembering that p divides n:

$$D^{(p)}D^{(-1)} = D^{p-1}(DD^{(-1)}) = D^{p-1}(G-N) = G-N.$$

By Lemma 2.3, this already gives the desired congruence (5.2). □

Theorem 5.2 is helpful both for constructing affine difference sets as well as in nonexistence proofs. In order to apply it, we need an affine analogue of Lemma 3.8 which is due to Hoffmann [96] in the cyclic case, and to the second author [112] in general. The proof of Lemma 3.8 carries over to give the following result.

Lemma 5.3 *If D is an abelian affine difference set, then there exists a translate $D+g$ fixed by all multipliers of D.*

Example 5.4 As in the planar case, multipliers can be used to simplify the construction of affine difference sets by trial and error. If $n = p^a$ is a prime power, we know by Theorem 5.1 that there exists a cyclic affine difference set D of order n. By Theorem 5.2 and Lemma 5.3 we may assume that D is fixed by the multiplier p, so $pD = D$. Thus D is the union of orbits of the automorphism $x \mapsto px$ of G. For instance: for $n = 4$, one easily finds that $D = \{1,2,4,8\} \subset \mathbb{Z}_{15}$. For $n = 8$, one orbit is $\{1,2,4,8,16,32\}$ which, together with $\{21,42\}$ (the unique orbit of length 2), forms the affine difference set $\{1,2,4,8,16,21,32,42\} \subset \mathbb{Z}_{63}$.

For nonexistence results the following affine analogue of Lemma 3.11 due to Ko and Ray-Chaudhuri [127] is useful; see also [112] for the simple proof.

Lemma 5.5 *Let t_1, t_2, t_3, t_4 be four multipliers of D such that $t_1 - t_2 = t_3 - t_4$. Then $\exp G$ divides the least common multiple of $t_1 - t_2$ and $t_1 - t_3$.*

Example 5.6 Assume the existence of an abelian affine difference set D of order n divisible by 10. Noting that 5, 4, 2 and 1 are multipliers of D satisfying $5-4 = 2-1$, we conclude from Lemma 5.5 that $\exp G$ divides 3. This implies that $n^2 - 1 = (n-1)(n+1) = 3^a$, which is clearly absurd. Thus no abelian affine difference set has order a multiple of 10.

Along the lines of Example 5.6, one may prove the following result due to Hoffmann [96] in the cyclic case and to the second author [112] in general.

Proposition 5.7 *There is no abelian affine difference set of order n whenever n is a multiple of one of the following numbers:*

$$\begin{aligned} 2p &\text{ for } p \in \{3,5,7,11,13,17,19,29,31,47,61\},\\ 3p &\text{ for } p \in \{5,7,11,13,17,19,29\},\\ 5p &\text{ for } p \in \{11,13,29\}. \end{aligned}$$

In the cyclic case, one can also exclude the cases $2p$ for $p \in \{23,67\}$.

This sort of reasoning becomes a lot simpler if we restrict attention to the cyclic case, since then $n^2 - 1$ has to divide the least common multiple of $t_1 - t_2$ and $t_1 - t_3$ which is impossible; see Ko and Ray-Chaudhuri [127] and Jungnickel [112, Theorem 3.7].

Theorem 5.8 *Let t_1, t_2, t_3, t_4 be multipliers for a cyclic affine difference set of order n and assume that $t_1 \not\equiv t_2, t_3 \pmod{n^2 - 1}$. Then $t_1 - t_2 \not\equiv t_3 - t_4$.*

Ko and Ray-Chaudhuri [127] used Theorem 5.8 to establish the PPC for cyclic affine difference sets of order $n \leqslant 5,000$. Later, the PPC was verified for abelian affine difference sets up to $n \leqslant 10,000$ by Jungnickel and Pott [116]. It should be no problem to extend the range of this result considerably - a nice project for a small thesis.

Next, we will discuss the affine analogues of the Mann test and of Wilbrink's theorem without giving proofs, for which we refer the reader to [112]. We begin with the Mann test. For simplicity, we shall only state the abelian case. Weak versions of this result are due to Hoffman [96] and Elliott and Butson [57], whereas the full result - including a non-abelian version - is due to the second author [112, Theorem 4.1]. The rather technical proof uses elementary but lengthy group ring computations, similar to the one given in the case of ordinary difference sets in [116].

Theorem 5.9 *Let D be an abelian affine difference set of order n in G with respect to N, and let U be a subgroup of index $u \neq 1$ of G which does not contain N. Put $H = G/U$ and write $u^* = \exp H$. Finally, let t be a multiplier for D and let p be a prime divisor of n satisfying $tp^f \equiv -1 \pmod{u^*}$ for some non-negative integer f. Then n is a perfect square.*

Corollary 5.10 *Let D be an abelian affine difference set of order n, and let p and q be primes dividing n and $n - 1$, respectively, where $q \neq 2$. If p has even order modulo q (in particular, if p is a quadratic non-residue modulo q), then n is a perfect square.*

One of the main results of Hiramine [87] says that $\mathrm{ord}_q(p)$ is odd whenever q and p are prime divisors of $n-1$ and the square-free part n^* of n, respectively. Note that this restriction - which was obtained using a different method - is weaker than the one provided by Corollary 5.10. Some applications of the non-abelian version of the Mann test are given in [112]. We mention the following result originally proved by Arasu, Davis, Jungnickel and Pott [7] which applies to an infinite class of non-abelian groups; see [112, Theorem 4.7].

Theorem 5.11 *Let D be a splitting affine difference set of order n in $G = N \oplus H$, where N is abelian. If q and p are primes dividing $n - 1$ and the square-free part of n, respectively, then p has odd order modulo q (and thus is a quadratic residue).*

The following affine analogue of Wilbrink's theorem is due to Pott [159]. In [112] the second author gave a more elementary proof using direct computations in $\mathbb{Z}G$ modulo p and p^2, respectively. This is similar to the proof of Wilbrink's theorem in [111] and our proof of Theorem 3.18.

Theorem 5.12 *Let D be an affine difference set of order n in an abelian group G (written multiplicatively, with forbidden subgroup N). Let p be a prime dividing n such that p^2 does not divide n, and assume, without loss of generality, that D is fixed under the multiplier p. Then the following identity holds in $\mathbb{Z}_pG$:*

$$D^{p-1} + (D^{(-1)})^{p-1} = 1 + 2G - N. \tag{5.3}$$

Note that the assumption that D is fixed by the multiplier p is no restriction in view of Theorem 5.2 and Lemma 5.3. Using Theorem 5.12, Pott [159] proved the following affine analogue of Corollary 5.13. We note that the proof requires several steps which are not just analogues of Wilbrink's arguments. As in the planar case, for $p \geqslant 5$ equation (5.3) becomes so complex that it has not (yet) been possible to derive from it any restrictions.

Corollary 5.13 *Let D be an abelian affine difference set of order n divisible by 3. Then either $n = 3$ or n is a multiple of 9.*

There are several papers concerning an affine analogue of Corollary 3.19, namely Ko and Ray-Chaudhuri [128], Jungnickel [111], Arasu [4] and Arasu and Jungnickel [10], where the following result was obtained.

Theorem 5.14 *Let D be an abelian affine difference set of even order n. Then $n = 2$, $n = 4$ or n is divisible by 8. In the last case, the following two conditions are satisfied.*

(i) *There exists a Paley-Hadamard difference set $\tilde{D}$ of order $n/4$, that is, an $(n-1, n/2, n/4)$-difference set, in the forbidden subgroup N of G which admits every prime divisor of n as a multiplier.*

(ii) *Either n is a square or each prime p dividing n is a quadratic residue for each prime q dividing $n - 1$.*

A considerably simplified proof of Theorem 5.14 can be found in [112]. We note that all known proofs are quite different from that of Corollary 3.19. Later, Arasu and Pott [12] considered the first open case, namely $n \equiv 8 \pmod{16}$, and proved the following further restrictions.

Theorem 5.15 *Let D be an abelian affine difference set of order $n \equiv 8 \pmod{16}$. Then $n - 1$ must be a prime power. If D is cyclic, then $n - 1$ must be a prime.*

In the proof of Theorems 5.14 and 5.15, the $(n-1, n/2, n/4)$-difference set $\tilde{D}$ in the forbidden subgroup N plays a crucial role. If the affine difference set D corresponds to the desarguesian plane $\mathrm{AG}(2,q)$, it can be described explicitly in the cyclic group $\mathrm{GF}(q^2)^*$ as $D = \{\alpha \in \mathrm{GF}(q^2) : \alpha + \alpha^q = 1\}$. Then $n = 2^t$ and $\tilde{D}$ is equivalent to the complement of a classical Singer difference set with parameters $(2^t - 1, 2^{t-1} - 1, 2^{t-2} - 1)$. If n is odd, it seems impossible to construct a difference set in N. However, for $n \equiv 3 \pmod 4$, one can construct a *difference family* of a very special type in N, see Yamada [178] and the discussion given in Pott [161, §5.2].

In [174] VanAken recalled previous work on cyclic affine difference sets of order $n \equiv 8 \pmod{16}$ and added some further technical restrictions concerning the orders of prime divisors of n. All these results together formed the basis of a computer search which showed that cyclic affine difference sets of even order $n < 100\,000$ have order a power of 2, except possibly for four cases which were still undecided. Also, Broad and VanAken [31] applied several known numerical constraints in the same case and used an extensive computer search to show that no cyclic affine difference set of order $8 < n \leqslant 1,000,000$ with $n \equiv 8 \pmod{16}$ exists.

Finally, we mention that Arasu and Jungnickel [10] also obtained a non-abelian version of Theorem 5.14 leading – together with the celebrated Feit-Thompson theorem [60] – to the following strong restriction which is still the only known general nonexistence result for the non-abelian case.

Theorem 5.16 *There is no splitting affine difference set of order* $n \equiv 2 \pmod 4$. *In particular, there exists no nilpotent affine difference set for any such order.*

One can also use more geometric arguments to study large abelian groups of finite projective planes and investigate certain associated *polarities* – isomorphisms of order two between the plane and its dual. This approach goes back to Ganley and Spence [68]. In the affine case, Arasu and Pott [11] proved the following result using polarities, strengthening a previous theorem of Arasu, Davis, Jungnickel and Pott [7] obtained by purely algebraic methods.

Theorem 5.17 *Let D be an abelian affine difference set in G relative to N. Then the Sylow 2-subgroup of G is cyclic. Moreover, the Sylow 2-subgroup of the multiplier group of D contains exactly one involution, namely the numerical multiplier n. Hence, if the multiplier group is abelian – in particular, if G is cyclic – then the Sylow 2-subgroup of the multiplier group is also cyclic.*

As noted by Arasu and Pott [11], the preceding result is enough to verify the PPC for orders up to $10,000$ in the abelian case by purely geometric reasoning with only nine exceptions. An alternative proof for the first part of Theorem 5.17 was given by Hiramine in [87] and some further, rather technical restrictions on the structure of the Sylow p-subgroups of G were provided by Dizon-Garciano and Hiramine [54].

6 Type (e): The Baer case

In this section Π will denote a plane of type (e), that is, a projective plane with an abelian (or, more generally, quasiregular) automorphism group G of order $n^2 - \sqrt{n}$ without fixed points and lines and with two point orbits, one of which is the set of points of a Baer subplane Π_0. Thus $\Delta = \Pi \setminus \Pi_0$ is a Baer semiplane of order n admitting G as a Singer group and may be represented by an $(n + \sqrt{n} + 1, n - \sqrt{n}, n, 1)$-RDS D in G relative to the stabiliser G_{Π_0}, as explained in Section 2. The point classes are the points of the lines of the omitted Baer subplane Π_0, and the line classes are the line pencils through the points of Π_0.

The only known abelian example is described by the $(7, 2, 4, 1)$–RDS $D = \{0, 1, 4, 6\}$ in $\mathbb{Z}_{14}$. It corresponds to $\Delta = \mathrm{PG}(2, 4) \setminus \mathrm{PG}(2, 2)$. There is also a non-abelian example for the Hughes plane of order 9 associated with an RDS in $\mathbb{Z}_{13} \times S_3$ which we will discuss next. Finally, there are two examples of the desarguesian planes of orders 9 and 16 which correspond to relative difference sets where the forbidden subgroup is not normal; see Jungnickel [106, Corollary 4.10].

As in Hughes [102], Denniston [51] and Martin [143], the Hughes plane Π of order 9 can be described as follows. The point set of Π is

$$\{X_i : X = A, B, C, D, E, F, G;\ i \in \mathbb{Z}_{13}\}$$

and the lines are obtained from the action of $\mathbb{Z}_{13}$ on the seven base lines $L_1, \dots, L_7$ displayed below. More precisely, the lines of Π are the 91 sets $L_i + k$, where $L_i + k$ is obtained from L_i by adding k to all subscripts ($i = 1, \dots, 7$; $k \in \mathbb{Z}_{13}$). As $\{0, 1, 3, 9\} \subset \mathbb{Z}_{13}$ is a difference set for $\mathrm{PG}(2, 3)$, it is clear that the point set $V = \{A_0, \dots, A_{12}\}$, together with the lines $(L_1 + k) \cap V$ ($k \in \mathbb{Z}_{13}$), form a Baer subplane Π_0 of Π. The corresponding Baer semiplane Δ then has as its lines the orbits of $\mathbb{Z}_{13}$ on the base lines $L_i' = L_i \setminus \{A_0\}$ ($i = 2, \dots, 7$).

$$\begin{aligned}
L_1 &= \{A_0, A_1, A_3, A_9, B_0, C_0, D_0, E_0, F_0, G_0\};\\
L_2 &= \{A_0, B_1, B_8, D_3, D_{11}, E_2, E_5, E_6, G_7, G_9\};\\
L_3 &= \{A_0, C_1, C_8, E_7, E_9, F_3, F_{11}, G_2, G_5, G_6\};\\
L_4 &= \{A_0, B_7, B_9, D_1, D_8, F_2, F_5, F_6, G_3, G_{11}\};\\
L_5 &= \{A_0, B_2, B_5, B_6, C_3, C_{11}, E_1, E_8, F_7, F_9\};\\
L_6 &= \{A_0, C_7, C_9, D_2, D_5, D_6, E_3, E_{11}, F_1, F_8\};\\
L_7 &= \{A_0, B_3, B_{11}, C_2, C_5, C_6, D_7, D_9, G_1, G_8\}.
\end{aligned}$$

Now Π admits $S_3 \times \mathbb{Z}_{13}$ as a quasiregular collineation group, and thus there exists an RDS with parameters $(13, 6, 9, 1)$ in this group. The first example of such a difference set is due to Carol Whitehead, see [68, p. 153]. We will present a different, but equivalent description given by the second author [109,

§8] which allows us to show the connection between the RDS-representation of Δ and the description just given. Put

$$D = \{(id, 2), (id, 5), (id, 6), (\tau, 1), (\tau, 8), (\sigma\tau, 7), (\sigma\tau, 9), (\tau\sigma, 3), (\tau\sigma, 11)\},$$

where the elements of S_3 are $\sigma = (0, 1, 2)$, $\sigma^2 = (0, 2, 1)$, $\tau = (1, 2)$, $\tau\sigma = (0, 1)$, $\sigma\tau = (0, 2)$ and id. Now write the coordinates in $\mathbb{Z}_{13}$ as subscripts and identify the elements of S_3 with B, C, D, E, F, G as follows:

$$B = \tau,\ C = \sigma^2,\ D = \tau\sigma,\ E = id,\ F = \sigma,\ G = \sigma\tau.$$

Then $L'_2 = D$, $L'_3 = D\sigma\tau$, $L'_4 = D\sigma$, $L'_5 = D\tau$, $L'_6 = D\tau\sigma$ and $L'_7 = D\sigma^2$. Thus we have the desired explicit representation $\Delta = \operatorname{dev} D$.

Unfortunately, this example is sporadic, and no larger Hughes plane allows anything comparable. Similarly, no desarguesian plane of order $q \neq 4$ allows a representation by an abelian Baer-RDS. Indeed, we want to offer the following conjecture.

Conjecture 6.1 *The only projective planes admitting a quasiregular collineation group of type (e) are* PG(2, 4) *and the Hughes plane of order* 9, *and an abelian group of this type exists only in the former case.*

While we have to admit that there is not much evidence for the non-abelian part of the preceding conjecture, there are overwhelming reasons to believe in the abelian part. In particular, we have the following strong restrictions due to Ganley and Spence [68].

Theorem 6.2 *Let* Π *be a projective plane of type (e). If* n *is a prime power, then* $n = 4$. *Otherwise* n *is odd and has no prime divisors* $\equiv 3 \pmod 4$.

In their proof of Theorem 6.2, Ganley and Spence used geometric arguments concerning the number of absolute points of polarities and a multiplier theorem (see [68, Theorem 7.1]) obtained by modifying the arguments given by Elliot and Butson [57].

Further evidence for Conjecture 6.1 is provided by the fact that the corresponding relative difference sets would be "liftings" of the complements of planar difference sets. If D is an $(n + \sqrt{n} + 1, n - \sqrt{n}, n, 1)$-RDS in G relative to N, then the canonical image of D in G/N is an $(n + \sqrt{n} + 1, n, n - \sqrt{n})$-difference set $\tilde{D}$, and thus the complement of $\tilde{D}$ is a planar difference set of order $\sqrt{n}$. In particular, the validity of the PPC for the planar case would imply that in case (e). Also, in the abelian case, Conjecture 6.1 holds for $\sqrt{n} \leqslant 2,000,000$ by the result of Gordon [79] already quoted.

We note that certain liftings of the complements of the classical Singer difference sets exist by results of Arasu et al. [8, 9] (see also Pott [161, Theorem 3.3.4]), but one can not lift them so far that the λ-value becomes 1. In any case, the existence question for planes of type (e) is related to how far we can extend the complement of a planar difference set to a relative difference set.

Relative difference sets representing Baer semiplanes have been also investigated by the second author in [109], who introduced a definition of arcs in divisible semiplanes. Arcs of $\Delta = \Pi \setminus \Pi_0$ are arcs of Π that interact in a prescribed manner with the omitted Baer subplane. The precise definition is chosen in such a way that Baer subplanes, and then the relative difference set D representing them, always contain a large collection of arcs related to D, namely $-D$ and all its translates. The existence problem for large arcs in Baer semiplanes poses interesting questions regarding the interaction of Baer subplanes and ovals (respectively conics) in both desarguesian and non-desarguesian planes. This leads to a possible geometric attack on the abelian part of Conjecture 6.1, as we shall briefly explain.

The study of arcs in Baer semiplanes turns out to be considerably more difficult than the study of arcs in symmetric nets and biaffine planes; compare Section 10 for results in these two cases. It is proved in [109] that any oval in a projective plane of order $n = m^2$ ($n \geqslant 7$) can only contain h-arcs of a corresponding Baer semiplane Δ with $h < m^2 - \sqrt{m}$. In particular, one never obtains an oval of Δ in this way, unless $n = 4$. Moreover, no desarguesian Baer semiplane of order $n \neq 4$ contains any oval. In fact, for $n \neq 4$, no example of an oval in any Baer semiplane is known to us. The reader is referred to [109] for further details and for some general constructions of relatively large complete arcs in Baer semiplanes, which can not arise from conics and regular hyperovals in $\mathrm{PG}(2, q)$. In [109] also arcs in the Baer semiplane associated with the Hughes plane of order 9 are considered. Apparently, the construction of large arcs in Baer semiplanes is a rather difficult problem. On the other hand, any Baer semiplane belonging to a relative difference set would have to contain an oval (many ovals in fact) and these ovals would not extend in the corresponding projective plane, unless $n = 4$. So it seems interesting to investigate arcs in non-desarguesian Baer semiplanes. In [109], the following conjecture is stated which would prove Conjecture 6.1 in the abelian case, and hence establish the PPC in case (e). Some evidence for this conjecture is given in [109, §9], where several restrictions on arcs belonging to a putative abelian Baer-RDS of odd order are obtained.

Conjecture 6.3 *The only Baer semiplane containing an oval is the semiplane* $\Delta = \mathrm{PG}(2,4) \setminus \mathrm{PG}(2,2)$.

7 Type (f): Direct product difference sets

In this section Π will denote a plane of type (f), that is, a projective plane with an abelian (or, more generally, quasiregular) automorphism group G of order $n(n-1)$ which fixes a double flag $(\infty_A, \infty_B, L_\infty)$, together with a further line L_A through ∞_A, and acts regularly on the $n(n-1)$ points not incident with either of the two special lines L_∞ and L_A and on the $n(n-1)$ lines not incident with either of the two special points. Planes of type (f) may be represented by a variant of relative difference sets, namely the *direct product difference sets* (DPDS) introduced by Ganley [65]. Using group ring notation, a DPDS of order n may be defined to be a subset D of a group G of order $n(n-1)$ with two normal subgroups A and B of orders n and $n-1$, respectively, which satisfies the equation

$$DD^{(-1)} = n + G - A - B \tag{7.1}$$

in $\mathbb{Z}G$. Thus every element not in the union of the two *forbidden subgroups* A and B has a unique "difference representation" from D. Note that $G = A \times B$ under our assumptions. There are also similar – but not quasiregular – examples in semidirect products, see Pott [161].

Let us give an explicit description for the associated projective plane Π which may be obtained from the semiplane $\Delta = \operatorname{dev} D$. Note that, in view of equation (7.1), D meets every coset of A and all but one coset of B exactly once. In particular, we may assume $D \cap B = \emptyset$. Under this assumption, one has the following simplified variant of the construction due to Ganley [65] first introduced by de Resmini and the present authors [52]. The points of Π are

- the $n(n-1)$ group elements $g \in G$;
- a point ∞_A and n points (a), where $a \in A$;
- a point ∞_B and $n-1$ points $((b))$, where $b \in B$.

The lines of Π are

- $n(n-1)$ lines $[a, b] = Dab \cup \{(a), ((b))\}$, where $a \in A$ and $b \in B$;
- a line L_∞ containing ∞_A, ∞_B and the $n-1$ points $((b))$, where $b \in B$;
- a line L_A containing ∞_A and the n points (a), where $a \in A$;
- $n-1$ lines $[Ab] = Ab \cup \{\infty_A\}$, where $b \in B$;
- n lines $[Ba] = Ba \cup \{\infty_B, (a)\}$, where $a \in A$.

It is a little tedious (but not difficult) to check that we do obtain a projective plane Π of order n in this way. Note that Π is both (∞_A, L_∞)- and (∞_B, L_A)-transitive and therefore at least in Lenz-Barlotti class II.2. Conversely, any plane admitting two such transitivities is of type (f) and can be described by a DPDS, see Ganley [65] or Pott [161, §5.3]. Here $G = A \times B$, where A is the group of (∞_A, L_∞)-elations and B the group of (∞_B, L_A)-homologies.

The only known abelian examples correspond to the desarguesian planes $\mathrm{PG}(2, q)$ which can be represented by the DPDS

$$D = \{(x, x) : x \in \mathrm{GF}(q)^*\} \subset (\mathrm{GF}(q), +) \times (\mathrm{GF}(q)^*, \cdot); \tag{7.2}$$

see Pott [161, §5.3]. The fact that (7.2) defines a DPDS was first noted by Spence; see Ganley [65, p. 323]. We remark that the same construction works if one replaces the field $\mathrm{GF}(q)$ by a nearfield K and uses the additive and multiplicative groups of K; see Hiramine [88, Example 4.2(ii)]. The resulting plane can easily be seen to be the corresponding nearfield plane. That these planes admit quasiregular groups of type (f) was already observed by Dembowski and Piper [48].

Loosely speaking, a proper nearfield may be thought of as a skewfield with only one distributive law. To be precise, a finite *nearfield* is a finite set K on which two operations, addition and multiplication $(\cdot)$, are defined with the following properties:

(N1) $(K, +)$ is an abelian group with identity 0;
(N2) $(K^*, \cdot)$ is a group, where $K^* = K \backslash \{0\}$;
(N3) $(a + b) \cdot c = (a \cdot c) + (b \cdot c)$ for all $a, b, c \in K$;
(N4) $a \cdot 0 = 0$ for all $a \in K$.

Any proper nearfield yields a non-abelian DPDS by Hiramine's example. We remark that the finite nearfields were completely classified by Zassenhaus [179], see [46, §5.2]. Regarding the abelian case, the following conjecture seems reasonable:

Conjecture 7.1 *Any finite projective plane admitting an abelian group of type (f) is desarguesian.*

Pott [160] characterised the abelian DPDS's describing desarguesian planes and used this to prove the following partial result towards Conjecture 7.1; see also Pott [161, §5.3].

Theorem 7.2 *Let Π be a projective plane of order p admitting an abelian collineation group G of type (f), where p is a prime. Then Π is the desarguesian plane* $\mathrm{PG}(2, p)$.

Recently, Jungnickel and de Resmini [114] established the PPC for abelian groups of type (f); in fact, their result shows a little more.

Theorem 7.3 *Let G be an abelian collineation group of type (f) for a projective plane Π of order n. Then n must be a power of a prime p and the p-part of G is elementary abelian.*

Proof We write G multiplicatively and use the group ring approach. Let D be the direct product difference set representing the plane Π, and let A and B be the two forbidden subgroups. As noted above, D meets every coset of A and all but one coset of B exactly once, and therefore we may assume $D \cap B = \emptyset$ in what follows. Thus

$$D = \sum_{b \in B} b f(b), \tag{7.3}$$

where $f: B \to A\backslash\{1\}$ is a bijection. As for Theorem 4.4, the proof proceeds via computing the group ring element $D^{(-1)}D^{(p)}$ modulo p, where p is any prime dividing n. First it is easily shown by induction that the following equations in $\mathbb{Z}_pG$ hold for all $m \in \mathbb{N}$:

$$\begin{aligned} GD^m &= (-1)^m G; \\ AD^m &= GD^{m-1} = (-1)^{m-1}G; \\ BD^m &= (G-B)D^{m-1} = (-1)^{m-1}mG + (-1)^m B. \end{aligned}$$

Using these auxiliary equations and (7.1), a short computation establishes

$$D^{(p)}D^{(-1)} = G - B \quad \text{in } \mathbb{Z}_pG. \tag{7.4}$$

But $|D^{(p)}D^{(-1)}| = |G-B| = (n-1)^2$, and so (7.4) implies the corresponding identity in $\mathbb{Z}G$. Together with (7.3), we obtain

$$\sum_{b,c\in B} c^p f(c)^p b^{-1} f(b)^{-1} = D^{(p)}D^{(-1)} = G - B \quad \text{in } \mathbb{Z}G. \tag{7.5}$$

If some element $f(c)^p$ equals one of the elements $f(b)$, we get the element $c^p b^{-1} \in B$ from the sum in (7.5), which is forbidden. Hence $f(c)^p = 1$ for all $c \in B$ and thus $a^p = 1$ for all elements $a \neq 1$ of A. Therefore A is an elementary abelian p-group and n is a power of p, as claimed. □

The even order case of Theorem 7.3 was already established much earlier by Ganley [65]. Also, Pott [160] had already proved – using entirely different methods – that n has to be a prime power or a perfect square.

Theorem 7.3 gives some evidence for the validity of Conjecture 7.1; in particular, the subgroup A of order n of G has the correct structure. Not much is known about the structure of the group B of order $n-1$ which should, of course, be cyclic. We just have the following partial result due to Pott [160].

Theorem 7.4 *Let G be an abelian collineation group of type (f) for a projective plane Π of odd order q. Then the Sylow 2-subgroup of G is cyclic.*

In Subsection 10.5, we will present the short proof of de Resmini, Ghinelli and Jungnickel [52] for this result. Finally, let us point out that Pott's [160] characterisation of the the abelian DPDS's describing desarguesian planes would imply the validity of Conjecture 7.1 if one could show that

- B is cyclic;
- the function f appearing in (7.3) is an isomorphism from B to $\mathrm{GF}(q)^*$, where we assume $A = (\mathrm{GF}(q), +)$ by Theorem 7.3.

We note that we may at least achieve $f(1) = 1$, after replacing D by a suitable translate.

8 Type (g): Neofields

In this section Π will denote a plane of type (g), that is, a projective plane with a quasiregular automorphism group G of order $|G| = (n-1)^2$, with $t = 7$ orbits on points and lines. The fixed structure F consists of the vertices O, X, Y of a triangle and their opposite sides L_O, L_X, L_Y, and Π is (O, L_O)-, (X, L_X)- and (Y, L_Y)-transitive. The desarguesian planes are examples of planes of type (g).

There are some closely related concepts which appeared in various places in the literature:

- finite projective planes admitting a collineation group of Lenz-Barlotti type I.4 and the corresponding algebraic structures, namely "neofields", which were studied quite intensively, see for example Hughes [97, 98], Kantor [122], Kantor and Pankin [125] and Pankin [150, 151];
- partially transitive planes of type (3) in the sense of Hughes [99];
- a certain type of abelian difference set relative to disjoint subgroups in the sense of Hiramine [88].

If one looks at the literature, a certain amount of confusion is likely to arise, as the connections between these notions have not yet been made really precise. This will be done in our forthcoming paper [77], as a detailed treatment of this topic takes considerable effort and is far beyond the scope of the present survey. We shall here restrict ourselves to a brief report on the most important concepts and results.

Using group ring notation, an *abelian neo-difference set* of order n may be defined to be a subset D of an abelian group G of order $(n-1)^2$ with three pairwise disjoint subgroups X, Y, and Z, of order $n-1$, which satisfies the equation

$$DD^{(-1)} = n + G - X - Y - Z \tag{8.1}$$

in $\mathbb{Z}G$. Thus every element g not in the union N of the *forbidden subgroups* X, Y, and Z, has a unique "difference representation" $g = de^{-1}$ with $d, e \in D$.

In the *Lenz-Barlotti classification*, collineation groups of projective planes are classified according to the configuration F formed by the point-line pairs (p, L) for which the given group G is (p, L)-transitive. In the special case $G = \operatorname{Aut}\Pi$, one speaks of the *Lenz-Barlotti class* of Π. Lenz [135] considered elations and distinguished seven types; by including homologies, Barlotti [20] refined this to 53 types. In this classification, the desarguesian planes – which are (p, L)-transitive for each point-line pair (p, L) – stand at one end; at the other end are the planes which are not (p, L)-transitive for any (p, L). The Coulter-Matthews planes which we met in Section 4 are in Lenz-Barlotti class II.1, where the configuration $F = \{(p, L)\}$ is a flag (∞, L_∞). For a group of Lenz-Barlotti type I.4, F is – by abuse of language – a triangle: if the vertices are O, X, Y and the opposite sides are L_O, L_X, L_Y as above, then Π is (O, L_O)-,

(X, L_X)- and (Y, L_Y)-transitive. It should be stressed that the desarguesian planes admit a group of Lenz-Barlotti type I.4, but that no finite examples of planes in Lenz-Barlotti class I.4 are known. We refer the reader to Demboski [46], Hughes and Piper [104] and the survey article by Yaqub [173] for accounts on the Lenz-Barlotti classification.

Loosely speaking, a proper neofield may be thought of as a field with non-associative addition. To be precise, a finite *neofield* – or a *planar division neo-ring* (PDNR) in the terminology of Hughes [97, 98] – is a finite set K on which two operations, addition and multiplication $(\cdot)$, are defined with the following properties:

(Ne1) $(K, +)$ is a loop with identity 0;
(Ne2) $(K^*, \cdot)$ is a group, where $K^* = K \backslash \{0\}$;
(Ne3) both distributive laws hold in $(K, +, \cdot)$,
$(a + b)c = ac + bc$, $c(a + b) = ca + cb$ for all $a, b, c \in K$.

In fact, finite neofields have further interesting properties, as shown by Hughes [97, 98] and Kantor and Pankin [125]:

(Ne4) $(K, +)$ is commutative;
(Ne5) $(a + b) + (-b) = a$ for all $a, b \in K$;
(Ne6) $(K^*, \cdot)$ is commutative.

Thus we see that a finite neofield satisfies all of the field axioms except for the associativity of addition, which has been replaced by condition (Ne5), the so-called *inverse property.* Given a neofield K, the incidence structure $\Sigma = \Sigma(K)$ defined as follows is an affine plane.

- The points are the ordered pairs (x, y) with $x, y \in K$.
- The lines are the point sets

$$[m, k] = \{(x, xm + k) \mid x \in K\}, \qquad [a] = \{(a, y) \mid y \in K\}.$$

- Incidence is set theoretic inclusion.

After these preparations, we can state the following equivalence theorem.

Theorem 8.1 *Let Π be a finite projective plane with a collineation group G. Then the following assertions are equivalent:*

(i) *G is of Lenz-Barlotti type I.4;*

(ii) *G is a quasiregular group of type* (g)*;*

(iii) *Π extends an affine plane coordinatised by a neofield K as above, and $G \cong K^* \times K^*$;*

(iv) *Π can be represented by an abelian neo-difference set in G.*

Under these conditions, the group G is necessarily abelian.

We refer the reader to our paper [77] for details. In particular, an explicit description of Π in terms of the abelian neo-difference set D may be found there. This is similar to the one presented here for case (f) but more involved, and therefore it is omitted from the present survey. However, we at least include the following description of $D \subset G = K^* \times K^*$ in terms of K, due to Hughes [97, 98]:

$$D = \{(\xi, \psi) \in K^* \times K^* \colon \xi + \psi = 1\}, \tag{8.2}$$

and the three forbidden subgroups are

$$X = K^* \times \{1\}, \quad Y = \{1\} \times K^*, \quad Z = \{(\xi, \xi) : \xi \in K^*\}.$$

The reader may esily check that (8.2) indeed yields a neo-difference set. In particular, if we take for K the finite field $\mathrm{GF}(q)$, we obtain a neo-difference set associated with the classical plane $\mathrm{PG}(2, q)$.

The basic problem concerning finite neofields is whether they must be fields. If this were so, finite planes in Lenz-Barlotti class I.4 would not exist. The next natural step after property (Ne6) would be to show that K^* is cyclic. All that is known in this direction is the following result due to Hughes [97, 98] and Kantor [122].

Theorem 8.2 *K^* has cyclic Sylow 2-subgroups and 3-subgroups.*

By the commutativity of K^* and the results of Hughes [97, 98], we have the following.

Theorem 8.3 *If $(K, +, \cdot)$ is a neofield of order n and p divides n, then the mapping $x \mapsto x^p$ is an automorphism of K.*

An integer p is called a *multiplier* if the mapping $\alpha \colon x \mapsto x^p$ (for $x \in K$) is an automorphism of K. Equivalently, α leaves the associated neo-difference set D defined in equation (8.2) invariant. In this formulation, a much shorter and also more transparent proof – similar to the cases of planar and affine difference sets discussed in this survey – is given in [77].

Cyclic neofields were studied by Pankin [150, 151] who obtained the following nonexistence theorem.

Theorem 8.4 *Let $n \geqslant 8$ have a divisor p such that the exponent r of p modulo $n-1$ satisfies*

- *r is not divisible by 3;*
- *$r > \lfloor (n-2)/6 \rfloor$ if r is odd and $r > 2\lfloor (n-2)/6 \rfloor$ if r is even.*

Then there is no cyclic neofield of order n.

Hughes [98], by hand, ruled out the possibility of proper cyclic neofields for the orders $9, 11, 13, 16, 27, 32$ and 64. Pankin [150] gave an algorithm for constructing any cyclic neofield. Using this algorithm with a computer search, he could add to the above list the orders $17, 19, 23, 25, 29, 31, 37, 41, 43, 47, 49$, 81 and 128. Thus the only cyclic neofields of the orders in the above two lists (which, by Theorem 8.2, are the only neofields of these orders) are fields. Therefore, no planes in Lenz-Barlotti class I.4 exist for these orders.

We now list some further theorems which have been used to eliminate various particular composite orders as possible orders of proper neofields. The first of these was used but not stated explicitly by Hughes [97, 98].

Theorem 8.5 *If t_1, t_2, t_3, t_4 are multipliers of an abelian neo-difference set of order n with $t_1 - t_2 \equiv t_3 - t_4 \pmod{n-1}$, then the exponent of G divides the least common multiple of $t_1 - t_2$ and $t_1 - t_3$.*

This leads to the following result, basically due to Hughes [97, p. 520], but with some improvements taken from [77].

Theorem 8.6 *There is no neofield whose order is divisible by any of the following pairs of primes:* $(2,3)$, $(2,5)$, $(2,7)$, $(2,13)$, $(2,17)$, $(2,19)$, $(2,31)$, $(3,5)$, $(3,7)$, $(3,11)$, $(3,13)$, $(3,17)$, $(3,19)$, $(5,7)$, $(5,11)$, $(5,13)$, $(7,13)$.

Pankin [150, p. 24] proved the following result.

Theorem 8.7 *No neofield of order $n > 4$ has -1 as a multiplier.*

Combining this with Theorem 8.3, one also obtains the following result.

Corollary 8.8 *If $n > 4$ has a divisor p such that $p^a \equiv -1 \pmod{n-1}$ for some a, then there is no neofield of order n.*

The following results are due to Kantor [122].

Theorem 8.9 *If a neofield of order n has a multiplier of even order, then n is a square.*

Theorem 8.10 *Assume the existence of a neofield of even order n. Then $n = 2$, $n = 4$, or n is a multiple of 8.*

The above results have been used together with the aid of a computer to show that every neofield of order $\leqslant 1,000$ has prime power order, see [151]. However, these techniques are clearly not sufficient to prove the PPC. A few further restrictions will be discussed in [77].

We finally remark that Hughes [99] and Hiramine [88, Example 4.2(iv)] noted that the same definition as in (8.2) also gives a solution of the group ring equation (8.1) if one uses for K a nearfield (compare with Section 7). Of course K^* is a non-abelian group, if K is a proper nearfield. Let us point out that this does not contradict the preceding statements, as one may check that the group G is not quasiregular in this case. In fact, non-abelian neo-difference sets for which two of the forbidden subgroups are normal correspond exactly to collineation groups of Lenz-Barlotti type I.3; see [77] for details.

9 Type (h): A sporadic case

In this section Π will denote a plane of type (h), that is, a projective plane with a quasiregular automorphism group G of order $|G| = (n - \sqrt{n} + 1)^2$, with $t = 2\sqrt{n}$ orbits on points and lines and without fixed elements. As mentioned in Section 1, in this case there are $t - 1$ disjoint subplanes $\Pi_1 \dots \Pi_{t-1}$ of order $\sqrt{n} - 1$ whose point sets constitute $t - 1$ orbits, each of length $n - \sqrt{n} + 1$. The remaining point orbit, on which G acts faithfully, is the set of all other points of Π; see Dembowski and Piper [48, Theorem 5].

If $n = 4$, there is a classical example, already mentioned in Dembowski and Piper [48], of an elementary abelian (and hence quasiregular) collineation group of order 9 in the desarguesian plane of order 4 with $t = 5$ point (and line) orbits. This group has four triangular point orbits while the fifth, of length 9, consists of the absolute points of a unitary polarity and may be regarded as the affine plane of order 3, embedded in an unusual way into the projective plane of order 4. If one is willing to consider a triangle as a "subplane of order 1", then this group provides an example for type (h). Note, however, that the condition $|G| > (n^2 + n + 1)/2$ is not satisfied for this group.

In 1975 Ganley and McFarland [67] proved the following theorem which implies that for $n \neq 4$ no example can exist even if G is not assumed to be abelian; in particular, the PPC holds for quasiregular groups of type (h).

Theorem 9.1 *A finite projective plane of order n admits a quasiregular collineation group G of order $(n - \sqrt{n} + 1)^2$ if and only if $n = 4$.*

The method of proof used in [67] is as follows. First, by a result of Hering, $p = n - \sqrt{n} + 1$ is a prime and G is an elementary abelian group of order p^2. This can be shown using simple counting arguments involving Sylow subgroups and orbit sizes, see [46, p. 183]. One then proves that the existence of G for $n > 4$ implies the existence of a DPDS-like subset D of G, which is then shown to be non-existent using Hering's result and the usual algebraic methods which apply for difference sets. Let us briefly sketch the main steps of the proof.

For each i, $1 \leqslant i \leqslant t - 1$, let G_i be the pointwise stabiliser of Π_i, and set $N = \bigcup_{i=1}^{t-1} G_i$. Let q and L be elements in the faithful point and line orbits, respectively. We write G multiplicatively in what follows. Then the following result holds, see [67, Lemma 3].

Lemma 9.2 *Let G be a quasiregular collineation group of order $(n - \sqrt{n} + 1)^2$ of a projective plane of order $n > 4$. With the notation above, consider the set $D = \{d \in G : q^d \in L\}$. Then every element of $G \setminus N$ can be represented exactly* *once in the form $d_i^{-1}d_j$ with $d_i, d_j \in D, d_i \neq d_j$, whereas no element of N can be represented in this way. In terms of the group ring $\mathbb{Z}G$, with $m = \sqrt{n} - 1$,*

$$D^{(-1)}D = m^2 + G - N. \tag{9.1}$$

By the result of Hering, $p = n - \sqrt{n} + 1 = m^2 + m + 1$ is a prime, and G is an elementary abelian group of order p^2. Thus there are $p+1$ subgroups of order p in G and every element $\neq 1$ of G is contained in exactly one of these subgroups. Now N is the set theoretic union of $t - 1 = 2m + 1$ of these subgroups, and so $G - N + 1$ is the set theoretic union of the remaining $p + 1 - (t - 1) = m^2 - m$ subgroups of order p, say $H_1, \dots, H_{m^2-m}$. By assumption, $m \neq 1$, and thus $p + 1 - (t - 1) \geqslant 2$. Now let $C = G/H_1$. Using the natural epimorphism $\alpha : \mathbb{Z}G \to \mathbb{Z}C$ induced by the group epimorphism $G \to C$ with kernel H_1, equation (9.1) implies that the group ring $\mathbb{Z}C$ contains an element $\tilde{D}$ with nonnegative coefficients satisfying

$$\tilde{D}^{(-1)}\tilde{D} = (m+1)^2 + (m^2 - m - 1)C. \tag{9.2}$$

Ganley and McFarland [67] then obtained a contradiction by showing that any element $\tilde{D}$ in $\mathbb{Z}C$ satisfying (9.2) must have a negative coefficient; for this final part, they used algebraic number theory and representation theory.

Hiramine [88] investigated a common generalisation of direct product difference sets as well as the variants appearing in the study of groups of type (g) and (h), namely "difference sets relative to disjoint subgroups". We will not give the precise definition, but we mention that Hiramine classified all possible types of such difference sets, see [88, Theorem 4.1].

10 Applications to geometry

Up to now, we have investigated the existence problem for planes with a quasiregular group and – so we hope – convinced the reader that the group ring approach to this problem is both useful and elegant. In this final section, we want to show that one can not only obtain existence criteria and sometimes characterisations in this way, but that the algebraic approach also allows some nice geometric applications to those planes which admit an abelian difference set representation. This applies mainly to the classical desarguesian planes, but also to commutative semifield planes. We begin with some results concerning Baer subplanes, unitals, and certain complete arcs in $\mathrm{PG}(2, q^2)$ and then discuss results dealing with the construction of ovals, hyperovals, and maximal arcs with interesting intersection patterns.

10.1 The geometry of $\mathrm{PG}(2, q^2)$

In this subsection we discuss some geometrical properties of the classical projective planes $\mathrm{PG}(2, q^2)$, giving constructions for which the representation by a Singer difference set is useful. In fact, these constructions work for arbitrary planar abelian difference sets of square order and will therefore be given in this slightly more general setting, even though it seems unlikely that this really constitutes a generalisation, see Conjecture 3.2. Thus $\Pi = \operatorname{dev} D$ will be a projective plane associated with a planar difference set D of order $n = m^2$ in an abelian group G throughout this subsection.

A first result was already provided in the proof of Theorem 3.14 where we noted that Π contains a Baer subplane. Using the notation there, the cosets of B also yield Baer subplanes. Thus one can say a lot more, as observed by Bruck [33] for the cyclic case and by the second author [109] in general.

Theorem 10.1 Π *admits a partition into Baer subplanes which is invariant under the Singer group G.*

There is a further interesting application of the proof of Theorem 3.14. For this, recall that a polarity of a projective plane of order m^2 with exactly m^3+1 absolute points is called a *unitary* polarity.

Theorem 10.2 Π *admits a unitary polarity.*

Proof We use the notation introduced in the proof of Theorem 3.14. By Lemma 3.8, we may assume that D is fixed under the multiplier $t = m^3$. It is easily checked that the correspondence

$$\pi : g \leftrightarrow D - tg \tag{10.1}$$

defines a polarity π of Π. Clearly g is an absolute point of π if and only if $2g^\beta = g + tg \in D$. Since A is the kernel of β, the set U of absolute points of π is given by $U = \{a + b : a \in A, 2b \in D \cap B\}$, and therefore π has exactly $(m^2 - m + 1)(m + 1) = m^3 + 1$ absolute points. □

By a theorem of Seib [167], any unitary polarity of Π induces a *unital* $\mathcal{U}$, that is, a resolvable Steiner system $S(2, m + 1, m^3 + 1)$. Thus the point set of $\mathcal{U}$ is the set U of the $m^3 + 1$ absolute points of π, and the lines are the intersections $U \cap L$, where L is a non-absolute line of π; all such intersections have cardinality $m+1$. Moreover, the m^2 non-absolute points on an arbitrary absolute line determine a resolution of the line set of $\mathcal{U}$ into parallel classes; see [23, Theorem VIII.5.26] for details. A brief historical survey on unitals, including a listing of important papers with short abstracts, may be found in the appendix of the thesis by Barwick [21].

Theorem 10.2 shows that Π contains unitals – a result due to Bose [30] for $\Pi = \mathrm{PG}(2, q^2)$, see also Ghinelli [70], and to Blokhuis, Brouwer and Wilbrink [25] in general – and its proof gives an explicit description of the point set U of the unital $\mathcal{U}$ associated with the polarity (10.1). But one can say even more, as the next result shows. Recall that a set K of k points in a projective plane Π is called a *k-arc* provided that every line intersects K in at most two points.

Theorem 10.3 *Both Π and $\mathcal{U}$ can be partitioned into arcs of size $m^2 - m + 1$ which are complete (that is, none of them is contained in any arc of larger size) for $m \neq 2$. Moreover, each of these arcs arises as the intersection of two unitals.*

The proof proceeds by showing that the subgroup A is an arc; then the desired partition of Π is given by the translates $A + b$, where $b \in B$, and the partition of $\mathcal{U}$ follows by restricting attention to the translates with $2b \in D \cap B$. Finally, counting arguments can be used to prove the completeness of these arcs for $m \geqslant 4$. The case $m = 3$ needs special arguments.

Theorem 10.3 is due to Blokhuis, Brouwer and Wilbrink [25], but the cyclic case was obtained earlier by Fisher, Hirschfeld and Thas [62] and Boros and Szőnyi [28] who only considered the special case $\Pi = \mathrm{PG}(2, q^2)$. In this case, the arcs in question had been constructed previously using different methods by Kestenband [126] who, however, did not note their completeness. This completeness is of particular interest, since it shows that a bound of Segre on the size of a complete arc in $\mathrm{PG}(2, q^2)$ for q even is best possible; see Hirschfeld [89, Theorem 10.3.3].

We remark that Storme and Van Maldeghem [169], Szőnyi [170] and Ho [93] considered the general problem under which conditions the orbit of a subgroup of a Singer group is an arc.

Finally, Blokhuis, Brouwer and Wilbrink in [25] used their difference set approach to prove the following beautiful characterisation of the classical *Hermitian unitals*, that is, unitals induced by a unitary polarity which can be described by a Hermitian matrix.

Theorem 10.4 *Let $\mathcal{U}$ be a unital embedded in $\Pi = \mathrm{PG}(2, q^2)$, where $q = p^r$. Then $\mathcal{U}$ is Hermitian if and only if it is contained in the $\mathbb{Z}_p$-code spanned by the lines of Π.*

Here the connection to difference sets and group rings is given by the fact that the $\mathbb{Z}_p$-code spanned by the lines of Π is nothing but the ideal generated by the difference set D in the group algebra $\mathbb{Z}_p G$.

10.2 Arcs and ovals from large abelian groups

Any k-arc in a projective plane of order n satisfies $k \leqslant n+2$. The $(n+2)$-arcs are usually called *hyperovals*, while an *oval* is an $(n + 1)$-arc. It is well-known that hyperovals can only exist in projective planes of even order; in planes of odd order, ovals are the largest possible arcs. We refer the reader to Hirschfeld [89] for background.

The study of ovals and hyperovals in (not necessarily desarguesian) projective planes has for many years been a topic of considerable interest in Finite Geometries. For the desarguesian case, only even orders q need to be considered, as any oval in $\mathrm{PG}(2, q)$, q odd, actually is a conic, by the famous theorem of Segre [166]. We refer the reader to the survey by Cherowitzo [41] for hyperovals in $PG(2, q)$, q even. For the non-desarguesian case, let us just mention a few references, namely the classification of hyperovals in the translation planes of order 16 due to Cherowitzo [40]; and the existence of ovals, respectively hyperovals, in the Figueroa planes (see Cherowitzo [39] and de Resmini and Hamilton [53]); in the Hughes planes (see Room [163]); and in commutative semifield planes (see Garner [69] and Jha and Wene [105]).

In the remainder of this section we discuss some constructions of ovals and hyperovals in projective planes with large abelian collineation groups of type (a), (b), (d) and (f), respectively, using their difference set representation. We shall follow the recent account by de Resmini and the present authors [52]. The basic result is as follows.

Proposition 10.5 *Let Π be a projective plane of order q admitting an abelian collineation group G of one of the types (a), (b) and (d), and let D denote a relative difference set in G associated with Π. Then the sets $A_g = -D + g$ are arcs in $\Delta = \operatorname{dev} D$, and the lines $D - 2d + g$ with $d \in D$ are tangents to A_g (with $-d+g$ as the tangency point). If G has type (a), A_g is an oval of $\Pi = \Delta$; and if G has type (b) or (d), the lines of Π through ∞ cannot be secants, hence A_g may be extended to an oval $O_g = A_g \cup \{\infty\}$ of Π.*

Proof We first show that any line $D+x$ intersects A_g at most twice. Assume that a is some point of intersection, say

$$a = d + x = -d' + g, \quad \text{hence} \quad d + d' = g - x,$$

with $d, d' \in D$. If b is a second point of intersection, we similarly obtain

$$b = e + x = -e' + g, \quad \text{hence} \quad e + e' = g - x,$$

with $e, e' \in D$. From these two equations, $d - e = e' - d'$ follows, therefore $d = e'$ and $d' = e$, since D is a relative difference set with $\lambda = 1$. This shows that b is uniquely determined by a, proving that $D + x$ intersects A_g at most twice. So, A_g is an arc in the divisible semiplane Δ. Since we may always rewrite $d + x = -d' + g$ as $d' + x = -d + g$, it is obvious that the line $D + x$ intersects A_g in two points unless $d = d'$. Thus $D + x$ is the (unique) tangent in Δ to A_g at the point $-d + g$ if and only if $d = d'$, hence $x = -2d + g$. If G is not of type (a), then $-D$ intersects any coset $N + g$ at most once (otherwise we would get a difference representation for an element in N); thus the corresponding lines of Π are either tangents or exterior lines, proving the final assertion. □

Special cases of Proposition 10.5 appeared much earlier, see [34, 107, 108, 110, 119] and [109], where the second author gave a systematic study of the arcs in divisible semiplanes associated with relative difference sets. In these papers, the emphasis was generally on exploiting arcs to obtain nonexistence results, whereas [52] is mainly concerned with using the relative difference set representation for constructing families of (hyper)ovals with interesting intersection patterns in an easy and systematic way. Regarding hyperovals, we have the following addendum to Proposition 10.5.

Proposition 10.6 *Let Π be a projective plane of even order q admitting an abelian collineation group G of one of the types (a), (b) and (d), and let D denote a relative difference set in G associated with Π. The oval O_g defined in Proposition 10.5 completes to a hyperoval H_g by adjoining its nucleus – that is, the common intersection point of all tangents – which is determined as follows.*

(i) *If G has type (a) or (d) assume, without loss of generality, that D is fixed under the multiplier 2, so that $2D = D$. Then the nucleus of O_g is the point g.*

(ii) *If G has type (b) assume, without loss of generality, that $0 \in D$. Then the nucleus of O_g is the point ∞_{Ng} on the line L_∞ which is determined by the parallel class of lines $D + g + n$, where $n \in N$.*

Proof For types (a) and (d), the tangents to O_g described in Proposition 10.5 take the form $D - d + g$, where d runs over the elements of D. Obviously, these tangents intersect in the point g. For type (b), $2D = N$ by Lemma 4.3. Hence the tangents to O_g described in Proposition 10.5 now take the form $D + n + g$, where n runs over the elements of N. □

10.3 Arcs and ovals in planes of type (a) and (d)

Some interesting families of ovals and hyperovals can be obtained from abelian planar and affine difference sets. Using Propositions 10.5 and 10.6, we get the following result for the planar case.

Proposition 10.7 *Let Π be a projective plane of type (a) associated with a planar difference set D of order q in an abelian group G. Then the sets $-D+g$ with $g \in G$ are $q^2 + q + 1$ ovals which pairwise intersect in a unique point; in other words, they form the lines of another projective plane on the point set G which is isomorphic to Π.*

If q is even, we also obtain hyperovals as described in Proposition 10.6. In the affine case, only even orders yield really interesting configurations of ovals. Here Proposition 10.6 easily gives the first part of the following general result; the second, more difficult assertion was proved by the second author in [110] as a consequence of a result by Arasu and Jungnickel [10].

Proposition 10.8 *Let Π be a projective plane of type (d) associated with an affine difference set D of even order q in an abelian group G, and assume $2D = D$. Then the sets $O_g = (-D + g) \cup \{g\}$ with $g \in G$ are $q^2 - 1$ ovals with common nucleus ∞ which can be partitioned into $q + 1$ families of $q - 1$*

ovals each such that any two ovals from different families meet in exactly one point, whereas the ovals in any of the $q+1$ families partition the point set of the divisible semiplane Δ, the set of affine points $\neq \infty$.

Moreover, the group G splits into a direct sum $G = H \oplus N$ for a suitable subgroup H of order $q+1$, as $(q-1, q+1) = 1$. The $q-1$ sets $H+n$ with $n \in N$ constitute a further family $\mathcal{O}(H)$ of $q-1$ ovals with common nucleus ∞ which partition the set of affine points $\neq \infty$. The group H acts regularly on each of these ovals, and N acts regularly on $\mathcal{O}(H)$.

In view of the conjecture that the only projective planes admitting an abelian group of type (a) or (d) are in fact the desarguesian planes, the preceding results probably only apply to $\mathrm{PG}(2,q)$ and to cyclic groups G. In this case, the ovals appearing in the general results are in fact conics; see [110, 119].

Proposition 10.9 *Let $\Pi = \mathrm{PG}(2,q)$. Then Π may be represented by a cyclic planar difference D for which the oval $-D$ is the conic with the equation $xz - y^2 + z^2 = 0$ (in homogeneous coordinates). Moreover, Π may also be represented by a cyclic affine difference D for which the oval $-D \cup \{\infty\}$ is the affine conic with the equation $dx^2 + y^2 + xy + x = 0$, where $x^2 + x + d$ is a primitive polynomial over $\mathrm{GF}(q)$. Similarly, the oval H is the affine conic with the equation $dx^2 + y^2 + xy = 1$.*

The preceding results may also be used to construct two interesting families of semisymmetric designs, see [52].

10.4 Arcs and ovals in planes of type (b)

From a constructive point of view, this is the most interesting case, as it also provides examples in non-desarguesian planes. But first we sketch the proof of Lemma 4.3. Assertions (i) and (ii) are obvious from the definition of an $(n,n,n,1)$-RDS. In order to prove assertion (iii), we consider the symmetric net $\Delta = \mathrm{dev}D$ associated with D and proceed via the following steps. By Proposition 10.5, $-D$ is an arc of Δ and the tangent to $-D$ at the point $-r$ is the line $D - 2r$ (for all $r \in D$). Next one shows that every point $g \notin -D$ is on some tangent. As the union of the tangents $D - 2r$ with $r \in D$ is G, the tangents form a parallel class of lines. Finally, $2D$ is a coset of N, hence $2D = N$ because $0 \in D$. For more details, we refer the reader to the original paper [108].

Now we turn our attention to constructing ovals. The basic observation is as follows.

Theorem 10.10 *Let Π be a projective plane of order q admitting an abelian collineation group G of type (b). Then Π contains a G-orbit of q^2 ovals sharing the common point ∞ which can be partitioned into q families of q ovals each such that any two ovals from different families meet in exactly two points* *(including ∞), whereas the ovals in any of the q families partition the affine plane $\Delta = \Pi \setminus L_\infty$.*

Proof With the notation of Proposition 10.5, we may define the desired q families $\mathcal{O}_h$ to consist of the q ovals O_{h+n}, where n runs over N and where the h's form a system of coset representatives for N in G. The intersection properties claimed in the assertion follow from the fact that $-D$ is again a $(q,q,q,1)$-RDS, which shows that the arcs $A_g = -D + g$ form a divisible semiplane isomorphic to Δ. □

As we have seen in Section 4, any projective plane defined over a commutative semifield or via an abelian planar function admits an abelian collineation group G of type (b), and hence Theorem 10.10 applies. In particular, one obtains the existence of ovals in arbitrary commutative semifield planes – which was known before, see Garner [69] and Jha and Wene [105] – but also in the Coulter-Matthews planes [44], a result first observed by de Resmini and the present authors [52]. Here the Coulter-Matthews planes are of particular interest, as they are associated with the only known planar functions which do not give rise to translation planes.

Next we present two general constructions for hyperovals and maximal arcs, respectively, in planes of even order with an abelian collineation group of type (b) given in [52]. As already noted in Section 4, as far as we are aware, the only known examples of such planes are those defined over a commutative semifield of even order.

Theorem 10.11 *Let Π be a projective plane of even order $q = 2^b$ admitting an abelian collineation group G of type (b). Then Π contains a G-orbit of q^2 hyperovals sharing the common point ∞ which can be partitioned into q families of q hyperovals each such that any two hyperovals from different families meet in exactly two points (namely, ∞ and a further point on the line L_∞), whereas the hyperovals in any of the q families partition the affine plane $\Sigma = \Pi \setminus L_\infty$. Moreover, these q^2 hyperovals together with the $q^2 + q$ points $\neq \infty$ of Π yield an embedding of the dual affine plane Σ^* into Π.*

Proof With the notation of Proposition 10.6, we may define the desired q families $\mathcal{H}_h$ to consist of the q hyperovals H_{h+n}, where n runs over N and where the h's form a system of coset representatives for N in G. Then the intersection properties claimed in the statement follow as in the proof of Theorem 10.10, using Proposition 10.6 instead of Proposition 10.5. Only the final assertion requires an additional argument which uses the fact that planes with an abelian collineation group of type (b) are self-dual; see Ganley [64]. The details can be found in [52, Theorem 4.1]. □

It is well-known that the exterior lines to a hyperoval H in a projective plane Π of even order q form a *maximal arc* of degree $q/2$ in the dual plane Π^*, that is, a set of points in Π^* which meets each line in either 0 or $q/2$ points. See, for instance, Hirschfeld [89] for background on maximal arcs. As noted above, planes with an abelian collineation group of type (b) are self-dual. Hence, the existence of hyperovals in Π also implies that of maximal arcs of degree $q/2$ in Π. The following result gives an explicit description of such a maximal arc in terms of the underlying relative difference set D (using group ring notation); for the proof, we refer the reader to [52, Theorem 4.2].

Theorem 10.12 *Let Π be a projective plane of even order $q = 2^b$ admitting an abelian collineation group G of type (b), and let D be an associated $(q, q, q, 1)$-*RDS. *Write G multiplicatively and assume, without loss of generality, $1 \in D$. Then the element $M \in \mathbb{Z}G$ defined by*

$$M = G - \frac{1}{2}(D^2 + N) \tag{10.2}$$

is a maximal arc of degree 2^{b-1} in Π. Moreover, the affine points together with the q^2 translates Mg of M (as blocks) form a symmetric design with parameters

$$(2^{2b}, 2^{2b-1} - 2^{b-1}, 2^{2b-2} - 2^{b-1}) \tag{10.3}$$

admitting G as a regular automorphism group.

In standard terminology, M is a Hadamard difference set of order 2^{2b-2} in G; see [23]. The existence of such objects in direct products of cyclic groups of order 4 is easy to prove; but the fact that some associated symmetric designs may be embedded into planes as sets of maximal arcs came as a surprise. We remark that we do not know if this property holds for every design of this kind.

In particular, Theorems 10.11 and 10.12 apply to the case where Π is defined over a commutative semifield S of order q. Using the setup of Section , the following result was obtained in [52]. Here a hyperoval H meeting the line at infinity of the affine semifield plane in two points is called a *translation oval* if there exists a group of q translations acting regularly on the q affine points of H; compare Cherowitzo [40].

Theorem 10.13 *Under the assumptions of Theorem 4.1, the arc associated with the* RDS *D as in Proposition 10.5 is the set*

$$A = \{(-x, -x + x^2) : x \in S\}. \tag{10.4}$$

In fact, A belongs to a translation oval, and the same holds for all q^2 translates Ag of A $(g \in G)$. Also A can be described as the affine conic with equation $y = x^2 + x$ over S.

In particular, each of the q families $\mathcal{H}_h$ of hyperovals defined in Theorem 10.11 yields a partition of the affine semifield plane Σ into translation ovals. This theorem of [52] considerably strengthens a result of Jha and Wene [105] who constructed $q-1$ pairwise disjoint translation ovals in any affine plane over a commutative semifield of even order q^n with middle nucleus of order q.

The derivation of the equation for the arc A is actually not dependent on the characteristic of S. Thus we also have the following general result which complements the analogous results for planes of types (a), (d) and (f) in Propositions 10.9 and 10.17.

Proposition 10.14 *Let Π be a commutative semifield plane of order q (in particular, $\Pi = \mathrm{PG}(2,q)$). Then Π may be represented by a $(q,q,q,1)$-RDS D for which the q-arc $-D$ is the affine conic with the equation $y = x^2 + x$.*

10.5 Arcs and ovals in planes of type (f)

For planes of type (f), the proof of Proposition 10.5 carries over to give the following result of [52]. Here we use multiplicative notation.

Proposition 10.15 *Let Π be a projective plane of order q with an abelian collineation group $G = A \times B$ of type (f), and let D denote an associated* DPDS *such that $D \cap B = \emptyset$. Then the $(q-1)$-sets $D^{-1}g$ are arcs in $\Delta = \mathrm{dev}\, D$, and the $q-1$ lines $Dd^{-2}g$ with $d \in D$ are tangents to $D^{-1}g$ (with $d^{-1}g$ as the tangency point). Moreover, the lines of Π corresponding to the cosets of A and B, respectively, can not be secants; hence, $D^{-1}g$ may be extended to an oval $O_g = D^{-1}g \cup \{\infty_A, \infty_B\}$ of Π.*

Next, let us consider the intersections of the tangents to D^{-1} with the special line L_A. We shall use the description of Π given in Section 7. By Proposition 10.15, each point $d = ab \in D$ leads to the tangent $[a^{-2}, b^{-2}]$ of $O = O_1$ which intersects L_A in the point (a^{-2}). If q is odd, no point can be on more than two tangents, and hence A can not contain more than one involution. Thus the Sylow 2-subgroup of G is cyclic, proving Theorem 7.4. In the even order case, an analogous argument shows that the Sylow 2-subgroup of G is elementary abelian, providing a third proof for this instance of Theorem 7.3. The original proofs of [160] and [65] use polarities and are considerably more cumbersome.

The argument about the nucleus of the oval O easily extends to give the following analogue of Proposition 10.6.

Corollary 10.16 *Under the hypotheses of Proposition 10.15, assume that q is even. Let $g \in G$, say $g = ab$ with $a \in A$ and $b \in B$. Then the nucleus of the oval O_g is the point $(a) \in L_A$.*

We observe that the ovals appearing in the preceding results are in fact conics for $\Pi = \mathrm{PG}(2,q)$, as expected. Using the DPDS given explicitly in (7.2), one gets $D^{-1} = \{(-x, x^{-1}) : x \in \mathrm{GF}(q)^*\}$, which proves the following result of [52].

Proposition 10.17 *Let $\Pi = \mathrm{PG}(2, q)$. Then Π may be represented by a direct product difference set D in a group $G \cong EA(q) \times \mathbb{Z}_{q-1}$ for which the arc D^{-1} is the affine hyperbola with the equation $y = -1/x$.*

As in the case of affine difference sets, even order planes of type (f) provide us with rather nice families of hyperovals, whereas the odd order case seems less interesting. Thus, we only note the following analogue of Theorem 10.11 and leave the odd order case to the reader.

Theorem 10.18 *Let Π be a projective plane of even order q admitting an abelian collineation group G of type (f). Then Π contains a G-orbit of $q(q-1)$ hyperovals sharing two common points ∞_A and ∞_B which can be partitioned into $q-1$ families of q hyperovals each such that any two hyperovals from different families meet in exactly three points (namely, ∞_A, ∞_B and a further point neither on L_∞ nor on L_A), whereas the hyperovals in any of the $q-1$ families partition the affine plane $\Sigma = \Pi \setminus L_\infty$.*

It is well-known that any projective plane which admits an abelian collineation group of type (f) is self-dual; see Ganley [65]. Hence, as for type (b) planes, the existence of hyperovals in Π in the even order case also implies that of maximal arcs of degree $q/2$. The following analogue of Theorem 10.12 gives an explicit description of such a maximal arc.

Proposition 10.19 *Let Π be a projective plane of even order $q = 2^b$ admitting an abelian collineation group $G = A \times B$ of type (f), and let D be an associated* DPDS. *Write G multiplicatively and assume, without loss of generality, that $D \cap B = \emptyset$. Then the element $M \in \mathbb{Z}G$ defined by*

$$M = G - \frac{1}{2}(D^2 + B) \tag{10.5}$$

is a maximal arc of degree 2^{b-1} in Π.

In contrast to the case of type (b) planes, the family of maximal arcs obtained from the arc M constructed in Proposition 10.19 does not seem interesting. It can be shown that three different intersection numbers occur, and thus we do not even get a partially symmetric design.

Finally, we remark that a similar study can also be made for planes with an abelian group of type (g); however, no particularly interesting configurations of ovals arise in this way. One obtains, however, a nice connection to projective triangles and hence to small blocking sets; for this, we refer the reader to our forthcoming paper [77].

Acknowledgements

This work was partially supported by GNSAGA, by the Italian Ministry for University, Research and Technology (project: *Strutture geometriche, combinatoria e loro applicazioni*) and the Università di Roma "La Sapienza" (project: *Gruppi, Grafi e Geometrie*). The paper started while the second author was a Visiting Research Professor at the University of Rome "La Sapienza"; he gratefully acknowledges the hospitality and financial support extended to him. The authors are indebted to Yutaka Hiramine and Bill Kantor for some helpful comments concerning planes of type (g) and to Alex Pott for reading a draft of this survey and making several helpful suggestions.

References

[1] I. Anderson, *Combinatorial designs: Construction methods*, Ellis Horwood, Chichester (1990).

[2] J. André, Über nicht-Desarguessche Ebenen mit transitiver Translationsgruppe, *Math. Z.* **62** (1954), 156–186.

[3] K.T. Arasu, Abelian projective planes of square orders, *European J. Combin.* **10** (1989), 207–209.

[4] K.T. Arasu, Cyclic affine planes of even order, *Discrete Math.* **76** (1989), 177–181.

[5] K.T. Arasu, Cyclic subplanes of cyclic affine planes, *Sankhyā Ser. A* **54** (1992), 31–34.

[6] K.T. Arasu, J.A. Davis, D. Jungnickel and A. Pott, A note on intersection numbers of difference sets, *European J. Combin.* **11** (1990), 95–98.

[7] K.T. Arasu, J.A. Davis, D. Jungnickel and A. Pott, Some nonexistence results on divisible difference sets, *Combinatorica* **11** (1991), 1–8.

[8] K.T. Arasu, J.F. Dillon, D. Jungnickel and A. Pott, The solution of the Waterloo problem, *J. Combin. Theory Ser. A* **71** (1995), 316–331.

[9] K.T. Arasu, J.F. Dillon, K.H. Leung and S.L. Ma, Cyclic relative difference sets with classical parameters, *J. Combin. Theory Ser. A* **94** (2001), 118–126.

[10] K.T. Arasu and D. Jungnickel, Affine difference sets of even order, *J. Combin. Theory Ser. A* **52** (1989), 188–196.

[11] K.T. Arasu and A. Pott, On quasiregular collineation groups of projective planes, *Des. Codes Cryptogr.* **1** (1990), 83–92.

[12] K.T. Arasu and A. Pott, Cyclic affine planes and Paley difference sets, *Discrete Math.* **106/107** (1992), 19–23.

[13] R.D. Baker, An elliptic semiplane, *J. Combin. Theory Ser. A* **25** (1978), 193–195.

[14] R.D. Baker, J. Dover, G.L. Ebert and K. Wantz, Perfect Baer subplane partitions and three-dimensional flag-transitive planes, *Des. Codes Cryptogr.* **21** (2000), 19–39.

[15] R.D. Baker, J. Dover, G.L. Ebert and K. Wantz, Baer subgeometry partitions, *J. Geom.* **67** (2000), 23–34.

[16] R.D. Baker and G.L. Ebert, Two-dimensional flag-transitive planes revisited, *Geom. Dedicata* **63** (1996), 1–15.

[17] R.D. Baker and G.L. Ebert, K.H. Leung and Q. Xiang, A trace conjecture and flag-transitive affine planes, *J. Combin. Theory Ser. A* **95** (2001), 158–168.

[18] S. Ball and M.R. Brown, The six semifields associated with a semifield flock, in preparation.

[19] S. Ball and M. Lavrauw, Commutative semifields of rank 2 over their middle nucleus, in *Finite fields and applications to coding theory, cryptography and related areas* (eds. G.L. Mullen, H. Stichtenoth & H. Tapia-Recillas), Springer, Berlin (2002), pp. 1–21.

[20] A. Barlotti, Le possibili configurazioni del sistema delle coppie punto-retta (A, a) per cui un piano grafico risulta (A, a) transitivo, *Boll. Unione Mat. Ital.* **12** (1957), 212–226.

[21] S.G. Barwick, Substructures of finite geometries, Ph.D. thesis, University of London, 1994.

[22] L.D. Baumert, *Cyclic difference sets*, Springer, New York (1971).

[23] T. Beth, D. Jungnickel & H. Lenz, *Design theory*, Cambridge University Press, Cambridge (1999).

[24] M. Biliotti, V. Jha & N.L. Johnson, *Foundations of translation planes*, Marcel Dekker, New York (2001).

[25] A. Blokhuis, A.E. Brower and H.A. Wilbrink, Hermitian unitals are code words, *Discrete Math.* **97** (1991), 63–68.

[26] A. Blokhuis, D. Jungnickel and B. Schmidt, On a class of symmetric divisible designs which are almost projective planes, in *Finite Geometries* (eds. A. Blokhuis, J.W.P. Hirschfeld, D. Jungnickel & J.A. Thas), Kluwer, Dordrecht (2001), pp. 27–34.

[27] A. Blokhuis, D. Jungnickel and B. Schmidt, Proof of the prime power conjecture for projective planes of order n with abelian collineation groups of order n^2, *Proc. Amer. Math. Soc.* **130** (2002), 1473–1476.

[28] E. Boros and T. Szőnyi, On the sharpness of a theorem of Segre, *Combinatorica* **6** (1986), 261–268.

[29] R.C. Bose, An affine analogue of Singer's theorem, *J. Indian Math. Soc.* **6** (1942), 1–15.

[30] R.C. Bose, On the application of finite projective geometry for deriving a certain series of balanced Kirkman arrangements, *Calcutta Math. Soc. golden jubilee commemoration* **II** ((),1958–59) 341–356.

[31] S.E. Broad and T.D. VanAken, Cyclic affine difference sets of order 8 modulo 16, *Congr. Numer.* **122** (1996), 203–206.

[32] R.H. Bruck, Difference sets in a finite group, *Trans. Amer. Math. Soc.* **78** (1955), 464–481.

[33] R.H. Bruck, Quadratic extensions of cyclic planes, *Proc. Sympos. Appl. Math.* **10** (1960), 15–44.

[34] R.H. Bruck, Circle geometry in higher dimensions, II, *Geom. Dedicata* **2** (1973), 133–188.

[35] R.H. Bruck and R.C. Bose, The construction of translation planes from projective spaces, *J. Algebra* **1** (1964), 85–102.

[36] R.H. Bruck and H.J. Ryser, The nonexistence of certain finite projective planes, *Canad. J. Math.* **1** (1949), 88–93.

[37] A.T. Butson, Relations among generalised Hadamard matrices, relative difference sets and maximal length linear recurring sequences, *Canad. J. Math.* **15** (1963), 42–48.

[38] C. Charnes and U. Dempwolff, The translation planes of order 49 and their automorphism groups, *Math. Comp.* **67** (1998), 1207–1224.

[39] W.E. Cherowitzo, Ovals in Figueroa planes, *J. Geom.* **37** (1990), 84–86.

[40] W.E. Cherowitzo, Hyperovals in the translation planes of order 16, *J. Combin. Math. Combin. Comput.* **9** (1991), 39–55.

[41] W.E. Cherowitzo, Hyperovals in desarguesian planes: An update, *Discrete Math.* **155** (1996), 31–38.

[42] S. Chowla and H.J. Ryser, Combinatorial problems, *Canad. J. Math.* **2** (1950), 93–99.

[43] S. D. Cohen and M.J. Ganley, Commutative semifields, two dimensional over their middle nuclei, *J. Algebra* **75** (1982), 373–385.

[44] R. Coulter and R. Matthews, Planar functions and planes of Lenz-Barlotti class II, *Des. Codes Cryptogr.* **10** (1997), 167–184.

[45] P. Dembowski, Gruppentheoretische Kenzeichnungen der endlichen desarguesschen Ebenen, *Abh. Math. Sem. Univ. Hamburg* **29** (1965), 92–106.

[46] P. Dembowski, *Finite geometries*, Springer, Berlin (1968).

[47] P. Dembowski and T. G. Ostrom, Planes of order n with collineation groups of order n^2, *Math. Z.* **103** (1968), 239–258.

[48] P. Dembowski and F.C. Piper, Quasiregular collineation groups of finite projective planes, *Math. Z.* **99** (1967), 53–75.

[49] U. Dempwolff, Translation planes of order 27, *Des. Codes Cryptogr.* **4** (1994), 105–121.

[50] U. Dempwolff and R. Reifart, The classification of the translation planes of order 16, I, *Geom. Dedicata* **15** (1983), 137–153.

[51] R.H.F. Denniston, Subplanes of the Hughes plane of order 9, *Math. Proc. Cambridge Philos. Soc.* **64** (1968), 589–598.

[52] M.J. de Resmini, D. Ghinelli and D. Jungnickel, Arcs and ovals from abelian groups, *Des. Codes Cryptogr.* **26** (2002), 213–228.

[53] M.J. de Resmini and N. Hamilton, Hyperovals and unitals in Figueroa planes, *European J. Combin.* **19** (1998), 215–220.

[54] A.V. Dizon-Garciano and Y. Hiramine, On Sylow subgroups of abelian affine difference sets, *Des. Codes Cryptogr.* **22** (2001), 157–163.

[55] G.L. Ebert, Partitioning problems and flag–transitive planes, *Rend. Circ. Mat. Palermo (2) Suppl.* **53** (1998), 27–44.

[56] E. Ellers and H. Karzel, Endliche Inzidenzgruppen, *Abh. Math. Sem. Univ. Hamburg* **27** (1964), 250–264.

[57] J.E.H. Elliott and A.D. Butson, Relative difference sets, *Illinois J. Math.* **10** (1966), 517–531.

[58] R.J. Evans and H.B. Mann, On simple difference sets, *Sanhkyā* **11** (1951), 357–364.

[59] W. Feit, Finite projective planes and a question about primes, *Proc. Amer. Math. Soc.* **108** (1990), 561–564.

[60] W. Feit and J.G. Thompson, Solvability of groups of odd order, *Pacific J. Math.* **13** (1963), 755–1029.

[61] J.B. Fink, A note on sharply flag-transitive projective planes, in *Finite geometries* (eds. N.L. Johnson, M.J. Kallaher & C.T. Long), Marcel Dekker, New York (1983), pp. 161–164.

[62] J.C. Fisher, J.W.P. Hirschfeld and J.A. Thas, Complete arcs in planes of square order, *Ann. Discrete Math.* **30** (1986), 243–250.

[63] C.I. Fung, M.K. Siu and S.L. Ma, On arrays with small off-phase binary autocorrelation, *Ars Combin.* **29A** (1990), 189–192.

[64] M.J. Ganley, On a paper of Dembowski and Ostrom, *Arch. Math. (Basel)* **27** (1976), 93–98.

[65] M.J. Ganley, Direct product difference sets, *J. Combin. Theory Ser. A* **23** (1977), 321–332.

[66] M.J. Ganley and F.C. Piper, Quasiregular collineation groups, *Rend. Istit. Mat. Univ. Trieste* **1** (1969), 112–122.

[67] M.J. Ganley and R.L. McFarland, On quasiregular collineation groups, *Arch. Math. (Basel)* **26** (1975), 327–331.

[68] M.J. Ganley and E. Spence, Relative difference sets and quasiregular collineation groups, *J. Combin. Theory Ser. A* **19** (1975), 134–153.

[69] C.W. Garner, Von Staudt conics in semifield planes, *Aequationes Math.* **11** (1974), 183–188.

[70] D. Ghinelli, Varietà Hermitiane e strutture finite, I, *Rend. Mat. Appl. (6)* **2** (1969), 23–62.

[71] D. Ghinelli(-Smit), On semisymmetric designs, report, Westfield College, University of London, 1980.

[72] D. Ghinelli(-Smit), Automorphisms and generalized incidence matrices of point-divisible designs, *Ann. Discrete Math.* **18** (1983), 377–400.

[73] D. Ghinelli(-Smit), On abelian projective planes, *Arch. Math. (Basel)* **44** (1985), 282–288.

[74] D. Ghinelli(-Smit), A new result on difference sets with -1 as multiplier, *Geom. Dedicata* **23** (1987), 309–317.

[75] D. Ghinelli(-Smit), Hall-Ryser type theorems for relative difference sets, *Ann. Discrete Math.* **37** (1988), 189–194.

[76] D. Ghinelli, A rational congruence for a standard orbit decomposition, *European J. Combin.* **11** (1990), 105–113.

[77] D. Ghinelli and D. Jungnickel, On finite projective planes in Lenz-Barlotti class at least I.3, *Adv. Geom.*, in press.

[78] D. Gluck, A note on permutation polynomial and finite geometries, *Discrete Math.* **80** (1990), 97–100.

[79] D.M. Gordon, The prime power conjecture is true for $n < 2,000,000$, *Electronic J. Combin.* **1 R6** (1994),

[80] D.M. Gordon, Some restrictions on orders of abelian planar difference sets, *J. Combin. Math. Combin. Comput.* **29** (1999), 241–246.

[81] M. Hall, Jr., Projective planes, *Trans. Amer. Math. Soc.* **54** (1943), 229–277.

[82] M. Hall, Jr., Cyclic projective planes, *Duke Math. J.* **14** (1947), 1079–1090.

[83] M. Hall, Jr., *Combinatorial theory (2nd edition)*, Wiley, New York (1986).

[84] Y. Hiramine, A conjecture on affine planes of prime order, *J. Combin. Theory Ser. A* **52** (1989), 44–50.

[85] Y. Hiramine, On planar functions, *J. Algebra* **133** (1990), 103–110.

[86] Y. Hiramine, Planar functions and related group algebras, *J. Algebra* **15** (1992), 135–145.

[87] Y. Hiramine, Affine difference sets and related factor sets, *Geom. Dedicata* **54** (1995), 13–29.

[88] Y. Hiramine, Difference sets relative to disjoint subgroups, *J. Combin. Theory Ser. A* **88** (1999), 205–216.

[89] J.W.P. Hirschfeld, *Projective geometries over finite fields (2nd edition)*, Oxford University Press, Oxford (1998).

[90] C.Y. Ho, Some remarks on orders of projective planes, planar difference sets and multipliers, *Des. Codes Cryptogr.* **1** (1991), 69–75.

[91] C.Y. Ho, On bounds for groups of multipliers of planar difference sets, *J. Algebra* **148** (1992), 325–336.

[92] C.Y. Ho, Subplanes of a tactical decomposition and Singer groups of a projective plane, *Geom. Dedicata* **53** (1994), 307–326.

[93] C.Y. Ho, Arc subgroups of planar Singer groups, in *Mostly finite geometries* (ed. N.L. Johnson), Marcel Dekker, New York (1997), pp. 227–233.

[94] C.Y. Ho, Finite projective planes with transitive abelian collineation groups, *J. Algebra* **208** (1998), 533–550.

[95] C.Y. Ho and A. Pott, On multiplier groups of planar difference sets and a theorem of Kantor, *Proc. Amer. Math. Soc.* **109** (1990), 803–808.

[96] A.J. Hoffmann, Cyclic affine planes, *Canad. J. Math.* **4** (1952), 295–301.

[97] D.R. Hughes, Planar division neo-rings, *Trans. Amer. Math. Soc.* **80** (1955), 502–527.

[98] D.R. Hughes, Planar division neo-rings, Ph.D. Thesis, University of Wisconsin, Madison, 1955.

[99] D.R. Hughes, Partial difference sets, *Amer. J. Math.* **78** (1956), 650–674.

[100] D.R. Hughes, A note on some partially transitive projective planes, *Proc. Amer. Math. Soc.* **8** (1957), 978–981.

[101] D.R. Hughes, Generalized incidence matrices over group algebras, *Illinois J. Math* **1** (1957), 545–551.

[102] D.R. Hughes, A class of non-desarguesian projective planes, *Canad. J. Math.* **9** (1957), 378–388.

[103] D.R. Hughes, On designs, in *Geometries and groups* (eds. M. Aigner & D. Jungnickel), Springer, Berlin (1981), pp. 43–67.

[104] D.R. Hughes & F.C. Piper, *Projective planes*, Springer-Verlag, New York (1982).

[105] V. Jha and G. Wene, An oval partition of the central units of certain semifield planes, *Discrete Math.* **155** (1996), 127–134.

[106] D. Jungnickel, On automorphism groups of divisible designs, *Canad. J. Math.* **34** (1982), 257–297.

[107] D. Jungnickel, A note on affine difference sets, *Arch. Math. (Basel)* **47** (1986), 279–280.

[108] D. Jungnickel, On a theorem of Ganley, *Graphs Combin.* **3** (1987), 141–143.

[109] D. Jungnickel, Divisible semiplanes, arcs, and relative difference sets, *Canad. J. Math.* **39** (1987), 1001–1024.

[110] D. Jungnickel, On the geometry of affine difference sets of even order, *Arab Gulf J. Scient. Res. A. Math. Phys. Sci.* **A7** (1989), 21–28.

[111] D. Jungnickel, An elementary proof of Wilbrink's theorem, *Arch. Math. (Basel)* **52** (1989), 615–617.

[112] D. Jungnickel, On affine difference sets, *Sankhyā Ser. A* **54** (1992), 219–240.

[113] D. Jungnickel, Difference sets, *Contemporary design theory: A collection of surveys*, (eds. J.H. Dinitz & D.R. Stinson), Wiley, New York (1992), pp. 241–324.

[114] D. Jungnickel and M.J. de Resmini, Another case of the prime power conjecture for finite projective planes, *Adv. Geom.* **2** (2002), 215–218.

[115] D. Jungnickel and A. Pott, Two results on difference sets, *Colloq. Math. Soc. János Bolyai* **52** (1988), 325–330.

[116] D. Jungnickel and A. Pott, Computational nonexistence results for abelian affine difference sets, *Congr. Numer.* **68** (1989), 91–98.

[117] D. Jungnickel and B. Schmidt, Difference sets: An update, in *Geometry, combinatorial designs and related structures* (eds. J.W.P. Hirschfeld, S.S. Magliveras & M.J. de Resmini), Cambridge University Press, Cambridge (1997), pp. 89–112.

[118] D. Jungnickel and B. Schmidt, Difference sets: A second update, *Rend. Circ. Mat. Palermo Ser. II Suppl.* **53** (1998), 89–118.

[119] D. Jungnickel and K. Vedder, On the geometry of planar difference sets, *European J. Combin.* **5** (1984), 143–148.

[120] M.J. Kallaher, *Affine planes with transitive collineation groups*, North-Holland, New York–Amsterdam (1982).

[121] M.J. Kallaher, Translation planes, in *Handbook of incidence geometry* (ed. F. Buekenhout), Elsevier, Amsterdam (1995), pp. 137–192.

[122] W. M. Kantor, Projective planes of type I-4, *Geom. Dedicata* **3** (1974), 335–346.

[123] W. M. Kantor, Linear groups containing a Singer cycle, *J. Algebra* **62** (1980), 232–234.

[124] W. M. Kantor, Primitive permutation groups of odd degree, and an application to finite projective planes, *J. Algebra* **106** (1987), 15–45.

[125] W. M. Kantor and M.D. Pankin, Commutativity in finite planes of type I.4, *Arch. Math. (Basel)* **23** (1972), 544–547.

[126] B. Kestenband, Unital intersections in finite projective planes, *Geom. Dedicata* **11** (1981), 107–117.

[127] H.P. Ko and D.K. Ray-Chaudhuri, Multiplier theorems, *J. Combin. Theory Ser. A* **30** (1981), 134–157.

[128] H.P. Ko and D.K. Ray-Chaudhuri, Intersection theorems for group divisible difference sets, *Discrete Math.* **39** (1982), 37–58.

[129] P.V. Kumar, On the existence of square dot-matrix patterns having a specific three-valued periodic-correlation function, *IEEE Trans. Inform. Theory* **34** (1988), 271–277.

[130] C.W.H. Lam, On relative difference sets, *Congr. Numer.* **20** (1977), 445–474.

[131] C.W.H. Lam, L. Thiel and S. Swiercz, The nonexistence of finite projective planes of order 10, *Canad. J. Math.* **41** (1989), 1117–1123.

[132] E.S. Lander, Topics in algebraic coding theory, Ph.D. Thesis, Oxford University, 1980.

[133] E.S. Lander, *Symmetric designs: An algebraic approach*, Cambridge University Press, Cambridge (1983).

[134] E.S. Lander, Restrictions upon multipliers of an abelian difference set, *Arch. Math. (Basel)* **50** (1988), 241–242.

[135] H. Lenz, Kleiner desarguesscher Satz und Dualität in projektiven Ebenen, *Jahresber. Deutsch. Math.-Verein.* **57** (1954), 20–31.

[136] K.H. Leung, S.L. Ma and V. Tan, Planar functions from Z_n to Z_n, *J. Algebra* **224** (2000), 427–436.

[137] J.H. van Lint & R.M. Wilson, *A course in combinatorics*, Cambridge University Press, Cambridge (2001).

[138] H. Lüneburg, *Translation planes*, Springer, Berlin (1980).

[139] S.L. Ma, Planar functions, relative difference sets, and character theory, *J. Algebra* **185** (1996), 342–356.

[140] S.L. Ma and A. Pott, Relative difference sets, planar functions and generalized Hadamard matrices, *J. Algebra* **175** (1995), 505–525.

[141] H.B. Mann, Some theorems on difference sets, *Canad. J. Math.* **4** (1952), 222–226.

[142] H.B. Mann, Balanced incomplete block designs and abelian difference sets, *Illinois J. Math.* **8** (1964), 252–261.

[143] G.E. Martin, On arcs in a finite projective plane, *Canad. J. Math.* **19** (1967), 376–393.

[144] R. Mathon, On a new divisible semiplane, Announcement, 11th British Combinatorial Conference, 1987.

[145] R. Mathon and G.F. Royle, The translation planes of order 49, *Des. Codes Cryptogr.* **5** (1995), 57-72.

[146] P. Müller, On the collineation group of cyclic planes, *J. Combin. Theory Ser. A* **65** (1994), 60–66.

[147] T.G. Ostrom, Concerning difference sets, *Canad. J. Math.* **5** (1953), 421–424.

[148] T.G. Ostrom and A. Wagner, On projective and affine planes with transitive collineation groups, *Math. Z.* **71** (1959), 186–199.

[149] U. Ott, Endliche zyklische Ebenen, *Math. Z.* **144** (1975), 195–215.

[150] M.D. Pankin, On finite planes of type I.4, Ph.D. thesis, University of Illinois at Chicago Circle, 1971.

[151] M.D. Pankin, On finite planes of type I.4, in *On projective planes* (eds. M.J. Kallaher & T.G. Ostrom), Washington State University Press, Washington (1973), pp. 215–218.

[152] G. Pickert, *Projektive Ebenen*, Springer, Berlin (1955,^{2}1975).

[153] F.C. Piper, The orbit structure of collineation groups of finite projective planes, *Math. Z.* **103** (1968), 318–332.

[154] F.C. Piper, On relative difference sets and projective planes, *Glasg. Math. J.* **15** (1975), 150–154.

[155] A. Pott, Applications of the DFT to abelian difference sets, *Arch. Math. (Basel)* **51** (1988), 283–288.

[156] A. Pott, A note on self-orthogonal codes, *Discrete Math.* **76** (1989), 283–284.

[157] A. Pott, On abelian difference sets with multiplier −1, *Arch. Math. (Basel)* **53** (1989), 510–512.

[158] A. Pott, On multiplier theorems, in *Coding Theory and Design Theory Part II* (ed. D. Ray-Chaudhuri), Springer, New York (1990), pp. 286–289.

[159] A. Pott, An affine analogue of Wilbrink's theorem, *J. Combin. Theory Ser. A* **55** (1990), 313–315.

[160] A. Pott, On projective planes admitting elations and homologies, *Geom. Dedicata* **52** (1994), 181–193.

[161] A. Pott, *Finite geometry and character theory*, Springer, Berlin (1995).

[162] A. Pott, A survey on relative difference sets, in *Groups, difference sets and the monster* (eds. K.T. Arasu et al.), Walter de Gruyter, Berlin (1996), pp. 195–232.

[163] T. G Room, Polarities and ovals in the Hughes plane, *J. Austral. Math. Soc.* **13** (1972), 196–204.

[164] L. Rónyai and T. Szőnyi, Planar functions over finite fields, *Combinatorica* **9** (1989), 315–320.

[165] B. Schmidt, Cyclotomic integers and finite geometry, *J. Amer. Math. Soc.* **12** (1999), 929–952.

[166] B. Segre, Ovals in a finite projective plane, *Canad. J. Math.* **7** (1955), 414–416.

[167] G. Seib, Unitäre Polaritäten endlicher projektiver Ebenen, *Arch. Math. (Basel)* **21** (1970), 103–112.

[168] J. Singer, A theorem in finite projective geometry and some applications to number theory, *Trans. Amer. Math. Soc.* **43** (1938), 377–385.

[169] L. Storme and H. Van Maldeghem, Cyclic arcs in $PG(2, q)$, *J. Algebraic Combin.* **3** (1994), 113–128.

[170] T. Szőnyi, On cyclic caps in projective spaces, *Des. Codes Cryptogr.* **8** (1996), 327–332.

[171] K. Thas, Finite flag-transitive projective planes: a survey and some remarks, *Discrete Math.*, in press.

[172] R.J. Turyn, Character sums and difference sets, *Pacific J. Math.* **15** (1965), 319–346.

[173] J.C.D.S. Yaqub, The Lenz-Barlotti classification, in *Projective Geometry* University of Illinois, Chicago (1967), pp. 129–160.

[174] T.D. VanAken, A search for affine difference sets of even order, *J. Statist. Plann. Inference* **62** (1997), 125–133.

[175] A. Wagner, On finite affine line transitive planes, *Math. Z.* **87** (1965), 1–11.

[176] W. Wei, S. Gao and Q. Xiang, Extraneous multipliers of abelian difference sets, in *Combinatorial designs and applications* (eds. W.D. Wallis, H. Shen, W. Wei & L. Zhu), Marcel Dekker, New York (1990), pp. 159–164.

[177] H.A. Wilbrink, A note on planar difference sets, *J. Combin. Theory Ser. A* **38** (1985), 94–95.

[178] M. Yamada, On a relation between a cyclic relative difference set associated with the quadratic extensions of a finite field and the Szekeres difference sets, *Combinatorica* **8** (1988), 207–216.

[179] H. Zassenhaus, Über endliche Fastkörper, *Abh. Math. Sem. Univ. Hamburg* **11** (1935), 187–220.

Dipartimento di Matematica
Università di Roma "La Sapienza"
2, Piazzale Aldo Moro
I–00185 Roma, Italy
Dina.Ghinelli@uniroma1.it

Lehrstuhl für Diskrete Mathematik,
Optimierung und Operations Research
Universität Augsburg
D–86135 Augsburg, Germany
jungnickel@math.uni-augsburg.de

Algorithmic aspects of graph homomorphisms

Pavol Hell

Abstract

Homomorphisms are a useful model for a wide variety of combinatorial problems dealing with mappings and assignments, typified by scheduling and channel assignment problems. Homomorphism problems generalize graph colourings, and are in turn generalized by constraint satisfaction problems; they represent a robust intermediate class of problems – with greater modeling power than graph colourings, yet simpler and more manageable than general constraint satisfaction problems. We will discuss various homomorphism problems from a computational perspective. One variant, with natural applications, gives each vertex a list of allowed images. Such list homomorphisms generalize list colourings, precolouring extensions, and graph retractions. Many algorithms for finding homomorphisms adapt well to finding list homomorphisms. Semi-homomorphisms are another variant; they generalize the kinds of partitions that homomorphisms induce, to allow both homomorphism type constraints, and constraints that correspond to homomorphisms of the complementary graphs. Surprisingly, semi-homomorphism partition problems cover a great variety of concepts arising in the study of perfect graphs. We illustrate some of the ideas leading to efficient algorithms for all these problems.

1 Introduction

Graphs we consider may be directed or undirected, and, correspondingly, uv can denote a directed arc or an undirected edge, depending on the context. Both kinds of graphs will be allowed to have loops, but no parallel edges, and all graphs will be assumed to be finite. A *homomorphism* of a graph G to a graph H is a mapping $f : V(G) \to V(H)$ such that $uv \in E(G)$ implies $f(u)f(v) \in E(H)$. Thus homomorphisms of undirected graphs preserve just the adjacency relation, while homomorphisms of directed graphs also preserve the directions of the arcs. We note that an undirected graph may be viewed as a directed graph where each edge is replaced by the two opposite arcs. It is easy to see that the two definitions of homomorphisms coincide for this class of graphs. Thus, whenever convenient, we shall consider undirected graphs to be a subclass of directed graphs.

For undirected graphs without loops, a homomorphism to the complete graph K_n is exactly an n-colouring. For this reason, a homomorphism of G to H is also called an *H-colouring* of G.

By analogy with classical colourings, we associate with each H-colouring f of G a partition of $V(G)$ into the sets $S_h = f^{-1}(h), h \in V(H)$. It is clear that a mapping $f : V(G) \to V(H)$ is a homomorphism of G to H if and only if the associated partition satisfies the following two constraints:

(H1) if hh is not a loop in H, then S_h is an independent set; and

(H2) if hh' is not an edge (arc) of H then there are no edges (arcs) from S_h to $S_{h'}$.

Thus for a graph G to admit an H-colouring is equivalent to admitting a partition satisfying (H1) and (H2).

Let H be a *fixed* graph. The *homomorphism problem* for H, also called the *H-colouring problem*, asks whether or not an input graph G admits a homomorphism to H, that is, a partition into sets $S_h, h \in V(H)$, satisfying (H1) and (H2). The same problem can be of course stated for more general structures H. When H is an arbitrary relational structure (a set with a number of relations of various arities), the problem of finding a mapping (of a relational structure G with the same numbers of relations of the same arities) preserving all the relations, is known as the *constraint satisfaction problem (or CSP)* with template H, [25, 36, 37, 45, 64, 65, 76, 80].

Two graphs are *homomorphically equivalent* if each admits a homomorphism to the other. Clearly two homomorphically equivalent graphs result in the same homomorphism problem. Specifically, if H, H' are homomorphically equivalent, then a graph G is H-colourable if and only if it is H'-colourable. (This easily follows from the fact the composition of two homomorphisms is again a homomorphism.) It turns out that any graph G is homomorphically equivalent to a unique (up to isomorphism) minimal subgraph of H which is called the *core* of G [41, 54]. Thus we may restrict our focus to H-colouring problems where H is a core.

2 A Simple Algorithm and Duality Result

Some H-colouring problems are easy to solve. Consider the directed three cycle $\vec{C}_3$ given in Fig. 1(a). For every vertex of $\vec{C}_3$ there is exactly one arc in and one arc out. Thus a homomorphism of a (weakly) connected directed graph G to $\vec{C}_3$ is determined by the image of any one vertex. (Each weak component of G can be mapped separately.) In fact, the following (folklore) algorithm can be used.

ALGORITHM 1
Input: A connected directed graph G.
Task: Find a $\vec{C}_3$-colouring of G, if one exists.
Action: Choose a vertex $v \in V(G)$ and set $f(v) = 0$. Then whenever a vertex x has been assigned an image i, set $f(y) = i - 1$ for all vertices y that dominate x, and set $f(z) = i + 1$ for all vertices z that are dominated by x. (Addition is taken modulo three.)

If the algorithm ever needs to change the colour of a vertex, it is said to *fail*; otherwise it is said to *succeed*.

The *net length* of an oriented walk is the difference between the number of forward and backward arcs.

Proposition 2.1 *The following statements are equivalent for a connected directed graph G.*

1. *Algorithm 1 applied to graph G succeeds.*

2. *There is a homomorphism of G to $\vec{C}_3$.*

3. *G does not contain a closed walk of net length not divisible by three.*

Proof Since the implications from 1 to 2 and from 2 to 3 are obvious, the equivalence is proved by showing that 3 implies 1 – by contrapositive, analyzing the situation at the moment of failure, when a vertex u that already has a colour needs to have another colour. By following the two walks from v to u that resulted in the different colours, we easily find a closed walk of net length not divisible by three. □

The equivalence of 1 and 2 implies the following:

Corollary 2.2 *Algorithm 1 is correct, that is, it succeeds if and only if G is $\vec{C}_3$-colourable.*

From the equivalence of 2 and 3 we derive the following example of what we shall call duality results (cf. section 6).

Corollary 2.3 ([21]) *G is $\vec{C}_3$-colourable if and only if every closed walk has net length divisible by three.*

Clearly, a similar technique will deal with any situation where mapping one vertex determines the images of all its neighbours. This happens when each vertex of H has a unique neighbour (in-neighbour and out-neighbour, in case of directed graphs); included here are all directed cycles $\vec{C}_k$ and directed paths $\vec{P}_k$. A special case of this is the undirected situation when $H = K_2$, which can be viewed as $H = \vec{C}_2$. Thus K_2-colouring of undirected graphs is solved in linear time, by a similar simple algorithm (essentially the Breadth First Search). The analogue of Corollary 2.3 in that case is the theorem of König which states that a graph is bipartite if and only if it does not contain an odd closed walk (or, equivalently, an odd cycle).

3 Dichotomy for Undirected Graphs

Recall that if H and H' are homomorphically equivalent then the H-colouring problem is the same as the H'-colouring problem. In particular, it is easy to see that the core of any (undirected) bipartite graph is K_1 or K_2. Similarly, the core of any graph (directed or undirected) which has a loop is the one-vertex graph with a loop, which we denote by K_1^*. Note that both the K_1-colouring and the K_1^*-colouring problems are trivial (for both undirected and directed graphs): All graphs admit a K_1^*-colouring, and precisely the edge-less graphs admit a K_1-colouring. In particular, for undirected graphs H, we have proved the first statement of the following *dichotomy* classification theorem:

Theorem 3.1 ([55]) *Let H be an undirected graph.*

- *If H is bipartite or contains a loop, then the H-colouring problem has a polynomial time algorithm.*
- *Otherwise the H-colouring problem is NP-complete.*

This was one of first known dichotomy results for general homomorphism (or constraint satisfaction) problems [29, 36, 37, 31, 39, 84, 88]. The reader should note that the it includes the well known dichotomy classification result that *the classical k-colouring problem is polynomial time solvable when $k \leqslant 2$ and NP-complete when $k > 2$.*

The proof of the second statement of Theorem 3.1 is too technical to be included in full (cf. [55, 57]). We illustrate the proof technique with the following example. It is clear that each H-colouring problem is in the class NP.

Proposition 3.2 *Let H be the five-cycle C_5. Then the H-colouring problem is NP-complete.*

Proof Here is a reduction from 5-COL (the problem of five-colourability) to the H-colouring problem: Given an instance G of 5-COL, we replace each edge of G with a path of length three, calling the resulting graph $^*G$. We now claim that G has a five-colouring if and only if $^*G$ has an H-colouring. Indeed, a five-colouring of G, with colours $0, 1, 2, 3, 4$, provides us with images for all the 'old' vertices of $^*G$ (those that were present in G). The 'new' vertices of $^*G$, lying on the added paths, can easily be mapped so that edges are preserved, since for any pair of *distinct* vertices of $H = C_5$ there is a homomorphism from the path of length three to H, which maps the endpoints to these prescribed vertices. Conversely, any homomorphism of $^*G$ to $H = C_5$ yields a mapping of the 'old' points in which two points joined by one of the added paths (thus adjacent in G) have distinct images. This is so, since H contains no triangle, and hence there is no homomorphism of the path of length three to H in which the endpoints have the same image. □

The construction can be easily generalized to prove

Proposition 3.3 *C_{2k+1}-COL is NP-complete, for any positive integer k.*

Our Theorem 3.1 has the following corollary.

Corollary 3.4 *Suppose H' is an induced subgraph of H. There is a polynomial time reduction of the H'-colouring problem to the H-colouring problem.*

Proof We may assume that H' is not bipartite otherwise the problem is polynomial time solvable and we can reduce each instance of H'-colouring to one of two fixed instances of H-colouring. But then H is also not bipartite and thus H-colouring is NP-complete. Since H'-colouring is a problem in NP, according to Cook's theorem, it admits a polynomial time reduction to any NP-complete problem, including H-colouring. □

It is ironic that we derived Corollary 3.4 in such a roundabout way (using Cook's theorem). After all, it is precisely the kind of statement which is routinely proved *directly* – by actually exhibiting a reduction. If we could find such a direct construction, we would have a much more satisfying proof of the above theorem - since every nonbipartite graph contains an odd cycle, and the NP-completeness of H-colouring when H is an odd cycle is proved above. On the other hand, there seem to be good reasons for the difficulties in trying to prove the corollary directly. For instance, the result fails to hold for directed graphs, cf. Fig. 1.

4 Directed Graph Homomorphisms

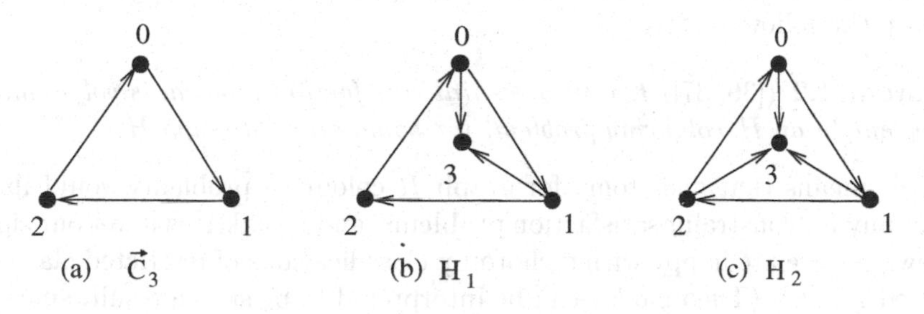

Figure 1: Graphs $\vec{C}_3, H_1$, and H_2

In Fig. 1, we have directed graphs H_1 (in part (b)), and H_2 (in part (c)), such that H_1 is a subgraph of H_2. Surprisingly, the smaller graph yields the harder problem.

Proposition 4.1 ([49])

- *The H_1-colouring problem is NP-complete.*
- *The H_2-colouring problem is polynomial time solvable.*

Proof We shall only prove the second statement. (The first statement is proved by a reduction from ONE-IN-THREE-SAT [49, 57].) In a graph G, no vertex of positive out-degree can be coloured 3, in any H_2-colouring of G. On the other hand, if a vertex of out-degree zero obtained a colour different from 3, in some H_2-colouring of G, we can re-colour it by 3, and still have an H_2-colouring of G. Thus we can test for the H_2-colourability of G by first colouring all vertices of out-degree zero by 3 and removing them from G. The original graph G is H_2-colourable if and only if the reduced graph is $\vec{C}_3$-colourable, which can be checked by Algorithm 1. □

There also exist graphs H_1, H_2 such that H_1-colouring is NP-complete and H_2-colouring is polynomial time solvable, where H_1 is an *induced* subgraph of H_2 [99, 57]. (On the other hand, if H_1 is a component of H_2, then there is a reduction of the H_1-colouring problem to the H_2-colouring problem [6].)

The H-colouring problem for directed graphs has received much attention, and yet no dichotomy classification has been obtained, or conjectured. In fact, it is an open problem whether or not this class of problems has dichotomy at all, in other words whether or not every homomorphism problem is NP-complete or polynomial time solvable. It is known [70] that if $P \neq NP$ then there are problems in NP which are neither NP-complete nor polynomial time solvable. However, it is a longstanding open problem whether or not such problems exist among constraint satisfaction problems [36, 37]. (See [20] for a new conjecture on what makes CSP problems easy; this distinction could turn out to be the sought after dichotomy classification of constraint satisfaction problems.) The role of graph homomorphism problems is central to this, as we have the following fact.

Theorem 4.2 ([36, 37]) *Every constraint satisfaction problem is polynomially equivalent to an H-colouring problem, for some directed graph H.*

This means that dichotomy for graph H-colouring problems would imply dichotomy for constraint satisfaction problems, and it makes us lower our sights somewhat – we are happy with dichotomy classifications of restricted classes of directed graphs. (Theorem 3.1 can be interpreted to be such a result, since the class of undirected graphs can be viewed as a subclass of the class of directed graphs.) Even so, dichotomy results seem hard to come by. It is, for instance, still open whether or not the class of oriented trees has dichotomy, cf. [58]. In fact, while oriented paths H all yield polynomial time solvable H-colouring problems [62, 50], dichotomy is not even known for trees H which consist of three oriented paths meeting at one central vertex, called *triads*.

For oriented cycles, dichotomy has been proved in [31], but it does not yield a classification. In other words, Feder proves that each homomorphism problem to a fixed oriented cycle is polynomial time solvable or NP-complete, without giving a full description of which oriented cycles yield polynomial time solvable problems and which yield NP-complete ones.

5 Tournaments and Graphs without Sources and Sinks

For tournaments, there is a dichotomy classification. To state it in greater generality, we define a *semi-complete digraph* to be a directed graph H which contains a spanning tournament; in other words, between any two vertices there is either one arc or both (opposite) arcs.

Theorem 5.1 ([7]) *Suppose H is a semi-complete digraph.*

- *If H contains at most one directed cycle, the H-colouring problem can be solved in polynomial time.*
- *Otherwise, the problem is NP-complete.*

Proof We prove the first assertion. First of all, if H has zero directed cycles, then it is the transitive tournament $H = \vec{T}_n$. Suppose we are trying to find an $\vec{T}_n$-colouring of an input graph G. This is a slightly more complex situation than we have previously encountered, as having mapped a vertex u of the input graph to i (say), does not uniquely determine the image of an out-neighbour of u. However, if we proceed in the right order, the choices will be easy to make. Note that if G contains a directed walk of length n (or more) then it cannot be homomorphic to $\vec{T}_n$. In particular, if G is $\vec{T}_n$-colourable it cannot contain directed cycles. Assume the vertices of $\vec{T}_n$ are $1, 2, \ldots, n$.

> **ALGORITHM 2**
> **Input:** A graph G.
> **Task:** Find a $\vec{T}_n$-colouring of G, if one exists.
> **Action:** Colour by n all vertices v of out-degree zero in G. After using the colour i, remove all vertices that have been coloured, and colour by $i - 1$ all vertices of out-degree zero in the remaining digraph.

If the algorithm colours all vertices, it is said to *succeed*; otherwise it is said to *fail*. The following proposition is also easily proved.

Proposition 5.2 *Let G be a graph. Then the following three statements are equivalent.*

1. *Algorithm 2 applied to graph G succeeds.*
2. *There is a homomorphism of G to $\vec{T}_n$.*

3. *G does not contain a directed walk of length n.*

Proof It is again easy to argue that 1 implies 2, and 2 implies 3. To see that 3 implies 1, note that the colour of vertex v is the maximum length of a directed path starting from v. □

The algorithm extends easily to semi-complete digraphs with a unique cycle. In fact, such a graph is either a directed cycle, or is obtained from a smaller semi-complete digraph by adjoining a *dominating vertex* (a vertex dominating all other vertices), or a *dominated vertex* (a vertex dominated by all other vertices). The technique of Algorithm 2 actually shows that adjoining a dominating or dominated vertex to a directed graph H preserves the polynomiality of the H-colouring problem (G admits a homomorphism to the extended graph if and only if G without the sources, or sinks respectively, admits a homomorphism to H). This completes the proof of the first part of Theorem 5.1. For the NP-completeness, we refer the reader to [7]. □

Directed cycles do seem to play an important role in this context. In fact, for the restricted class of directed graphs without sources and sinks, there is a conjectured dichotomy classification. (Sources and sinks are the vertices of out-degree respectively in-degree zero.)

Conjecture 5.3 ([6]) *Suppose H is a connected graph without sources and sinks. If the core of H is a directed cycle, then H-colouring is polynomial time solvable. Otherwise H-colouring is NP-complete.*

We note that it is easy to check whether or not the core of a given graph G is a directed cycle, since it would have to be a shortest directed cycle in G. Thus finding the directed girth k and testing for $\vec{C}_k$-colourability, as in Algorithm 1, will accomplish the check. (On the other hand, we remark in passing that it is NP-complete to decide whether or not the core of G is an oriented tree [60]. It is also NP-complete to decide if a given graph G is equal to its core [54].)

The conjecture has been verified for a number of graph classes [6, 7, 8, 9, 77, 79]. It has, for instance, been verified for vertex transitive graphs H [75]. We may also view Theorem 3.1 as confirming a special case of the conjecture, since a connected undirected graph H has a core which is a directed cycle if and only if it is bipartite or has a loop.

6 Duality

We now note the similarity of many results accompanying our polynomial time algorithms. Each offers the same kind of certificate of non-colourability of G by the G-colourability of certain other graphs. To express these concisely, we shall write $G \to H$ to mean there exists a homomorphism of G to H and

$G \not\rightarrow H$ to mean that there is no such homomorphism. Note that the existence of walks and closed walks is expressible in terms of homomorphisms from paths and cycles, for instance $\vec{C}_n \rightarrow G$ if and only if G has a directed walk of length n. Thus our previous results can be summarized as follows: (Most of these are folklore; some are explicitly stated for the first time in [21].)

Proposition 6.1

1. *$G \not\rightarrow \vec{C}_n$, for a digraph G, if and only if $C \rightarrow G$ for some oriented cycle C of net length not divisible by n.*
2. *$G \not\rightarrow K_2$, for a graph G, if and only if $C_k \rightarrow G$ for some odd k.*
3. *$G \not\rightarrow \vec{P}_n$, for a digraph G, if and only if $P \rightarrow G$ for some oriented path P of net length at least n.*
4. *$G \not\rightarrow \vec{T}_n$, for a digraph G, if and only if $\vec{P}_{n+1} \rightarrow G$.*

Proof We denoted by $\vec{C}_n, \vec{P}_n, \vec{T}_n$ the directed cycle, path, and transitive tournament on n vertices, respectively. Statement 1 is discussed, in the special case $n = 3$, in Corollary 2.3. Statement 2 is a special case of Statement 1, as remarked before. Statement 3 is easy to prove using the same technique. Finally, statement 4 follows from the discussion following Algorithm 2. □

These results suggest the following terminology: Let $\mathcal{T}$ denote the class of all oriented trees. We say that the graph H *has tree duality*, if there exists a family $\mathcal{F} \subseteq \mathcal{T}$ such that

- $G \not\rightarrow H$ if and only if $T \rightarrow G$ for some $T \in \mathcal{F}$.

Statements 3 and 4 of the above Proposition show that $\vec{P}_n$ and $\vec{T}_n$ have tree duality.

Let $\mathcal{T}_k$ denote the class of all oriented graphs of treewidth k. We say that the graph H *has treewidth k duality*, if there exists a family $\mathcal{F} \subseteq \mathcal{T}_k$ such that

- $G \not\rightarrow H$ if and only if $T \rightarrow G$ for some $T \in \mathcal{F}$.

Since cycles have treewidth two, statements 1 and 2 above show that $\vec{C}_n, K_2$ have treewidth two duality.

A graph H has *bounded treewidth duality*, if there is an integer k such that H has treewidth k duality. It turns out that bounded treewidth duality is closely related to the existence of polynomial time colouring algorithms.

Theorem 6.2 ([58]) *If H has bounded treewidth duality, then the H-colouring problem has a polynomial time algorithm.*

We remark that it is a well known open question whether or not each problem in NP which is also in co-NP belongs to P. Theorem 6.2 answers this question in the affirmative, for the special case of H-colouring problems which are in co-NP by virtue of having bounded treewidth duality. (It can be shown that the certificate T of non-H-colourability, implied by the definition of treewidth k duality, can be made small enough, placing the problem in co-NP, [58].)

7 Consistency

There is an inference procedure, well known in the Artificial Intelligence community [25, 45, 76, 80], which is closely related to tree duality. Before making that relation explicit, we take time to introduce lists. Given a graph G, input to the H-colouring problem, we may have (or introduce) *lists* $L(g) \subseteq V(H), g \in V(G)$. For the time being we assume that G is always given with all initial lists $L(g) = V(H)$. (This will change when we discuss list homomorphisms.) We aim to modify the initial lists, to satisfy the following property.

Lists L are *consistent* if for any $gg' \in E(G)$ and any $h \in L(g)$ there exists an $h' \in L(g')$ with $hh' \in E(H)$, and (in the case of directed graphs), for any $gg' \in E(G)$ and any $h' \in L(g')$ there exists an $h \in L(g)$ with $hh' \in E(H)$.

The following algorithm reduces the given lists L to consistent lists L^*.

ALGORITHM 3 - Consistency Test
Input: A graph G, with lists $L(g) = V(H), g \in V(G)$.
Task: Reduce the lists to $L^*(g) \subseteq V(H), g \in V(G)$, which are consistent.
Action: Initially set all lists $L^*(g) = L(g)$, and then, as long as changes occur, process $gg' \in E(G)$ repeatedly as follows: Remove from $L^*(g)$ any h for which no element $h' \in L^*(g)$ has $hh' \in E(H)$, and remove from $L^*(g')$ any h' for which no $h \in L^*(g)$ has $hh' \in E(H)$.

We say that the consistency test *fails* if some list $L^*(g)$ becomes empty; otherwise we say that the consistency test *succeeds*.

Edges may have to be re-processed as the changes propagate in the graph G; however, the procedure is linear in the size of G [45, 58]. It is easy to see that if a list homomorphism of G to H with respect to the lists L exists, then the consistency test succeeds. (If f is such a list homomorphism, then $f(g)$ will never be removed from $L^*(g)$.)

Suppose we seek an H-colouring of an input graph G. If the consistency test fails, then G is not H-colourable. Of course, if the test succeeds, we still cannot be sure a homomorphism exists. For instance, suppose $H = K_2$ and $G = K_3$: The initial lists $L(g) = \{1, 2\}$ are consistent, so the consistency test will not modify them; yet no homomorphism of G to H exists. The reason that tree duality (and bounded treewidth duality) leads to polynomial algorithms is the following fact:

Theorem 7.1 ([58]) *H has tree duality if and only if $G \to H$ whenever the consistency test applied to G succeeds.*

(Thus the above example shows that K_2 does not have tree duality. Recall that it does have treewidth two duality.)

We illustrate the use of the technique by discussing a polynomial time algorithm for the H-colouring problem, where H is any oriented path. This problem was first solved in polynomial time in [50].

Theorem 7.2 ([62]) *If H is an oriented path, then H has tree duality.*

Proof In fact, [62] shows that the family $\mathcal{F}$ from the definition of tree duality (the family of certificates of non-colourability) can be chosen to consist of oriented *paths*. We will prove the theorem by showing that if the consistency test succeeds then an H-colouring exists (via Theorem 7.1). Thus suppose that the vertices of the oriented path H are consecutively named $1, 2, \ldots, n$. If the consistency test for an input graph G (with all lists equal to $V(H)$) succeeds, then each final list $L^*(g)$ will be nonempty, and will consist of some integers between 1 and n. We now claim that choosing for each $f(g)$ the smallest integer in $L^*(g)$ will define a homomorphism of G to H. If this were not the case, then some edge $gg' \in E(G)$ would have $i = \min(L^*(g)), i' = \min(L^*(g'))$, and $ii' \notin E(H)$. Recall also that the lists L^* are consistent, thus some $j' > i'$ has $ij' \in E(H)$, and some $j > i$ has $ji' \in E(H)$. This cannot happen: If $i < i'$, then $i < j' - 1$ so $ij' \in E(H)$ is impossible; if $i > i'$, then $j > i' + 1$ and $ji' \in E(H)$ is impossible; finally, if $i = i'$ then $ji' \in E(H)$ implies $j = i+1$ and $ij' \in E(H)$ implies $j' = i + 1$ – but both $i(i+1) \in E(H)$ and $(i+1)i \in E(H)$ cannot hold. This contradiction shows that f is indeed a homomorphism of G to H. In particular, we have a polynomial time algorithm for the H-colouring problem, when H is an oriented path. □

This idea can be obviously extend to any graph H which admits an ordering of the vertices $1, 2, \ldots, n$ with the above property, namely, if ij', ji' are edges of H and if $i < j, i' < j'$, then ii' is also an edge of H.

This property is called the *X-underbar property* [50], and can be nicely restated as follows: If ij and $i'j'$ are edges of H then $\min(i, i') \min(j, j')$ is also an edge of H.

Thus for any fixed graph H with the X-underbar property we have an efficient H-colouring algorithm.

ALGORITHM 4
Input: A graph G with lists $L(g) = V(H), g \in V(G)$.
Task: Find a $\vec{T}_n$-colouring of G, if one exists.
Action: Apply consistency test (Algorithm 3). If the test fails, no homomorphism exists. If the test succeeds, let each $f(g)$ be the smallest element of $L^*(g)$.

Corollary 7.3 *If H has the X-underbar property, then H has tree duality.*

There are several equivalent mechanisms for defining graphs which have tree duality [3, 37, 58]. We focus on the following property. Let $\mathcal{P}(H)$ be the graph whose vertices are nonempty subsets of $V(H)$, and two subsets S, S' of $V(H)$ are adjacent vertices of $\mathcal{P}(H)$ if for each $s \in S$ there is an $s' \in S'$, and for each $s' \in S'$ an $s \in S$, such that $ss' \in E(H)$.

Proposition 7.4 ([37]) *H has tree duality if and only if $\mathcal{P}(H) \to H$.*

Proof Suppose ϕ is a homomorphism of $\mathcal{P}(H)$ to H. We can use ϕ to define a homomorphism f of G to H whenever the consistency test succeeds: For each $g \in V(G)$, let $f(g) = \phi(L^*(g))$. It follows from the definitions that f is a homomorphism. Conversely, suppose H has tree duality. Let $G = \mathcal{P}(H)$, and let each list $L(S) = V(H)$. It is easy to see that after applying the consistency test each $L^*(S)$ will contain the set S (adjacency in $\mathcal{P}(H)$ is defined in such a way that the lists $L'(S) = S$ are consistent, so nothing in these lists will be eliminated during the consistency test). Since the consistency test succeeds, we know from Theorem 7.1 that there is a homomorphism from $G = \mathcal{P}(H)$ to H. □

Proposition 7.4 is a decision procedure for tree duality. Even though much of the above discussion can be extended to bounded treewidth duality [37], no decision procedure for treewidth k duality is known apart from the above case of $k = 1$. We have already seen examples H which do not have tree duality but do have treewidth two duality. No examples H are known, at this time, which have bounded treewidth duality but no treewidth two duality, and it would be interesting to construct such examples. More importantly, no examples H are known where H-colouring is polynomial time solvable without H having bounded treewidth duality. It seems conceivable that such examples do not exist. Other attempts to define a general technique to solve graph homomorphism problems in polynomial time seem to lead to the same class of problems. For instance, Feder and Vardi [36, 37] introduce a class of H-colouring problems solvable by Datalog programs. (These programs are guaranteed to run in polynomial time.) It turns out [37] that an H-colouring problem can be solved by Datalog if and only if it has bounded treewidth duality. Similar results link bounded treewidth duality to fractional homomorphisms and proof systems [3, 40].

We remark in passing that Theorem 6.2 allows us to 'predict' theorems, using the general expectation that $P \neq NP$. For instance, we have seen that undirected nonbipartite graphs H yield NP-complete H-colouring problems. This means, in view of Theorem 6.2, that we should not expect undirected nonbipartite graphs to have bounded treewidth duality. The predicted theorem has in fact been proved by Nešetřil and Zhu.

Theorem 7.5 ([82]) *Let H be an undirected nonbipartite graph. Then H does not have bounded treewidth duality.*

Proof We seek, for any integer k, a graph G_k which is not H-colourable, but which has the property that all G_k-colourable graphs of treewidth k are H-colourable. To assure that G_k is not H-colourable, it suffices take G_k to be a graph of chromatic number greater than H. Since H is nonbipartite, it contains an odd cycle, say $C_{2\ell+1}$. It turns out that by taking G_k of high girth, we can assure that all G_k-colourable graphs of treewidth k are $C_{2\ell+1}$-colourable, and hence H-colourable. (Recall that there exist graphs of arbitrarily high girth and chromatic number, [30].) The following result, of independent interest, completes the proof. □

Proposition 7.6 ([82]) *There is a function $g(k, \ell)$ such that if G is a graph of girth greater than $g(k, \ell)$, then any graph of treewidth k which is G-colourable is also $C_{2\ell+1}$-colourable.*

Note that the result implies, for any fixed ℓ, that a graph whose girth is sufficiently high, with respect to its treewidth, is $C_{2\ell+1}$-colourable. This is interesting even for C_3-colourability, since we have just recalled that there are graphs of arbitrarily high girth that are not C_3-colourable [30] (and so we may conclude they must have high treewidth as well).

Since we have seen that a bipartite graph has treewidth two duality, Theorem 7.5 completely classifies undirected graphs which have bounded treewidth duality. The corresponding problem for directed graphs is wide open.

8 Pair Consistency and Majority Functions

Let us now return to consistency, and bounded treewidth duality. Consistency could be called 'single consistency', in the sense that we are testing whether lists of single vertices are consistent over pairs of vertices. (Indeed, only adjacent pairs pose any restrictions, so checking for all pairs amounts to the same thing as checking over edges.) There is a natural extension of the concept, called *k-tuple consistency*, in which k-tuples of vertices are tested for consistency over $(k+1)$-tuples [25, 36, 37, 45, 76, 80]. For our purposes, we shall focus on the case $k = 2$, and talk about *pair consistency*. (Everything we discuss extends in an obvious way to higher values of k.) Assume that each pair of distinct vertices g, g' of G has a pair list $L(g, g') \subseteq V(H) \times V(H)$. We will always assume that if $(h, h') \in L(g, g')$ then the mapping taking g to h and g' to h' is a homomorphism of the subgraph of G induced by $\{g, g'\}$ to H. (In other words, any edges and loops present between g and g' yield corresponding edges and loops between h and h'.) We say that the pair lists L are *consistent* if for any three vertices $g, g', g'' \in V(G)$ and any $(h, h') \in L(g, g')$ there exists an $h'' \in V(H)$ with $(h, h'') \in L(g, g'')$ and $(h', h'') \in L(g', g'')$. In this case we will take the following initial pair lists $L(g, g')$: Consider the set

F of all homomorphisms of the subgraph of G induced by g, g' to H, and let $L(g, g')$ consist of all pairs $f(g)f(g')$ with $f \in F$. Note that, if there are no edges of G on g, g', then $L(g, g') = V(H) \times V(H)$; if g has a loop in G, then in any $(h, h') \in L(g, g')$, the vertex h has a loop in H; and so on. The following generalization of the consistency test reduces the pair lists L to consistent pair lists L^*.

ALGORITHM 5 - Pair Consistency Test
Input: A graph G, with initial pair lists $L(g, g')$, as described.
Task: Reduce the pair lists to $L^*(g, g')$, which are consistent.
Action: Initially set all lists $L^*(g, g') = L(g, g')$, and then, as long as changes occur, process triples g, g', g'' of vertices of G as follows: Remove from $L^*(g, g')$ any (h, h') for which no element $h'' \in V(H)$ has $(h, h'') \in L^*(g, g'')$ and $(h', h'') \in L(g', g'')$.

We say that the consistency test *fails* if some list $L^*(g, g')$ becomes empty; otherwise we say that the consistency test *succeeds*. It is again the case that the pair consistency test can be performed in polynomial time, and that the test cannot fail if a homomorphism exists. In the AI literature, consistency is usually called 'arc consistency' and pair consistency 'path consistency', [25].

Theorem 8.1 ([58]) *H has treewidth two duality if and only if $G \to H$ whenever the pair consistency test applied to G succeeds.*

We see that for graphs H which enjoy treewidth two duality we can efficiently test H-colourability of input graphs by running the polynomial time pair consistency test. If the test fails, we know that the input graph is not H-colourable. If the test succeeds, we know that the input graph is H-colourable; but how may we find an H-colouring ?

At this point we have, for each vertex $g \in V(G)$, a list of candidate images in H: Consider any other vertex $g' \in V(G)$ and let

$$L^*(g) = \{h \in V(H) : (h, h') \in L^*(g, g') \text{ for some } h' \in V(H)\}.$$

(Because the pair lists L^* are consistent, it follows that $L^*(g)$ does not depend on which g' is chosen). Since the pair consistency test succeeded, each set $L^*(g)$ is nonempty. At this point it would be good if we could concisely describe how to choose a member of each $L^*(g)$ to obtain a homomorphism of G to H. (We were able to do that after the single consistency test using the homomorphism $F : \mathcal{P}(H) \to H$.) Unfortunately, no general technique for this is known, and all algorithms to find an H-colouring are somewhat more involved.

We first describe a class of graphs with treewidth two duality which admit a particular kind of algorithm to find an H-colouring.

A *majority function* μ on H is a mapping $V(H) \times V(H) \times V(H) \to V(H)$ such that the following two conditions are satisfied.

- If $aa', bb', cc' \in E(H)$ then $\mu(a,b,c)\mu(a',b',c') \in E(H)$.
- If at least two of the vertices a, b, c are equal to h then $\mu(a,b,c) = h$.

The first property says that μ is a homomorphism of the categorical product $H \times H \times H$ to H. The second property says that the μ-value of a triple with repetition is equal to the majority (that is, repeated) value.

Theorem 8.2 ([36, 37]) *If H admits a majority function then H has treewidth two duality.*

Corollary 8.3 ([36, 37]) *If H admits a majority function, then H-colouring is polynomial time solvable.*

In this case there is a canonical way to find a homomorphism following the success of the pair consistency check. Specifically, suppose $L^*(g, g'), g, g' \in V(G)$, $g \neq g'$, are the final pair lists produced by the pair consistency test, on an input graph G. If f is a mapping of some subset U of $V(G)$ to $V(H)$, we say that f is *consistent*, if $f(g)f(g') \in L^*(g, g')$ for all $g, g' \in U, g \neq g'$. It turns out [37, 36] that H admits a majority function if and only if for any graph G, any set U of vertices of G, any consistent mapping $f : U \to V(H)$, and any vertex $u' \in V(G) - U$, there exists a consistent mapping f' of $U \cup \{u'\}$ to $V(H)$ which extends the mapping f. (A detailed proof of this fact may be found in [57]). Note that a consistent mapping defined on $U = V(G)$ is a homomorphism of G to H.

Therefore if H has a majority function, we can use the following algorithm to actually find an H-colouring when G passed the pair consistency test.

ALGORITHM 6
Input: A graph G with vertices $g_1, g_2, \ldots, g_n$.
Task: Find an H-colouring of G, if one exists.
Action: Perform the pair consistency test (Algorithm 5), obtaining final pair lists $L^*(g, g')$. If the test fails (some pair list is empty), G is not H-colourable. Otherwise find an H-colouring of G as follows: Define $f(g_1), f(g_2)$ so that the pair $(f(g_1), f(g_2))$ is in $L^*(g_1, g_2)$. Then, having defined f consistently on $g_1, g_2, \ldots g_{i-1}$, for $3 \leqslant i \leqslant n$, find a vertex $h \in V(H)$, so that letting $f(g_i) = h$ extends the definition of f consistently to $g_1, g_2, \ldots, g_i$.

According to the above remark, the existence of a majority function on H assures that this procedure is correct, so that at each stage one can actually find a required vertex h.

Some concepts related to the proofs of these results [36, 37] were anticipated in [25], and subsequently applied to majority functions in [65]. Majority functions are just one of a variety of similar functions whose existence makes H-colouring CSP problems polynomial time solvable. In [20, 64, 65, 71] and several other related papers, an algebraic approach to H-colouring problems is developed, in the context of constraint satisfaction problems and universal algebra. The crux of the connection is the observation that if G and H have the same set of homomorphisms of $G^k \to G$ and $H^k \to H$ for every positive integer k, then there is a reduction from G-colouring to H-colouring. In other words, if G-colouring is NP-complete then so is H-colouring, and if H-colouring is polynomial time solvable then so is G-colouring. A general conjecture arises as to which CSP problems are polynomial time solvable, and could turn out to give the dichotomy classification for constraint satisfaction problems, and hence for digraph H-colouring and other problems discussed here.

We now return to the general case of a graph H with treewidth two duality and give a self-reduction procedure, obtained by modifying Algorithm 6, that will still find an H-colouring of the input graph, if one exists. (However, the modified algorithm is more complex and less efficient.) According to the last paragraph of Section 1, we may assume that H is a core. The **Action** begins as above, by performing the pair consistency test, and declaring there is no H-colouring if the test fails. If the test succeeds, we proceed to assign images $f(g_1), f(g_2), \dots$. Say that an assignment of $f(g_1) = h_1, f(g_2) = h_2, \dots f(g_{i-1}) = h_{i-1}$ is *admissible*, if there is an H-colouring of the input graph extending this assignment. Having a consistent assignment $f(g_1) = h_1, f(g_2) = h_2, \dots f(g_{i-1})$, we test all $h \in L^*(g_i)$ until we find an h such that the assignment $f(g_1) = h_1, f(g_2) = h_2, \dots f(g_{i-1}) = h_{i-1}$, and $f(g_i) = h$, is also admissible. Since we know an H-colouring of the input graph exists, we will always be able to extend the assignment so that it remains admissible, until an H-colouring of G is found. It remains to explain how to test whether an assignment $f(g_1) = h_1, f(g_2) = h_2, \dots f(g_i) = h_i$, is admissible. Here is where we use the fact that H is a core. Let G' be obtained from the input graph G and a copy of the graph H by identifying each vertex g_j with the corresponding vertex h_j, for all $j = 1, 2, \dots, i$. Then $G' \to H$ if and only if there is a homomorphism of G to H extending the assignment $f(g_1) = h_1, f(g_2) = h_2, \dots f(g_i) = h_i$. Indeed, any homomorphism of G to H extending the given assignment is easily extended to G'. Conversely, if there is a homomorphism of G' to H, then there is such a homomorphism which maps the vertices of the copy of H included in G' identically to the vertices of H. (Since H is a core, the restriction to the copy of H in G' is an automorphism a of H, and we can compose the homomorphism with the inverse of a.) Therefore the restriction of this homomorphism to G extends the given assignment. Of course, we can test the existence of a homomorphism $G' \to H$ by performing a pair consistency test, since H has treewidth two duality.

9 List Homomorphisms

We now return to lists. Let H be a fixed graph. Assume that for each vertex g of the input graph G we are given a list $L(g) \subseteq V(H)$. A *list homo-*

morphism of G to H, or a *list H-colouring of* G, with respect to the lists L, is a homomorphism f of G to H, such that $f(g) \in L(g)$ for all $g \in V(G)$. The *list H-colouring problem* asks whether or not an input graph G with lists L admits a list homomorphism to H with respect to L. We note that when $H = K_n$ a list homomorphism of G to H is a list colouring of G, and hence the list K_n-colouring problem is essentially the list colouring problem (with a restricted set of colours). The complexity of list colouring problems has been studied by J. Kratochvíl and Zs. Tuza [66]. We also note that each H-colouring problem is a list H-colouring problem restricted to inputs G with all lists $L(g) = V(H)$.

List H-colourings tend to be more manageable than H-colourings, since they offer a natural way to recurse: Seeking a list H-colouring of G we may choose a value for $g \in V(G)$ (from the list $L(g)$), modify correspondingly the lists of all neighbours of g (they can no longer use colours not adjacent to the value assigned to g) and then remove g from consideration. Similarly, if the list H-colouring problem is NP-complete, then so is any list H'-colouring problem where H is an induced subgraph of H'. (Restrict the inputs G to have their lists contained in $V(H)$.) Also, many natural applications of homomorphisms, such as frequency assignment, scheduling, and so on, tend to have additional constraints expressible by lists. Finally, it turns out that many algorithms for graph homomorphisms adapt very naturally to lists. This is the case for virtually all the algorithms we have discussed, and is particularly plain for consistency tests, which introduce lists even if lists were not originally present.

For list H-colouring problems for undirected graphs we have the following dichotomy classification.

Theorem 9.1 ([34]) *Let H be a fixed graph.*

- *If H is a bi-arc graph then the list H-colouring problem has a polynomial time algorithm.*
- *Otherwise the problem is NP-complete.*

Bi-arc graphs are defined as follows: Let C be a fixed circle with two specified points p and q on C. A *bi-arc* is an ordered pair of arcs (N, S) on C such that N contains p but not q, and S contains q but not p. A graph H is a *bi-arc graph* if there is a family of bi-arcs $\{(N_x, S_x) : x \in V(H)\}$ such that, for any $x, y \in V(H)$, not necessarily distinct, the following hold.

- If x and y are adjacent, then neither N_x intersects S_y nor N_y intersects S_x.
- If x and y are not adjacent, then both N_x intersects S_y and N_y intersects S_x.

(Note that we are not allowing bi-arcs $(N, S), (N', S')$ such that N intersects S' but S does not intersect N'.)

This result is much simplified if we restrict our focus to graphs in which every vertex has a loop; such graphs are called *reflexive graphs.*

Reflexive graphs arise naturally in applications of list homomorphisms. Suppose for instance that G is a graph whose vertices are jobs to be executed, and H a graph whose vertices are the processors to be used. Two vertices in H are adjacent if the processors communicate quickly, say, by a direct connection. Two vertices in G are adjacent if the jobs need a lot of cross-referencing during their execution. Moreover, each job comes with a list of processors that it can be executed on. Clearly, the problem calls for a list H-colouring of G, and both the graphs G, H are naturally reflexive and undirected.

An *interval graph* is a graph G whose vertices can be represented by intervals such that two vertices are adjacent in G if and only if the corresponding intervals intersect. Note that by our definition all interval graphs are reflexive. (It is more common to define adjacency by intersection only for distinct vertices, but for our purposes we need reflexive interval graphs.)

Theorem 9.2 ([32]) *Let H be an undirected reflexive graph.*

- *If H is an interval graph then the list H-colouring problem has a polynomial time algorithm.*
- *Otherwise the problem is NP-complete.*

For irreflexive graphs (graphs without loops), the list H-colouring problem is polynomial time solvable if H is a bipartite graph whose complement of H is a *circular arc graph*; and is NP-complete otherwise [33].

Proof A well known result due to Lekkerkerker and Boland [72] characterizes interval graphs by the absence of induced subgraphs which are cycles of length greater than three, and graphs with structures called *asteroidal triples.* According to the discussion above, it then suffices to prove the NP-completeness of the list H-colouring problem for cycles of length greater than three and graphs asteroidal triples. This is done in [32].

To show that list H-colouring problems for interval graphs H have polynomial time algorithms, one can take advantage of the the interval representation of H, using a 2-satisfiability algorithm, [32]. Alternately, [34], one can prove that each interval graph H has a *conservative* majority function, that is, one satisfying the additional constraint that $\mu(a, b, c) \in \{a, b, c\}$ for all $a, b, c \in V(H)$, and use the following analogue of Theorem 8.2 for list homomorphisms. □

Theorem 9.3 ([37, 34]) *If H admits a conservative majority function then the list H-colouring problem has a polynomial time algorithm.*

It is shown in [34] that every bi-arc graph admits a conservative majority function, but the functions for arbitrary (not necessarily reflexive or irreflexive) graphs do not seem to admit a concise description, except for the class of trees. Amongst trees, bi-arc graphs are characterized by the absence of certain induced subgraphs [34].

Here is another theorem predicted by an NP-completeness proof: Since every graph H which admits a conservative majority function defines a polynomial time solvable H-colouring problem, and since the H-colouring problem for graphs which are not bi-arc graphs is NP-complete, we expect that a graph which is not a bi-arc graph does not admits a conservative majority function. Indeed, we were able to prove:

Theorem 9.4 ([15]) *H admits a conservative majority function if and only if it is a bi-arc graph.*

A graph is *chordal* if it does not contain an induced cycle other than triangle. We close this section with the following dichotomy classification:

Proposition 9.5 ([32]) *Let H be a reflexive undirected graph, and consider the list H-colouring problem restricted to inputs G with connected lists L (so each $L(g), g \in V(G)$, induces a connected subgraph of H).*

- *If H is a chordal graph, then the problem is polynomial time solvable.*
- *Otherwise, the problem is NP-complete.*

A similar classification for irreflexive graphs is given in [63, 95].

10 Variants and Restrictions

Many well known colouring problems are in fact specialized H-colouring problems. This includes, in addition to classical colourings (the graphs H are complete), also circular colourings (the graphs H are the 'circular graphs' G^p_q [100]), multicolourings (the graphs H are the 'Kneser graphs' $K(k,n)$ [89]), T-colourings (again the graphs H are of a certain special form [73]), and so on.

Counting the number of homomorphisms is of interest in statistical physics [16, 17, 29]. For instance, a well known problem from statistical physics is to count the number of independent sets in a graph G; this is simply the number of homomorphisms of G to the graph H consisting of two vertices $0, 1$ with the edge 01 and the loop 11. There is a dichotomy of polynomial versus $\#P$-complete counting problems for undirected graphs [29]: *When every component of H is a reflexive complete graph or an irreflexive complete bipartite graph, the problem of counting H-colourings is polynomial time solvable. Otherwise, the problem is $\#P$-complete.*

Surprisingly, exactly the same classification describes the dichotomy of polynomial versus NP-complete *equitable H-colouring* problems. In this problem we are asking whether or not the input graph G admits an H-colouring f in which the numbers $f^{-1}(v), v \in V(G)$, differ from each other by at most one [3, 27].

A dichotomy classification of the complexity of the counting problem for directed graphs, and general constraint satisfaction problems, is conjectured in [19].

Homomorphism and list homomorphism problems restricted to graphs of bounded degree are discussed in [35, 46, 52, 56]. It turns out that homomorphism problems that are intractable in general may become polynomial time solvable for graphs of bounded degree. The best known example of this is 3-colouring – which becomes polynomial time solvable when the maximum degree of the input graphs is restricted to be three (or less). On the other hand, the problem is again NP-complete for the class of graphs with maximum degree four or less. In [46, 52] both positive and negative results are given, indicating that the situation is very complex. No dichotomy is known or conjectured. The classification of the complexity of list H-colourings (Theorem 9.1) does not change when input graphs are restricted to have bounded degree (except the trivial case of degrees bounded by two cf. below) [35]. However [35], there are certain special list C_n-colouring problems (C_n-retraction problems, cf. below) which become polynomial time solvable for progressively more and more graphs H, as the degree constraint is tightened from 6 (this degree constraint makes no new problems easy) to 5 (some new problems become easy), to 4 (more additional easy problems) and 3 (even more problems become easy).

Other restrictions that have been investigated include planarity [81], vertex and edge transitivity [75], and generalization of chordality [74]. (Observe that restricting the input graphs to be chordal makes little sense for the problem of H-colouring, since a chordal G is homomorphic to H if and only if the clique number of G does not exceed that of H.)

The *H-retraction problem* is the list H-colouring problem restricted to graph G whose lists are either singletons or all of $V(H)$. Equivalently, the H retraction problem asks whether or not an input graph G which contains H as a subgraph admits a homomorphism $f : G \to H$ such that $f(h) = h$ for all vertices of the subgraph H. There is no dichotomy known for H-retraction problems, and one is unlikely to be easily found, as Feder and Vardi have shown [36, 37] that dichotomy here would imply dichotomy for all constraint satisfaction problems. The complexity of H-retraction problems has also been studied in [4, 5, 15, 16, 17, 32, 33, 34, 35, 56, 61, 74].

Counting problems for degree constrained graphs are discussed in [28, 29, 56]. It has been conjectured in [29] that the same classification proved for counting H-colourings (above), also applies to counting H-colourings for graphs of bounded degrees (except when the degree bound is two, cf. below). Dichotomy (polynomial versus $\#P$-complete) is established for the complexity of counting list H-colourings for graphs of bounded degree [56].

It is not hard to see (by standard arguments) that homomorphism problems can be solved in polynomial time when restricted to graphs of bounded treewidth. Essentially, a tree decomposition of the graph G can be used to streamline the execution of the k-consistency test, as well as to choose images for the vertices of G from the final lists L^*, if the test succeeds. This has already been observed in the AI literature (at least for a related notions of width) since the mid-seventies [25, 45, 76], and is explicitly mentioned in [58] (above Theorem 2.5) and in [83]. The topic is explored in depth in [26, 27, 28], and applied to homomorphisms, list homomorphisms, counting homomorphisms, and various combinations thereof. By exploiting *nice tree decompositions* the computations are made more efficient. Note that graphs with degrees bounded by two have treewidth bounded by two as well; thus all the variants of H-colourings for these problems are polynomial time solvable (cf. above). The effect of fixing the treewidth (or certain other parameters) on the complexity of H-colouring problems is investigated, from the perspective of fixed parameter tractability, in [28].

Recall that for a graph G to admit an H-colouring is equivalent to admitting a partition satisfying (H1) and (H2). We say that G admits a *surjective* H-colouring if the associated partition satisfies (H1), (H2), and the following two additional constraints:

(H3) each set S_h is nonempty; and

(H4) if hh' is an edge (arc) of H then there is at least one edge (arc) from S_h to $S_{h'}$.

A surjective H-colouring is also called a *compaction* to H. Compaction to undirected cycles H is classified in [94]. No general dichotomy is known, and Feder and Vardi have shown that dichotomy for compaction problems would imply dichotomy for all constraint satisfaction problems [36, 37]. These problem remain, of course, tractable for bounded treewidth. In fact, [28] they remain fixed parameter tractable.

By way of comparison, a graph G admits a *contraction* to H, if it has a partition of $V(G)$ into sets $S_h, h \in V(H)$, which satisfy (H2), (H3), (H4), and, instead of (H1), the property requiring that each part S_h induce a *connected* subgraph of G. The complexity of deciding if G admits a contraction to H is studied in [18]. The complexity of deciding if a *subgraph* of G admits a contraction to H is, of course, the well known problem of deciding if G admits H as a minor, solved in polynomial time in [86].

An H-colouring f of G is called an H-*cover* of G if it is a 'local isomorphism', that is, if for each vertex $g \in V(G)$, the mapping f is a bijection between the set of neighbours of g in G and the set of neighbours of $f(g)$ in H. (For instance, identifying antipodal vertices is a C_n-cover of C_{2n}.) The study of the complexity of H-cover problems was initiated by [1]. In [68, 67], many polynomially solvable and NP-complete cases are presented. There are interesting polynomial time algorithms, in particular, using linear equations modulo two. (Similar situations arise in the algorithms in [51].) All small graphs H are classified in [69]. No dichotomy is known. It is conjectured that all H-cover problems where H is a regular graph (of degree at least three) are NP-complete; some special cases have been verified [68].

11 Semi-homomorphism Partitions

Recall the kind of vertex partition problems that can be expressed by homomorphism problems: We can find a partition of $V(G)$ into parts S_h, where certain parts S_h are required to be independent, and certain pairs of parts $S_h, S_{h'}$ are required to have no edges (or arcs) from S_h to $S_{h'}$. Note that by requiring the complement of the input graph G to admit a homomorphism to the complement of H, we may also express the problems of finding vertex a partition into parts S_h, where certain parts S_h are required to be cliques, and certain pairs of parts $S_h, S_{h'}$ are required to have all edges (or arcs) from S_h to $S_{h'}$. In both cases, we may allow the input graphs G to be equipped with lists restricting the parts to which the vertices of G can be placed.

A natural extension of the homomorphism problem seeks vertex partitions in which some parts may be required to be either independent or cliques, and some ordered pairs of parts may be required to have either no edges or all edges. These are called *semi-homomorphism partitions*, since they extend homomorphism partitions with respect to complementation, as suggested above. We shall illustrate the surprisingly strong modeling power of this simple generalization of the homomorphism framework.

To introduce the problem formally, let M be a fixed k-by-k matrix over $0, 1, *$. An *M-partition* of a graph G is a partition of the vertex set $V(G)$ into k parts $S_1, S_2, \ldots, S_k$, such that S_i is independent if $M_{i,i} = 0$, or a clique if $M_{i,i} = 1$ (with no restriction if $M_{i,i} = *$), and such that there is no edge from S_i to S_j if $M_{i,j} = 0$, or all edges from S_i to S_j if $M_{i,j} = 1$ (with no restriction if $M_{i,j} = *$). When k is small, we usually refer to parts $A, B, C, \ldots$ instead of $A_1, A_2, A_3, \ldots$ and write, for example, $A = 0$ to mean $M_{A,A} = 0$ or $AB = 1$ instead of $M_{A,B} = 1$.

In all our examples we shall take M a symmetric matrix and the inputs G to be undirected graphs; nevertheless there are interesting examples of non-symmetric M-partition problems applied to directed graphs [93].

When M is a $0, *$-matrix, an M-partition of a graph G is an H-colouring of G, where H is the graph whose adjacency matrix is M with $*$'s replaced by 1's.

Thus we define the *M-partition problem* to ask whether or not an input graph G admits an M-partition (for any fixed matrix M). In addition to H-colouring problems, this framework includes, for instance, the following natural problems.

- Is G a split graph [48]? This is the M-partition problem with the following matrix M:

$$\begin{pmatrix} 1 & * \\ * & 0 \end{pmatrix}$$

- Does G have a clique cutset [90, 97, 98]? This problem is an M-partition problem which requires that all parts be nonempty; the matrix M is

$$\begin{pmatrix} 1 & * & * \\ * & * & 0 \\ * & 0 & * \end{pmatrix}$$

- Does G have an independent cutset [92]? This problem also requires all parts to be nonempty. The matrix M is

$$\begin{pmatrix} 0 & * & * \\ * & * & 0 \\ * & 0 & * \end{pmatrix}$$

- Does G have a skew cutset [24]? The parts are required to be nonempty and the matrix M is

$$\begin{pmatrix} * & 0 & * & * \\ 0 & * & * & * \\ * & * & * & 1 \\ * & * & 1 & * \end{pmatrix}$$

- Does G have a Winkler partition [94]? This is again an M-partition into nonempty parts, with M equal to

$$\begin{pmatrix} * & * & 0 & * \\ * & * & * & 0 \\ 0 & * & * & * \\ * & 0 & * & * \end{pmatrix}$$

- Does G have a homogeneous set [48]? M is the matrix

$$\begin{pmatrix} * & * & 1 \\ * & * & 0 \\ 1 & 0 & * \end{pmatrix}$$

  and there are additional constraints requiring that certain parts have at least a certain number of elements (cf. below).

These well known classes of graphs (and many others that fit the framework) have been investigated from the perspective of perfect graphs and combinatorial optimization. For instance, clique cutsets are related to chordal graphs; they can be found in polynomial time [90, 97, 98], and are the basis of a decomposition algorithm [90, 98], which allows efficient solution of many optimization problems for the class of decomposable graphs. Independent cutsets are of interest because a result of Tucker [92] asserts that a minimal imperfect graph other than an odd cycle cannot contain a independent cutset; this M-partition problem has been proved NP-complete in [42].) Chvátal conjectured [24] that a minimal imperfect graph cannot contain a skew cutset. Skew cutsets played an important role in the proof of the Strong Perfect Graph Conjecture recently announced by Chudnovsky et al. [23]. (Several of the other important concepts are also M-partitions of certain kinds.) Chvátal also asked whether skew cutsets can be found in polynomial time. An information theoretic sub-exponential algorithm was given in [38], and recently a polynomial algorithm was found [43]. Winkler's partition problem has been proved NP-complete by N. Vikas, in response to a problem posed by Peter Winkler [94]. Homogeneous sets also define a decomposition (the 'modular decomposition') which facilitates the recognition of comparability graphs (and other graph classes) [22, 78]. Homogeneous sets (and modular decompositions) can be found efficiently [78].

Consider more carefully the last example, of homogeneous sets. Specifically, a *homogeneous set* in a graph G is a set C of vertices of G such that each vertex outside of C is adjacent either to all or to none of the vertices in C. It is easy to see that the existence of such a set is equivalent to an M-partition with the above matrix M. However, and this is typical of several of the above examples, homogeneous sets have additional constraints designed to avoid the trivial cases of a homogeneous sets consisting of a single vertex or the entire vertex set. We also require that C contains at least two but not all vertices. Similarly, all the cutset problems above require all three parts nonempty, as does Winkler's partition problem [94]. It is easy to model these additional requirements (that certain parts be non-empty, or have at least a fixed number of vertices, or have at least some edges joining them, etc.) by allowing the input graphs G come equipped with lists, restricting which parts a vertex can belong to. This gives us a variety of options in restricting the contents of the individual parts or of their connections. For instance, in the case of homogeneous sets, we may ensure that C has at least two but not all vertices by choosing three vertices x, y, z of the input graph and specifying that the lists of x, y only consist of C and the list of z does not contain C. Thus the problem of finding a homogeneous set in a graph with n vertices is reduced to n^3 list partition problems. (A homogeneous set exists if and only if at least one of the n^3 choices of x, y, z has a desired list partition.)

12 Lists of Size at most Two

The most basic technique for solving list M-partition problems is the 2-satisfiability algorithm of [2].

Suppose first that M is a 2-by-2 matrix seeking to partition the input graph into two parts, say A, B. We can solve the list M-partition problem by introducing a boolean variable x_v for each vertex v of the input graph G; we think of the value of x_v as encoding whether or not the vertex v belongs to the part A of the partition ($x_v = 1$ means $v \in A$, $x_v = 0$ means $v \notin A$). It is then easy to see that all the constraints, and lists, of the list M-partition problem can be stated by polynomially many clauses with at most two literals each. For instance, if A is to be an independent set ($A = 0$), we impose the constraint $\overline{x}_u \vee \overline{x}_v$ for each edge uv of G. Similarly, if, say, every edge from A to B is to be present ($AB = 1$), we impose the constraint $x_u \vee \overline{x}_v$ for each non-edge uv of G. Finally, if the list of v is, say, B, we impose the constraint $\overline{x}_v$. Hence the problem can now be solved by the 2-satisfiability algorithm [2].

For instance, we can solve this way the list split partition problem. The two parts are respectively a clique and an independent set, and we correspondingly denote them by C and I. Each vertex v corresponds to a variable x_v, and $x_v = 1$ means $v \in C$, $x_v = 0$ means $v \in I$. We obtain one constraint for each edge $vw \in E(G)$, namely $x_v \vee x_w$, and one constraint for each non-edge $vw \notin E(G)$, namely $\overline{x}_v \vee \overline{x}_w$. Moreover, if a vertex v has a restricted list, we add the constraint x_v or $\overline{x}_v$.

The same technique applies any time we have an instance in which every list has size at most two.

Proposition 12.1 *There is a polynomial-time algorithm which solves any list M-partition problem restricted to instances in which the list of every vertex of the input graph has size at most two.*

13 List Split Partitions

The encoding of the list split partition problem as a 2-satisfiability problem results in $O(n^2)$ constraints, and hence an $O(n^2)$-time algorithm. On the other hand, there are several natural split graph recognition algorithms which run in linear time [48]. As usual, several of them adapt easily to lists. We present an algorithm from [53], which takes advantage of the fact that split graphs are necessarily chordal.

A *perfect elimination ordering* of a graph G is an ordering $v_1, v_2, \ldots, v_n$ of the vertices of G such that when $a < b, a < c$, and v_a is adjacent to both v_b and v_c, then v_a and v_b are adjacent. A graph admits a perfect elimination scheme if and only if it is chordal [48]. There are linear time algorithms which for an input graph G either find a perfect elimination ordering (if G is chordal), or find an induced cycle other that triangle (if G is not chordal) [87, 91]. It is easy to see that a split graph is necessarily chordal [48].

We now give our algorithm for the list version of the split partition problem. Thus we wish to partition the vertices of the input graph G into sets I, C where I is independent and C is a clique, conforming to the lists $L(g) \subseteq \{I, C\}, g \in V(G)$. Since there are only two parts, the lists are either unrestricted ($L(g) = \{I, C\}$), or singleton lists. We shall view a vertex g with singleton list $L(g) = \{X\}$ as being precoloured by X ($X = I$ or $X = S$), and speak of extending the given precolouring. We may assume that the precolouring is consistent, so no two vertices precoloured I are adjacent, and any two vertices precoloured C are adjacent.

ALGORITHM 7

Input: A graph G, and lists $L(g) \subseteq \{I, S\}, g \in V(G)$.

Task: Colour the vertices of G by I and S, respecting the precolouring, and so that no two vertices coloured I are adjacent and any two vertices coloured C are adjacent.

Action:

- Find a perfect elimination ordering $v_1, v_2, \ldots, v_n$ of G.
- In the order $v_1, v_2, \ldots,$ colour the vertices by I as long as possible. (It is possible to colour v_i by I if it is not precoloured by C, and is nonadjacent to the vertices $v_1, v_2, \ldots, v_{i-1}$, previously coloured by I, as well as to all vertices precoloured by I.) Let j be the first subscript such that we cannot colour all of $v_1, v_2, \ldots, v_j$ by I.
- In the order $v_j, v_{j+1}, \ldots,$ colour the vertices by giving a preference to C, i.e, for $i \geq j$,
  - colour v_i by C if possible (if it is not precoloured by I, and is adjacent to all vertices previously coloured by C and all vertices precoloured by C), otherwise (if it is not precoloured by I and is adjacent to v_j)
  - colour v_i by I if possible (it is nonadjacent to all vertices previously coloured by I and all vertices precoloured by I).

If the algorithm colours all vertices of G it is said to *succeed*; otherwise it is said to *fail*. The algorithm can fail because there is no perfect elimination ordering. In that case, the perfect elimination algorithm will exhibit an induced $C_k, k \geq 4$, certifying that the graph G is not chordal, and hence not split [91]. The algorithm can also fail by reaching a vertex v_i which cannot be coloured by I or by C.

We first observe that if v_i cannot be coloured by C then v_i is nonadjacent to v_j. Otherwise v_i would be nonadjacent to some $v_k, k > j$, coloured or precoloured C, with v_j, v_k adjacent (since both have colour C), contradicting the definition of a perfect elimination ordering. On the other hand, there are two possible reasons v_i cannot be coloured by I: (1) v_i is adjacent to some vertex v_k previously coloured by I; or (2) v_i is adjacent to some vertex v_k precoloured by I. Recall that there are three possible reasons v_j could not be coloured by I: (a) v_j is adjacent to some vertex v_ℓ precoloured by I; (b) v_j is adjacent to some vertex v_ℓ previously coloured by I; or (c) v_j is precoloured by C. (There are fewer reasons for v_i because we are only colouring it by I when we failed to colour it by C, hence it is not precoloured by C.)

Thus we assume that v_i is nonadjacent to v_j, and one of (1), (2), and one of (a), (b), (c) holds.

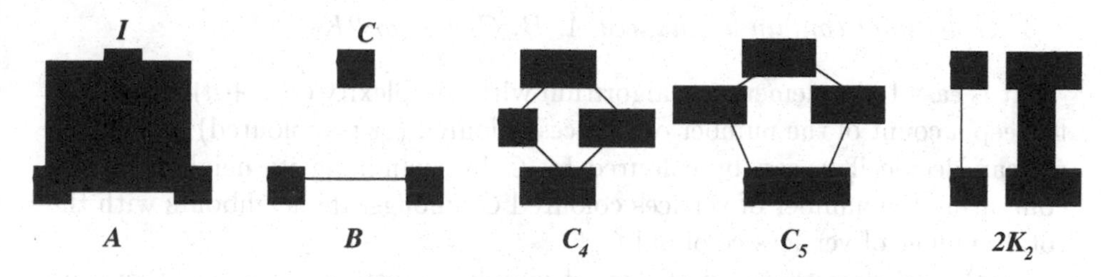

Figure 2: Obstructions for list split partition

Suppose first that (1) holds, so that v_i is adjacent to some vertex v_k previously coloured by I.

We claim that v_j and v_k must be nonadjacent. If $k < j$, this follows from the properties of a perfect elimination ordering. If $k > j$, then if v_k were adjacent to v_j, the algorithm would have coloured it by C. (The perfect elimination ordering assures that v_k is then adjacent to all preceding vertices coloured C.)

Suppose moreover that case (a) applies, so that v_j was coloured by C because it was adjacent to a vertex v_ℓ previously coloured by I. Since $\ell < j < i$, and v_i, v_j are nonadjacent, v_ℓ must be nonadjacent to v_i. Additionally, the two vertices v_k, v_ℓ are nonadjacent, as they are both coloured by I. We conclude that v_i, v_j, v_k, v_ℓ form an induced $2K_2$ in G. It is easy to see that this certifies that G is not a split graph.

In case (b), when v_j was coloured by C because it was adjacent to a vertex v_ℓ precoloured by I, then v_ℓ and v_k are still nonadjacent. If v_ℓ is also nonadjacent to v_i, we have a $2K_2$ as before. Otherwise the three vertices v_ℓ, v_j, v_i induce a copy of the precoloured graph A, which also certifies that the precolouring cannot be extended.

Finally, in case (c), if v_j was precoloured by C, then the vertices v_j, v_i, v_k, induce a copy of the precoloured graph B, also certifying that the precolouring cannot be extended.

Suppose second, that (2) holds, i.e, that v_i cannot be coloured by I because it is adjacent to some vertex v_k precoloured by I. Then v_k, v_i, v_j induce A provided v_j is adjacent to v_k. Otherwise (v_j and v_k are nonadjacent), if (a) or (b) hold, we argue about v_ℓ as in the previous case. Finally, if (c) holds, then v_i, v_j, v_k form an induced B.

Since each $C_k, k \geq 6$, contains an induced $2K_2$, we have proved the following fact.

Proposition 13.1 ([53]) *The following statements, for a graph G precoloured by I, C, are equivalent.*

1. *Algorithm 7 succeeds.*

2. *The precolouring can be extended.*

3. *G does not contain an induced A, B, C_4, C_5, or $2K_2$.*

It is easy to implement the algorithm with complexity $O(m+n)$. It suffices to keep a count of the number of vertices coloured (or precoloured) by C; then we can decide if v_i can be coloured by C by examining its neighbours and comparing the number of vertices coloured C amongst its neighbours with the total number of vertices coloured C.

Both the algorithm and the proof simplify substantially when there are no precoloured vertices. In particular, we obtain the following well known characterization of split graphs.

Corollary 13.2 ([48]) *A graph G is a split graph if and only if it does not contain an induced $2K_2, C_4$ or C_5.*

The following generalization of split graphs has been defined by Brandstädt (who also proposed the first polynomial time algorithms [11, 12, 13]). An (a, b)-*graph* is a graph whose vertices can be partitioned into a independent sets and b cliques. Let M be the symmetric $(a+b) \times (a+b)$ matrix with all off-diagonal entries $*$, and having a 0's and b 1's on the diagonal.

Proposition 13.3 ([11, 38]) *If both $a \leqslant 2$ and $b \leqslant 2$, then the list M-partition problem is polynomial-time solvable. Otherwise the list M-partition problem is NP-complete.*

Proof First we note that if $a \geq 3$ then the M-partition problem is NP-complete, since we can decide whether or not an input graph G is 3-colourable by endowing all its vertices with the list $\{1, 2, 3\}$, and then asking whether or not it has a list M-partition. (If $b \geq 3$ the proof is similar.)

The are polynomial time algorithms for the $(1,1)-, (2,1)-, (1,2)-, (2,2)-$ problems which are similar to each other. We illustrate the technique (from [38]) on the case of $(2,1)-$graphs. A $(2,1)$-*partition* of G is a partition of the vertices of G into an induced bipartite subgraph and a clique. Of course, G has a $(2,1)$-partition if and only if it is a $(2,1)$-graph: it has a partition into two independent sets and a clique. By focusing on the union of the two independent sets (the bipartite subgraph) we emphasize the fact that we shall treat their vertices in the same way. In fact, we shall repeatedly use the obvious fact that a bipartite graph and a clique can meet in at most two vertices.

We claim that a graph G on n vertices has at most n^4 different $(2,1)$-partitions, and all these partitions can be found in time proportional to n^8.

Let $V(G) = B \cup C$ be a particular $(2,1)$-partition. Then any other $(2,1)$-partition $V(G) = B' \cup C'$ has $|B' \cap C| \leqslant 2$ and $|B \cap C'| \leqslant 2$, so B' is obtained from B by deleting at most two vertices and inserting at most two new vertices. In fact, if we allow ourselves to insert back a vertex that has just been deleted, we can say that we make exactly two deletions and exactly two insertions. Each of these at most four operations can be made in at most n ways. This observation proves the first part of the claim and allows us to find *all* $(2,1)$–partitions if one such partition is known. It amounts to a 4-local search (the current S is changed in at most four vertices), and can be performed in time n^4.

It remains to explain how to find the first $(2,1)$-partition. The algorithm proceeds in two phases. The first phase attempts to find as large a bipartite subgraph as possible. This is based on the observation that if $V(G) = B \cup C$ is a $(2,1)$-partition and B' a bipartite graph smaller than B, then $B' \cap C$ has at most two vertices, and hence, as above, B' can be enlarged by removing some two vertices and inserting some three new vertices. Thus, starting with any bipartite subgraph of G, we can increase its size by performing a 5-local search (making all possible two deletions and three insertions and testing if the result is bipartite) in time n^5 times the time to check whether a graph is bipartite. After performing this operation at most n times, we reach a situation where the current bipartite subgraph can no longer be enlarged in this way. Clearly, at this point our current bipartite graph B' has the same size as the (unknown) B.

The second phase of the algorithm attempts to change B', without changing its size, until $V(G) - B$ is a clique. This is accomplished by a 4-local search, based on a very similar principle – namely, if $V(G) = B \cup C$ is a $(2,1)$-partition and $|B| = |B'|$, then B is obtained from B' by a deletion of two vertices and the insertion of two other vertices. Thus we can test all n^4 possible new sets B' for being bipartite and the corresponding $V(G) - B'$ for being a clique, and if no $(2,1)$-partition is found we can be sure none exists.

The most time-consuming operation is the first phase of the algorithm, finding one $(2,1)$-partition – taking time n^8. □

To summarize the steps of the algorithm:

ALGORITHM 8
Input: A graph G.
Task: Find a $(2,1)$-partition of G, if one exists.
Action: Start with any induced bipartite subgraph B of G (say, a single edge). While the size of the current B can be increased by deleting two current vertices and inserting three new vertices, do so.
When the size of B can no longer be increased in this way, test whether the complement of B is a clique, and if not, delete two current elements of B and insert two new elements to B, keeping B bipartite, until the complement of the current B is a clique (or until all possible ways have been tried).

It would be worthwhile to obtain a significantly faster algorithm.

14 List Clique Cutset Partitions

Our second illustration concerns the clique cutset problem. Here we seek a partition of the vertex set of the input graph G into three sets A, B, C where A is a clique of G and there are no edges of G between the parts B and C. Additionally, vertices of G have lists, which are subsets of $\{A, B, C\}$, restricting the parts to which they may belong.

We begin by recalling that there is a polynomial time algorithm to find a minimal chordal extension H of a graph G [87]. Thus we obtain a graph H which is a chordal supergraph of G, with the same vertex set, and with $E = E(H') - E(H)$ inclusion minimal in the sense that $G \cup E'$ is not chordal for any $E' \subset E, E' \neq E$. It is easy to see that for any M-partition A, B, C of G, the graph H will not contain any edges between the sets B and C [90]. (Indeed, H with all these edges removed will still be a chordal extension of G, since any cycle in it that contains both a vertex of B and a vertex of C goes twice through the clique A and hence has a chord.)

Let $v_1, v_2, \ldots, v_n$ be a perfect elimination ordering of H. Let F_i be the *forward clique* of H, consisting of v_i and all adjacent $v_j, j > i$. The fact that each F_i is a clique follows from the properties of a perfect elimination ordering. Moreover, each clique A of H (and hence also each clique A of G) is a subset of some F_i – it is enough to let i be the lowest subscript of any $v_i \in A$. In conclusion, the sets F_i have the following property: For any M-partition A, B, C of $V(G)$, there exists a set F_i that contains A and is disjoint from either B or C. (Indeed, F_i, being a clique of H, cannot contain both a vertex of B and a vertex of C.)

We can now state our algorithm.

ALGORITHM 9
Input: A graph G, with lists $L(g) \subset \{A, B, C\}, g \in V(G)$.
Task: Find a partition $V(G) = A \cup B \cup C$ where A is a clique, there are no edges between B and C, and each $g \in L(g)$.
Action:

- Compute a minimal chordal extension H of G.
- For each forward clique F_i of H try to find a list M-partition A, B, C in which F_i contains A and is disjoint from B, or contains A and is disjoint from C. (The former is tested by deleting B from all lists of vertices in F_i and deleting A from all lists of vertices not in F_i; the latter is tested by deleting C from all lists of vertices in F_i and deleting A from all lists of vertices not in F_i. In both cases we have lists of size at most two and we can use Proposition 12.1.)
- If all tests fail, no list M-partition exists.

Corollary 14.1 *The list clique partition problem is solvable in polynomial time.*

15 Small Matrices M

Since most of our motivational examples have few parts (and are symmetric), it may be of interest to classify the complexity of list M-partition problems for all symmetric matrices M of small size k. (For small non-symmetric matrices, see [93].)

Theorem 15.1 ([38]) *Let M be a symmetric $k \times k$ matrix.*

When $k = 2$, all list M-partition problems are polynomial time solvable.

When $k = 3$, all list M-partition problems are polynomial time solvable, except for the 3-colouring problem and the independent cutset problems and their complements, which are NP-complete even as M-partition problems (that is, without lists).

For a discussion of the case $k = 4$, see [38]. There seem to be only three minimal NP-complete list partition problems (up to complementation): 3-colouring, independent cutset, and Winkler partition. In any event, all other problems admit a sub-exponential algorithm, and thus are not expected to be NP-complete. Some very recent work by C. Hoang et al., in progress, seems to be making good progress towards finding polynomial time algorithms for these remaining problems.

References

[1] J. Abello, M.R. Fellows and J.C. Stillwell, On the complexity and combinatorics of covering finite complexes, *Australas. J. Combin.* **4** (1991), 103–112.

[2] B. Aspvall, F. Plass and R.E. Tarjan, A linear time algorithm for testing the truth of certain quantified Boolean formulas, *Inform. Process. Lett.* **8** (1979), 121–123.

[3] R. Bačík, Graph homomorphisms and semidefinite programming, Ph.D. thesis, Simon Fraser University, 1996.

[4] H-J. Bandelt, A. Dahlmann and H. Schutte, Absolute retracts of bipartite graphs, *Discrete Appl. Math.* **16** (1987), 191–215.

[5] H-J. Bandelt, M. Farber and P. Hell, Absolute reflexive retracts and absolute bipartite retracts, *Discrete Appl. Math.* **44** (1993), 9–20.

[6] J. Bang-Jansen and P. Hell, The effect of two cycles on the complexity of colourings by directed graphs, *Discrete Appl. Math.* **26** (1990), 1–23.

[7] J. Bang-Jansen, P. Hell and G. MacGillivray, The complexity of colourings by semicomplete digraphs, *SIAM J. Discrete Math.* **1** (1988), 281–289.

[8] J. Bang-Jensen, P. Hell and G. MacGillivray, On the complexity of colouring by superdigraphs of bipartite graphs, *Discrete Math.* **109** (1992), 27–44.

[9] J. Bang-Jensen, P. Hell and G. MacGillivray, Hereditarily hard colouring problems, *Discrete Math.* **138** (1995), 75–92.

[10] J.A. Bondy & U.S.R. Murty, *Graph theory with applications*, American Elsevier, New York (1976).

[11] A. Brandstädt, Partitions of graphs into one or two stable sets and cliques, *Discrete Math.* **152** (1996), 47–54.

[12] A. Brandstädt, Corrigendum, *Discrete Math.* **186** (1998), 295.

[13] A. Brandstädt, Partitions of graphs into one or two stable sets and cliques, Informatik-Berichte Nr. 105, 1/1991, FernUniversität Hagen, 1991.

[14] R.C. Brewster, Vertex colourings of edge coloured graphs, Ph.D. thesis, Simon Fraser University, 1993.

[15] R.C. Brewster, T. Feder, P. Hell, J. Huang and G. MacGillivray, Near-unanimity functions and varieties of graphs, preprint, 2002.

[16] G.R. Brightwell and P. Winkler, Graph homomorphisms and phase transitions, *J. Combin. Theory Ser. B* **77** (1999), 221–262.

[17] G.R. Brightwell and P. Winkler, Gibbs measures and dismantlable graphs, *J. Combin. Theory Ser. B* **78** (2000), 141–166.

[18] A.E. Brouwer and H.J. Veldman, Contractibility and NP-completeness, *J. Graph Theory* **11** (1987), 71–79.

[19] A.A. Bulatov and V. Dalmau, Towards a dichotomy for the counting constraint satisfaction problem, preprint, 2003.

[20] A.A. Bulatov and P. Jeavons, Algebraic structures in combinatorial problems, preprint, 2002.

[21] S. Burr, P. Erdős and L. Lovász, On graphs of Ramsey type, *Ars Combin.* **1** (1976), 167–190.

[22] A. Cournier and M. Habib, A new linear algorithm for modular decomposition, in *CAAP'94: International Colloquium, Lectures Notes in Computer Science* (ed. Sophie Tison), Springer, Edinburgh (1994), pp. 68–82.

[23] M. Chudnovsky, N. Robertson, P. Seymour and R. Thomas, The strong perfect graph theorem, preprint, 2002.

[24] V. Chvátal, Star-cutsets and perfect graphs, *J. Combin. Theory Ser. B* **39** (1985), 189–199.

[25] R. Dechter, From local to global consistency, *Artificial Intelligence* **55** (1992), 87–107.

[26] J. Diaz, M. Serna and D.M. Thilikos, Counting H-colourings of partial k-trees, *Theoret. Comput. Sci.* **281** (2002), 291–309.

[27] J. Diaz, Counting H-colourings and variants, preprint, 2002.

[28] J. Diaz, M. Serna and D.M. Thinlikos, The complexity of parametrized H-colourings: A survey, preprint, 2002.

[29] M. Dyer and C. Greenhill, The complexity of counting graph homomorphisms, *Random Structures Algorithms* **17** (2000), 260–289.

[30] P. Erdös, Graph theory and probability, *Canad. J. Math.* **11** (1959), 34–38.

[31] T. Feder, Classification of homomorphisms to oriented cycles and of k-partite satisfiability, *SIAM J. Discrete Math.* **14** (2001), 471–480.

[32] T. Feder and P. Hell, List homomorphisms to reflexive graphs, *J. Combin. Theory Ser. B* **72** (1998), 236–250.

[33] T. Feder, P. Hell and J. Huang, List homomorphisms and circular arc graphs, *Combinatorica* **19** (1999), 487–505.

[34] T. Feder, P. Hell and J. Huang, Bi-arc graphs and the complexity of list homomorphisms, *J. Graph Theory* **42** (2003), 61–80.

[35] T. Feder, P. Hell and J. Huang, List homomorphisms of graphs with bounded degrees, preprint, 2003.

[36] T. Feder and M.Y. Vardi, Monotone monadic SNP and constraint satisfaction, in *Proceedings of the Twenty-fifth Annual ACM Symposium on the Theory of Computing* (1993), pp. 612–622.

[37] T. Feder and M.Y. Vardi, The computational structure of monotone monadic SNP and constraint satisfaction: a study through Datalog and group theory, *SIAM J. Comput* **28** (1998), 57–104.

[38] T. Feder, P. Hell, S. Klein and R. Motwani, Complexity of graph partition problems, in *Proceedings of the Thirty-first Annual ACM Symposium on the Theory of Computing* (1999), pp. 464–472.

[39] T. Feder, F. Madelaine, I.A. Stewart, Dichotomies for classes of homomorphism problems involving unary functions, preprint, 2002.

[40] U. Feige and L. Lovász, Two-prover one-round proof systems: Their power and their problems, in *Proceedings of the Twenty-fourth Annual ACM Symposium on the Theory of Computing* (1992), pp. 733-744.

[41] W.D. Fellner, On minimal graphs, *Theoret. Comput. Sci.* **17** (1982), 103–110.

[42] C. M. H. de Figueiredo and S. Klein, The NP-completeness of multipartite cutset testing, *Congr. Numer.* **119** (1996), 217–222.

[43] C.M.H. de Figueiredo, S. Klein, Y. Kohayakawa and B. Reed, Finding skew partitions efficiently, *J. Algorithms* **37** (2000), 505–521.

[44] S. Foldes and P.L. Hammer, Split Graphs, in *8th Southeastern Conference on Combinatorics, Graph Theory and Computing* (ed. F. Hoffman et al), Louisiana State Univ, Baton Rouge, Louisiana pp. 311–315.

[45] E.C. Freuder, A sufficient condition for backtrack-free search, *J. ACM* **29** (1982), 24–32.

[46] A. Galluccio, P. Hell and J. Nešetřil, The complexity of H-colouring of bounded degree graphs, *Discrete Math.* **222** (2000), 101–109.

[47] M.R. Garey and D.S. Johnson, *Computers and Intractability*, W.H. Freeman and Company, San Francisco (1979).

[48] M.C. Golumbic, *Algorithmic graph theory and perfect graphs*, Academic Press, New York (1980).

[49] W. Gutjahr, Graph colourings, Ph.D. thesis, Free University Berlin, 1991.

[50] W. Gutjahr, E. Welzl and G. Woeginger, Polynomial graph-colorings, *Discrete Appl. Math.* **35** (1992), 29–45.

[51] L. Haddad, P. Hell and E. Mendelsohn, On the complexity of colouring areflexive h-ary relations, *Ars Combin.* **48** (1998), 111 - 128.

[52] R. Häggkvist and P. Hell, Universality of *A*–mote graphs, *European J. Combin.* **14** (1993), 23 - 27.

[53] P. Hell, S. Klein, L.T. Nogueira and F. Protti, On generalized split graphs, *Electron. Notes Discrete Math.* **7** (2001),

[54] P. Hell and J. Nešetřil, The core of a graph, *Discrete Math.* **109** (1992), 117–126.

[55] P. Hell and J. Nešetřil, On the complexity of H-colouring, *J. Combin. Theory Ser. B* **48** (1990), 92–110.

[56] P. Hell and J. Nešetřil, Counting list homomorphisms and graphs with bounded degrees, *Discrete Math.*, in press.

[57] P. Hell and J. Nešetřil, *Graphs and Homomorphisms*, Clarendon Press, Oxford (2003).

[58] P. Hell, J. Nešetřil and X. Zhu, Duality and polynomial testing of tree homomorphisms, *Trans. Amer. Math. Soc.* **348** (1996), 147–156.

[59] P. Hell, J. Nešetřil and X. Zhu, Complexity of tree homomorphisms, *Discrete Appl. Math.* **70** (1996), 23–36.

[60] P. Hell, J. Nešetřil and X. Zhu, Duality of Graph Homomorphisms, in *Combinatorics, Paul Erdös is Eighty, Vol. 2* (ed. A. Hajnal et al), Bolyai Society Mathematical Studies, Budapest, Hungary (1993),

[61] P. Hell and I. Rival, Absolute retracts and varieties of reflexive graphs, *Canad. J. Math.* **39** (1987), 544–567.

[62] P. Hell and X. Zhu, Homomorphisms to oriented paths, *Discrete Math.* **132** (1994), 107-114.

[63] J. Huang, personal communication, 1996.

[64] P.G. Jeavons, On the algebraic structure of combinatorial problems, *Theoret. Comput. Sci.* **200** (1998), 185–204.

[65] P.G. Jeavons, D. Cohen and M. Gyssens, Closure properties of constraints, *J. ACM* **44** (1997), 527–548.

[66] J. Kratochvíl and Zs. Tuza, Algorithmic complexity of list colorings, *Discrete Appl. Math.* **50** (1994), 297–302.

[67] J. Kratochvíl and J.A. Telle, Complexity of coloured graph covers, Technical Report 97-354, KAM-DIMATIA Series, Charles University, 1997.

[68] J. Kratochvíl, A. Proskurowski and J.A. Telle, Covering regular graphs, *J. Combin. Theory Ser. B* **71** (1997), 1–16.

[69] J. Kratochvíl, A. Proskurowski and J.A. Telle, *Nordic J. Comput.* **5** (1998), 173–195.

[70] R.E. Ladner, On the structure of polynomial time reducibility, *J. ACM* **22** (1975), 155-171.

[71] B. Larose and L. Zádori, The complexity of the extendability problem for finite posets,, preprint, 2002.

[72] C.G. Lekkerkerker and J.Ch. Boland, Representation of a finite graph by a set of intervals on the real line, *Fund. Math.* **51** (1962), 45–64.

[73] D. Liu, T-colorings and chromatic number of distance graphs, *Ars Combin.* **56** (2000), 65–80.

[74] C. Loten, Retractions and generalizations of chordality, Ph.D. thesis, Simon Fraser University, 2003.

[75] G. MacGillivray, On the complexity of colourings by vertex-transitive and arc-transitive digraphs, *SIAM J. Discrete Math.* **4** (1991), 297–408.

[76] A.K. Mackworth, Consistency in networks of relations, *Artificial Intelligence* **8** (1977), 99–118.

[77] H.A. Maurer, J.H. Sudborough and E. Welzl, On the complexity of the general colouring problem, *Inform. and Control* **51** (1981), 123-145.

[78] R.M. McConnell and J.P. Spinrad, Linear-time modular decomposition and efficient transitive orientation of comparability graphs, in *Proceedings of the ACM-SIAM SODA* (1994), pp. 536–545.

[79] B. Mohar, P. Hell and T. Feder, Acyclic homomorphisms and circular colorings of digraphs, preprint, 2003.

[80] U. Montanari, Networks of constraints: Fundamental properties and applications to picture processing, *Inform. Sci.* **7** (1974), 95–132.

[81] R. Naserasr and C. Tardif, Homomorphisms of planar graphs, preprint, 2003.

[82] J. Nešetřil and X. Zhu, On Bounded Treewidth Duality of Graphs, *J. Graph Theory* **23** (1996), 151–162.

[83] A. Proskurowski and S. Arnborg, Linear time algorithms for NP-hard problems restricted to partial k-trees, *Discrete Appl. Math.* **23** (1989), 11–24.

[84] A. Puricella and I.A. Stewart, Stewart, Greedy algorithms, H-colourings, and a complexity-theoretic dichotomy, *Theoret. Comput. Sci.*, in press.

[85] N. Robertson and P.D. Seymour, Graph minors, II: Algorithmic aspects of treewidth, *J. Algorithms* **7** (1986), 309–322.

[86] N. Robertson and P.D. Seymour, Graph minors—a survey, *London Math. Soc. Lecture Note Ser.* **103** (1985), 153–171..

[87] D.J. Rose, R.E. Tarjan and G.S. Lueker, Algorithmic aspects of vertex elimination on graphs, *SIAM J. Comput.* **5** (1976), 266–283.

[88] T.J. Schaeffer, The complexity of satisfiability problems, in *Proceedings of the Tenth Annual ACM Symposium on the Theory of Computing* (1978), pp. 216–226.

[89] S. Stahl, The multichromatic numbers of some Kneser graphs, *Discrete Math.* **185** (1998), 287–291.

[90] R.E. Tarjan, Decomposition by clique separators, *Discrete Math.* **55** (1985), 221–232.

[91] R.E. Tarjan and M. Yannakakis, Addendum: simple linear-time algorithms to test chordality of graphs, test acyclicity of hypergraphs, and selectively reduce acyclic hypergraphs, *SIAM J. Comput.* **14** (1985), 254–255.

[92] A. Tucker, Coloring graphs with stable sets, *J. Combin. Theory Ser. B* **34** (1983), 258–267.

[93] K. Tucker, List M-partitions of directed graphs, M.Sc. thesis, Simon Fraser University, 2003.

[94] N. Vikas, Computational complexity of graph compaction, in *Proceedings of the ACM-SIAM SODA* (1999),

[95] N. Vikas, Connected and loosely connected list homomorphisms, preprint, 2002.

[96] E. Welzl, Symmetric graphs and interpretations, *J. Combin. Theory Ser. B* **37** (1984), 235-244.

[97] S. Whitesides, An algorithm for finding clique cutsets, *Inform. Process. Lett.* **12** (1981), 31–32.

[98] S. Whitesides, A method for solving certain graph recognition and optimization problems, with applications to perfect graphs, *Annals of Discrete Mathematics* **21** (1984), 281–297.

[99] G. Woeginger, personal communication, 1988.

[100] X. Zhu, Circular chromatic number: A survey, *Discrete Math.* **229** (2001), 371–410.

School of Computing Science
Simon Fraser University
Burnaby, B.C.
Canada V5A 1S6
pavol@cs.sfu.ca

Counting Lattice Triangulations

Volker Kaibel and Günter M. Ziegler

Abstract

We discuss the problem to count, or, more modestly, to estimate the number $f(m,n)$ of unimodular triangulations of the planar grid of size $m \times n$.

Among other tools, we employ recursions that allow one to compute the (huge) number of triangulations for small m and rather large n by dynamic programming; we show that this computation can be done in polynomial time if m is fixed, and present computational results from our implementation of this approach.

We also present new upper and lower bounds for large m and n, and we report about results obtained from a computer simulation of the random walk that is generated by flips.

1 Introduction

An innocent little combinatorial counting problem asks for the number of triangulations of a finite grid of size $m \times n$. That is, for $m, n \geq 1$ we define $P_{m,n} := \{0,1,\ldots,m\} \times \{0,1,\ldots,n\}$, "the grid". Equivalently, the point configuration $P_{m,n}$ consists of all points of the integer lattice $\mathbf{Z}^2$ in the lattice rectangle $\operatorname{conv}(P_{m,n}) = [0,m] \times [0,n]$ of area mn. Every triangulation of this rectangle point set that uses all the points in $P_{m,n}$ has $(m+1)(n+1) = |P_{m,n}|$ vertices, $2mn$ facets/triangles, and $3mn + m + n$ edges, $2(m+n)$ of them on the boundary, the other $3mn - m - n$ ones in the interior. All the triangles are minimal lattice triangles of area $\frac{1}{2}$ (that is, of determinant 1), which are referred to as *unimodular* triangles. The grid triangulations that use all the points are thus called *unimodular triangulations.* The number of unimodular triangulations of the grid $P_{m,n}$ will be denoted by $f(m,n)$.

As an example, our first figure shows one unimodular triangulation of the 5×6 grid:

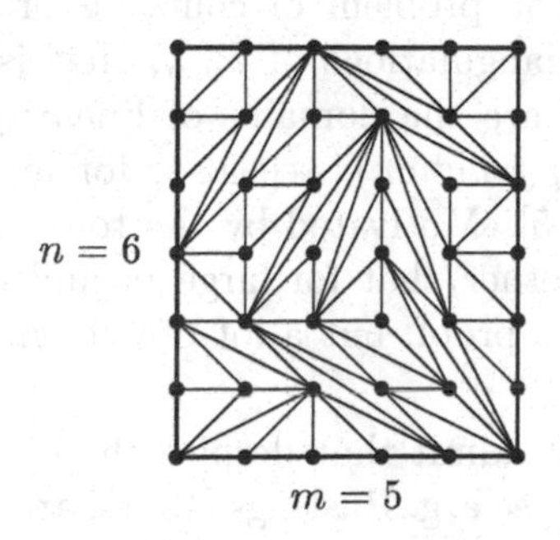

To get started, one notes that of course $f(m,n) = f(n,m)$, one discovers with pleasure that $f(1,n) = \binom{2n}{n}$, and one works out by hand with a bit of pain that $f(2,2) = 64$ and $f(2,3) = 852$. Furthermore, one observes that the special triangulations that decompose into m vertical strips of width 1 yield the lower bound

$$f(m,n) \ \geq \ f(1,n)^m \ = \ \binom{2n}{n}^m \tag{1.1}$$

so for larger m and n the numbers $f(m,n)$ get huge *very soon*; for example, the bound yields $f(5,6) > 6 \cdot 10^{14}$.

It is equally interesting to study/enumerate more general types of triangulations, such as triangulations of finite point sets that do not necessarily use all the points, triangulations of general convex or non-convex lattice polygons, triangulations of general position point sets, triangulations of higher-dimensional point sets, etc. None of these will appear here, but we refer interested readers to Lee [22] and De Loera, Rambau & Santos [12].

Lattice triangulations are basic combinatorial objects, and they are fundamental discrete geometric structures; so it is no surprise that they appear in various computational geometry contexts. However, lattice triangulations have also been studied intensively from different algebraic geometry angles. So, triangulations of a convex lattice polygon

- provide the data for Viro's [31] famous construction method of plane algebraic curves with prescribed combinatorics and topology, related to Hilbert's sixteenth problem;
- appear in Gel'fand–Kapranov–Zelevinsky's [16] theory of discriminants, where "regular" triangulations are in bijection with the vertices of the secondary polytope of the point configuration, and
- model torus-equivariant crepant resolution of singularities for toric threefolds, where "regular" triangulations correspond to projective desingularizations; see e. g., Kempf et al. [21, Chap. 3] and Dais [10].

The last two points pose the problem of counting or estimating the number $f^{\text{reg}}(m,n)$ of *regular* triangulations of $P_{m,n}$, that is, of triangulations of $\text{conv}(P_{m,n})$ whose triangles are the domains of linearity for a piecewise linear convex function (*lifting function*); see also, for example, Sturmfels [30, Chap. 8], Ziegler [32, Lect. 5]. Motivated by the toric variety considerations, one would like to know whether/that for large m and n most triangulations are non-regular. There is no proof, but a lot of evidence that this should be true (see, e.g., Table 2).

In this context, we must admit that despite the effort that has been put into studying this question (see e. g. Hastings [19, Chap. 2]), it has not become clear what non-regular triangulations really "look like." The "mother of all examples" is the whirlpool triangulation of the 3×3 grid:

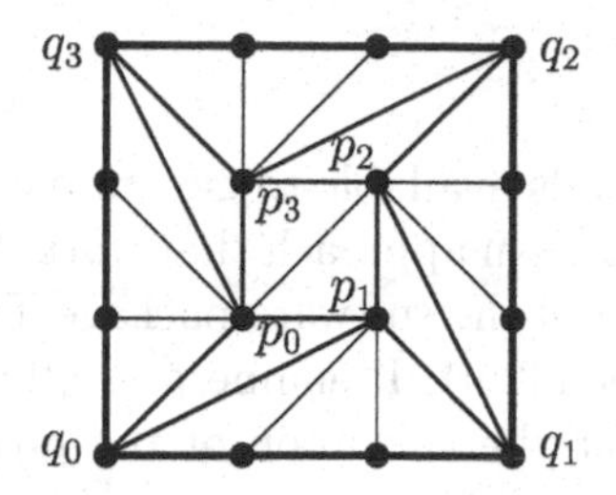

Indeed, this unimodular triangulation is irregular, since a lifting function h would have to satisfy $\frac{3}{4}h(p_i) + \frac{1}{4}h(q_{i-1}) < \frac{3}{4}h(p_{i-1}) + \frac{1}{4}h(q_i)$ for $i = 0, \ldots, 3$ (all indices taken modulo 4), implying the contradiction $h(q_0) + \cdots + h(q_3) < h(q_0) + \cdots + h(q_3)$.

For $m = n = 3$, except for three symmetric copies this is the only example of a non-regular triangulation: We have $f(3,3) - f^{\text{reg}}(3,3) = 4$. This may lead one to the conjecture that some kind of "generalized whirlpools" are responsible for non-regularity. However, the pictures of "non-regular triangulations" that we present below do not support this intuition.

The plan for this paper is as follows: In Section 2 we face the challenge to ***count*** explicitly, trying to cope with the "combinatorial explosion." For this, we present a simple dynamic programming technique by which we get surprisingly far, and which on the specific problem of grid triangulations outruns the much more sophisticated general techniques such as Avis & Fukuda's [6] reverse search, Aichholzer's [1] path of a triangulation, or the oriented matroid technique of Rambau [27].

In Section 3 we ***estimate*** $f(m, n)$ for large m and n, trying to narrow the bounds for the asymptotics. It is interesting to compare with the situation for $N = (m+1)(n+1)$ points in the plane in general position, where the currently best available upper bound seems to be Santos & Seidel's [28] estimate that there are $o(59^N)$ triangulations. However, in our problem the N points are not at all in general position — and the upper bounds that we have to offer are much better: We report on a neat $O(2^{3N})$ upper bound by Anclin [5], which substantially improves on a previous $O(2^{4N})$ upper bound by Orevkov [25]. Based on explicit enumeration results from Section 2, we get a lower bound of $2^{2.055\,mn}$ when both m and n get large; note that (1.1) yields already a lower bound $2^{(1-o(1))2mn}$.

Finally, in Section 4 we ***sample*** lattice triangulations for large parameters m and n, and thus try to understand what typical lattice triangulations, as well as typical regular lattice triangulations, "look like." While we have some pictures to offer, proofs seem harder to come by. Indeed, the pictures display some long-range order; while this may make lattice triangulations interesting as a statistical physics model, it generates serious obstacles for any proof that the obvious Markov chain is rapidly mixing, and thus to application of the (by now) standard theory [8].

2 Explicit Values

There are several methods available to generate or count all triangulations of a finite set of points in $\mathbf{R}^2$. An approach that works for point sets in arbitrary dimensions is implemented in the software package TOPCOM by Rambau [27] (see also Pfeifle and Rambau [26]). It enumerates all triangulations in a purely combinatorial manner after the chirotope of the point set (oriented matroid data) has been computed.

The *reverse search algorithm* proposed by Avis and Fukuda [6] is a rather general enumeration scheme that can be specialized to triangulations of point configurations in $\mathbf{R}^2$ (see also Bespamyatnikh [7]). Since it was used to obtain some of the results reported in Section 4, and because it is based upon some structural properties that are relevant for our treatment later, we briefly describe the method here.

Let $\mathcal{T}$ be a *fine triangulation* of a point set $S \subset \mathbf{R}^2$; i.e., $\mathcal{T}$ is a set of triangles, for which the set of vertices equals S, such that the union of all triangles is the convex hull of S, and any two triangles intersect in a common (possibly empty) face. The unimodular triangulations of $P_{m,n}$ are precisely its fine triangulations. An edge of some triangle in $\mathcal{T}$ is *flippable* if it is contained in two triangles of $\mathcal{T}$ whose union is a strictly convex quadrangle. Replacing these two triangles by the two triangles into which the other diagonal cuts that quadrangle (*flipping* the edge) yields another fine triangulation of S. The graph on the fine triangulations of S defined via flipping is the *flip graph* of S.

Let us fix an arbitrary ordering of the points in S, inducing via lexicographical ordering a total order on the set of pairs of points, and thus, again via lexicographical ordering, a total order on the fine triangulations of S, which are identified with their sets of edges here. There is a distinguished fine triangulation $\mathcal{T}_0$ of S with respect to that ordering, namely the smallest *Delaunay triangulation* (i.e., a triangulation characterized by the condition that for every triangle the circumcircle contains no point from S in its interior).

Furthermore, for each fine triangulation $\mathcal{T} \neq \mathcal{T}_0$, there is a distinguished flippable edge (computable in $O(|S|)$ steps) such that, starting from any fine triangulation of S, iterated flipping of the respective distinguished edges eventually yields $\mathcal{T}_0$. This algorithmically defines a spanning tree in the flip graph of S, rooted at $\mathcal{T}_0$; in particular, the flip graph of a two dimensional finite point set is connected.

The basic idea of the reverse search method is to traverse that spanning tree from its root $\mathcal{T}_0$. At each iteration one chooses a leaf $\mathcal{T}$ of the current partial tree, and determines those among the triangulations adjacent to $\mathcal{T}$ on whose path to $\mathcal{T}_0$ in the spanning tree $\mathcal{T}$ lies. Properly implemented, the reverse search algorithm generates all fine triangulations in $O(|S| \cdot f(S))$ steps, where $f(S)$ is the number of fine triangulations of S; see [6].

Via the "secondary polytope" of a finite point set $S \subset \mathbf{R}^2$ [16] one can design a variant of the reverse search algorithm that generates *all regular* triangulations of S in $O(|S| \cdot F^{\text{reg}}(S))$ steps, where $F^{\text{reg}}(S)$ is the number of such triangulations. It is, unclear, however, if one can also generate all *regular fine* triangulations of S in a number of steps that is bounded by a polynomial in the number of such triangulations.

If one is interested in the number of triangulations of a two-dimensional set of points rather than in the explicit generation of all of them, then the *path of a triangulation method* due to Aichholzer [1] is more efficient than the reverse search algorithm.

For counting the unimodular (fine) triangulations of the very special point sets $P_{m,n}$, however, different methods are much more efficient. These are described in the following sections.

2.1 Narrow Strips

Strips of width $m = 1$. For any lattice trapezoid of width 1, whose parallel vertical sides have lengths a and b, the number of unimodular triangulations is $g_1(a,b) = \binom{a+b}{a} = \binom{a+b}{b}$; indeed, a bijection between these triangulations and the a-subsets of $\{1,\ldots,a+b\}$ is established by top-down numbering the triangles of a triangulation $\mathcal{T}$ by $1,\ldots,a+b$, and mapping $\mathcal{T}$ to the a-set of all numbers of triangles whose vertical edges are on the left. In particular, we have

$$f(1,n) \;=\; \binom{2n}{n}. \tag{2.1}$$

Strips of width $m = 2$. For $f(2,n)$ we have no explicit formula, and we cannot evaluate the asymptotics precisely, but we have a "quadratic" recursion that can be evaluated efficiently: For this we enumerate the triangulations according to the highest "width 2 diagonal," which (if it exists) decomposes the rectangle into two width 1 strips, a single triangle, and a trapezoid of width 2 (see the left-hand figure below). Let $g_2(A,B)$ denote the number of triangulations of a trapezoid of width 2 with vertical edges of lengths A and B and horizontal base line, as in the right-hand figure below — where $A + B \equiv 1 \bmod 2$ implies that the midpoint of the diagonal is not a lattice point, and where we may assume $A < B$ by symmetry:

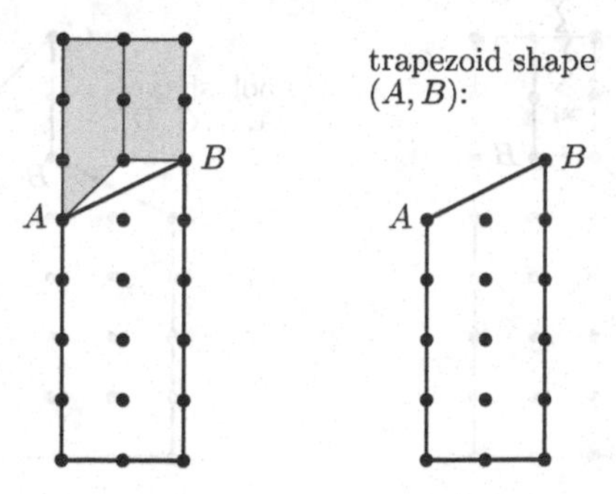

Thus we get

$$f(2,n) = \binom{2n}{n}^2 + 2 \sum_{\substack{0 \le A < B \le n \\ A+B \equiv 1 \bmod 2}} g_2(A,B) \binom{2n - \frac{3A+B+1}{2}}{n-A} \binom{2n - \frac{A+3B+1}{2}}{n-B}.$$

The binomial coefficients in this recursion correspond to triangulations of width 1 strips. So they could be rewritten in terms of g_1, as $g_1(n-A, n-\frac{A+B+1}{2})$ resp. $g_1(n-B, n-\frac{A+B+1}{2})$. A similar remark applies to the binomial coefficients that appear in the following.

For $g_2(A,B) = g_2(B,A)$ we also get a recursion by considering the highest diagonal of width 2:

$$\begin{aligned} g_2(A,B) &= \binom{\frac{3A+B-1}{2}}{A} \binom{\frac{A+3B-1}{2}}{B} \\ &\quad + \sum_{\substack{0 \le a \le A,\ 0 \le b \le B \\ a+b \equiv 1 \bmod 2,\ a+b < A+B}} g_2(a,b) \binom{\frac{3A+B-3a-b}{2} - 1}{A-a} \binom{\frac{A+3B-a-3b}{2} - 1}{B-b}. \end{aligned}$$

Here the parameters A, B, a, b may be interpreted as the y-coordinates of certain lattice points. The shaded parts of the figure consist of strips of width 1, whose triangulations are counted by the binomial coefficients $g_1(\cdot, \cdot)$.

Strips of width $m = 3$. For $f(m,3)$ we have a recursion of order 4; it relies on the observation that if we screen the middle strip from the top for diagonals of width at least 2, then the first diagonal to find will be of width exactly 2, since any width 3 diagonal is flippable, and contained in a parallelogram that is bounded by two width 2 diagonals. The corresponding decomposition of our rectangle is indicated in the left figure below:

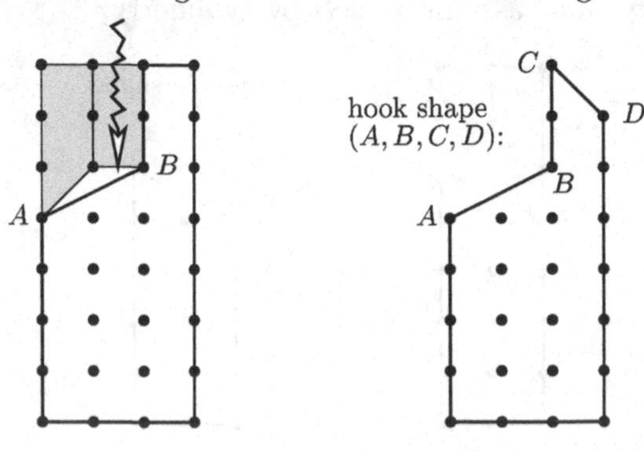

Therefore, we obtain

$$f(3,n) = \binom{2n}{n}^3 + 2 \sum_{\substack{0\le A,B\le n \\ A+B\equiv 1 \bmod 2}} h(A,B,n,n) \binom{2n - \frac{3A+B+1}{2}}{n-A} \binom{2n - \frac{A+3B+1}{2}}{n-B},$$

where $h(A,B,C,D)$ counts the number of triangulations in a "hook" shape as given in the right drawing in the figure above, which depends on four parameters $0 \le A,B,C,D \le n$, with $A+B \equiv 1 \bmod 2$ and $B \le C$. For the number of triangulations of such a hook shape we get a recursion

$$\begin{aligned} h(A,B,C,D) &= \binom{\frac{3A+B-1}{2}}{A}\binom{\frac{A+3B-1}{2}}{B}\binom{C+D}{C} \\ &+ \sum_{\substack{0\le a\le A,\ 0\le b\le B \\ a+b\equiv 1 \bmod 2,\ a+b<A+B}} h(a,b,C,D) \binom{\frac{3A+B-3a-b}{2}-1}{A-a}\binom{\frac{A+3B-a-3b}{2}-1}{B-b} \\ &+ \sum_{\substack{0\le a\le D,\ 0\le b\le \frac{A+B-1}{2} \\ a+b\equiv 1 \bmod 2,\ \frac{a+b+1}{2}\le B}} h(a,b,\tfrac{A+B-1}{2},A) \binom{D+C-\frac{3a+b+1}{2}}{D-a}\binom{\frac{A+3B-a-3b}{2}-1}{B-\frac{a+b+1}{2}} \\ &+ h(\tfrac{3B-A-1}{2},\tfrac{A+B-1}{2},\tfrac{A+B-1}{2},A)\binom{C+D-\frac{5B-A-1}{2}}{C-B} \quad \text{if } D \ge \tfrac{3B-A-1}{2} \ge 0. \end{aligned}$$

The four terms in this recursion correspond to the four cases depicted in the figure below, where the fourth case — of a long diagonal of width 3 — occurs only in the case where the second endpoint of the diagonal, which may be computed to have y-coordinate $\frac{3B-A-1}{2}$, comes to lie within the hook.

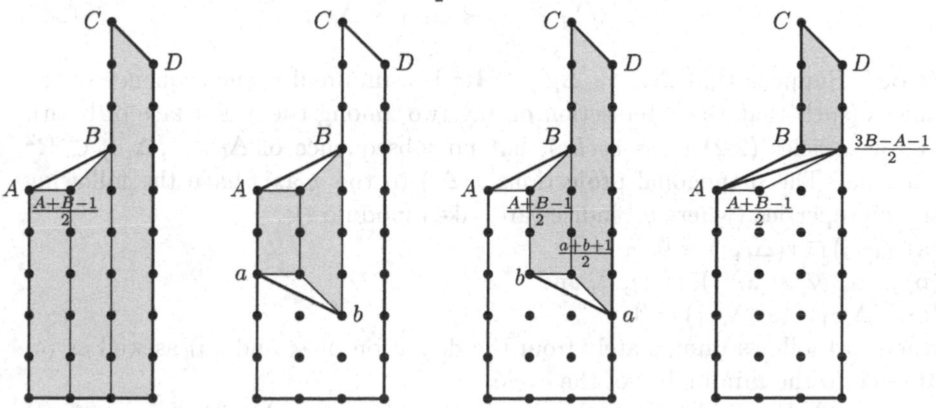

2.2 Strips of (fixed) width m.

We now describe a recursive strategy for the enumeration of unimodular triangulations of grids of arbitrary size. The method is applicable for triangulations of general finite point sets — but it is effective only in the special case where the points lie on a small family of parallel (vertical, say) lines; in our case this is the situation of small (fixed) m and variable n. The key observation is that any triangulation may be dismantled by removing triangles from the upper boundary, while maintaining a lattice triangulation of a y-convex lattice polygon. Since for fixed m the number of such polygons in $P_{m,n}$ is bounded by a polynomial in n, this yields an efficient dynamical programming algorithm in this case.

Let $\Delta_1, \Delta_2 \subset \mathbf{R}^2$ be two triangles whose intersection $\Delta_1 \cap \Delta_2$ is a common (empty, zero-, or one-dimensional) face of both Δ_1 and Δ_2. We say that Δ_2 *lies above* Δ_1 ($\Delta_1 \prec \Delta_2$) if there are two points $(x, y_1) \in \Delta_1 \setminus \Delta_2$ and $(x, y_2) \in \Delta_2 \setminus \Delta_1$ on a vertical line with $y_1 < y_2$. For example, in our figure the shaded triangle lies above the other one; the other two pairs of triangles are incomparable.

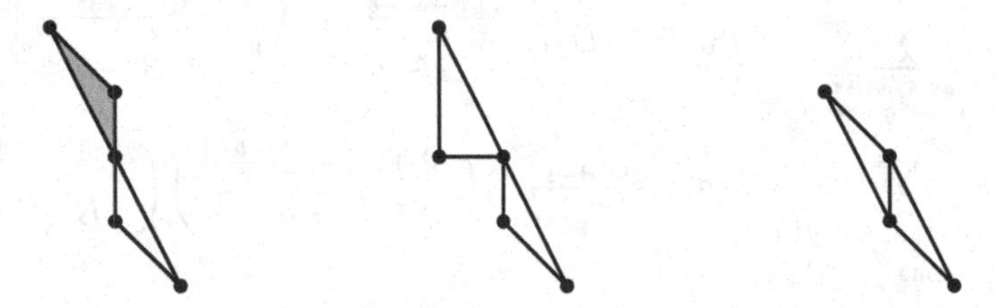

Due to the convexity of the triangles and the intersection condition imposed on them, this is a well-defined asymmetric and irreflexive relation.

Lemma 2.1 *There is no sequence $\Delta_0, \ldots, \Delta_{t-1} \subset \mathbf{R}^2$ of triangles (such that the intersection of any two among them is a face of both) satisfying*

$$\Delta_0 \prec \Delta_1 \prec \ldots \prec \Delta_{t-1} \prec \Delta_0 \ . \tag{2.2}$$

Proof Suppose that $\Delta_0, \ldots, \Delta_{t-1} \subset \mathbf{R}^2$ is a minimal cyclic sequence of triangles (such that the intersection of any two among them is a face of both), i.e., it satisfies (2.2) (it is *cyclic*), but no subsequence of $\Delta_0, \ldots, \Delta_{t-1} \subset \mathbf{R}^2$ is cyclic. The orthogonal projections $x(\Delta_i)$ to the x-axis have the following three properties (where all indices are taken modulo t):
(a) $x(\Delta_i) \cap x(\Delta_{i+1}) \neq \emptyset$,
(b) $x(\Delta_i) \not\subseteq x(\Delta_{i-1}), x(\Delta_{i+1})$, and
(c) $x(\Delta_{i-1}) \cap x(\Delta_{i+1}) = \emptyset$,
where (a) follows immediately from the definition of $\prec$ and (b) as well as (c) are due to the minimality of the cycle.

But (a), (b), and (c) together imply that the intervals $x(\Delta_0), \ldots, x(\Delta_{t-1})$ "either run left-to-right or right-to-left"; in particular, we have $x(\Delta_0) \cap x(\Delta_i) = \emptyset$ for $i \in \{2, \ldots, t-1\}$, contradicting $\Delta_{t-1} \prec \Delta_0$. □

Of course, both in the definition of $\prec$ as well as in Lemma 2.1 one can replace "triangle" by "compact convex set."

The relation $\prec$ was defined with respect to parallel projection here. If one defines an ordering with respect to central rather than to parallel projection, then the analog to Lemma 2.1 (for arbitrary centers of projection) does not hold. For Delaunay triangulations, however, there is an analog to Lemma 2.1. (See De Floriani et al. [11] for dimension 2, and Edelsbrunner [13] for arbitrary dimensions.)

Due to Lemma 2.1, the relation $\prec$ induces a partial order on the set of triangles in $\mathbf{R}^2$, which we will also denote by $\prec$.

A sequence $\mathcal{T}_1, \ldots, \mathcal{T}_{2mn}$ of sets of triangles will be called an *admissible sequence* for $P_{m,n}$ if $\mathcal{T}_1$ is a unimodular triangulation of $P_{m,n}$, and if, for each $i = 2, 3, \ldots, 2nm$, we have $\mathcal{T}_i = \mathcal{T}_{i-1} \setminus \{\Delta\}$ for some $\prec$-maximal triangle Δ in $\mathcal{T}_{i-1}$. A subset $S \subset \mathbf{R}^2$ is called an *admissible shape* (of $P_{m,n}$) if it can be obtained as a union $S = \bigcup_{\Delta \in \mathcal{T}_i} \Delta$ for an admissible sequence $\mathcal{T}_1, \ldots, \mathcal{T}_{2mn}$ and some $i \in \{1, \ldots, 2mn\}$. Every admissible shape is y-convex (i.e., its intersection with any vertical line is connected). It is determined by its *upper boundary segments*, i.e., the sequence of line segments $[l^{(1)}, r^{(1)}], \ldots, [l^{(t)}, r^{(t)}]$ with $l_x^{(1)} = 0$, $l_x^{(t)} = n$, and $r_x^{(j-1)} = l_x^{(j)}$ for $j \in \{2, \ldots, t\}$, such that, for each point p in the relative interior of any of the segments, $p + (0, \varepsilon) \notin S$ holds for all $\varepsilon > 0$.

Let S be an admissible shape. We denote by $\mathcal{T}_{\max}(S)$ the set of all $\prec$-maximal unimodular triangles in S, that is, the finite set of all unimodular triangles that could be $\prec$-maximal in *some* unimodular triangulation of S. For example, the figure below indicates the 12 $\prec$-maximal triangles of the shaded admissible shape.

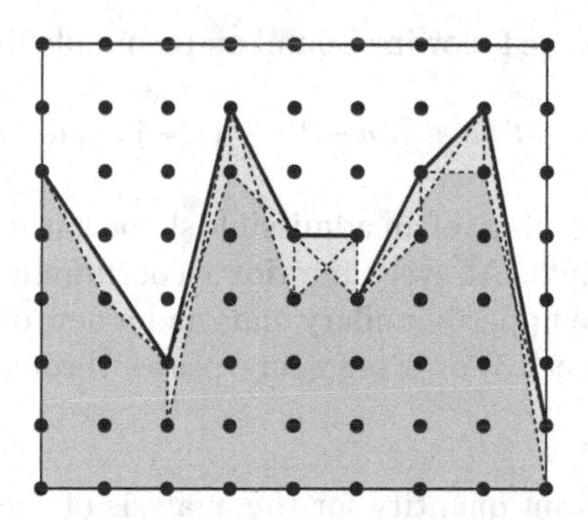

Any admissible shape S' that arises from S by removing some triangles contained in $\mathcal{T}_{\max}(S)$ is called an *admissible subshape* of S. These triangles have disjoint interiors, and they are uniquely determined for each admissible subshape S' of S (compare the proof of Lemma 2.5). Their number is denoted by $\#(S', S)$.

Since every unimodular triangulation of S contains at least one triangle from $\mathcal{T}_{\max}(S)$, we obtain the following inclusion-exclusion formula for the numbers $f(S)$ of unimodular triangulations of admissible shapes S.

Lemma 2.2 *Every admissible shape* S *has*

$$f(S) = \sum_{S'} (-1)^{\#(S',S)-1} f(S') \tag{2.3}$$

unimodular triangulations, where the sum is taken over all admissible proper subshapes S' *of* S.

Lemma 2.2 allows us to compute $f(m,n) = f([0,m] \times [0,n])$ via a dynamic programming approach: Determine the numbers $f(S)$ via (2.3) in some order such that every admissible shape appears after all its admissible subshapes. In order to analyze the running time of such an algorithm, we first need to estimate the number of admissible shapes.

Lemma 2.3 *Let* $[l^{(1)}, r^{(1)}], \dots, [l^{(t)}, r^{(t)}]$ *be the sequence of upper boundary segments of an admissible shape* S. *Then*

$$l_y^{(j)} \in \{r_y^{(j-1)} - 1, r_y^{(j-1)}, r_y^{(j-1)} + 1\}$$

holds for each $2 \le j \le t$.

Proof This follows by induction on the number of triangles removed in order to obtain S: Every vertical edge of a triangle in a unimodular triangulation has length one, and thus removing a $\prec$-maximal triangle from an admissible shape never creates a vertical boundary part of height more than 1. □

Lemma 2.3 implies the following bound on the number of admissible shapes.

Lemma 2.4 *There are at most* $(3n+2)^{m-1}(n+1)^2$ *admissible shapes of* $P_{m,n}$.

Proof The upper boundary of an admissible shape has $n+1$ possible start and $n+1$ possible end points. At every interior x-coordinate ($x \in \{1, \dots, m-1\}$) either a segment of the upper boundary ends and a new one starts ($3(n+1)-2$ possibilities, by Lemma 2.3) or a segment "passes through" (one possibility). □

The second important quantity for the analysis of the running time of the dynamic programming algorithm proposed above is the maximal number of summands that may occur in (2.3).

Lemma 2.5 *Every admissible shape* S *of* $P_{m,n}$ *has at most*

$$\left(\frac{3+\sqrt{13}}{2}\right)^m < 3.31^m$$

admissible subshapes.

Proof Let $(B_1, \ldots, B_t)$ be the sequence (from left to right) of upper boundary segments of S. Each triangle in $\mathcal{T}_{\max}(S)$ contains at least one of the edges $\{B_1, \ldots, B_t\}$.

Let $B \in \{B_1, \ldots, B_t\}$ be a segment of the upper boundary of S. There are at most two triangles in $\mathcal{T}_{\max}(S)$ containing B and one of its adjacent segments. Each other triangle in $\mathcal{T}_{\max}(S)$ that contains B must have its third vertex v in $B + \mathbf{R} \cdot (0, -1)$ on the line that is parallel to B at distance $\ell_2(B)^{-1}$, where $\ell_2(B)$ denotes the Euclidean length of B (because the triangle has area $\frac{1}{2}$). Since v must be integral and there is no integral point in the relative interior of B, there are at most two possibilities for v. Let us call one of them *the first triangle below* B, and the other one, *the second triangle below* B (if they exist).

For every admissible subshape T of S, define a word $w(T) \in \{0, \alpha, \beta, \gamma\}^\star$ by replacing B_i by

'α' if $S \setminus T$ contains the first triangle below B_i,
'β' if $S \setminus T$ contains the second triangle below B_i,
'γ' if $S \setminus T$ contains the triangle formed by B_i and B_{i-1}, and
'0' otherwise.

Clearly, every 'γ' in $w(T)$ has a '0' as its left neighbor; furthermore, $w(\cdot)$ is an injective mapping. Therefore, the number of admissible subshapes of S is bounded from above by the function $\varphi(m)$ defined recursively via

$$\varphi(0) = 1\,, \qquad \varphi(1) = 3\,, \qquad \varphi(m) = 3\varphi(m-1) + \varphi(m-2) \quad (m \geq 2).$$

Using standard techniques (see, e.g., [29, Thm. 4.1.1]), one derives from this recursion that $\varphi(m) < \left(\frac{3+\sqrt{13}}{2}\right)^m$ for $m \geq 1$. $\square$

Lemmas 2.4 and 2.5 imply that (for every m) the dynamic programming algorithm needs at most

$$3.31^m (3n+2)^{m-1} (n+1)^2 \; < \; 10^m (n+1)^{m+1}$$

arithmetic operations. The actual running time of an implementation heavily depends on the data structures for storing the admissible shapes and on the way by which one determines the admissible subshapes in that data structure. Therefore, here we include only the following rough statement.

Theorem 2.6 *For every fixed m, the function $f(m, n)$ can be computed in time bounded by a polynomial in n.*

In our implementation, we organize the admissible shapes as the leaves of a tree whose nodes are the prefixes of the sequences of upper boundaries segments of admissible shapes. The data structure allows quite efficient access to the admissible subshapes while not wasting too much memory. Nevertheless, the bottleneck in the computations is always memory. It is crucial to use an ordering of the admissible shapes in which for each shape S the shapes of which it is an admissible subshape come as soon as possible after it; this allows one to keep in memory only a subset of the admissible shapes at each point of time.

Furthermore, in (2.3) there is no need to sum over *all* admissible subshapes S'. It suffices to consider those S' that arise from removing triangles in $\mathcal{T}_{\max}(S)$ that are contained in $\{(x, y) \in \mathbf{R}^2 : x \geq c\}$, where c is the maximal x-coordinate where $l_y^{(j)} = r_y^{(j-1)} + 1$ for some j and the upper boundary segments $[l^{(1)}, r^{(1)}], \ldots, [l^{(t)}, r^{(t)}]$ of S (if that maximum exists).

Of course, when computing the number of unimodular triangulations of $P_{m,n}$ by our method, we obtain as a byproduct the number of unimodular triangulations of several interesting polygons inside $P_{m,n}$, including $P_{m,1}, \ldots, P_{m,n-1}$.

2.3 Explicit values

We have implemented the algorithms described in Subsection 2.1 in C++, using the `gmp` library [17] for exact arithmetic, with the interface to it provided by the `polymake` system [15].

Results obtained by our code are compiled in the Appendix (see Tables 3, 4, 5, 6, and 7). The number of unimodular triangulations of $P_{m,n}$ asymptotically grows exponentially with mn (see also Section 3). Therefore, it is more convenient to view the function $f(m, n)$ on a logarithmic scale, normalized by mn.

Definition The *capacity* of the $m \times n$ grid is

$$c(m, n) \quad := \quad \frac{\log_2 f(m, n)}{mn}.$$

The following Figure shows the capacity functions $c(m, n)$ for $m \in \{1, \ldots, 6\}$. The largest capacity we found is 2.055792 (for $m = 4$ and $n = 32$, see Table 5).

To give an impression of the amount of (machine) resources required for the calculations: The run of the admissible shape algorithm for $m = 6$ and $n = 7$ needed about three gigabytes of memory. The grid $P_{6,7}$ has 370252552 admissible shapes. We generated them in lexicographical order with respect to the pairs of starting heights and volumes. This way, never more than 15% of the admissible shapes had to be kept in memory simultaneously. Notice that more than 400 megabytes (of the 3 gigabytes in total) where needed just to store the *numbers* of triangulations of the admissible shapes in the memory. The CPU time used for the computation was about 25 hours. Our computations were performed on a SUN UltraAX MP machine equipped with four 448 MHz UltraSPARC-II processors (of which we used only one) and 4 gigabytes main memory.

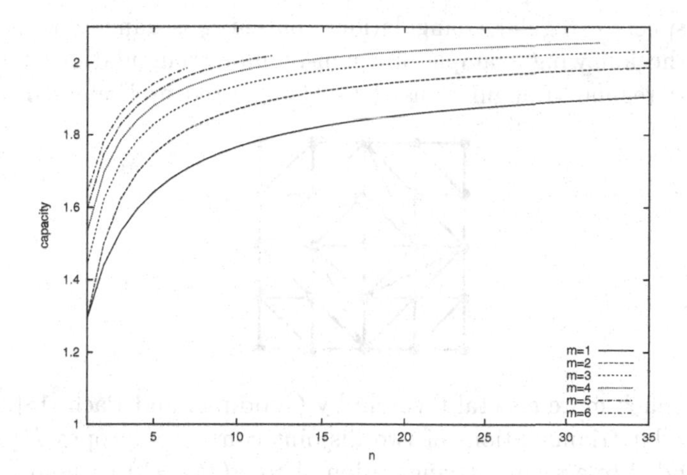

For very small parameters, Meyer [24] has enumerated all unimodular triangulations by Avis and Fukuda's reverse search method (sketched at the beginning of this section) and checked them for regularity. Table 1 shows the results. While for these small parameters irregular triangulations are quite rare, the picture changes drastically when m and n get larger, see Section 4.

$m \times n$	# triangulations	# irregular	fraction
3×3	46456	4	.000086
3×4	2822648	502	.000178
3×5	182881520	63528	.000347
4×4	736983568	1553020	.002107

Table 1: Number of regular triangulations of small grids.

3 Bounds

3.1 Patching

Any two unimodular triangulations of P_{m,n_1} and P_{m,n_2} can be patched to a unimodular triangulation of P_{m,n_1+n_2}. Thus we have the follwing supermultiplicativity relation, where $f^{\text{irreg}}(m,n)$ denotes the number of irregular unimodular triangulations of $P_{m,n}$.

Lemma 3.1 *For $m, n_1, n_2 \geq 1$ the following relations hold.*

(i) $$f(m, n_1 + n_2) \geq f(m, n_1) f(m, n_2)$$

(ii) $$f^{\text{irreg}}(m, n_1 + n_2) \geq f(m, n_1) f^{\text{irreg}}(m, n_2)$$

With respect to regular triangulations, patching is dangerous, as demonstrated by the following example of a non-regular triangulation of $P_{4,4}$ composed of four regular triangulations of $P_{2,2}$ (suggested by Francisco Santos).

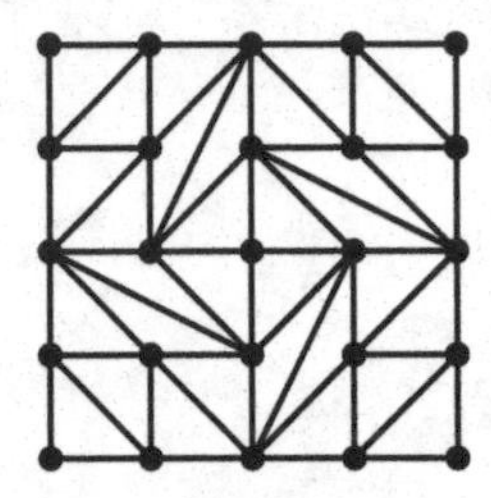

However, a much more general theorem by Goodman and Pach [18] says that any two regular triangulations of two disjoint convex polytopes $P_1, P_2 \subset \mathbf{R}^d$ can be extended to a regular triangulation of $\operatorname{conv}(P_1 \cup P_2)$ without additional vertices. Thus also for the regular case we get a (slightly weaker) supermultiplicativity relation.

Lemma 3.2 *For $m, n_1, n_2 \geq 1$ we have*

$$f^{\mathrm{reg}}(m, n_1 + n_2 + 1) \geq f^{\mathrm{reg}}(m, n_1) f^{\mathrm{reg}}(m, n_2) .$$

The following figure illustrates the patching of Lemma 3.2:

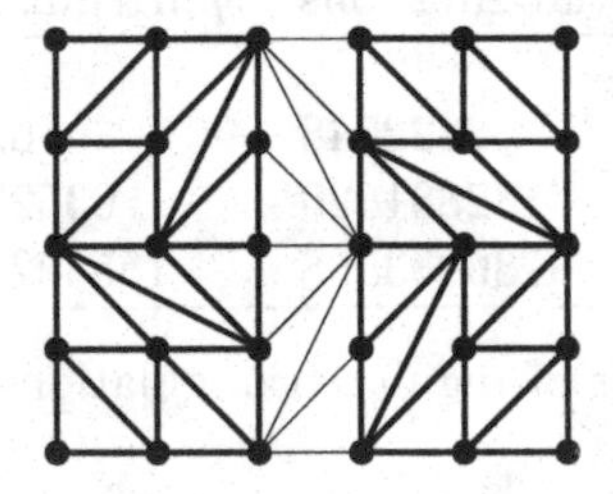

For $n_2 = 1$ we will (Lemma 3.5) strengthen the inequality in Lemma 3.2.

Let us fix some notations first. For any function $h : P_{m,n} \longrightarrow \mathbf{R}$ we denote by $H : P_{m,n} \longrightarrow \mathbf{R}^3$ the function with $H(x, y) = (x, y, h(x, y))$. The function h is called a *lifting function* of a triangulation $\mathcal{T}$ of $P_{m,n}$, if $\mathcal{T}$ is the image of the set of "lower facets" of the 3-polytope $\operatorname{conv}\{H(x, y) : (x, y) \in P_{m,n}\}$ under orthogonal projection to the x, y-plane (deletion of the third coordinate).

A function h is a lifting function of $\mathcal{T}$ if and only if h is convex and piecewise linear, and its (maximal) domains of linearity are the triangles in $\mathcal{T}$. In particular, one may add to h any convex piecewise linear function whose domains of linearity are unions of triangles of $\mathcal{T}$ in order to obtain another lifting function for $\mathcal{T}$.

A triangulation is regular if and only if it has a lifting function.

The following result shows that all unimodular triangulations of a strip of width 1 are regular; moreover one has a lot of freedom in choosing the respective lifting functions, which we will exploit below.

Lemma 3.3 *Let $\mathcal{T}$ be a unimodular triangulation of a lattice trapezoid with two parallel vertical or horizontal sides S_0 and S_1 at distance one. Every piecewise linear function $h_0 : S_0 \longrightarrow \mathbf{R}$ that is strictly convex on $S_0 \cap \mathbf{Z}^2$ can be extended to a lifting function for $\mathcal{T}$.*

Proof Let $p_0, \ldots, p_r$ be the integral points on S_0, and let $\{e_{i,1}, \ldots, e_{i,k(i)}\}$ be the edges of $\mathcal{T}$ connecting p_i to S_1. Let $S_0^+(i)$ and $S_0^-(i)$ be the closures of the two components of $S_0 \setminus \{p_i\}$. Then we may decompose the function h_0 as

$$h_0(x) = \sum_{i=0}^{r} \sum_{j=1}^{k(i)} h_{i,j}(x) ,$$

where each $h_{i,j}$ is a convex function on S_0, linear both on $S_0^+(i)$ and on $S_0^-(i)$, and having its unique break-point at p_i.

Now we extend each $h_{i,j}$ to a convex, piecewise-linear function defined on the entire trapezoid such that it has a break-line at the edge $e_{i,j}$ and is linear above and below this line. Then the sum of all $h_{i,j}$ is a lifting function for $\mathcal{T}$. □

Proposition 3.4 *For $n \geq 1$ the following relations hold.*

(i) $$f^{\text{reg}}(1,n) = f(1,n) = \binom{2n}{n}$$

(ii) $$f^{\text{reg}}(2,n) = f(2,n)$$

Proof Part (i) follows immediately from Lemma 3.3(i) (and equation (2.1)).

Since patching two regular triangulations along a *single* edge preserves regularity, it suffices for the proof of part (ii) to show that every unimodular triangulation of shapes of one of the forms

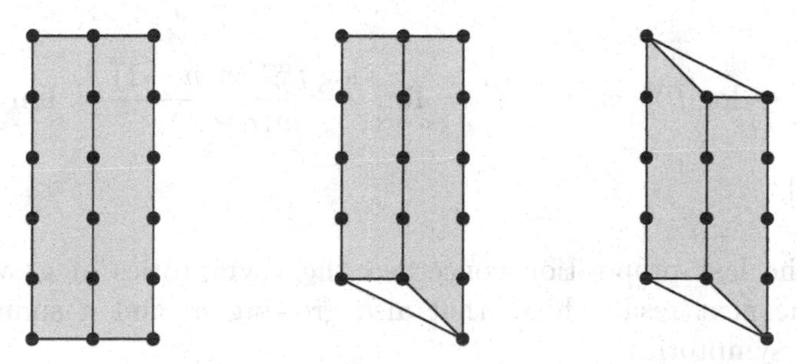

is regular. But this can be derived from Lemma 3.3. One starts from an arbitrary prescribed strictly convex function on the middle column, and after the extensions to the two shaded vertical trapezoids of width one obtained from Lemma 3.3, one adds a piecewise linear function that is constant on the left strip, linear on the right strip, and sufficiently large on the right column of vertices. □

Similarly, one proves the following strengthening of Lemma 3.2 for $n_2 = 1$, as announced earlier.

Lemma 3.5 *For $m, n \geq 1$, we have*

$$f^{\text{reg}}(m, n+1) \geq f^{\text{reg}}(m, n) \cdot \binom{2n}{n} .$$

3.2 Limit capacities

In the following, we will show that the capacities $c(m,n)$ and $c^{\text{reg}}(m,n)$ (with $f(m,n) = 2^{c(m,n)mn}$ and $f^{\text{reg}}(m,n) = 2^{c^{\text{reg}}(m,n)mn}$) asymptotically behave well, which allows us to focus on their limits subsequently. Note that all capacities are bounded (see Theorem 3.9).

Proposition 3.6 *Let $m \geq 1$.*

(i) *The limit* $c_m := \lim_{n\to\infty} c(m,n)$ *exists.*

(ii) *The limit* $c_m^{\text{reg}} := \lim_{n\to\infty} c^{\text{reg}}(m,n)$ *exists.*

Proof Lemmas 3.1(i) and 3.2 imply by Fekete's lemma [23, Lemma 11.6] that

$$\lim_{n\to\infty} f(m,n)^{\frac{1}{n}} \qquad \text{and} \qquad \lim_{n\to\infty} f^{\text{reg}}(m,n-1)^{\frac{1}{n}}$$

exist. Therefore,

$$\lim_{n\to\infty} \frac{1}{m} \log f(m,n)^{\frac{1}{n}} = \lim_{n\to\infty} \frac{\log f(m,n)}{mn} = \lim_{n\to\infty} c(m,n)$$

and

$$\lim_{n\to\infty} \frac{n}{m(n-1)} \log f^{\text{reg}}(m,n-1)^{\frac{1}{n}} = \lim_{n\to\infty} \frac{\log f^{\text{reg}}(m,n-1)}{m(n-1)} = \lim_{n\to\infty} c^{\text{reg}}(m,n)$$

exist as well. □

While the last proposition concerned the asymptotics of growing n for fixed m, the next result shows that also growing m and n simultaneously yields nice asymptotics.

Proposition 3.7 *Let $m \geq 1$.*

(i) *The limit* $c := \lim_{m\to\infty} c(m,m)$ *exists. It satisfies*

$$c = \lim_{m\to\infty} c_m \qquad \textit{and} \qquad c_{m_0} \leq c \quad (m_0 \in \mathbf{N}) .$$

(ii) *The limit* $c^{\mathrm{reg}} := \lim_{m\to\infty} c^{\mathrm{reg}}(m,m)$ *exists. It satisfies*

$$c^{\mathrm{reg}} = \lim_{m\to\infty} c_m^{\mathrm{reg}} \qquad \textit{and} \qquad c_{m_0}^{\mathrm{reg}} \le c^{\mathrm{reg}} \quad (m_0 \in \mathbf{N}) .$$

Proof From Lemma 3.1(i) one derives (for $m_0, n_0 \ge 1$) the inequality

$$f(m,n) \ge f(m_0,n_0)^{\lfloor \frac{m}{m_0} \rfloor \lfloor \frac{n}{n_0} \rfloor}. \tag{3.1}$$

For integers $p, q \ge 1$ we define $\Phi(p,q) := 1 - \frac{p \bmod q}{p}$. We have $\Phi(q,q) = 1$ and $\lim_{p\to\infty} \Phi(p,q) = 1$ for all $q \in \mathbf{N}$.

Equation(3.1) then implies

$$c(m,n) \ge \Phi(m,m_0)\Phi(n,n_0)c(m_0,n_0) , \tag{3.2}$$

and, in particular,

$$c(m_0,n) \ge \Phi(n,n_0)c(m_0,n_0) . \tag{3.3}$$

Inequality (3.3) (together with $\lim_{n\to\infty} \Phi(n,n_0) = 1$) yields

$$c_m \ge c(m,n_0) . \tag{3.4}$$

Inequality (3.2) implies (together with $\lim_{m\to\infty} \Phi(m,m_0)\Phi(m,n_0) = 1$)

$$\liminf_{m\to\infty} c(m,m) \ge c(m_0,n_0) ,$$

and therefore,

$$\liminf_{m\to\infty} c(m,m) \ge c_{m_0} . \tag{3.5}$$

Finally, we obtain the following chain of inequalities, which, together with (3.5) proves part (i) of the proposition. The middle inequality is from (3.5), and the outer ones are due to (3.4).

$$\liminf_{m\to\infty} c_m \ge \liminf_{m\to\infty} c(m,m) \ge \limsup_{m\to\infty} c_m \ge \limsup_{m\to\infty} c(m,m)$$

Part (ii) is proved similarly, starting from Lemma 3.2. □

Note that a similar proof yields

$$c = \lim_{m\to\infty} c(\alpha m, \beta n) \qquad \text{and} \qquad c^{\mathrm{reg}} = \lim_{m\to\infty} c^{\mathrm{reg}}(\alpha m, \beta n)$$

for each pair $\alpha, \beta > 0$.

By Lemma 3.1(ii), the corresponding "irregular limit capacities" do exist as well, and they are equal to c_m ($m \ge 3$) and c, respectively. Therefore, we do not treat them explicitly.

3.3 Lower bounds

Proposition 3.8 *The following estimates hold.*

(i) $c_1 = c_1^{\text{reg}} = 2$

(ii) $c_2 = c_2^{\text{reg}} > 2.044$

(iii) $c_3 > 2.051,\ c_4 > 2.055$

(iv) $c_m > 2.048$ *for* $m \geq 5$

(v) $c_m \geq c_m^{\text{reg}} > 2$ *for* $m \geq 3$

Proof Part (i) follows from

$$f(1,n) = f^{\text{reg}}(1,n) = \binom{2n}{n} \approx \frac{2^{2n}}{\sqrt{2n}} .$$

Parts (ii) and (iii) are results of the computer calculations reported in Section 2, combined with Proposition 3.4(ii).

Lemma 3.1(i) applied to the "transposed grids" implies

$$c(m_1+m_2,n) \geq \frac{m_1}{m_1+m_2}c(m_1,n) + \frac{m_2}{m_1+m_2}c(m_2,n) , \tag{3.6}$$

and thus

$$c_{m_1+m_2} \geq \frac{m_1}{m_1+m_2}c_{m_1} + \frac{m_2}{m_1+m_2}c_{m_2} .$$

With (ii) and (iii), this yields (iv).

Similarly, Lemma 3.5 leads to

$$c_{m+1}^{\text{reg}} \geq \frac{m}{m+1}c_m^{\text{reg}} + \frac{1}{m+1} \cdot 2 .$$

With (ii) this proves part (v). □

Equation (3.6) implies that $c_{kn} \geq c_n$ for $k \geq 2$; for example, this implies that $c_4 \geq c_2$ — but it is not obvious that $c_3 \geq c_2$. Even stronger, one would assume that

$$2 = c_1 < c_2 < c_3 < \ \cdots \ \leq c ,$$

and

$$2 = c_1^{\text{reg}} < c_2^{\text{reg}} < c_3^{\text{reg}} < \ \cdots \ \leq c^{\text{reg}} ,$$

but neither monotonicity is proved.

3.4 Upper bounds

From general principles (see Ajtai et al. [4]) one gets that the capacity for any configuration of N points in the plane is finite. In the general position case (no three points on a line) the currently best upper bound is $o(2^{59N})$, due to Santos & Seidel [28]. However, in the very "degenerate" case of lattice triangulations, there are far fewer triangulations: Orevkov [25] obtained the bound $f(m,n) \leq 4^{3mn} = 2^{6mn}$. Very recently, this has been substantially improved by Anclin [5], as follows.

Theorem 3.9 (Anclin [5]) *For all $m, n \geq 1$,*

$$f(m,n) \ < \ 2^{3mn-m-n}.$$

Proof Our sketch relies on the essential ideas of Anclin's proof.

The first, crucial observation is that the midpoints of the edges in any unimodular triangulation are exactly the half-integral, not integral points in $\mathrm{conv}(P_{m,n})$. (Clearly the midpoint of every edge is half-integral; the converse may be derived from Pick's theorem, or from the fact that all unimodular triangles are equivalent to $\mathrm{conv}\{(0,0),(1,0),(0,1)\}$.) The number of these half-integral points in the interior of $P_{m,n}$ is $e = 3mn - m - n$.

Now any triangulation is built as follows: The half-integral points are processed in an order that is given by a parallel *sweep*. (See de Berg et al. [9, Sect. 2.1] for a discussion and many further sweeping applications of this fundamental technique.)

Whenever a point is processed, we add to the partial triangulation a new edge with the given midpoint. The key *claim* is that at each such step, when a half-integral point v is processed, there are (at least one and) at most two possibilities for the new edge with midpoint v to be added. If this claim is true, then the number of triangulations is bounded by 2^e.

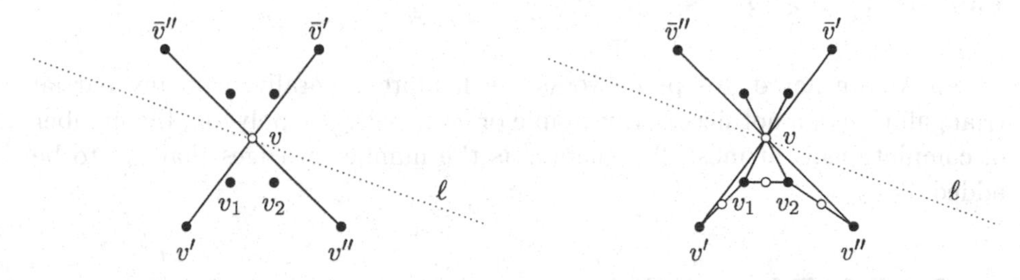

To prove the claim, one can verify the following: Let $[v', \bar{v}']$ and $[v'', \bar{v}'']$ be two potential edges with midpoint v that could be added, where v' and v'' are the endpoints below the sweep-line ℓ; let Q be the convex hull of the integral points in the triangle $[v, v', v'']$; then all the $k \geq 2$ vertices $v' = v_0, v_1, \ldots, v_{k-1}, v_k = v''$ of Q are "visible" from v, and any one of the edges $[v_i, \bar{v}_i]$ with midpoint v could potentially be added at this step. Furthermore, the midpoints $\frac{1}{2}(v_{i-1} + v_i)$ lie below the sweep-line, and thus the edges $[v_{i-1}v_i]$ are present in the current partial triangulation.

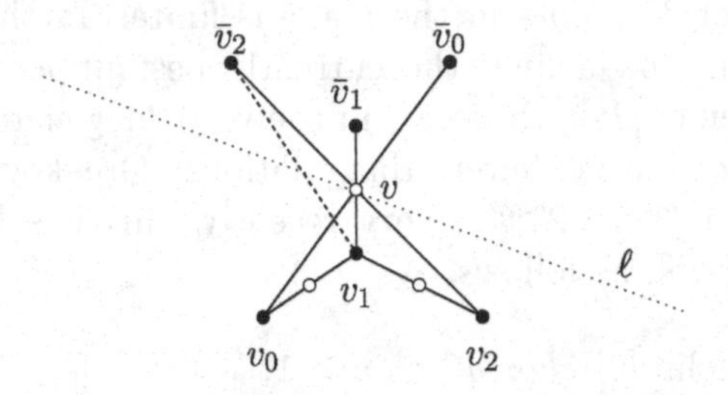

Now assume that $[v_0, \bar{v}_0]$, $[v_1, \bar{v}_1]$, $[v_2, \bar{v}_2]$ are three *adjacent* edges with midpoint v that could be added when processing v. By central symmetry with respect to v we may assume that the midpoint w of $[v_1, \bar{v}_2]$ lies below the sweep-line. But the triangle $[v_1, v, \bar{v}_2]$ is then an empty triangle of area $\frac{1}{4}$, just like the triangle $[v_1, v, v_2]$. From this we conclude that in the current partial triangulation, the edge $[v_1, \bar{v}_2]$ must be present — which creates a crossing with the potential edge $[v_0, \bar{v}_0]$, and thus a contradiction. □

The following upper bounds on the limit capacities follow immediately from Theorem 3.9.

Corollary 3.10 *For all $m, n \geq 1$, the following inequalities hold:*

(i) $c^{\text{reg}}(m,n) \leq c(m,n) \leq 3 - \frac{1}{m} - \frac{1}{n}$

(ii) $c_m^{\text{reg}} \leq c_m \leq 3 - \frac{1}{m}$ (in particular, $c_2^{\text{reg}} \leq c_2 \leq 2.5$)

(iii) $c^{\text{reg}} \leq c \leq 3$

As Anclin noted, his proof works much more generally: For any partial triangulation of a not necessarily simple or convex lattice polygon, the number of completions is at most $2^{e'}$, where e' is the number of edges that are to be added.

4 Explicit Triangulations

For small grids, one can enumerate all unimodular triangulations by the reverse search algorithm sketched in Section 2. For larger grids, it is desirable to obtain from the huge set of unimodular triangulations "random" ones. There are ways to produce them, however, in most cases the probability distribution from which they are chosen is unknown.

4.1 Generating Random Triangulations

A standard way to compute complex random objects such as triangulations is to set up a random walk. In our case of unimodular triangulations of $P_{m,n}$ the method is described easily: First one determines any starting triangulation $\mathcal{T}$ of $P_{m,n}$. Then, the following operation is performed τ times: Choose one of the (inner) edges of $\mathcal{T}$ uniformly at random; if this edge is flippable (see Section 2), then with probability $\frac{1}{2}$ the current triangulation $\mathcal{T}$ is replaced by the one obtained from it by flipping that edge.

As the flip graph of the triangulations is connected (see Section 2), it follows from general principles that, with τ tending to infinity, the probability distribution defined by the output of this algorithm converges to the uniform distribution (with respect to the "total variation distance"); see, e.g., Jerrum and Sinclair [20] or Behrends [8]. Unfortunately, not much is known about the speed of convergence of the distribution. In particular, for general m and n it is not known whether there is a polynomial bound (in $n + m + \varepsilon^{-1}$) on the number τ of steps needed to guarantee that the total variation distance between the produced and the uniform distribution is at most ε (i.e., if the associated Markov chain is *rapidly mixing*). The only exception is the case $m = 1$: Here it follows from results of Felsner and Wernisch [14] that the Markov chain is indeed rapidly mixing.

Despite this lack of knowledge on the distribution of the output of the random walk algorithm, one still can use it in order to produce examples of "interesting" triangulations.

4.2 Empirical Results

For each $n \in \{10, \ldots, 20\}$, Meyer [24] generated 1000 random unimodular triangulations of $P_{n,n}$ by running the random walk described in Subsection 4.1 10^9 steps, recording every 10^6th triangulation; see Table 2.

The results support the conjecture that, for n tending to infinity, a random unimodular triangulation of $P_{n,n}$ is irregular with probablity one. A second observation is that the (expected value) of the average length of an edge in a random triangulation seems to grow very slowly with n.

4.3 Obstructions to Regularity

All figures shown below have been produced by Meyer [24], who implemented procedures for checking regularity by solving linear programs using the CPLEX 6.6.1 library.

Proposition 3.4 shows that $P_{2,n}$ has only regular triangulations. The grid $P_{3,3}$ has precisely the following four (pairwise congruent) irregular unimodular triangulations.

$m \times n$	irregularity	max. edge length	av. edge length
10×10	.355	5.538	1.614
11×11	.435	5.843	1.630
12×12	.559	6.118	1.645
13×13	.696	6.397	1.659
14×14	.782	6.650	1.670
15×15	.875	6.911	1.681
16×16	.927	7.151	1.690
17×17	.965	7.391	1.700
18×18	.971	7.618	1.708
19×19	.992	7.821	1.713
20×20	.997	8.060	1.723

Table 2: Results for random unimodular triangulations of large grids. The first column shows the (empirical) probability of irregularity, the second and third columns contain the (empirical) expected values of the maximal and the average edge length.

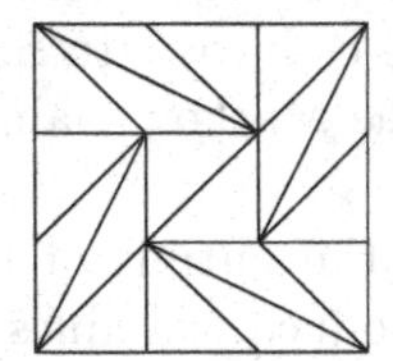 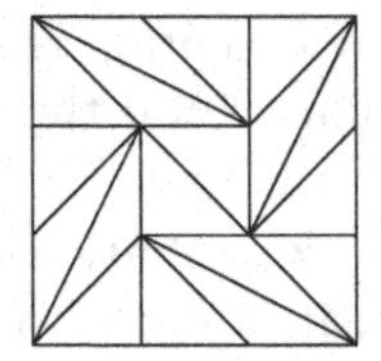 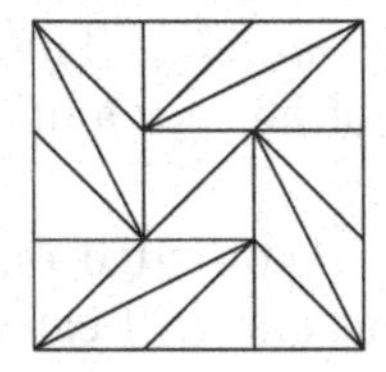 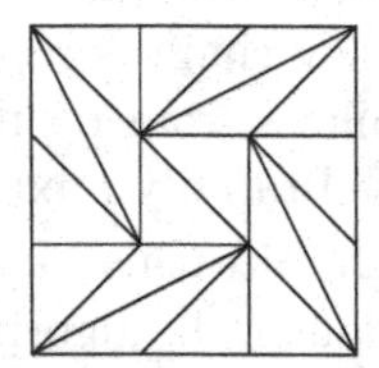

When trying to understand the reasons for irregularity, it seems useful to consider (smallest) forbidden patterns for regular triangulations. Let $\mathcal{T}$ be a set of unimodular triangles of $P_{m,n}$ (such that any two of them intersect in a common face). We denote by $S \subset P_{m,n}$ the set of all grid points covered by triangles in $\mathcal{T}$. The set $\mathcal{T}$ is called *regular* if there is a height function $h : S \longrightarrow \mathbf{R}$ such that for each triangle $\Delta \in \mathcal{T}$, all h-lifted points in $S \backslash \Delta$ lie strictly above the affine hull of the h-lifting of Δ. A subset of $\mathcal{T}$ is called a *minimal irregular configuration* if it is not regular, but all its proper subsets are regular. Clearly, a regular unimodular triangulation of $P_{m,n}$ cannot contain any minimal irregular configuration.

Examples for minimal irregular configurations of $P_{3,3}$ are the following:

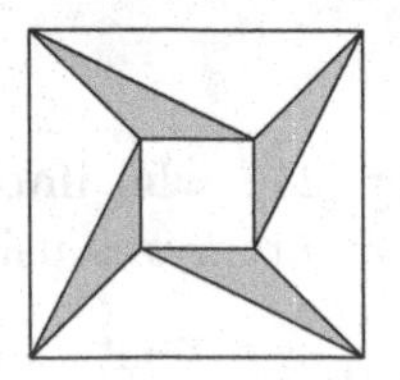 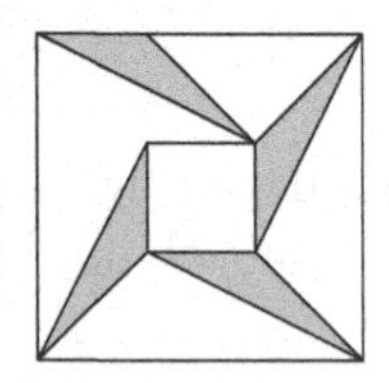 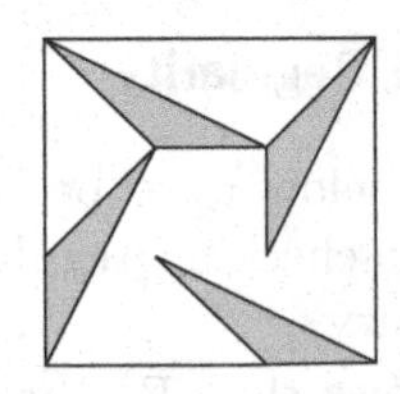 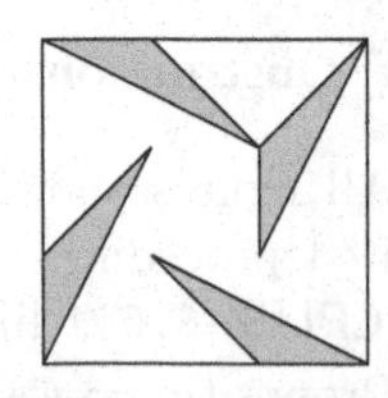

Already for $P_{3,4}$ many other minimal irregular configurations occur; some are depicted here:

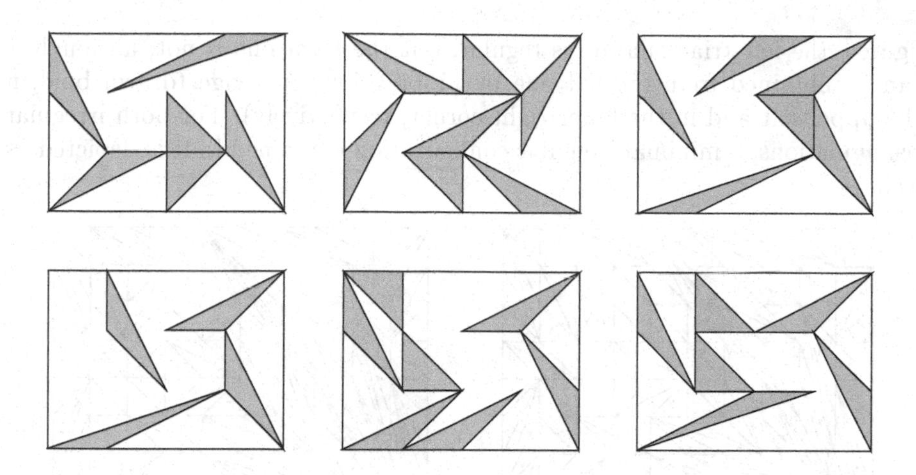

While these figures still have some similarities with the nice "whirlpools" ones for $P_{3,3}$, the picture gets more and more complicated with growing grid sizes, as the following examples demonstrate:

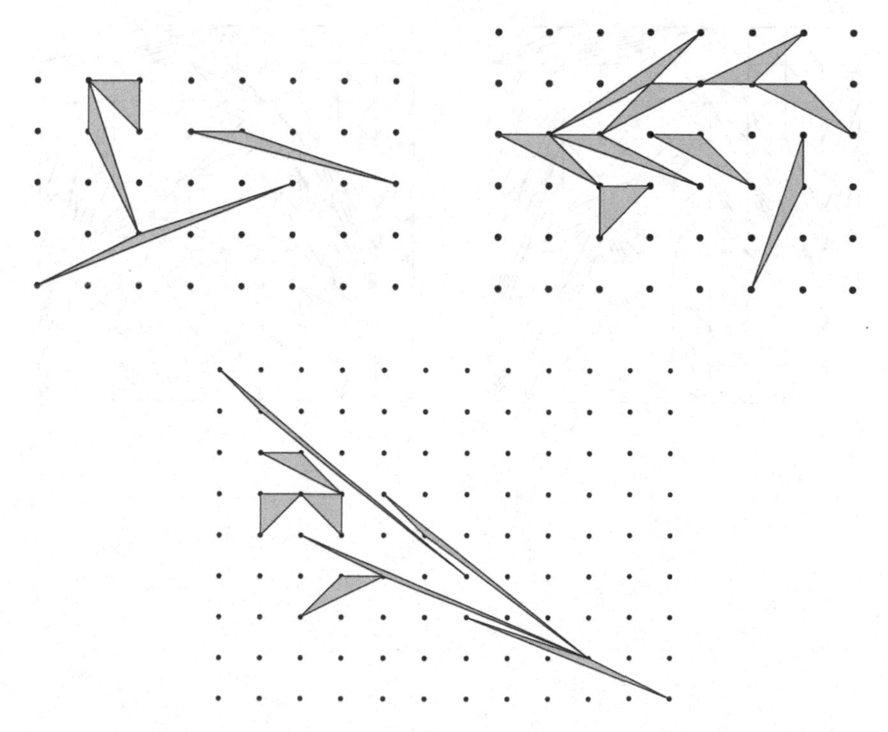

Viewing these figures, it seems unlikely that one can find any compact characterization of regularity for unimodular triangulations of $P_{m,n}$ in terms of forbidden substructures.

We close our zoo of "explicit triangulations" with some pairs of triangulations, found by Meyer's implementation of the random walk. In each of the figures, the left triangulation is regular, but the right one is not, although it can be obtained from the left one by flipping just one edge (drawn bold in the upper left and in the lower right corner, respectively). For both irregular triangulations, a minimal irregular configuration contained in it is depicted as well.

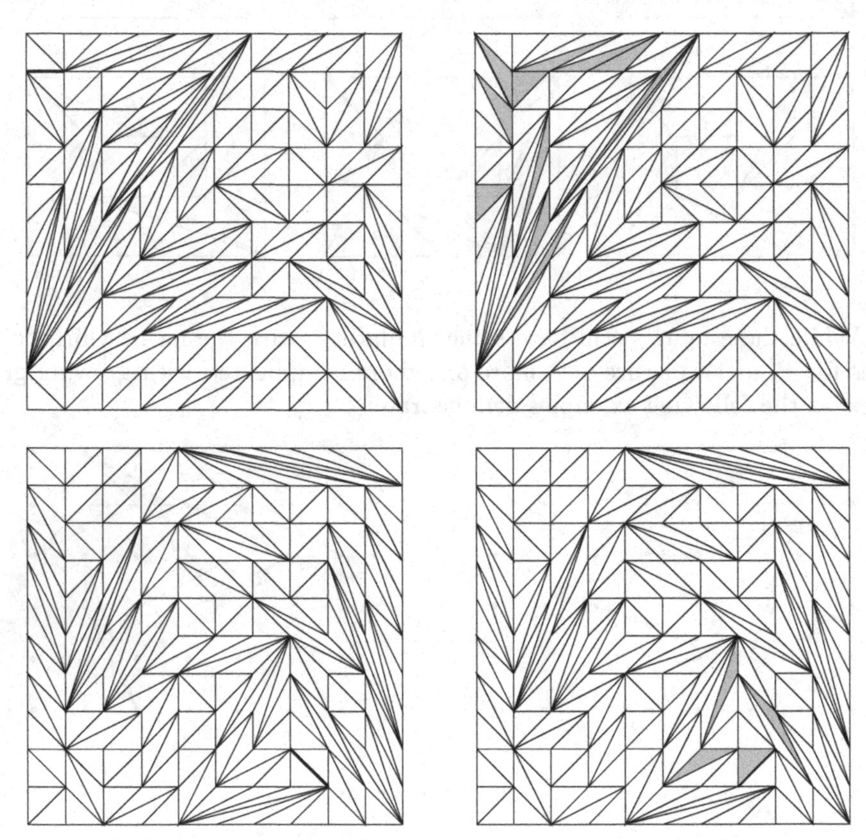

Acknowledgements.

We are grateful to Oswin Aichholzer, Émile Anclin, Stephan Meyer, and Jörg Rambau for valuable discussions. Thanks to Francisco Santos and to the referee for helpful comments on an earlier version of the manuscript.

References

[1] O. Aichholzer, The path of a triangulation, in *Proceedings of the 15th Annual ACM Workshop Computational Geometry (Miami Beach, Florida, June 13-15, 1999)* ACM, New York (1999), pp. 14–23.

[2] O. Aichholzer, Counting triangulations – Olympics, WWW page, created Oct. 5, 1999, last modified Nov. 20, 2001 `http://www.cis.TUGraz.at/igi/oaich/triangulations/counting/counting.html`.

[3] M. Aigner & G. M. Ziegler, *Proofs from THE BOOK*, Springer-Verlag, Heidelberg (1998; second edition, 2001).

[4] M. Ajtai, V. Chvátal, M. Newborn & E. Szemerédi, Crossing-free subgraphs, *Annals of Discrete Mathematics*, 12, (1982), pp. 9–12.

[5] E. Anclin, An upper bound for the number of planar lattice triangulations, preprint, TU Berlin, 2002. ArXiv: `math.CO/0212140`

[6] D. Avis & K. Fukuda, Reverse search for enumeration, *Discrete Applied Math.* **65** (1996), 21–46.

[7] S. Bespamyatnikh, An efficient algorithm for enumeration of triangulations, *Comput. Geom.* **23** (2002), 271–279.

[8] E. Behrends, *Introduction to Markov Chains, with Special Emphasis on Rapid Mixing*, *Advanced Lectures in Mathematics*, Friedr. Vieweg & Sohn, Braunschweig/Wiesbaden (2000).

[9] M. de Berg, M. van Kreveld, M. Overmars & O. Schwarzkopf, *Computational Geometry. Algorithms and Applications*, Springer-Verlag, Berlin (second edition, 2000).

[10] D.I. Dais, Resolving 3-dimensional toric singularities, *Sémin. Congr.* **6** (2002), 155–186.

[11] L. De Floriani, B. Falcidieno, G. Nagy & C. Pienovi, On sorting triangles in a Delaunay tessellation, *Algorithmica* **6** (1991), 522–532.

[12] J. De Loera, J. Rambau & F. Santos, *Triangulations*, book in preparation.

[13] H. Edelsbrunner, An acyclicity theorem for cell complexes in *d* dimensions, *Combinatorica* **10** (1990), 251–260.

[14] S. Felsner & L. Wernisch, Markov chains for linear extensions, the two-dimensional case, in *Proceedings of the Eighth Annual ACM-SIAM Symposium on Discrete Algorithms (New Orleans, LA, 1997)* ACM, New York (1997), pp. 239–247.

[15] E. Gawrilow & M. Joswig, polymake: An Approach to Modular Software Design in Computational Geometry, in *Proceedings of the Seventeenth Annual Symposium on Computational Geometry, 2001* ACM, New York pp. 222–231.
Software package at `http://www.math.tu-berlin.de/polymake/`, version 1.5, September 2002.

[16] I.M. Gel'fand, M.M. Kapranov & A.V. Zelevinsky, *Discriminants, Resultants, and Multidimensional Determinants*, Birkhäuser, Boston (1994).

[17] GNU library for arbitrary precision arithmetic, `http://www.swox.com/gmp/`

[18] J.E. Goodman & J. Pach, Cell decomposition of polytopes by bending, *Israel J. Math.* **64** (1988), 129–138.

[19] R.E. Hastings, Triangulations of point configurations and polytopes, Dissertation, Cornell University, 1998.
`ftp://cam.cornell.edu/pub/theses/hastings.ps.gz`

[20] M.R. Jerrum & A. Sinclair, The Markov Chain Monte Carlo Method: An Approach to Approximate Counting and Integration, in *Approximation Algorithms for NP-hard Problems* (ed. D. Hochbaum), PWS Publishing, (1996), pp. 482–520.
`http://www.dcs.ed.ac.uk/home/mrj/hochbaum.ps`

[21] G. Kempf, F. Knudsen, D. Mumford & B. Saint-Donat, *Toroidal Embeddings I*, *Lecture Notes in Math.*, 339, Springer-Verlag, Berlin (1973).

[22] C.W. Lee, Subdivisions and triangulations of polytopes, in *Handbook on "Discrete and Computational Geometry"* (eds. J.E. Goodman & J. O'Rourke), CRC Press, Boca Raton FL (1997), pp. 271–290.

[23] J.H. van Lint & R.M. Wilson, *A Course in Combinatorics*, Cambridge University Press, Cambridge (1992, second edition 2001).

[24] S. Meyer, Enumeration von Triangulierungen, Diplomarbeit, TU Berlin, 2002.

[25] S.Yu. Orevkov, Asymptotic number of triangulations with vertices in $\mathbb{Z}^2$, *J. Combin. Theory, Ser. A* **86** (1999), 200-203.

[26] J. Pfeifle & J. Rambau, Counting triangulations using oriented matroids, in *Algebra, Geometry, and Software Systems* (eds. M. Joswig & N. Takayama), Springer-Verlag, Berlin (2003). pp. 49-76.

[27] J. Rambau, TOPCOM — Triangulations Of point configurations and oriented matroids, in *Proceedings of the First International Congress of Mathematical Software ICMS 2002 (Beijing, China, August 17-19, 2002)* World Scientific, Singapore (2002), pp. 330-340.
Software package at `http://www.zib.de/rambau/TOPCOM/`, version 0.11.1, July 2002.

[28] F. Santos & R. Seidel, A better upper bound on the number of triangulations of a planar point set, *J. Combin. Theory, Ser. A*, in press. ArXiv: `math.CO/0204045`

[29] R.P. Stanley, *Enumerative Combinatorics I*, *Cambridge Studies in Advanced Mathematics*, 49, Cambridge University Press, Cambridge (second edition 1997).

[30] B. Sturmfels, *Gröbner Bases and Convex Polytopes*, *University Lecture Series*, 8, Amer. Math. Soc., Providence, RI (1995).

[31] O.Ya. Viro, Real plane algebraic curves: Constructions with controlled topology, *Leningrad Math. J.* **1** (1990), 1059–1134.

[32] G.M. Ziegler, *Lectures on Polytopes*, *Graduate Texts in Mathematics*, 152, Springer-Verlag, New York (1995; revised edition 1998).

Appendix

Below we report on some values $f(m, n)$ and $c(m, n)$ that we have computed by the algorithms for narrow strips described in Subsection 2.1. Note that Aichholzer's results referred to in the captions of the tables have been obtained by a code that works for *general* point sets in the plane.

n	# unimodular triangulations of $P_{2,n}$	capacity
2	64	1.500000
3	852	1.622451
4	12170	1.696380
5	182132	1.747462
6	2801708	1.784822
7	43936824	1.813494
8	698607816	1.836244
9	11224598424	1.854774
10	181815529916	1.870184
11	2964167665340	1.883216
12	48580814410080	1.894393
13	799696199314500	1.904094
14	13212398835196240	1.912597
15	218976668040908248	1.920118
16	3639020246503687098	1.926820
17	60616163842958990268	1.932833
18	1011775545312594580868	1.938260
19	16918718677672553292440	1.943185
20	283368129709983000763876	1.947675
21	4752924784523774226889308	1.951787
22	79824154012907603962950312	1.955568
23	1342199498257069824064033644	1.959057
24	22592402326314503187343665228	1.962288
25	380653341141186360494812030908	1.965287
⋮	⋮	⋮
375	32237946979502385047429500335320672196108255805142308488024462249323984026708677398252396626798849612418104980258822126509337232116231190005654057238750599232186906105955093046025120869939338489702534179125985366638163193491909033373598455910903465733725460855531008615951205123532313200039339081267563585858626886602111651589423454725534616624385434519062538407256710064644710448005559085336858456557439009515491637998852932253379794927887762995003200075959780	2.044130

Table 3: Results for $m = 2$ (up to $n = 15$ by Aichholzer [2]).)

n	# unimodular triangulations of $P_{3,n}$	capacity
3	46456	1.722619
4	2822648	1.785718
5	182881520	1.829755
6	12244184472	1.861743
7	839660660268	1.886238
8	58591381296256	1.905656
9	4140106747178292	1.921429
10	295372308876234428	1.934510
11	21234538315776214604	1.945546
12	1535939689343151109944	1.954989
13	111655493479477379881272	1.963164
14	8150727077307189203809876	1.970314
15	597087996550303632801161860	1.976623
16	43871350204895836758556369212	1.982234
17	3231797978935266793268797809260	1.987258
18	238606105193380387756570932194588	1.991783
19	17651135152017098450035730535703808	1.995882
20	1308029292984065630362694842042395056	1.999613
21	97080539975603502667567153853690549804	2.003024
22	7215158047881650609075575773153609553148	2.006154
23	536905685776901371485436849505792415847140	2.009039
24	39997858254082097021224132017959794867440460	2.011705
25	2982752306685557862989393328648927558138800612	2.014178
26	222638546950211814977693932477091620801626551100	2.016478
27	16632293481947394846909242053460530217349594447732	2.018623
28	1243490745851056260557782821562507156418923407432920	2.020627
29	93034737749193459157244717739574844241159902101217660	2.022506
30	6965244882542454937020619818702059053741377290255068284	2.024269
31	521789556367416753405244328934612259884552790211207421244	2.025929
32	39111402471791798530057405675011481922023421912454706904712	2.027493
⋮	⋮	⋮
60	140299226506605674595297010256056870425972772255987818065754 82481045782421605312026256653301411517672578559472224	2.051236

Table 4: Results for $m = 3$ (up to $n = 10$ by Aichholzer [2].)

n	# unimodular triangulations of $P_{4,n}$	capacity
4	736983568	1.841066
5	208902766788	1.880202
6	61756221742966	1.908818
7	18792896208387012	1.930751
8	5831528022482629710	1.948080
9	1835933384812941453312	1.962138
10	584455230176565718869688	1.973785
11	187686028049755013528577884	1.983601
12	60685901262618326775192700244	1.991986
13	19731268926382148037209063600412	1.999235
14	6444884828545542240332780129017164	2.005567
15	2113222804656668311309302902100087020	2.011147
16	695163898467233943317499868644974218294	2.016103
17	229316915701559537858641762255000442720116	2.020535
18	75827610389461537709077484409103543784585710	2.024522
19	25126215170054967918515172517611569017605311400	2.028130
20	8341120564526486621411516194118239406742548614820	2.031409
21	2773492141234111587868660829757660194821668552146720	2.034405
22	923542836393484391634763396019705303948974557455679944	2.037151
23	307927840785137540445620416718491375007439099052396877524	2.039680
24	102788976776576952654837601573736966649525369087792041774754	2.042015
25	34348007173983906887977536976143285540616055151406837464386952	2.044178
26	11488735157368185493550332714289865763761071970295733672739011182	2.046188
27	3846114798702837748429637799631405899720860204464245538618538080224	2.048061
28	1288605098545745425026895981169173452565572617561190100930952999826626	2.049811
29	432053057804721606144370507673814019049397228524243323674624721189513152	2.051449
30	144960362040664009850310097114550102266012390064271163400409174321129109970	2.052986
31	48666988587816100417429461522083834520238266167053837281579839713435133670880	2.054431
32	16348321592766160525928861562545873132577437301251385744803248045014265044217096	2.055792

Table 5: Results for $m = 4$ (up to $n = 8$ by Aichholzer [2].)

n	# unimodular triangulations of $P_{5,n}$	capacity
1	252	1.595455
2	182132	1.747462
3	182881520	1.829755
4	208902766788	1.880202
5	260420548144996	1.915513
6	341816489625522032	1.941533
7	464476385680935656240	1.961547
8	645855159466371391947660	1.977388
9	913036902513499041820702784	1.990240
10	1306520849733616781789190513820	2.000871
11	1887591165891651253904039432371172	2.009821
12	2747848427721241461905176361078147168	2.017461

Table 6: Results for $m = 5$ (up to $n = 6$ by Aichholzer [2].)

n	# unimodular triangulations of $P_{6,n}$	capacity
1	924	1.641958
2	2801708	1.784822
3	12244184472	1.861743
4	61756221742966	1.908818
5	341816489625522032	1.941533
6	1999206934751133055518	1.965553
7	12169409954141988707186052	1.984082

Table 7: Results for $m = 6$.

Institut für Mathematik, MA 6–2
Technische Universität Berlin
D–10623 Berlin
Germany
kaibel@math.tu-berlin.de
ziegler@math.tu-berlin.de

Partition regular equations

Imre Leader

Abstract

A finite or infinite matrix A with rational entries is called partition regular if whenever the natural numbers are finitely coloured there is a monochromatic vector x with $Ax = 0$. Many of the classical theorems of Ramsey theory, such as van der Waerden's theorem or Schur's theorem, may naturally be interpreted as assertions that particular matrices are partition regular.

While in the finite case partition regularity is well understood, very little is known in the infinite case. In this survey paper we will review some finite results and then proceed to discuss some features of the infinite case.

1 Introduction

Let A be an $m \times n$ matrix with rational entries. We say that A is *partition regular* if for every finite colouring of the natural numbers $\mathbb{N} = \{1, 2, \ldots\}$ there is a monochromatic vector $x \in \mathbb{N}^n$ with $Ax = 0$. In other words, A is partition regular if for every positive integer k, and every function $c : \mathbb{N} \to \{1, \ldots, k\}$, there is a vector $x = (x_1, \ldots, x_n) \in \mathbb{N}^n$ with $c(x_1) = \ldots = c(x_n)$ such that $Ax = 0$. We may also speak of the 'system of equations $Ax = 0$' being partition regular.

Many of the classical results of Ramsey theory may naturally be considered as statements about partition regularity. For example, Schur's theorem [11], that in any finite colouring of the natural numbers we may solve $x+y = z$ in one colour class, is precisely the assertion that the 1×3 matrix $(1, 1, -1)$ is partition regular. As another example, the theorem of van der Waerden [13] that, for any m, every finite colouring of the natural numbers contains a monochromatic arithmetic progression with m terms, is (with the strengthening that we may also choose the common difference of the sequence to have the same colour) exactly the statement that the $(m-1) \times (m+1)$ matrix

$$\begin{pmatrix} 1 & 1 & -1 & 0 & \ldots & 0 & 0 \\ 1 & 0 & 1 & -1 & \ldots & 0 & 0 \\ 1 & 0 & 0 & 1 & \ldots & 0 & 0 \\ \vdots & \vdots & \vdots & \vdots & \ddots & \vdots & \vdots \\ 1 & 0 & 0 & 0 & \ldots & 1 & -1 \end{pmatrix}$$

is partition regular.

Note that not all matrices are partition regular. For example, the 1×2 matrix $(2, -1)$ is not partition regular. Indeed, if it were then in any finite colouring of the natural numbers there would exist an x such that x and $2x$ had the same colour, and this is certainly not the case (for example, colour x by the parity of the largest n such that 2^n divides x).

The partition regular matrices were characterised by Rado [10]. To state Rado's result, we need a small amount of notation. Let A be an $m \times n$ rational matrix, with columns $a^{(1)}, \ldots, a^{(n)}$. We say that A has the *columns property* if there is a partition of $[n] = \{1, \ldots, n\}$, say $[n] = D_0 \cup D_1 \cup \ldots \cup D_k$, such that

$$\sum_{i \in D_0} a^{(i)} = 0$$

and for every $r = 1, \ldots, k$ we have

$$\sum_{i \in D_r} a^{(i)} \in \langle a^{(i)} \ : \ i \in D_0 \cup \ldots \cup D_{r-1} \rangle,$$

where $\langle \cdot \rangle$ denotes rational linear span. For example, a $1 \times n$ matrix has the columns property if and only if some of its (non-zero) entries sum to zero, so that the above matrix $(1, 1, -1)$ certainly has the columns property. It is also easy to see that the $(m-1) \times (m+1)$ van der Waerden matrix above has the columns property.

Rado [10] proved that a matrix is partition regular if and only if it has the columns property. This reduces partition regularity to a property that is very tangible and, moreover, can be checked in finite time for any particular matrix. One of the remarkable features of Rado's theorem is that neither direction is obvious. For general background about Rado's theorem, see the excellent survey of Deuber [2]. For background on all aspects of Ramsey theory, see Graham, Rothschild and Spencer [4]. However, in this survey we will be concentrating on some different features of partition regularity, so that no background knowledge of this sort is needed. Indeed, our survey will be complementary to the Deuber survey, so that there is very little overlap.

This seminal work of Rado in the 1930s initiated the study of partition regularity. Another viewpoint, and another characterisation, were introduced by Deuber in the 1970s, and used by him to clarify and answer some questions left open by Rado. Roughly speaking, Deuber proved that the only partition regular systems consist of iterated versions of arithmetic progressions. More precisely, for positive integers m, p, c, and positive integers $u_1, \ldots, u_m$, the (m, p, c)*-set* generated by $u_1, \ldots, u_m$ consists of all sums of the form

$$\sum_{i=k}^{m} \lambda_i u_i,$$

where $k \in \{1, 2, \ldots, m\}$, $\lambda_k = c$, and $\lambda_i \in \{1, 2, \ldots, p\}$ for all $i > k$. Thus for example a $(2, p, 1)$-set is just an arithmetic progression of p terms, together with its common difference. What Deuber showed in [1] is that a matrix A is partition regular if and only if there exist m, p, c such that every (m, p, c)-set contains a solution of $Ax = 0$. See [2] or [4] for a full discussion of this, but again we stress that no background knowledge is needed here.

Let us mention one key closure property of the finite partion regular matrices, which will in fact be one of the main themes of this paper. This is the most important way of making 'new systems from old', and is known as 'consistency': if A and B are partition regular then so is the diagonal sum of A and B. Putting it another way, if we can guarantee to solve $Ax = 0$ in a colour class and we can guarantee to solve $By = 0$ in a colour class then we can guarantee to solve them both in the *same* colour class. Note that this is trivial by the columns property. However, because of its importance, let us give a direct proof. This proof is based on what is called, for obvious reasons, a 'product argument'.

Theorem 1.1 *Let A and B be finite partition regular matrices. Then the diagonal sum* $\begin{pmatrix} A & 0 \\ 0 & B \end{pmatrix}$ *is also partition regular.*

Proof Let c be a k-colouring of $\mathbb{N}$. Fix n large enough that, in any k-colouring of $\{1, \ldots, n\}$, there is a monochromatic solution of $Ax = 0$. (If there is no such n then for each n there is a bad k-colouring of $\{1, \ldots, n\}$, and we may put these together, by picking infinitely many that agree at 1, and infinitely many of those that agree at 2, and so on, to form a bad colouring of the whole of $\mathbb{N}$.)

Now consider, for each natural number a, the colours of the first n multiples of a, in other words the sequence $c(a), c(2a), \ldots, c(na)$. This determines a k^n-colouring of the natural numbers (by colouring a with this sequencc), and so we may find a vector y, with the entries of y monochromatic for this k^n-colouring, such that $By = 0$.

Write $y = (y_1, \ldots, y_m)$. Considering the sequence $c(y_1), c(2y_1), \ldots, c(ny_1)$, we see by the definition of n that there is an $x = (x_1, \ldots, x_l)$ with $Ax = 0$ such that $c(x_1 y_1) = c(x_2 y_1) = \ldots = c(x_l y_1)$. But then the vectors $y_1 x$ and $x_1 y$ satisfy $A(y_1 x) = 0$ and $B(x_1 y) = 0$ and have all entries the same colour, as required. □

This turns out to have many consequences (see [2]). For example, it follows immediately that, whenever the natural numbers are finitely coloured, some colour class contains solutions to *all* finite partition regular systems of equations. Although this sounds an extremely strong statement, it is very easy to deduce from the above consistency theorem. Indeed, suppose that $\mathbb{N} = D_1 \cup \ldots \cup D_k$, such that for each $1 \leqslant i \leqslant k$ there is a partition regular matrix A_i with D_i not containing a solution to $A_i x = 0$. Let A be the diagonal sum of the A_i. Then A is partition regular, by Theorem 1.1, but clearly no D_i contains a solution of $Ax = 0$, a contradiction.

Let us now turn from finite to infinite systems of equations. Are there any such systems at all? If A is an infinite matrix, with rational entries and only finitely many non-zero entries in each row, we say as before that A is *partition regular* if whenever the natural numbers are finitely coloured there is a monochromatic vector x with $Ax = 0$.

First of all, there are some rather trivial examples, that come directly from Ramsey's theorem itself. For example, let the natural numbers be finitely coloured, by a colouring c say, and let us induce a colouring of the pairs by giving the pair $\{x, y\}$ the colour $c(x + y)$. Ramsey's theorem states precisely that there is an infinite set all of whose pairs have the same colour, and in terms of the original colouring c this states that there is an infinite sequence $x_1, x_2, \ldots$ (with the x_i all distinct) such that the set $\{x_i + x_j : i < j\}$ is monochromatic.

Another example of the same nature may be obtained by colouring the pair $\{x, y\}$ (for $x < y$) with the colour $c(2x + y)$: we then obtain an infinite sequence $x_1, x_2, \ldots$ such that all $x_i + 2x_j$ $(i < j)$ have the same colour.

One can of course obtain similar examples using other coefficients, or using Ramsey's theorem for triples instead of pairs, and so on. But it is worth remarking that one cannot *vary* the coefficients. For example, could we guarantee a sequence $x_1, x_2, \ldots$ with all $x_i + 2x_j$ $(i \neq j)$ having the same colour? In such an (injective) sequence, it is clear that, as n tends to infinity, the ratio of $2x_1 + x_n$ to $x_1 + 2x_n$ tends to 2, and this may be defeated by a suitable 3-colouring, rather like the 2-colouring earlier that defeated $x = 2y$. We leave this as a simple exercise for the reader who wishes to get practice in colourings.

Let us pause for a moment to point out that the careful reader may have noticed that, as they stand, these examples from Ramsey's theorem are technically not given as the kernels of linear systems. Indeed, not only are the x_i required to be distinct, but also the x_i are allowed to have any colours themselves. However it is easy to rewrite the equations in the 'official' partition regular form (and if it were not, we would surely modify the definition of 'partition regular' to include such manifestly linear structures as these). For example, we may first rewrite $x_1, x_2, \ldots$ as $y_1, y_1 + y_2, y_1 + y_2 + y_3, \ldots$. Then, among the given expressions that are to be the same colour, choose a maximal linearly independent set and use these as the new variables. Now that we know that this can be done, we may completely ignore the precise manner in which this rewriting occurred, and we urge the reader to give it no further thought.

The first example of a (non-trivial) infinite partition regular system of equations was constructed in 1974 by Hindman [5], proving a conjecture of Graham and Rothschild: in any finite colouring of the natural numbers there is a sequence $x_1, x_2, \ldots$ of natural numbers such that the set

$$FS(x_1, x_2, \ldots) = \left\{ \sum_{i \in I} x_i \ : \ I \subseteq \mathbb{N},\ I \text{ finite and non-empty} \right\}$$

is monochromatic. This is also known as the Finite Sums theorem. (It is worth remarking that the finite analogue of this, known as Folkman's theorem, stating that, for any m, in any finite colouring of the natural numbers we may find

$x_1, \ldots, x_m$ with $FS(x_1, \ldots, x_m)$ monochromatic, follows easily from Rado's theorem).

A few other infinite systems are known. The first such result after Hindman's theorem was the Milliken-Taylor theorem [9, 12]. For a k-tuple $a = (a_1, \ldots, a_k)$ of natural numbers, and a sequence $x_1, x_2, \ldots$ of natural numbers, we write $FS_a(x_1, x_2, \ldots)$ for the set of all sums of the form

$$\sum_{i \in I_1} a_1 x_i + \sum_{i \in I_2} a_2 x_i + \cdots + \sum_{i \in I_k} a_k x_i \ ,$$

where $I_1, \ldots, I_k$ are non-empty finite subsets of $\mathbb{N}$ such that $\max I_t < \min I_{t+1}$ for all t. Then the Milliken-Taylor theorem asserts that, for any a, whenever the natural numbers are finitely coloured there exists a sequence $x_1, x_2, \ldots$ such that $FS_a(x_1, x_2, \ldots)$ is monochromatic.

Let us briefly remark that, roughly speaking, the Milliken-Taylor theorem may be proved by copying down the proof of Ramsey's theorem, replacing the appeals to the pigeonhole principle with appeals to Hindman's theorem. We also remark that one cannot allow any of the I_t to be empty, or allow the I_t to interleave in any way: this may be shown by some simple bad colourings, just as mentioned above.

While some other infinite partition regular systems are now known, the general problem of deciding which systems are partition regular looks hopelessly out of reach at present.

Our aim in this survey is to give an introduction to some of the ideas and methods that are used in attacking infinite systems. A large part of the modern approach to partition regularity is based on ultrafilters, or more precisely the structure of the space $\beta\mathbb{N}$, the Stone-Čech compactification of the natural numbers. For example, Hindman's theorem follows immediately from the existence of an idempotent for addition on $\beta\mathbb{N}$. We assume that the reader has no experience of such things. So in Section 2 we shall introduce ultrafilters and the Stone-Čech compactification $\beta\mathbb{N}$, the space of all ultrafilters on $\mathbb{N}$. This is a very big and very strange space, but we will only need to understand a little bit about it to be able to read off Hindman's theorem. It is extremely surprising that ultrafilters are at all relevant.

In Section 3, we look at consistency for infinite systems. It was proved in [3] that consistency now fails: different Milliken-Taylor systems are not consistent. We will explain here how this is established.

Finally, in Section 4, we turn to a question that arises naturally from the failure of consistency for Milliken-Taylor systems, namely the question of consistency for the Ramsey-type examples we started with. Interestingly, it turns out that we *do* have consistency here. The original proof [6] involved some intricate work in $\beta\mathbb{N}$, or to be more precise a careful examination of the structure of some ideals in $\beta\mathbb{N}$. In that paper, it was asked if there was a direct proof (not using ultrafilters) of this result. Very recently, such a proof has been found by Leader and Russell [8], and we present this short proof here.

2 Ultrafilters and Hindman's theorem

We start with some definitions that seem very far away from Hindman's theorem. Intuitively, a filter will be a rough way of selecting some subsets of $\mathbb{N}$ as 'large', while an ultrafilter will be a more precise way of doing so.

A *filter* on $\mathbb{N}$ is a collection $\mathcal{F} \subseteq \mathcal{P}(\mathbb{N})$ such that

(i) $\emptyset \notin \mathcal{F}$, $\mathbb{N} \in \mathcal{F}$;

(ii) if $A \in \mathcal{F}$ and $B \supset A$ then $B \in \mathcal{F}$;

(iii) if $A, B \in \mathcal{F}$ then $A \cap B \in \mathcal{F}$.

For example, $\{A \subseteq \mathbb{N} : 1, 2 \in A\}$ is a filter. So is $\{A \subseteq \mathbb{N} : A^c \text{ finite }\}$, the *cofinite filter*. (Here and elsewhere we write A^c for the complement of A). We may 'forget about the odd numbers', by taking $\{A \subseteq \mathbb{N} : E \setminus A \text{ finite }\}$, where E denotes the set of even numbers.

An *ultrafilter* is a maximal filter. For example, none of the above is an ultrafilter. Indeed, the second is contained in the third, and in a similar way the third is not maximal, while the first is contained in the set $\{A \subseteq \mathbb{N} : 1 \in A\}$. This set is clearly an ultrafilter, and for any $n \in \mathbb{N}$ we write $\tilde{n}$ for $\{A \subseteq \mathbb{N} : n \in A\}$, the *principal ultrafilter* at n.

Are there any non-principal ultrafilters? The reader may spend a few minutes seeing that it seems rather hard to specify which sets should be in the ultrafilter and which should not: the even numbers or the odd numbers, and so on. Indeed, it is easy to check that a filter $\mathcal{F}$ is an ultrafilter if and only if for every $A \subseteq \mathbb{N}$ we have either $A \in \mathcal{F}$ or $A^c \in \mathcal{F}$.

However, by Zorn's Lemma it is clear that every filter is contained in a maximal filter, and so there do exist non-principal ultrafilters: any ultrafilter extending the cofinite filter is non-principal. Some (weak) form of the Axiom of Choice *is* needed here, so non-principal ultrafilters are somewhat elusive.

If $\mathcal{U}$ is an ultrafilter and $A \in \mathcal{U}$ and $A = B \cup C$ then $B \in \mathcal{U}$ or $C \in \mathcal{U}$, since otherwise we would have $B^c, C^c \in \mathcal{U}$, whence $A^c \in \mathcal{U}$. It follows from this that an ultrafilter is non-principal if and only if it extends the cofinite filter.

We now step back from considering a single ultrafilter and instead focus on the set of *all* ultrafilters on $\mathbb{N}$. This set is denoted $\beta\mathbb{N}$. There is a natural topology on $\beta\mathbb{N}$, defined by taking as basic open sets all sets C_A, for $A \subseteq \mathbb{N}$, where

$$C_A = \{\mathcal{U} \in \beta\mathbb{N} : A \in \mathcal{U}\}.$$

This really is a base for a topology, because $C_A \cap C_B = C_{A \cap B}$ (since $A, B \in \mathcal{U}$ if and only if $A \cap B \in \mathcal{U}$).

Note that $\beta\mathbb{N} \setminus C_A = C_{A^c}$, so that the C_A also form a basis for the *closed* sets in $\beta\mathbb{N}$.

We may view $\mathbb{N}$ as living inside $\beta\mathbb{N}$, by identifying $n \in \mathbb{N}$ with the principal ultrafilter $\tilde{n}$. Each point of $\mathbb{N}$ is isolated in $\beta\mathbb{N}$, since $C_{\{n\}} = \{\tilde{n}\}$. Also, $\mathbb{N}$

is dense in $\beta\mathbb{N}$, because every non-empty C_A (in other words, every C_A with $A \neq \emptyset$) contains all the $\widetilde{n}$ with $n \in A$.

The most important property of $\beta\mathbb{N}$ is that it is compact. This may be deduced from Tychonov's theorem (that a product of compact spaces is compact), since, viewing an ultrafilter as a function from $\mathcal{P}(\mathbb{N})$ to $\{0,1\}$, the topology on $\beta\mathbb{N}$ is the restriction of the product topology (and $\beta\mathbb{N}$ is closed in this product). However, for completeness let us give a direct proof.

Theorem 2.1 *The space $\beta\mathbb{N}$ is a compact Hausdorff space.*

Proof To see that $\beta\mathbb{N}$ is Hausdorff, choose distinct ultrafilters $\mathcal{U}, \mathcal{V}$. Then there is some $A \in \mathcal{U}$ with $A \notin \mathcal{V}$. But then $A^c \in \mathcal{V}$, so that $\mathcal{U} \in C_A$ and $\mathcal{V} \in C_{A^c}$.

To show that $\beta\mathbb{N}$ is compact, let the $F_i, i \in I$ be a family of closed sets with the finite intersection property (all finite intersections are non-empty): our task is to find a point belonging to all the F_i. We may assume without loss of generality that each F_i is basic: say $F_i = C_{A_i}$.

Now the sets A_i must themselves have the finite intersection property because

$$C_{A_{i_1}} \cap \ldots \cap C_{A_{i_n}} = C_{A_{i_1} \cap \ldots \cap A_{i_n}}.$$

We may thus define a filter $\mathcal{F}$ generated by the A_i:

$$\mathcal{F} = \{A \subseteq \mathbb{N} \,:\, A \supseteq A_{i_1} \cap \ldots \cap A_{i_n}, \text{ some } i_1, \ldots, i_n \in I\}.$$

Let $\mathcal{U}$ be an ultrafilter extending $\mathcal{F}$. Then $A_i \in \mathcal{U}$ for all i, so $\mathcal{U} \in C_{A_i}$ for all i, as required. □

This topological space $\beta\mathbb{N}$ is known as the Stone-Čech compactification of $\mathbb{N}$. We mention in passing that $\beta\mathbb{N}$ is the 'largest' compact Hausdofff space in which $\mathbb{N}$ is dense. More precisely, if K is any compact Hausdorff space and $f : \mathbb{N} \to K$ is any function then there is a unique continuous function $\widetilde{f} : \beta\mathbb{N} \to K$ that extends f.

Our aim is going to be to define and use an addition operation on $\beta\mathbb{N}$, and for that it is very useful to introduce 'ultrafilter quantifiers'. It is possible to define addition without these quantifiers, but their use renders intuitive and simple some otherwise complicated concepts.

The idea is that we think the sets in an ultrafilter $\mathcal{U}$ as being 'almost all' of $\mathbb{N}$, or equivalently we view the sets not in $\mathbb{N}$ as being 'of measure zero'. More precisely, let $p(x)$ be some property of a natural number x. We write $(\forall_{\mathcal{U}} x) p(x)$ to mean that $\{x \in \mathbb{N} : p(x)\} \in \mathcal{U}$. We read the symbol $(\forall_{\mathcal{U}} x)$ as 'for most x' or 'for $\mathcal{U}$-most x'. For example, for any non-principal $\mathcal{U}$ we have $(\forall_{\mathcal{U}} x) x > 10$. As another example, if $\mathcal{U} = \widetilde{n}$ then we have $(\forall_{\mathcal{U}} x) p(x)$ if and only if $p(n)$.

The remarkable thing about ultrafilter quantifiers is that they behave in many ways *better* than the usual logical quantifiers. In the following proposition, part (i) is just as for usual quantifiers but parts (ii) and (iii) are rather striking.

Proposition 2.2 *Let $\mathcal{U}$ be an ultrafilter, and $p(x)$ and $q(x)$ statements. Then*

(i) $(\forall_{\mathcal{U}} x)(p(x)$ *and* $q(x))$ *if and only if* $((\forall_{\mathcal{U}} x)p(x)$ *and* $(\forall_{\mathcal{U}} x)q(x))$;

(ii) $(\forall_{\mathcal{U}} x)(p(x)$ *or* $q(x))$ *if and only if* $((\forall_{\mathcal{U}} x)p(x)$ *or* $(\forall_{\mathcal{U}} x)q(x))$;

(iii) *if not* $(\forall_{\mathcal{U}} x)p(x)$ *then* $(\forall_{\mathcal{U}} x)($*not* $p(x))$.

Proof Let $A = \{x : p(x)\}$ and $B = \{x : q(x)\}$. Then part (i) simply states that $A \cap B \in \mathcal{U}$ if and only if $A \in \mathcal{U}$ and $B \in \mathcal{U}$, while part (ii) asserts that $A \cup B \in \mathcal{U}$ if and only if $A \in \mathcal{U}$ or $B \in \mathcal{U}$. Finally, for part (iii), note that if $(\forall_{\mathcal{U}} x)p(x)$ does not hold then $A \notin \mathcal{U}$, whence $A^c \in \mathcal{U}$. □

Let us remark that, despite these good properties, there is one cautionary note: $(\forall_{\mathcal{U}} x)$ and $(\forall_{\mathcal{V}} y)$ do not in general commute – even if $\mathcal{U} = \mathcal{V}$. For example, let $\mathcal{U}$ be any non-principal ultrafilter. Then we certainly have $(\forall_{\mathcal{U}} x)(\forall_{\mathcal{U}} y) x < y$, while it is not the case that $(\forall_{\mathcal{U}} y)(\forall_{\mathcal{U}} x) x < y$.

Armed with ultrafilter quantifiers, we are now ready to define an addition on $\beta\mathbb{N}$. For $\mathcal{U}, \mathcal{V} \in \beta\mathbb{N}$, define

$$\mathcal{U} + \mathcal{V} = \{A \subseteq \mathbb{N} : (\forall_{\mathcal{U}} x)(\forall_{\mathcal{V}} y)\ x + y \in A\}.$$

Using Proposition 2.2, it is easy to check that $\mathcal{U} + \mathcal{V}$ is indeed an ultrafilter. For example, if $A \notin \mathcal{U} + \mathcal{V}$ then $\{x : (\forall_{\mathcal{V}} y) x + y \in A\} \notin \mathcal{U}$, whence $(\forall_{\mathcal{U}} x)($not $(\forall_{\mathcal{V}} y) x + y \in A)$, whence in turn $(\forall_{\mathcal{U}} x)(\forall_{\mathcal{V}} y) x + y \notin A$, so that $A^c \in \mathcal{U} + \mathcal{V}$. (It is an amusing exercise to write down the definition of $\mathcal{U} + \mathcal{V}$ without the quantifiers: the verification of even the simplest facts becomes very convoluted!)

It is also clear from the definition that the operation $+$ on $\beta\mathbb{N}$ is associative:

$$\mathcal{U} + (\mathcal{V} + \mathcal{W}) = (\mathcal{U} + \mathcal{V}) + \mathcal{W} = \{A \subseteq \mathbb{N} : (\forall_{\mathcal{U}} x)(\forall_{\mathcal{V}} y)(\forall_{\mathcal{W}} z)\ x + y + z \in A\}.$$

Another important property of $+$ is that it is left-continuous – in other words, for each fixed $\mathcal{V}$ the mapping $\mathcal{U} \mapsto \mathcal{U} + \mathcal{V}$ is continuous. Indeed, if we fix $\mathcal{V}$ and a basic open set C_A, then $\mathcal{U} + \mathcal{V} \in C_A$ if and only if $A \in \mathcal{U} + \mathcal{V}$, which is the same as $\{x : (\forall_{\mathcal{V}} y) x + y \in A\} \in \mathcal{U}$, which in turn is the same as $\mathcal{U} \in C_B$, where $B = \{x : (\forall_{\mathcal{V}} y) x + y \in A\}$.

We mention in passing that $+$ is neither commutative nor right-continuous: these make pleasant exercises.

The key to Hindman's theorem will turn out to be finding an ultrafilter $\mathcal{U}$ that is *idempotent*, meaning that $\mathcal{U} + \mathcal{U} = \mathcal{U}$. Indeed, once we have established the existence of an idempotent then Hindman's theorem will follow almost immediately. The following folklore result, known as the Idempotent Lemma, asserts that an idempotent ultrafilter does indeed exist. Very conveniently, it is true in just enough generality that it does apply to $\beta\mathbb{N}$.

Lemma 2.3 *Let X be a non-empty compact Hausdorff space, and let $+ : X \times X \to X$ be an associative and left-continuous binary operation on X. Then there exists $x \in X$ with $x + x = x$.*

Proof As we are seeking a single element x with $x + x = x$, it clearly makes sense to look for a 'small' subset M with $M + M \subseteq M$. (Here we write $A + B$ for $\{a + b : a \in A, b \in B\}$.) So let us say that a subset M of X is *good* if it is compact, non-empty and satisfies $M + M \subseteq M$. By Zorn's Lemma, there is a minimal good set M. We will now show that every $x \in M$ satisfies $x + x = x$.

So fix $x \in M$. The set $M + x$ is compact, being the image of M under a continuous function, and also

$$(M + x) + (M + x) = (M + x + M) + x \subseteq M + x.$$

So $M + x$ is good, whence by the minimality of M we must have $M + x = M$. It follows that there exists $y \in M$ with $y + x = x$. So let $N = \{y \in M : y + x = x\}$. Note that N is closed, being the inverse image of a singleton under a continuous function, and so is compact. Also, if $y, z \in N$ then

$$(y + z) + x = y + (z + x) = y + x = x,$$

so that $y + z \in N$. Thus N is good, and again it follows by the minimality of M that $N = M$. In particular, we have $x \in N$, as required. □

After all this preparation, we are at last ready for Hindman's theorem.

Theorem 2.4 *Whenever $\mathbb{N}$ is finitely coloured, there exist $x_1, x_2, \ldots$ with $FS(x_1, x_2, \ldots)$ monochromatic.*

Proof Fix an idempotent ultrafilter $\mathcal{U}$. Given a finite colouring on $\mathbb{N}$, choose a colour class A belonging to $\mathcal{U}$. Intuitively, we are thinking of A as the 'large' colour class.

We know that $(\forall_{\mathcal{U}} y) y \in A$, and also $(\forall_{\mathcal{U}} x)(\forall_{\mathcal{U}} y) x + y \in A$, because $\mathcal{U}$ is idempotent. It follows (from Proposition 2.2) that $(\forall_{\mathcal{U}} x)(\forall_{\mathcal{U}} y) FS(x, y) \subseteq A$, so we may choose x_1 with $(\forall_{\mathcal{U}} y) FS(x_1, y) \subseteq A$.

Now we just repeat. Formally, suppose we have chosen $x_1, \ldots, x_n$ with $(\forall_{\mathcal{U}} y)\ FS(x_1, \ldots, x_n, y) \subseteq A$. Then for each $z \in FS(x_1, \ldots, x_n)$ we have $(\forall_{\mathcal{U}} y) z + y \in A$, so that $(\forall_{\mathcal{U}} x)(\forall_{\mathcal{U}} y) z + x + y \in A$. Thus, by Proposition 2.2 again, we have $(\forall_{\mathcal{U}} x)(\forall_{\mathcal{U}} y) FS(x_1, \ldots, x_n, x, y) \subseteq A$, and now we may set x_{n+1} to be such an x. □

To end this section, let us just remark that the space $\beta\mathbb{N}$ is a very strange space: much remains unknown about its structure. In a sense, it is fortunate that one is able to prove just the things about it (above) that are needed for Hindman's theorem.

The original proof of Hindman's theorem in [5] was much longer. The beautiful ultrafilter viewpoint is due to Galvin, Glazer and Hindman. For much more information about $\beta\mathbb{N}$ see [7].

3 Inconsistency for Milliken-Taylor systems

In this section we will look at the first inconsistency result for partition regularity, from [3]. This paper shows that, apart from trivial special cases, Milliken-Taylor systems are *always* inconsistent, in the sense that the diagonal sum of two Milliken-Taylor systems is not partition regular. The construction of the colourings is by no means an easy task, and our aim here is to show what these colourings are like. We shall adopt a rather informal description of the colourings, often being content with clear descriptions rather than formulas, because the formulas soon become unreadable. However, the reader who wishes may easily 'formalise' our descriptions (or see the paper [3]).

We will refer to the particular k-tuple $(a_1, \ldots, a_k)$ in the Milliken-Taylor theorem as a *pattern*. There is clearly no point in ever considering patterns with a value repeated in consecutive places, so we shall restrict our attention to *compressed* patterns, meaning ones for which $a_i \neq a_{i+1}$ for all $1 \leqslant i \leqslant k-1$. Given two compressed patterns a and b, when could we hope that they are consistent? In other words, when could we hope that in any finite colouring of $\mathbb{N}$ there are sequences $x_1, x_2, \ldots$ and $y_1, y_2, \ldots$ such that the set

$$FS_a(x_1, x_2, \ldots) \cup FS_b(y_1, y_2, \ldots)$$

is monochromatic?

If a and b are rational multiples of one another, then this is certainly the case. But it turns out that, with this one exception, the Milliken-Taylor systems are *never* consistent. For simplicity, here we confine ourselves to the case when all entries of a and b are powers of 2. The general case has no essential new ideas, but just a large amount of extra notation.

We start with some general remarks. We will gradually build up some colourings, adding more and more colours in such a way that more and more is demanded of the x_i in a monochrome Milliken-Taylor system. Eventually, we will have enough restrictions that we will be able to show that there is no one colour class able to contain both a set $FS_a(x_1, x_2, \ldots)$ and a set $FS_b(y_1, y_2, \ldots)$. While building up our colourings, we will often bear in mind one 'example' Milliken-Taylor system with pattern $a = (a_1, \ldots, a_n) = (1, 2, 4)$.

When a number is written in binary, its *start* will be the position of its most significant 1 (the units digit is position 0, etc.), its *end* will be the position of its least significant 1, and its *gap sequence* will be the sequence of gaps between consecutive 1s. For example, the number 11001000100 has start 10, end 2, and gap sequence $0, 2, 3$.

Pick a number k, much larger than a_1, a_2, a_3. Suppose we have a sequence $x_1, x_2, \dots$ such that all of $FS_a(x_1, x_2, \dots)$ is monochromatic in some given colouring. By taking linear combinations, we may as well assume that each x_i starts far to the left of x_{i-1} – say the end of x_i is at least k greater than the start of x_{i-1}. From now on, any reference to an x_i will be restricted to $i > n$, so that each x_i could occur in a sum with coefficient $1, 2$ or 4.

Let us colour by start (mod k) and end (mod k) (thus we have k^2 colours so far). Then all the $4x_i$ start in the same place (mod k) (as $i > n$), so all the x_i start in the same place (mod k). Similarly, all the x_i end in the same place (mod k). Thus the gap between an x_i and an x_{i-1} is fixed (mod k): say it is g (so, (mod k), g is the difference between the end of x_i and the start of x_{i-1}, with 1 subtracted).

Now let us colour by the number, (mod k), of gaps of length congruent to j (mod k), for each $0 \leqslant j < k$. So we have k^k new colours (and hence k^{k+2} colours in total). Given a number z in $FS_a(x_1, x_2, \dots)$, say, ending with $4x_{i-1}$ (in other words the largest x in the sum forming z is x_{i-1}, with of course coefficient 4), we are free to add $4x_i$ or not, as we please – both z and $z + 4x_i$ must have the same colour. Now, adding $4x_i$ to z puts in some new gaps: one of length g (mod k), and also all the gaps inside x_i. We conclude that the distribution of gap-lengths (mod k) inside each x_i is the same, namely: the number of gaps in the gap sequence of x_i which are congruent to j (mod k) is -1 if $j \equiv g$ (mod k), and 0 otherwise (of course, '-1' and '0' are meant (mod k)).

We remark that at this stage we can already distinguish various patterns. For example, what is the gap-distribution for our pattern $(1, 2, 4)$? Any z in a monochromatic $FS_a(x_1, x_2, \dots)$ must have gap-distribution as follows: -3 gaps of length g mod k, 2 gaps of length $g + 1$ (mod k), and 0 gaps of length each other j (mod k). So, having coloured $\mathbb{N}$ like this, and found our x_i, we can read off the value of g – it is that value (mod k) for which the number of gaps congruent to it is -3. Thus to distinguish $(1, 2, 4)$ from $(2, 4, 16)$, say, we pick our large k (larger than the numbers in both patterns, of course), colour $\mathbb{N}$ as above, and find our supposed x_i for $(1, 2, 4)$ and y_i for $(2, 4, 16)$. Then the distribution vector for the colour class of the Milliken-Taylor system formed by the x_i is of the form $(0, 0, \dots, 0, -3, 2, 0, \dots, 0)$ (the jth entry in this vector denotes the number of gaps of length j (mod k)). However, the distribution vector for the other system is of the form $(0, 0, \dots, 0, -3, 1, 1, 0, \dots, 0)$. No vector is of both these forms, so we are done.

Of course, we cannot yet distinguish $(1, 2, 8)$ from $(1, 4, 8)$, and we certainly cannot distinguish $(1, 2, 1, 4, 1)$ from $(1, 4, 1, 2, 1)$. So let us now proceed to more colours. Our next construction will accomplish both of these. We remark that if we added colours to see where (mod k) each gap started (how many gaps of length j start in position h, and so on), we would indeed be able to distinguish $(1, 2, 8)$ from $(1, 4, 8)$. However, we would still not be able to distinguish $(1, 2, 1, 4, 1)$ from $(1, 4, 1, 2, 1)$, so we will not bother to add these colours: instead, the next argument manages to complete the inconsistency proof by itself.

Introduce new colours as follows (keeping all the old colours as well, of course). For each ordered pair (h, j) of numbers (mod k), count how many times in the gap sequence of x (the number we are colouring) we have an h followed later by a k. This gives k^{k^2} new colours. Note that we do not insist that the j must be immediately after the h, and each h and j can be counted lots of times. For example, if x has gap sequence $(2, 1, 3, 2, 7, 1, 7)$ then the ordered pair $(2, 1)$ occurs 3 times, while the ordered pair $(3, 7)$ occurs twice, and the pair $(2, 2)$ occurs once.

What happens when we consider a z in the Milliken-Taylor system, say ending with $4x_{i-1}$? We know that z and $z + 4x_i$ must have the same colour. But what have we done to the ordered-pair-counts when we put on this $4x_i$? Thinking of $y + 4x_i$ as being made up of $4x_i$, then a 'dividing gap' (of length g, of course) between the $4x_i$ and the z, and then the z, we have three places new gap-pairs could come in: (A) pairs of gaps both in the $4x_i$, (B) pairs of gaps of which the first is in the $4x_i$ and the second is the dividing gap, and (C) pairs of gaps of which the first is either in $4x_i$ or the dividing gap and the second is in the z. Now (C) contributes nothing at all to any pair (h, j), for the simple reason that the gaps of x_i, together with one g, have distribution vector $(0, \ldots, 0)$ (as we saw a few paragraphs ago). And (B) is easy: it adds -1 copies of (g, g) and 0 copies of everything else (again, this is by the distribution vector of x_i). We conclude that the pair-counts inside x_i must be: 1 for (g, g), 0 for each other (h, j).

So we know that each x_i has gap count of: -1 gaps of length g, 0 of each other length, and we also know that each x_i has gap pair count of: 1 ordered pair of gaps of lengths (g, g), 0 of each other pair of lengths. Thus, for example, a z in a Milliken-Taylor system with pattern $1, 2$ will have gap-pairs counts as follows (for some g): 3 of type (g, g), -1 of type $(g, g+1)$, -1 of type $(g+1, g)$, 0 of each other type. Note also that in a pattern with $n = 3$, for example $(1, 8, 2)$, the only gap-pair count which is not 0 and is not of a type (a, g) or (g, b) (any a and b) is the type $(g-2, g+3)$, which gets a count of 1.

Now we look at triples (ordered triples of gaps). Just as above, when we add in colours for these counts, we will force the triples-distribution of an x_i to be: -1 of type (g, g, g) and 0 of each other type. Also, in a pattern with $n = 4$, like $(1, 2, 8, 1)$, the only triple which does not contain a g and does not get count 0 is the triple $(-3, 2, 1)$, which gets a count of 1. Then on to quadruples, where each x_i will have 1 of type (g, g, g, g) and 0 of each other type; and so on.

Thus, in total, here is how to distinguish any two patterns (of powers of 2), say the patterns $(2^{a_1}, \ldots, 2^{a_m})$ and $(2^{b_1}, \ldots, 2^{b_n})$, where $m \leqslant n$. Pick a k much larger than all the 2^{a_i} and 2^{b_i} and n. Colour $\mathbb{N}$ by start, end, gap distribution, gap-pair distribution, and so on up to $(n-1)$-tuple distribution (all (mod k)). Suppose that we have x_i for the first pattern and y_i for the second pattern so that both Milliken-Taylor systems lie in the same colour class. We can assume that each x_i starts far to the left to x_{i-1}, and similarly for the y_i. Then the distribution of gaps of z, a typical element of the first system, has a $-n$ in exactly one place (that of length g, say), and all other coefficients are between 0 and $n-1$. Similarly for w, a typical member of the second system. Thus $m = n$.

So now we also know g. There is a unique $(n-1)$-tuple (in the distribution of z) which does not contain a g and does not get coefficient 0, namely

$$(g + a_n - a_{n-1},\ g + a_{n-1} - a_{n-2},\ \ldots,\ g + a_2 - a_1)\,.$$

Hence, as this must be the same for w, we have $a_i - a_{i-1} = b_i - b_{i-1}$ for every i. Thus one pattern is a multiple of the other, as required.

What makes the above complicated appears to be that it is hard to find colourings that 'mesh' with addition in a good enough way for us to be able to draw consequences. There are many other plausible colouring methods, but none seems to help.

4 Consistency for Ramsey systems

In this very short final section, we turn to a question left open by [3], namely whether or not the systems mentioned above like $x_i + x_j$ or $y_i + 2y_j$, whose partition regularity follows trivially from Ramsey's theorem, are consistent. This was answered in [6], where it was proved that, in contrast to the situation with Milliken-Taylor systems, these systems are consistent. This is of some interest as, besides being a very natural question, the diagonal sum of these two systems is then a partition regular system that is *not* contained in any Milliken-Taylor system.

The proof in [6] used a large amount of machinery on the structure of $\beta\mathbb{N}$, and in particular on its 'minimal ideal' (see [7] for a general discussion of this and similar notions). There it was asked if there is a simpler proof. Actually, various 'absoluteness' results from logic imply that as the theorem is provable in ZFC it in fact *must* be provable in ZF (this is from the logical form of the statement), but the question was whether there was a short proof in ZF, or at any rate a proof that did not rely on $\beta\mathbb{N}$.

One difficulty is that such a proof must *not* extend to giving a proof of the false result that the corresponding two Milliken-Taylor systems are consistent. This seems to rule out many approaches.

Very recently, Leader and Russell [8] have discovered such a proof. We present it here, to have an example of a positive consistency result for infinite systems. Just for simplicity, we write it out just for the two systems $x_i + x_j$

and $y_i + 2y_j$, although it will be clear that the method applies equally well to any other such pair of systems coming from Ramsey's theorem.

Theorem 4.1 *Whenever* $\mathbb{N}$ *is finitely coloured, there exists a pair of sequences* $x_1 < x_2 < \dots$ *and* $y_1 < y_2 < \dots$ *such that the set*

$$\{x_i + x_j \ : \ i \neq j\} \cup \{y_i + 2y_j \ : \ i < j\}$$

is monochromatic.

Proof Given a finite colouring of $\mathbb{N}$, let us induce a finite colouring of the 4-sets from $\mathbb{N}$ by giving the set $\{i, j, k, l\}$, where $i < j < k < l$, the colour of $i + 2j + k + 2l$. By Ramsey's theorem, there is an infinite monochromatic set for this colouring. In other words, there is a sequence $z_1 < z_2 < \dots$ such that all $z_i + 2z_j + z_k + 2z_l$ $(i < j < k < l)$ lie in the same colour class.

The choice of the x_i is now clear: we may take $x_i = z_{2i} + 2z_{2i+1}$. The choice of the y_i, on the other hand, proceeds as follows. Let us fix some $a < b$ for which z_a and z_b are congruent mod 3. We may then take as the y_i, not the z_i themselves, but rather $y_i = z_i + (z_a + 2z_b)/3$, each $i > b$. Then for $b < i < j$ we have $y_i + 2y_j = z_a + 2z_b + z_i + 2z_j$. □

To end, we would like to mention some unsolved problems. The most important of all, of course, is: which infinite systems of equations are partition regular? All known examples are either mentioned in this paper or else are very close in spirit. In particular, they are all in some way or other built up from Milliken-Taylor systems, or from combinations of them. And in a sense the Milliken-Taylor theorem is not too far from Hindman's theorem itself. Is there some general theorem here? It seems at first to be very unlikely that *all* infinite systems should in some way be related to Hindman's theorem, but the absence of different examples is suspicious!

As a concrete example of this question, we may consider the following. In every known partition regular system, any one variable can occur with coefficients only from a bounded set (although the set of all coefficients, as the variables vary, can be unbounded). Must this always be the case? In other words, is there a partition regular system in which some variable occurs with an unbounded set of coefficients? If such a system were found, it would certainly not come from Milliken-Taylor-type systems. (Needless to say, one also requires that the system cannot be rewritten to avoid this happening – so for example one would not be allowed to construct an example from the Schur system by rewriting the equation $x + y = z$ as the set of equations $x + y = z, 2x + 2y = 2z, 3x + 3y = 3z, \dots$!) It seems easy to think up plausible systems, but they all have a habit of failing to some or other unpleasant colouring.

References

[1] W.A. Deuber, Partitionen und lineare Gleichungssysteme, *Math. Z.* **133** (1973), 109–123.

[2] W.A. Deuber, Developments based on Rado's dissertation 'Studien zur Kombinatorik', in *Surveys in Combinatorics, 1989* (ed. J. Siemons), *London Math. Soc. Lecture Note Ser.*, 141, Cambridge Univ. Press, Cambridge (1989), pp. 52–74.

[3] W.A. Deuber, N. Hindman, I. Leader and H. Lefmann, Infinite partition regular matrices, *Combinatorica* **15** (1995), 333–355.

[4] R.L. Graham, B.L. Rothschild & J.H. Spencer, *Ramsey Theory*, Wiley, New York (1990).

[5] N. Hindman, Finite sums from sequences within cells of a partition of $\mathbb{N}$, *J. Combin. Theory Ser. A A*, **17** (1974), 1–11.

[6] N. Hindman, I. Leader and D. Strauss, Infinite partition regular matrices – solutions in central sets, *Trans. Amer. Math. Soc.*, in press.

[7] N. Hindman & D. Strauss, *Algebra in the Stone-Čech Compactification: Theory and Applications*, de Gruyter, Berlin (1998).

[8] I. Leader and P.A. Russell, Consistency for partition regular equations, preprint, University of Cambridge, 2003.

[9] K.R. Milliken, Ramsey's theorem with sums or unions, *J. Combin. Theory Ser. A* **18** (1975), 276–290.

[10] R. Rado, Studien zur Kombinatorik, *Math. Zeit.* **36** (1933), 424–480.

[11] I. Schur, Über die kongruenz $x^m + y^m = z^m \pmod{p}$, *Jahresbericht der Deutschen Math.-Verein.* **25** (1916), 114–117.

[12] A.D. Taylor, A canonical partition relation for finite subsets of ω, *J. Combin. Theory Ser. A* **21** (1976), 137–146.

[13] B.L. van der Waerden, Beweis einer baudetschen vermutung, *Nieuw Arch. Wiskd.* **15** (1927), 212-216.

Department of Pure Mathematics and Mathematical Statistics
Centre for Mathematical Sciences
Wilberforce Road
Cambridge CB3 0WB
England
I.Leader@dpmms.cam.ac.uk

Kostka–Foulkes polynomials and Macdonald spherical functions

Kendra Nelsen and Arun Ram

Abstract

Generalized Hall–Littlewood polynomials (Macdonald spherical functions) and generalized Kostka–Foulkes polynomials (q-weight multiplicities) arise in many places in combinatorics, representation theory, geometry, and mathematical physics. This paper attempts to organize the different definitions of these objects and prove the fundamental combinatorial results from "scratch", in a presentation which, hopefully, will be accessible and useful for both the nonexpert and researchers currently working in this very active field. The combinatorics of the affine Hecke algebra plays a central role. The final section of this paper can be read independently of the rest of the paper. It presents, with proof, Lascoux and Schützenberger's positive formula for the Kostka–Foulkes poynomials in the type A case.

Introduction

The classical theory of Hall–Littlewood polynomials and Kostka–Foulkes polynomials appears in the monograph of I.G. Macdonald [11]. The Hall–Littlewood polynomials form a basis of the ring of symmetric functions and the Kostka–Foulkes polynomials are the entries of the transition matrix between the Hall–Littlewood polynomials and the Schur functions.

This theory enters in many different places in algebra, geometry and combinatorics. Many of these connections appear in [11].

(a) [11, Ch. II] explains how this theory describes the structure of the Hall algebra of finite $\mathfrak{o}$-modules, where $\mathfrak{o}$ is a discrete valuation ring.

(b) [11, Ch. IV] explains how the Hall–Littlewood polynomials enter into the representation theory of $GL_n(\mathbb{F}_q)$ where $\mathbb{F}_q$ is a finite field with q elements.

(c) [11, Ch, V] shows that the Hall–Littlewood polynomials arise as spherical functions for $GL_n(\mathbb{Q}_p)$ where $\mathbb{Q}_p$ is the field of p-adic numbers.

(d) [11, Ch. III §6 Ex. 6] explains how the Kostka–Foulkes polynomials relate to the intersection cohomology of unipotent orbit closures for $GL_n(\mathbb{C})$ and [11, Ch. III §8 Ex. 8] explains how the Kostka–Foulkes polynomials describe the graded decomposition of the representations of the symmetric groups S_n on the cohomology of Springer fibers.

(e) [11, Ch. I App. A §8 and Ch. III §6] shows that the Kostka–Foulkes polynomials are q-analogues of the weight multiplicities for representations of $GL_n(\mathbb{C})$.

(f) [11, Ch. III (6.5)] explains how the Kostka–Foulkes polynomials encode a subtle statistic on column strict Young tableaux.

Macdonald [12, (4.1.2)] showed that there is a formula for the spherical functions for the Chevalley group $G(\mathbb{Q}_p)$ which generalizes the formula for Hall–Littlewood symmetric functions. This combinatorial formula is in terms of the root system data of the Chevalley group G. In [10] Lusztig showed that Macdonald's spherical function formula can be seen in terms of the affine Hecke algebra and that the "q-weight multiplicities" or generalized Kostka–Foulkes polynomials coming from these spherical functions are Kazhdan–Lusztig polynomials for the affine Weyl group. Kato [5] proved the "partition function formula" for the q-weight multiplicities which was conjectured by Lusztig. The partition function formula has led to continuing analysis of the connection between the q-weight multiplicities, functions on nilpotent orbits, filtrations of weight spaces by the kernels of powers of a regular nilpotent element, and degrees in harmonic polynomials (see [4] and the references there).

The connection between Hall–Littlewood polynomials and $\mathfrak{o}$-modules has seen generalizations in the theory of representations of quivers, the classical case being the case where the quiver is a loop consisting of one vertex and one edge. This theory has been generalized extensively by Ringel, Lusztig, Nakajima and many others and is developing quickly; fairly recent references are [15] and [16].

The connection to Springer representations of Weyl groups and the representations of Chevalley groups over finite fields has been developed extensively by Lusztig, Shoji and others; a good survey of the current theory is in [18] and the recent papers [19] show how this theory is beginning to extend its reach outside Lie theory into the realm of complex reflection groups.

Since the theory of Macdonald spherical functions (the generalization of Hall–Littlewood polynomials) and q-weight multiplicities (the generalization of Kostka–Foulkes polynomials) appears in so many important parts of mathematics it seems appropriate to give a survey of the basics of this theory. This paper is an attempt to collect together the fundamental combinatorial results analogous to those which are found for the type A case in [11]. The presentation here centers on the role played by the affine Hecke algebra. Hopefully this will help to illustrate how and why these objects arise naturally from a combinatorial point of view and, at the same time, provide enough underpinning to the algebra of the underlying algebraic groups to be useful to researchers in representation theory.

Using the terms *Hall–Littlewood polynomial* and *Macdonald spherical function* interchangeably, and using the words *Kostka–Foulkes polynomial* and q-

weight multiplicity interchangeably, the results that we prove in this paper are as follows.

(1) The interpretation of the Hall–Littlewood polynomials as elements of the affine Hecke algebra (via the Satake isomorphism).

(2) Macdonald's spherical function formula.

(3) The expansion of the Hall Littlewood polynomial in terms of the standard basis of the affine Hecke algebra.

(4) The triangularity of transition matrices between Macdonald spherical functions and other bases of symmetric functions.

(5) The straightening rules for Hall–Littlewood polynomials.

(6) The orthogonality of Macdonald spherical functions.

(7) The raising operator formula for Kostka–Foulkes polynomials.

(8) The partition function formula for q-weight multiplicities.

(9) The identification of the Kostka–Foulkes polynomial as a Kazhdan–Lusztig polynomial.

All of these results are proved here in general Lie type. They are all previously known, spread throughout various parts of the literature. The presentation here is a unified one; some of the proofs may (or may not) be new.

Section 4 is designed so that it can be read independently of the rest of the paper. In Section 4 we give the proof of Lascoux-Schützenberger's positive combinatorial formula [9] (see also [11, Ch. III (6.5)]) for Kostka–Foulkes polynomials in type A. Versions of this proof have appeared previously in [17] and in [2]. This proof has a reputation for being difficult and obscure. After finally getting the courage to attack the literature, we have found, in the end, that the proof is not so difficult after all. Hopefully we have been able to explain it so that others will also find it so.

Acknowledgements. A portion of this paper was written during a stay of A. Ram at the Newton Institute for the Mathematical Sciences at Cambridge University. A. Ram thanks them for their hospitality and support during Spring 2001. The preparation of this paper has been greatly aided by handwritten lecture notes of I.G. Macdonald from lectures he gave at the University of California, San Diego, in Spring 1991. In several places we have copied rather unabashedly from them. Over many years Professor Macdonald has generously given us lots of handwritten notes. We cannot thank him enough, these notes have opened our eyes to many beautiful things and shown us the "right way" many times when we were going astray.

1 Weyl groups, affine Weyl groups, and the affine Hecke algebra

This section sets up the definitions and notations. Good references for this preliminary material are [1], [20] and [14].

1.1 The root system and the Weyl group

Let $\mathfrak{h}_{\mathbb{R}}^*$ be a real vector space with a nondegenerate symmetric bilinear form $\langle\, ,\,\rangle$. The basic data is a reduced irreducible root system R (defined below) in $\mathfrak{h}_{\mathbb{R}}^*$. Associated to R are the *weight lattice*

$$P = \{\lambda \in \mathfrak{h}_{\mathbb{R}}^* \mid \langle \lambda, \alpha^\vee \rangle \in \mathbb{Z} \text{ for all } \alpha \in R\}, \quad \text{where } \alpha^\vee = \frac{2\alpha}{\langle \alpha, \alpha \rangle}, \tag{1.1}$$

and the *Weyl group* $W = \langle s_\alpha \mid \alpha \in R \rangle$ generated by the reflections

$$\begin{array}{rccc} s_\alpha : & \mathfrak{h}_{\mathbb{R}}^* & \longrightarrow & \mathfrak{h}_{\mathbb{R}}^* \\ & \lambda & \longmapsto & \lambda - \langle \lambda, \alpha^\vee \rangle \alpha \end{array} \tag{1.2}$$

in the hyperplanes

$$H_\alpha = \{x \in \mathfrak{h}_{\mathbb{R}}^* \mid \langle x, \alpha^\vee \rangle = 0\}, \qquad \alpha \in R. \tag{1.3}$$

With these definitions R is a reduced irreducible root system if it is a subset of $\mathfrak{h}_{\mathbb{R}}^*$ such that

(a) R is finite, $0 \notin R$ and $\mathfrak{h}_{\mathbb{R}}^* = \mathbb{R}\text{-span}(R)$,

(b) W permutes the elements of R, that is, $w\alpha \in R$ for $w \in W$ and $\alpha \in R$,

(c) W is finite,

(d) $R \subseteq P$,

(e) if $\alpha \in R$ then the only other multiple of α in R is $-\alpha$,

(f) $\mathfrak{h}_{\mathbb{R}}^*$ is an irreducible W-module.

The choice of a fundamental region C for the action of W on $\mathfrak{h}_{\mathbb{R}}^*$ is equivalent to a choice of *positive roots* R^+ of R,

$$R^+ = \{\alpha \in R \mid \langle x, \alpha^\vee \rangle > 0 \text{ for all } x \in C\}$$

and

$$C = \{x \in \mathfrak{h}_{\mathbb{R}}^* \mid \langle x, \alpha^\vee \rangle > 0 \text{ for all } \alpha \in R^+\}.$$

Example 1.1 If $\mathfrak{h}_{\mathbb{R}}^* = \mathbb{R}^2$ with orthonormal basis $\varepsilon_1 = (1,0)$ and $\varepsilon_2 = (0,1)$, $P = \mathbb{Z}\text{-span}\{\varepsilon_1, \varepsilon_2\}$, and $W = \{1, s_1, s_2, s_1s_2, s_2s_1, s_1s_2s_1, s_2s_1s_2, s_1s_2s_1s_2\}$ is the dihedral group of order 8 generated by the reflections s_1 and s_2 in the hyperplanes H_{α_1} and H_{α_2}, respectively, where

$$\begin{aligned} \alpha_1 &= 2\varepsilon_1, & \alpha_1^\vee &= \varepsilon_1, \\ \alpha_2 &= \varepsilon_2 - \varepsilon_1, & \alpha_2^\vee &= \alpha_2, \end{aligned}$$

then

$$R = \{\pm\alpha_1, \pm\alpha_2, \pm(\alpha_1+\alpha_2), \pm(\alpha_1+2\alpha_2)\}.$$

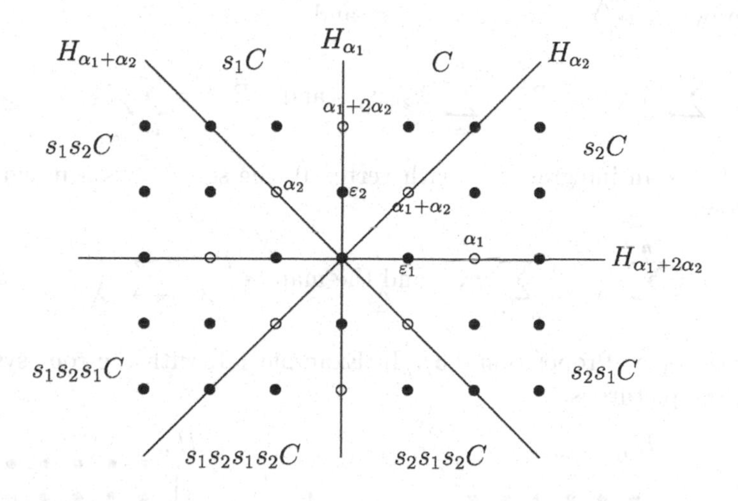

This is the root system of *type* C_2.

For each $\alpha \in R^+$ define the *raising operator* $R_\alpha \colon P \to P$ by $R_\alpha\mu = \mu + \alpha$. The *dominance order* on P is given by

$$\mu \leqslant \lambda \qquad \text{if} \qquad \lambda = R_{\beta_1} \cdots R_{\beta_\ell}\mu \tag{1.4}$$

for some sequence of positive roots $\beta_1, \ldots, \beta_\ell \in R^+$.

The various fundamental chambers for the action of W on $\mathfrak{h}_{\mathbb{R}}^*$ are the $w^{-1}C$, $w \in W$. The *inversion set* of an element $w \in W$ is

$$\begin{aligned} R(w) &= \{\alpha \in R^+ \mid H_\alpha \text{ is between } C \text{ and } w^{-1}C\}, \quad \text{and} \\ \ell(w) &= \mathrm{Card}(R(w)) \end{aligned} \tag{1.5}$$

is the *length* of w. If $R^- = -R^+ = \{-\alpha \mid \alpha \in R^+\}$ then

$$R = R^+ \cup R^- \ \text{ and } \ R(w) = \{\alpha \in R^+ \mid w\alpha \in R^-\}, \qquad \text{for } w \in W.$$

The weight lattice, the set of *dominant integral weights*, and the set of *strictly dominant integral weights*, are

$$\begin{aligned} P &= \{\lambda \in \mathfrak{h}_{\mathbb{R}}^* \mid \langle \lambda, \alpha^\vee \rangle \in \mathbb{Z} \text{ for all } \alpha \in R\}, \\ P^+ = P \cap \overline{C} &= \{\lambda \in \mathfrak{h}_{\mathbb{R}}^* \mid \langle \lambda, \alpha^\vee \rangle \in \mathbb{Z}_{\geqslant 0} \text{ for all } \alpha \in R^+\}, \\ P^{++} = P \cap C &= \{\lambda \in \mathfrak{h}_{\mathbb{R}}^* \mid \langle \lambda, \alpha^\vee \rangle \in \mathbb{Z}_{>0} \text{ for all } \alpha \in R^+\}, \end{aligned} \tag{1.6}$$

where $\overline{C} = \{x \in \mathfrak{h}_{\mathbb{R}}^* \mid \langle x, \alpha^\vee \rangle \geqslant 0 \text{ for all } \alpha \in R^+\}$ is the closure of the fundamental chamber C.

The *simple roots* are the positive roots $\alpha_1, \ldots, \alpha_n$ such that the hyperplanes H_{α_i}, $1 \leqslant i \leqslant n$, are the *walls* of C. The *fundamental weights*, $\omega_1, \ldots, \omega_n \in P$, are given by $\langle \omega_i, \alpha_j^\vee \rangle = \delta_{ij}$, $1 \leqslant i, j \leqslant n$, and

$$P = \sum_{i=1}^n \mathbb{Z}\omega_i, \qquad P^+ = \sum_{i=1}^n \mathbb{Z}_{\geqslant 0}\omega_i, \quad \text{and} \quad P^{++} = \sum_{i=1}^n \mathbb{Z}_{>0}\omega_i. \tag{1.7}$$

The set P^+ is an integral cone with vertex 0, the set P^{++} is a integral cone with vertex

$$\rho = \sum_{i=1}^n \omega_i = \tfrac{1}{2} \sum_{\alpha \in R^+} \alpha, \quad \text{and the map} \quad \begin{array}{ccc} P^+ & \longrightarrow & P^{++} \\ \lambda & \longmapsto & \lambda + \rho \end{array} \tag{1.8}$$

is a bijection (see Proposition 2.3). In Example 1.1, with the root system of type C_2, the picture is

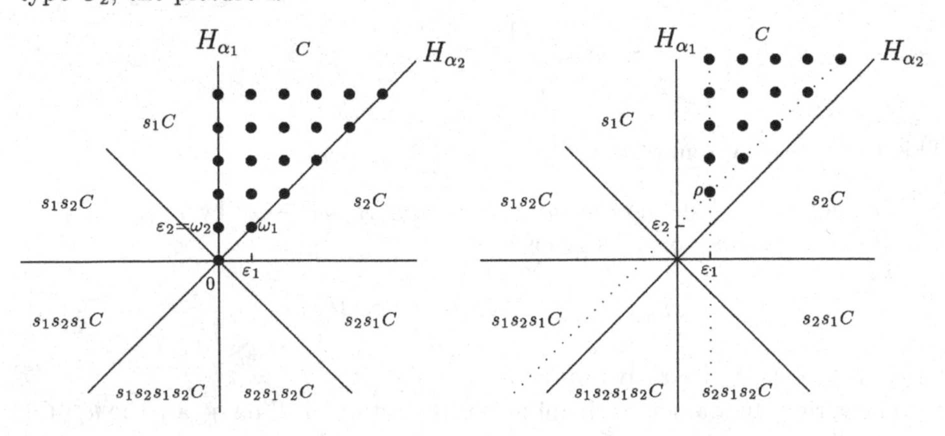

The set P^+ The set P^{++}

The *simple reflections* are $s_i = s_{\alpha_i}$, for $1 \leqslant i \leqslant n$. The Weyl group W has a presentation by generators $s_1, \ldots, s_n$ and relations

$$\begin{aligned} s_i^2 &= 1, & &\text{for } 1 \leqslant i \leqslant n, \\ \underbrace{s_i s_j s_i \cdots}_{m_{ij} \text{ factors}} &= \underbrace{s_j s_i s_j \cdots}_{m_{ij} \text{ factors}}, & &i \neq j, \end{aligned} \tag{1.9}$$

where π/m_{ij} is the angle between the hyperplanes H_{α_i} and H_{α_j}. A *reduced word* for $w \in W$ is an expression $w = s_{i_1} \cdots s_{i_p}$ for w as a product of simple reflections which has p minimal. The following lemma describes the inversion set in terms of the simple roots and the simple reflections and shows that if $w = s_{i_1} \cdots s_{i_p}$ is a reduced expression for w then $p = \ell(w)$.

Lemma 1.2 ([1, VI §1 no. 6 Cor. 2 to Prop. 17]) *Let $w = s_{i_1} \cdots s_{i_p}$ be a reduced word for w. Then*

$$R(w) = \{\alpha_{i_p}, s_{i_p}\alpha_{i_{p-1}}, \ldots, s_{i_p} \cdots s_{i_2}\alpha_{i_1}\}.$$

The *Bruhat order*, or *Bruhat-Chevalley order* (see [20, §8 App., p. 126]), is the partial order on W such that $v \leqslant w$ if there is a reduced word for v, $v = s_{j_1} \cdots s_{j_k}$, which is a subword of a reduced word for w, $w = s_{i_1} \cdots s_{i_p}$, (that is, $s_{j_1}, \ldots, s_{j_k}$ is a subsequence of the sequence $s_{i_1}, \ldots, s_{i_p}$).

1.2 The affine Weyl group

For $\lambda \in P$, the *translation in* λ is

$$\begin{array}{rccc} t_\lambda: & \mathfrak{h}^*_{\mathbb{R}} & \longrightarrow & \mathfrak{h}^*_{\mathbb{R}} \\ & x & \longmapsto & x + \lambda. \end{array} \tag{1.10}$$

The *extended affine Weyl group* $\tilde{W}$ is the group

$$\tilde{W} = \{wt_\lambda \mid w \in W, \lambda \in P\}, \tag{1.11}$$

with multiplication determined by the relations

$$t_\lambda t_\mu = t_{\lambda+\mu}, \quad \text{and} \quad t_{w\lambda}w = wt_\lambda, \tag{1.12}$$

for $\lambda, \mu \in P$ and $w \in W$, and so $\tilde{W}$ is a semidirect product of W and the group of translations $\{t_\lambda \mid \lambda \in P\}$. It is the group of transformations of $\mathfrak{h}^*_{\mathbb{R}}$ generated by the s_α, $\alpha \in R^+$, and t_λ, $\lambda \in P$. The *affine Weyl group* W_{aff} is the subgroup of $\tilde{W}$ generated by the reflections

$$\begin{gathered} s_{\alpha,k}\colon \mathfrak{h}^*_{\mathbb{R}} \to \mathfrak{h}^*_{\mathbb{R}} \text{ in the hyperplanes} \\ H_{\alpha,k} = \{x \in \mathfrak{h}^*_{\mathbb{R}} \mid \langle x, \alpha^\vee \rangle = k\}, \quad \alpha \in R^+, k \in \mathbb{Z}. \end{gathered} \tag{1.13}$$

The reflections $s_{\alpha,k}$ can be written as elements of $\tilde{W}$ via the formula

$$s_{\alpha,k} = t_{k\alpha}s_\alpha = s_\alpha t_{-k\alpha}. \tag{1.14}$$

The *highest short root* of R is the unique element $\varphi \in R^+$ such that the *fundamental alcove*

$$A = C \cap \{x \in \mathfrak{h}^*_{\mathbb{R}} \mid \langle x, \varphi^\vee \rangle < 1\} \tag{1.15}$$

is a fundamental region for the action of W_{aff} on $\mathfrak{h}^*_{\mathbb{R}}$. The various fundamental chambers for the action of W_{aff} on $\mathfrak{h}^*_{\mathbb{R}}$ are $\tilde{w}^{-1}A$, $\tilde{w} \in W_{\text{aff}}$. The *inversion set* of $\tilde{w} \in \tilde{W}$ is

$$R(\tilde{w}) = \{H_{\alpha,k} \mid H_{\alpha,k} \text{ is between } A \text{ and } \tilde{w}^{-1}A\} \quad \text{and} \quad \ell(\tilde{w}) = \text{Card}(R(\tilde{w}))$$

is the *length* of $\tilde{w}$. If $w \in W$ and $\lambda \in P$ then

$$\ell(wt_\lambda) = \sum_{\alpha \in R^+} |\langle \lambda, \alpha^\vee \rangle + \chi(w\alpha)|, \tag{1.16}$$

where, for a root $\beta \in R$, set $\chi(\beta) = 0$, if $\beta \in R^+$, and $\chi(\beta) = 1$, if $\beta \in R^-$.

Continuing Example 1.1, we have the picture

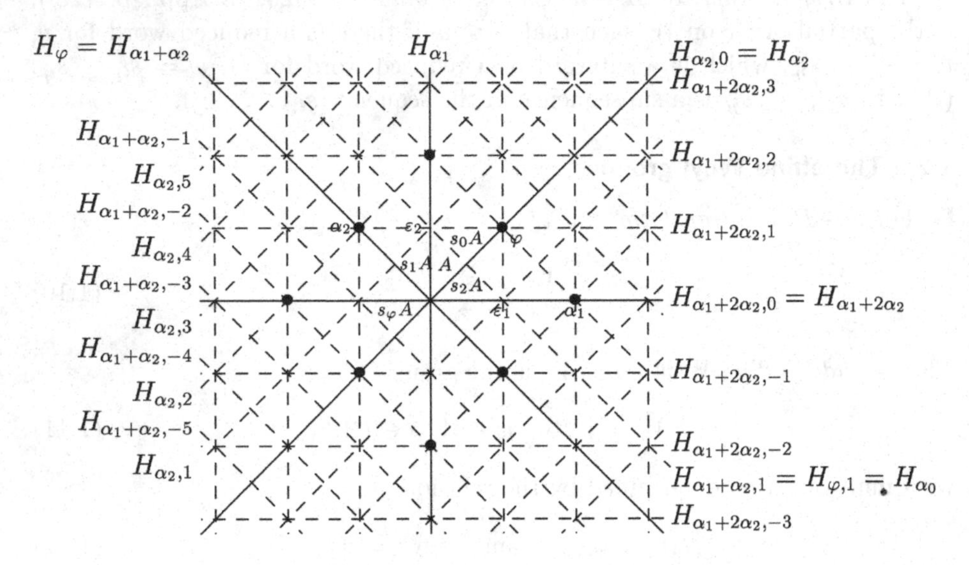

Let

$$H_{\alpha_0} = H_{\varphi,1} \quad \text{and} \quad s_0 = s_{\varphi,1} = t_\phi s_\phi = s_\phi t_{-\phi}, \tag{1.17}$$

and let $H_{\alpha_1}, \ldots, H_{\alpha_n}$ and $s_1, \ldots, s_n$ be as in (1.9). Then the walls of A are the hyperplanes $H_{\alpha_0}, H_{\alpha_1}, \ldots, H_{\alpha_n}$ and the group W_{aff} has a presentation by generators $s_0, s_1, \ldots, s_n$ and relations

$$\begin{aligned} s_i^2 &= 1, & \text{for } 0 \leqslant i \leqslant n, \\ \underbrace{s_i s_j s_i \cdots}_{m_{ij} \text{ factors}} &= \underbrace{s_j s_i s_j \cdots}_{m_{ij} \text{ factors}}, & i \neq j, \end{aligned} \tag{1.18}$$

where π/m_{ij} is the angle between the hyperplanes H_{α_i} and H_{α_j}.

Let w_0 be the longest element of W and let w_i be the longest element of the subgroup $W_{\omega_i} = \{w \in W \mid w\omega_i = \omega_i\}$. Let $\varphi^\vee = c_1\alpha_1^\vee + \cdots + c_n\alpha_n^\vee$. Then let

$$\Omega = \{g \in \tilde{W} \mid \ell(g) = 0\} = \{1\} \cup \{g_i \mid c_i = 1\}, \quad \text{where} \quad g_i = t_{\omega_i} w_i w_0, \tag{1.19}$$

(see [1, VI §2 no. 3 Prop. 6]). Each element $g \in \Omega$ sends the alcove A to itself and thus permutes the walls $H_{\alpha_0}, H_{\alpha_1}, \dots, H_{\alpha_n}$ of A. Denote the resulting permutation of $\{0, 1, \dots, n\}$ also by g. Then

$$gs_ig^{-1} = s_{g(i)}, \quad \text{for } 0 \leqslant i \leqslant n, \tag{1.20}$$

and the group $\tilde{W}$ is presented by the generators $s_0, s_1, \dots, s_n$ and $g \in \Omega$ with the relations (1.18) and (1.20). In the setting of Example 1.1, $W_{\omega_1} = \{1, s_2\}$, $W_{\omega_2} = \{1, s_1\}$, $w_1 = s_2$, $w_2 = s_1$ and $w_0 = s_1s_2s_1s_2$, and $\varphi^\vee = 2\alpha_1^\vee + \alpha_2^\vee$ so that $c_1 = 2$, $c_2 = 1$ and $\Omega = \{1, g_2\} \cong \mathbb{Z}/2\mathbb{Z}$, where $g_2 = t_{\omega_2}s_2s_1s_2$.

1.3 The affine Hecke algebra

Let q be an indeterminate and let $\mathbb{K} = \mathbb{Z}[q, q^{-1}]$. The affine Hecke algebra $\tilde{H}$ is the algebra over $\mathbb{K}$ given by generators T_i, $1 \leqslant i \leqslant n$, and x^λ, $\lambda \in P$, and relations

$$\begin{aligned}
&\underbrace{T_iT_jT_i\cdots}_{m_{ij}\text{ factors}} = \underbrace{T_jT_iT_j\cdots}_{m_{ij}\text{ factors}}, && \text{for all } i \neq j,\\
&T_i^2 = (q - q^{-1})T_i + 1, && \text{for all } 1 \leqslant i \leqslant n,\\
&x^\lambda x^\mu = x^\mu x^\lambda = x^{\lambda+\mu}, && \text{for all } \lambda, \mu \in P,\\
&x^\lambda T_i = T_i x^{s_i\lambda} + (q - q^{-1})\frac{x^\lambda - x^{s_i\lambda}}{1 - x^{-\alpha_i}}, && \text{for all } 1 \leqslant i \leqslant n,\ \lambda \in P.
\end{aligned} \tag{1.21}$$

An alternative presentation of $\tilde{H}$ is by the generators T_w, $w \in \tilde{W}$, and relations

$$\begin{aligned}
T_{w_1}T_{w_2} &= T_{w_1w_2}, && \text{if } \ell(w_1w_2) = \ell(w_1) + \ell(w_2),\\
T_{s_i}T_w &= (q - q^{-1})T_w + T_{s_iw}, && \text{if } \ell(s_iw) < \ell(w) \quad (0 \leqslant i \leqslant n).
\end{aligned}$$

With notations as in (1.10–1.20) the conversion between the two presentations is given by the relations

$$\begin{aligned}
T_w &= T_{i_1}\cdots T_{i_p}, && \text{if } w \in W_{\text{aff}} \text{ and } w = s_{i_1}\cdots s_{i_p} \text{ is a reduced word},\\
T_{g_i} &= x^{\omega_i}T^{-1}_{w_0w_i}, && \text{for } g_i \in \Omega \text{ as in (1.19)},\\
x^\lambda &= T_{t_\mu}T^{-1}_{t_\nu}, && \text{if } \lambda = \mu - \nu \text{ with } \mu, \nu \in P^+,\\
T_{s_0} &= T_{s_\phi}x^{-\phi}, && \text{where } \phi \text{ is the highest short root of } R,
\end{aligned} \tag{1.22}$$

1.4 The Kazhdan–Lusztig basis

The algebra $\tilde{H}$ has bases

$$\{x^\lambda T_w \mid w \in W, \lambda \in P\} \quad \text{and} \quad \{T_w x^\lambda \mid w \in W, \lambda \in P\}.$$

The Kazhdan–Lusztig basis $\{C'_w \mid w \in \tilde{W}\}$ is another basis of $\tilde{H}$ which plays an important role. It is defined as follows.

The *bar involution* on $\tilde{H}$ is the $\mathbb{Z}$-linear automorphism $\overline{} : \tilde{H} \to \tilde{H}$ given by

$$\overline{q} = q^{-1} \qquad \text{and} \qquad \overline{T_w} = T_{w^{-1}}^{-1}, \quad \text{for } w \in \tilde{W}.$$

For $0 \leqslant i \leqslant n$, $\overline{T_i} = T_i^{-1} = T_i - (q - q^{-1})$ and the bar involution is a $\mathbb{Z}$-algebra automorphism of $\tilde{H}$. If $w = s_{i_1} \cdots s_{i_p}$ is a reduced word for w then, by the definition of the Bruhat order (defined after Lemma 1.2),

$$\begin{aligned} \overline{T_w} &= \overline{T_{i_1} \cdots T_{i_p}} = \overline{T_{i_1}} \cdots \overline{T_{i_p}} = T_{i_1}^{-1} \cdots T_{i_p}^{-1} \\ &= (T_{i_1} - (q - q^{-1})) \cdots (T_{i_p} - (q - q^{-1})) = T_w + \sum_{v<w} a_{vw} T_v, \end{aligned}$$

with $a_{vw} \in \mathbb{Z}[(q - q^{-1})]$.

Setting $\tau_i = qT_i$ and $t = q^2$, the second relation in (1.21)

$$T_i^2 = (q - q^{-1})T_i + 1 \quad \text{becomes} \quad \tau_i^2 = (t-1)\tau_i + t. \tag{1.23}$$

Let $\tau_w = q^{\ell(w)} T_w$ for $w \in \tilde{W}$. The *Kazhdan–Lusztig basis* $\{C'_w \mid \tilde{w} \in \tilde{W}\}$ of $\tilde{H}$ is defined [6] by

$$\overline{C}'_w = C'_w \quad \text{and} \quad C'_w = t^{-\ell(w)/2} \left(\sum_{y \leqslant w} P_{yw} \tau_y \right), \tag{1.24}$$

subject to $P_{yw} \in \mathbb{Z}[t^{\frac{1}{2}}, t^{-\frac{1}{2}}]$, $P_{ww} = 1$, and $\deg_t(P_{yw}) \leqslant \frac{1}{2}(\ell(w) - \ell(y) - 1)$. If

$$p_{yw} = q^{-(\ell(w)-\ell(y))} P_{yw} \tag{1.25}$$

then

$$C'_w = q^{-\ell(w)} \sum_{y \leqslant w} P_{yw} q^{\ell(y)} T_y = \sum_{y \leqslant w} P_{yw} q^{-(\ell(w)-\ell(y))} T_y = \sum_{y \leqslant w} p_{yw} T_y, \tag{1.26}$$

with

$$p_{yw} \in \mathbb{Z}[q, q^{-1}], \quad p_{ww} = 1, \quad \text{and} \quad p_{yw} \in q^{-1}\mathbb{Z}[q^{-1}], \tag{1.27}$$

since $\deg_q(P_{yw}(q) q^{-(\ell(w)-\ell(y))}) \leqslant \ell(w) - \ell(y) - 1 - (\ell(w) - \ell(y)) = -1$. The following proposition establishes the existence and uniqueness of the C'_w and the p_{yw}.

Proposition 1.3 *Let $(\tilde{W}, \leqslant)$ be a partially ordered set such that for any $u, v \in \tilde{W}$ the interval $[u, v] = \{z \in \tilde{W} \mid u \leqslant z \leqslant v\}$ is finite. Let M be a free $\mathbb{Z}[q, q^{-1}]$-module with basis $\{T_w \mid w \in \tilde{W}\}$ and with a $\mathbb{Z}$-linear involution $\overline{} : M \to M$ such that*

$$\overline{q} = q^{-1} \quad \text{and} \quad \overline{T_w} = T_w + \sum_{v<w} a_{vw} T_v.$$

Then there is a unique basis $\{C'_w \mid w \in \tilde{W}\}$ of M such that

(a) $\overline{C'_w} = C'_w$,

(b) $C'_w = T_w + \sum_{v<w} p_{vw} T_v$, *with $p_{vw} \in q^{-1}\mathbb{Z}[q^{-1}]$ for $v < w$.*

Proof The p_{vw} are determined by induction as follows. Fix $v, w \in W$ with $v \leq w$. If $v = w$ then $p_{vw} = p_{ww} = 1$. For the induction step assume that $v < w$ and that p_{zw} are known for all $v < z \leqslant w$.

The matrices $A = (a_{vw})$ and $P = (p_{vw})$ are upper triangular with 1's on the diagonal. The equations

$$T_w = \overline{\overline{T_w}} = \sum_v \overline{a_{vw} T_v} = \sum_{u,v} a_{uv} \overline{a_{vw}} T_u \quad \text{and}$$

$$\sum_u p_{uw} T_u = C'_w = \overline{C'_w} = \sum_v \overline{p_{vw} T_v} = \sum_{u,v} \overline{p_{vw}} a_{uv} T_u,$$

imply $A\overline{A} = \mathrm{Id}$ and $P = A\overline{P}$. Then

$$f = \sum_{u<z\leqslant w} a_{uz} \overline{p}_{zw} = ((A-1)\overline{P})_{uw} = (A\overline{P} - \overline{P})_{uw} = (P - \overline{P})_{uw} = p_{uw} - \overline{p}_{uw},$$

is a known element of $\mathbb{Z}[q, q^{-1}]$;

$$f = \sum_{k\in\mathbb{Z}} f_k q^k \quad \text{such that} \quad \overline{f} = \overline{(p_{uw} - \overline{p}_{uw})} = \overline{p}_{uw} - p_{uw} = -f.$$

Hence $f_k = -f_{-k}$ for all $k \in \mathbb{Z}$ and p_{uw} is given by $p_{uw} = \sum_{k\in\mathbb{Z}_{<0}} f_k q^k$. □

The *finite Hecke algebra* H and the *group algebra of* P are the subalgebras of $\tilde{H}$ given, respectively, by

$$\begin{aligned} H &= (\text{subalgebra of } \tilde{H} \text{ generated by } T_1, \ldots, T_n), \\ \mathbb{K}[P] &= \mathbb{K}\text{-span } \{x^\lambda \mid \lambda \in P\}, \qquad \text{where } \mathbb{K} = \mathbb{Z}[q, q^{-1}], \end{aligned} \tag{1.28}$$

and $\mathbb{K}$-span$\{x^\lambda \mid \lambda \in P\}$ denotes the set of $\mathbb{K}$-linear combinations of elements x^λ in $\tilde{H}$. The Weyl group W acts on $\mathbb{K}[P]$ by

$$wf = \sum_{\mu\in P} c_\mu x^{w\mu}, \quad \text{for } w \in W \text{ and } f = \sum_{\mu\in P} c_\mu x^\mu \in \mathbb{K}[P]. \tag{1.29}$$

Theorem 1.4 *The center of the affine Hecke algebra is the ring*

$$Z(\tilde{H}) = \mathbb{K}[P]^W = \{f \in \mathbb{K}[P] \mid wf = f \text{ for all } w \in W\}$$

of symmetric functions in $\mathbb{K}[P]$.

Proof If $z \in \mathbb{K}[P]^W$ then by the fourth relation in (1.21), $T_i z = (s_i z)T_i + (q - q^{-1})(1 - x^{-\alpha_i})^{-1}(z - s_i z) = zT_i + 0$, for $1 \leqslant i \leqslant n$, and by the third relation in (1.21), $zx^\lambda = x^\lambda z$, for all $\lambda \in P$. Thus z commutes with all the generators of $\tilde{H}$ and so $z \in Z(\tilde{H})$.

Assume

$$z = \sum_{\lambda \in P, w \in W} c_{\lambda,w} x^\lambda T_w \in Z(\tilde{H}).$$

Let $m \in W$ be maximal in Bruhat order subject to $c_{\gamma,m} \neq 0$ for some $\gamma \in P$. If $m \neq 1$ there exists a dominant $\mu \in P$ such that $c_{\gamma+\mu-m\mu,m} = 0$ (otherwise $c_{\gamma+\mu-m\mu,m} \neq 0$ for every dominant $\mu \in P$, which is impossible since z is a finite linear combination of $x^\lambda T_w$). Since $z \in Z(\tilde{H})$ we have

$$z = x^{-\mu} z x^\mu = \sum_{\lambda \in P, w \in W} c_{\lambda,w} x^{\lambda-\mu} T_w x^\mu.$$

Repeated use of the fourth relation in (1.21) yields

$$T_w x^\mu = \sum_{\nu \in P, v \in W} d_{\nu,v} x^\nu T_v$$

where $d_{\nu,v}$ are constants such that $d_{w\mu,w} = 1$, $d_{\nu,w} = 0$ for $\nu \neq w\mu$, and $d_{\nu,v} = 0$ unless $v \leqslant w$. So

$$z = \sum_{\lambda \in P, w \in W} c_{\lambda,w} x^\lambda T_w = \sum_{\lambda \in P, w \in W} \sum_{\nu \in P, v \in W} c_{\lambda,w} d_{\nu,v} x^{\lambda-\mu+\nu} T_v$$

and comparing the coefficients of $x^\gamma T_m$ gives $c_{\gamma,m} = c_{\gamma+\mu-m\mu,m} d_{m\mu,m}$. Since $c_{\gamma+\mu-m\mu,m} = 0$ it follows that $c_{\gamma,m} = 0$, which is a contradiction. Hence $z = \sum_{\lambda \in P} c_\lambda x^\lambda \in \mathbb{K}[P]$.

The fourth relation in (1.21) gives

$$zT_i = T_i z = (s_i z)T_i + (q - q^{-1})z'$$

where $z' \in \mathbb{K}[P]$. Comparing coefficients of x^λ on both sides yields $z' = 0$. Hence $zT_i = (s_i z)T_i$, and therefore $z = s_i z$ for $1 \leqslant i \leqslant n$. So $z \in \mathbb{K}[P]^W$.

□

2 Symmetric and alternating functions and their q-analogues

Let $\mathbf{1}_0$ and ε_0 be the elements of the finite Hecke algebra H which are determined by

$$\begin{array}{lll} \mathbf{1}_0^2 = \mathbf{1}_0 & \text{and} \quad T_i\mathbf{1}_0 = q\mathbf{1}_0, & \text{for all } 1 \leqslant i \leqslant n, \\ \varepsilon_0^2 = \varepsilon_0 & \text{and} \quad T_i\varepsilon_0 = (-q^{-1})\varepsilon_0, & \text{for all } 1 \leqslant i \leqslant n. \end{array}$$

In terms of the basis $\{T_w \mid w \in W\}$ of H these elements have the explicit formulae

$$\mathbf{1}_0 = \frac{1}{W_0(q^2)} \sum_{w\in W} q^{\ell(w)} T_w, \quad \text{and} \quad \varepsilon_0 = \frac{1}{W_0(q^{-2})} \sum_{w\in W} (-q)^{-\ell(w)} T_w, \tag{2.1}$$

where $W_0(t) = \sum_{w\in W} t^{\ell(w)}$. (To define these elements one should adjoin the element $W_0(q^2)^{-1}$ to $\mathbb{K}$ or to H.) The elements $\mathbf{1}_0$ and ε_0 are q-analogues of the elements in the group algebra of W given by

$$\mathbf{1} = \frac{1}{|W|} \sum_{w\in W} w \quad \text{and} \quad \varepsilon = \frac{1}{|W|} \sum_{w\in W} (-1)^{\ell(w)} w, \tag{2.2}$$

and the vector spaces $\mathbf{1}_0\tilde{H}\mathbf{1}_0$ and and $\varepsilon_0\tilde{H}\mathbf{1}_0$ are q-analogues of the vector spaces (more precisely, free $\mathbb{K} = \mathbb{Z}[q, q^{-1}]$-modules) of *symmetric functions* and *alternating functions*,

$$\begin{aligned} \mathbb{K}[P]^W &= \{f \in \mathbb{K}[P] \mid wf = f \text{ for all } w \in W\} = \mathbf{1}\,\mathbb{K}[P], \\ \mathcal{A} &= \{f \in \mathbb{K}[P] \mid wf = (-1)^{\ell(w)} f \text{ for all } w \in W\} = \varepsilon\,\mathbb{K}[P], \end{aligned} \tag{2.3}$$

respectively, where the action of W on $\mathbb{K}[P]$ is as defined in 1.29.

For $\mu \in P$ let the orbit $W\mu$ and the stabilizer W_μ of μ be defined by

$$W\mu = \{w\mu \mid w \in W\} \qquad \text{and} \qquad W_\mu = \{w \in W \mid w\mu = \mu\}.$$

Then define

$$\begin{aligned} m_\mu &= \sum_{\gamma\in W\mu} x^\gamma = \frac{|W|}{|W_\mu|} \mathbf{1}\, x^\mu, & a_\mu &= \sum_{w\in W} (-1)^{\ell(w)} w x^\mu = |W|\,\varepsilon\, x^\mu, \\ M_\mu &= \mathbf{1}_0 x^\mu \mathbf{1}_0, & A_\mu &= \varepsilon_0 x^\mu \mathbf{1}_0. \end{aligned} \tag{2.4}$$

Theorem 2.2 below shows that the elements in (2.4) which are indexed by elements of P^+ and P^{++} form bases (over $\mathbb{K}$) of $\mathbb{K}[P]^W$, $\mathcal{A}$, $\mathbf{1}_0\tilde{H}\mathbf{1}_0$, and $\varepsilon_0\tilde{H}\mathbf{1}_0$. This will be a consequence of the following *straightening rules*. The straightening law for the M_μ given in the following Proposition is a generalization of [11, III §2 Ex. 2].

Proposition 2.1 *For $\gamma \in P$ let m_γ, a_γ, M_γ, and A_γ be as defined in (2.4). Let α_i be a simple root and let $\mu \in P$ be such that $d = \langle \mu, \alpha_i^\vee \rangle \geqslant 0$. Then*

$$m_{s_i\mu} = m_\mu, \qquad a_{s_i\mu} = -a_\mu, \qquad \text{and} \qquad A_{s_i\mu} = -A_\mu.$$

Letting $t = q^{-2}$, $M_\mu = M_{s_i\mu}$ if $d = 0$, and if $d > 0$ then

$$\begin{aligned} M_{s_i\mu} &= tM_\mu + \left(\sum_{j=1}^{\lfloor d/2-1 \rfloor} (t^2-1)t^{j-1} M_{\mu - j\alpha_i} \right) \\ &\quad + \begin{cases} (t-1)t^{d/2-1} M_{\mu-(d/2)\alpha_i}, & \text{if } d \text{ is even,} \\ 0, & \text{if } d \text{ is odd.} \end{cases} \end{aligned}$$

Proof The first two equalities follow from the definitions of m_λ and a_μ and the fact that $\ell(s_i) = 1$.

Let $\mu \in P$ such that $d = \langle \mu, \alpha_i^\vee \rangle \geqslant 0$. Since $x^\mu + x^{s_i\mu}$ is in the center of the tiny little affine Hecke algebra generated by T_i and the x^γ, $\gamma \in P$,

$$\begin{aligned} A_\mu + A_{s_i\mu} &= \varepsilon_0(x^\mu + x^{s_i\mu})\mathbf{1}_0 = q^{-1}\varepsilon_0(x^\mu + x^{s_i\mu})T_i\mathbf{1}_0 \\ &= q^{-1}\varepsilon_0 T_i (x^\mu + x^{s_i\mu})\mathbf{1}_0 = -q^{-2}\varepsilon_0(x^\mu + x^{s_i\mu})\mathbf{1}_0 \\ &= -q^{-2}(A_\mu + A_{s_i\mu}). \end{aligned}$$

Thus $A_\mu + A_{s_i\mu} = 0$ which establishes the third statement.

If $d = 0$ then, by definition, $M_\mu = M_{s_i\mu}$. If $d > 0$ then multiplying the fourth relation in (1.21) by $\mathbf{1}_0$ on both the left and the right (and then multiplying by q^{-1}) gives

$$\mathbf{1}_0(x^{s_i\mu} - x^\mu)\mathbf{1}_0 = q^{-1}(q - q^{-1})\mathbf{1}_0 \left(\frac{x^{s_i\mu} - x^\mu}{1 - x^{-\alpha_i}} \right) \mathbf{1}_0.$$

Subtracting the same relation with μ replaced by $\mu - \alpha_i$ gives

$$\begin{aligned} &\mathbf{1}_0(x^{s_i\mu} - x^\mu)\mathbf{1}_0 - \mathbf{1}_0(x^{s_i\mu+\alpha_i} - x^{\mu-\alpha_i})\mathbf{1}_0 \\ &= (1-q^{-2})\mathbf{1}_0 \left(\frac{x^{s_i\mu} - x^\mu - x^{s_i\mu+\alpha_i} + x^{\mu-\alpha_i}}{1 - x^{-\alpha_i}} \right) \mathbf{1}_0 \\ &= (1-q^{-2})\mathbf{1}_0(-x^{s_i\mu+\alpha_i} - x^\mu)\mathbf{1}_0. \end{aligned}$$

So

$$\mathbf{1}_0 x^{s_i\mu}\mathbf{1}_0 = q^{-2}\mathbf{1}_0 x^\mu \mathbf{1}_0 - \mathbf{1}_0 x^{\mu-\alpha_i}\mathbf{1}_0 + q^{-2}\mathbf{1}_0 x^{s_i\mu+\alpha_i}\mathbf{1}_0.$$

Inductively applying this relation yields the result. The first cases are

$$M_{s_i\mu} = \begin{cases} M_\mu, & \text{if } \langle \mu, \alpha_i^\vee \rangle = 0, \\ q^{-2}M_\mu, & \text{if } \langle \mu, \alpha_i^\vee \rangle = 1, \\ q^{-2}M_\mu + (q^{-2}-1)M_{\mu-\alpha_i}, & \text{if } \langle \mu, \alpha_i^\vee \rangle = 2, \\ q^{-2}M_\mu + (q^{-4}-1)M_{\mu-\alpha_i}, & \text{if } \langle \mu, \alpha_i^\vee \rangle = 3, \\ q^{-2}M_\mu + (q^{-4}-1)M_{\mu-\alpha_i} + q^{-2}(q^{-2}-1)M_{\mu-2\alpha_i}, & \text{if } \langle \mu, \alpha_i^\vee \rangle = 4. \end{cases}$$

□

Proposition 2.1 implies that, for all $\mu \in P$ and $w \in W$,

$$m_{w\mu} = m_\mu, \quad a_{w\mu} = (-1)^{\ell(w)} a_\mu, \quad \text{and} \quad A_{w\mu} = (-1)^{\ell(w)} A_\mu. \tag{2.5}$$

Theorem 2.2 *Let* $\mathbb{K} = \mathbb{Z}[q, q^{-1}]$. *As free* $\mathbb{K}$*-modules*

$$\mathbb{K}[P]^W \ \ \textit{has basis}\ \{m_\lambda \mid \lambda \in P^+\}, \quad \mathbf{1}_0\tilde{H}\mathbf{1}_0 \ \ \textit{has basis}\ \{M_\lambda \mid \lambda \in P^+\},$$

$$\mathcal{A} \quad \textit{has basis}\ \{a_\mu \mid \mu \in P^{++}\}, \quad \varepsilon_0\tilde{H}\mathbf{1}_0 \ \ \textit{has basis}\ \{A_\mu \mid \mu \in P^{++}\}.$$

Proof Since $\{x^\mu T_w \mid \mu \in P, w \in W\}$ form a basis of $\tilde{H}$ the elements $M_\mu = \mathbf{1}_0 x^\mu \mathbf{1}_0 = q^{-\ell(w)}\mathbf{1}_0 x^\mu T_w \mathbf{1}_0$, $\mu \in P$, span $\mathbf{1}_0\tilde{H}\mathbf{1}_0$. By Proposition 2.1, if μ is on the negative side of a hyperplane H_{α_i}, that is, if $\langle \mu, \alpha_i^\vee\rangle < 0$, then M_μ can be rewritten as a linear combination of M_γ such that all terms have γ on the nonnegative side of H_{α_i}. By repeatedly applying the relation in Proposition 2.1, M_μ can be rewritten as a linear combination of M_γ such that all terms have γ on the nonnegative side of $H_{\alpha_1}, \ldots, H_{\alpha_n}$, that is, $\gamma \in P^+ = P \cap \overline{C}$, where $\overline{C} = \{x \in \mathbb{R}^n \mid \langle x, \alpha_i^\vee\rangle \geqslant 0 \text{ for all } 1 \leqslant i \leqslant n\}$.

If $\lambda \in P^+$, using the fourth relation in (1.21),

$$\begin{aligned} M_\lambda &= \mathbf{1}_0 x^\lambda \mathbf{1}_0 = \frac{1}{W_0(q^2)} \sum_{w \in W} q^{\ell(w)} T_w x^\lambda \mathbf{1}_0 = \frac{1}{W_0(q^2)} \sum_{\gamma, v, w} q^{\ell(w)} d_{v,\gamma} x^\gamma T_v \mathbf{1}_0 \\ &= \frac{1}{W_0(q^2)} \sum_{\gamma, v, w} q^{\ell(w)} d_{v,\gamma} x^\gamma q^{\ell(v)} \mathbf{1}_0 = \frac{1}{W_0(q^2)} \sum_\gamma d_\gamma x^\gamma \mathbf{1}_0, \end{aligned}$$

where $d_{v,\gamma}$ and d_γ are some polynomials in $\mathbb{Z}[q, (q - q^{-1})]$ such that $d_{v,v\lambda} = 1$ so that $d_{w_0\lambda} = 1$. Furthermore $d_\gamma = 0$ unless γ is in the convex hull of the points in the orbit $W\lambda$. Thus the coefficient of $x^{w_0\lambda}$ in M_λ is $W_0(q^2)^{-1} q^{2\ell(w_0)}$ and the coefficient of $x^\gamma T_v$ can be nonzero only if $\gamma \geqslant w_0\lambda$. Thus the M_λ, $\lambda \in P^+$, are linearly independent.

The proof for the cases of m_μ, a_μ and A_μ is easier, following directly from (2.5), the fact that $C = \{x \in \mathbb{R}^n \mid \langle x, \alpha_i^\vee\rangle > 0 \text{ for all } 1 \leqslant i \leqslant n\}$ is a fundamental chamber for the action of W, and that if $\mu \in P^+ \backslash P^{++}$ then $\langle \mu, \alpha_i^\vee\rangle = 0$ and $a_\mu = -a_{s_i\mu} = -a_\mu$, in which case $a_\mu = 0$ (similarly for A_μ). □

For $\lambda \in P$ define the *Schur function*, or *Weyl character*, by

$$s_\lambda = \frac{a_{\lambda+\rho}}{a_\rho}, \quad \text{where} \quad \rho = \tfrac{1}{2} \sum_{\alpha \in R^+} \alpha. \tag{2.6}$$

The straightening law for a_μ in (2.5) implies the following straightening law for the Schur functions. If $\mu \in P$ and $w \in W$ then, by (2.5) and the definition of s_μ,

$$(-1)^{\ell(w)} s_\mu = \frac{(-1)^{\ell(w)} a_{\mu+\rho}}{a_\rho} = \frac{a_{w(\mu+\rho)-\rho+\rho}}{a_\rho} = s_{w \circ \mu}, \tag{2.7}$$

$$\text{where} \quad w \circ \mu = w(\mu + \rho) - \rho.$$

The *dot action* of the Weyl group W on $\mathfrak{h}_{\mathbb{R}}^*$ which is appearing here, $w \circ \mu = t_{-\rho} w t_\rho \mu = (t_\rho^{-1}) w t_\rho \mu$, is the ordinary action of W on $\mathfrak{h}_{\mathbb{R}}^*$ except with the "center" shifted to $-\rho$. For the root system of type C_2, in Example 1.1, the picture is

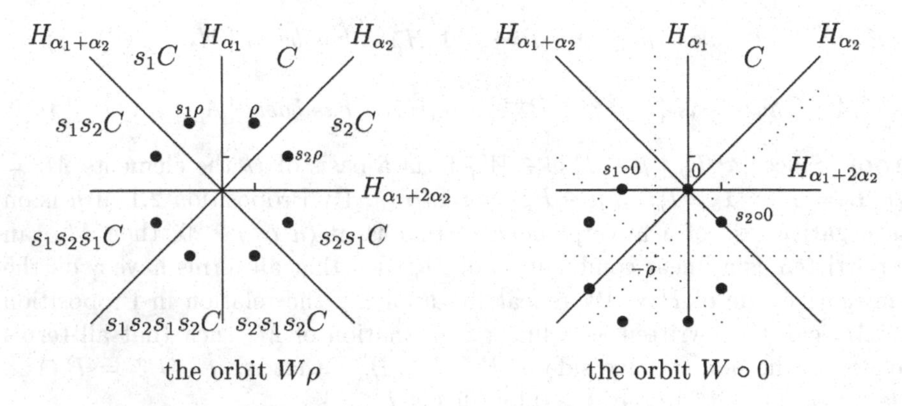

the orbit $W\rho$ the orbit $W \circ 0$

The following proposition shows that the Weyl characters s_λ are elements of $\mathbb{K}[P]^W$. The equality in part (a) is the *Weyl denominator formula*, a generalization of the factorization of the Vandermonde determinant $\det(x_i^{n-j}) = \prod_{1 \leqslant i,j \leqslant n}(x_i - x_j)$. In the remainder of this section we shall abuse language and use the term "vector space" in place of "free $\mathbb{K} = \mathbb{Z}[q, q^{-1}]$ module".

Proposition 2.3 *Let P^+, P^{++}, $\mathbb{K}[P]^W$ and $\mathcal{A}$ be as in (1.7) and (2.4) and let ρ be as in (1.8).*

(a) *If $f \in \mathcal{A}$ then f is divisible by a_ρ and*

$$a_\rho = x^\rho \prod_{\alpha \in R^+} (1 - x^{-\alpha})$$

(b) *The set $\{s_\lambda \mid \lambda \in P^+\}$ is a basis of $\mathbb{K}[P]^W$.*

(c) *The maps*

$$\begin{array}{ccc} P^+ & \longrightarrow & P^{++} \\ \lambda & \longmapsto & \lambda + \rho \end{array} \quad \text{and} \quad \begin{array}{cccc} \Phi : & \mathbb{K}[P]^W & \longrightarrow & \mathcal{A} \\ & f & \longmapsto & a_\rho f \\ & s_\lambda & \longmapsto & a_{\lambda+\rho} \end{array}$$

are a bijection and a vector space isomorphism, respectively.

Proof Since s_i takes α_i to $-\alpha_i$ and permutes the other elements of R^+,

$$\rho - \langle \rho, \alpha_i^\vee \rangle \alpha_i = s_i \rho = \rho - \alpha_i,$$

and so

$$\langle \rho, \alpha_i^\vee \rangle = 1, \quad \text{for all } 1 \leqslant i \leqslant n.$$

Thus the map $P^+ \to P^{++}$ given by $\lambda \mapsto \lambda + \rho$ is well defined and it is a bijection since it is invertible.

Let $d = x^\rho \prod_{\alpha \in R^+}(1 - x^{-\alpha}) = \prod_{\alpha \in R^+}(x^{\alpha/2} - x^{-\alpha/2})$. Since s_i takes α_i to $-\alpha_i$ and permutes the other elements of R^+, $s_i d = -d$ for all $1 \leqslant i \leqslant n$ and so $wd = (-1)^{\ell(w)} d$ for all $w \in W$. Thus d is an element of $\mathcal{A}$.

If $\alpha \in R^+$ and $f = \sum_{\mu \in P} c_\mu x^\mu \in \mathcal{A}$ then

$$\sum_{\mu \in P} c_\mu x^\mu = f = -s_\alpha f = \sum_{\mu \in P} -c_\mu x^{s_\alpha \mu},$$

and

$$f = \sum_{\langle \mu, \alpha^\vee \rangle \geqslant 0} c_\mu (x^\mu - x^{s_\alpha \mu}).$$

Since $(1 - x^{-\langle \mu, \alpha^\vee \rangle \alpha})$ is divisible by $(1 - x^{-\alpha})$ it follows that $x^\mu - x^{s_i \mu} = x^\mu(1 - x^{-\langle \mu, \alpha^\vee \rangle \alpha})$ is divisible by $(1 - x^{-\alpha})$ and thus that f is divisible by $(1 - x^{-\alpha})$ for all $\alpha \in R^+$. Since the elements $(1 - x^{-\alpha})$ are relatively prime in the Laurent polynomial ring $\mathbb{K}[P]$ and x^ρ is a unit in $\mathbb{K}[P]$, f is divisible by d. Since both f and d are in $\mathcal{A}$, the quotient f/d is an element of $\mathbb{K}[P]^W$.

The monomial x^ρ appears in a_ρ with coefficient 1 and it is the unique term x^μ in a_ρ with $\mu \in P^+$. Since d has highest term x^ρ with coefficient 1 and a_ρ is divisible by d it follows that $a_\rho / d = 1$. Thus $a_\rho = d$, the inverse of the map Φ in (c) is well defined, and Φ is an isomorphism.

Since $\{a_{\lambda+\rho} \mid \lambda \in P^+\}$ is a basis of $\mathcal{A}$ and the map Φ is an isomorphism it follows that $\{s_\lambda \mid \lambda \in P^+\}$ is a $\mathbb{K}$-basis of $\mathbb{K}[P]^W$. □

2.1 The Satake isomorphism

The following theorem establishes a q-analogue of the isomorphism Φ from Proposition 2.3(c). The map Φ_1 in the following theorem is the *Satake isomorphism*. We shall continue to abuse language and use the term "vector space" in place of "free $\mathbb{K} = \mathbb{Z}[q, q^{-1}]$ module".

Theorem 2.4 *The vector space isomorphism Φ in Proposition 2.3(c) generalizes to a vector space isomorphism*

$$\begin{array}{rccccc} \tilde{\Phi}: & Z(\tilde{H}) = \mathbb{K}[P]^W & \xrightarrow{\Phi_1} & Z(\tilde{H})\mathbf{1}_0 = \mathbf{1}_0 \tilde{H} \mathbf{1}_0 & \xrightarrow{\Phi_2} & \varepsilon_0 \tilde{H} \mathbf{1}_0 \\ & f & \longmapsto & f\mathbf{1}_0 & \longmapsto & A_\rho f \mathbf{1}_0 \\ & s_\lambda & \longmapsto & s_\lambda \mathbf{1}_0 & \longmapsto & A_{\lambda+\rho}. \end{array}$$

Proof Using the third equality in (2.5),

$$\varepsilon_0 a_\lambda \mathbf{1}_0 = \varepsilon_0 \left(\sum_{w \in W} (-1)^{\ell(w)} x^{w\lambda} \right) \mathbf{1}_0 = \sum_{w \in W} (-1)^{\ell(w)} A_{w\lambda} = |W| \, A_\lambda.$$

By Proposition 2.3(c) and Theorem 1.4, $s_\lambda \in \mathbb{K}[P]^W = Z(\tilde{H})$, and so

$$A_\rho s_\lambda \mathbf{1}_0 = \frac{1}{|W|}\varepsilon_0 a_\rho \mathbf{1}_0 s_\lambda \mathbf{1}_0 = \frac{1}{|W|}\varepsilon_0 a_\rho s_\lambda \mathbf{1}_0^2 = \frac{1}{|W|}\varepsilon_0 a_{\lambda+\rho}\mathbf{1}_0 = A_{\lambda+\rho}.$$

Since $\{s_\lambda \mid \lambda \in P^+\}$ is a basis of $\mathbb{K}[P]^W = Z(\tilde{H})$ and $\{A_{\lambda+\rho} \mid \lambda \in P^+\}$ is a basis of $\varepsilon_0\tilde{H}\mathbf{1}_0$, the composite map

$$\begin{array}{ccccccc} Z(\tilde{H}) & \xrightarrow{\mathbf{1}_0} & Z(\tilde{H})\,\mathbf{1}_0 & \hookrightarrow & \mathbf{1}_0\tilde{H}\mathbf{1}_0 & \xrightarrow{A_\rho} & \varepsilon_0\tilde{H}\mathbf{1}_0 \\ f & \longmapsto & f\,\mathbf{1}_0 & \mapsto & f\,\mathbf{1}_0 & \longmapsto & A_\rho\, f\,\mathbf{1}_0 \\ s_\lambda & \longmapsto & s_\lambda\,\mathbf{1}_0 & \mapsto & s_\lambda\,\mathbf{1}_0 & \longmapsto & A_{\lambda+\rho} \end{array}$$

is a vector space isomorphism. □

If $\mu \in P$ let

$$W_\mu = \{w \in W \mid w\mu = \mu\} \quad \text{and} \quad W_\mu(t) = \sum_{w\in W_\mu} t^{\ell(w)}. \tag{2.8}$$

In particular, if $\mu = 0$, then $W_0 = W$ and $W_0(t)$ is the polynomial that appears in (2.1).

The *Hall–Littlewood polynomials*, or *Macdonald spherical functions*, are defined by

$$P_\mu(x;t) = \frac{1}{W_\mu(t)}\sum_{w\in W} w\Big(x^\mu \prod_{\alpha\in R^+}\frac{1-tx^{-\alpha}}{1-x^{-\alpha}}\Big), \quad \text{for } \mu \in P. \tag{2.9}$$

Then $m_\mu = P_\mu(x;1)$ and, using the Weyl denominator formula,

$$\begin{aligned} P_\mu(x;0) &= \sum_{w\in W} w\Big(\frac{x^\rho x^\mu}{x^\rho \prod_{\alpha\in R^+}(1-x^{-\alpha})}\Big) \\ &= \frac{1}{a_\rho}\sum_{w\in W}(-1)^{\ell(w)} w x^{\mu+\rho} = \frac{a_{\mu+\rho}}{a_\rho} = s_\mu, \end{aligned} \tag{2.10}$$

and so, conceptually, the spherical functions $P_\mu(x;t)$ interpolate between the Schur functions s_μ and the monomial symmetric functions m_μ.

The double cosets in $W\backslash\tilde{W}/W$ are $Wt_\lambda W$, $\lambda \in P^+$. If $\lambda \in P^+$ let n_λ and m_λ be the maximal and minimal length elements of $Wt_\lambda W$, respectively. Theorem 2.9 below will show that under the Satake isomorphism the Weyl characters s_λ correspond to Kazhdan Lusztig basis elements C'_{n_λ} and the polynomials $P_\mu(x;q^{-2})$ correspond to the elements $M_\mu = \mathbf{1}_0 x^\mu \mathbf{1}_0$. More precisely, we have the following diagram:

$$\begin{array}{llcl} \Phi_1: & Z(\tilde{H}) = \mathbb{K}[P]^W & \longrightarrow & Z(\tilde{H})\mathbf{1}_0 = \mathbf{1}_0\tilde{H}\mathbf{1}_0 \\ & f & \longmapsto & f\mathbf{1}_0 \\ & q^{-\ell(w_0)}W_0(q^2)\,s_\lambda & \longmapsto & C'_{n_\lambda} \\ & \dfrac{W_\mu(q^{-2})}{W_0(q^{-2})}\,P_\mu(x;q^{-2}) & \longmapsto & M_\mu \end{array} \tag{2.11}$$

where w_0 is the longest element of W. The following three lemmas (of independent interest) are used in the proof of Theorem 2.9.

Lemma 2.5 *Let t_α, $\alpha \in R^+$, be commuting variables indexed by the positive roots. For $\lambda \in P^+$ let $P_\lambda(x;t)$ be as in (2.9), W_λ as in (2.8), and define*

$$R_\lambda(x;t_\alpha) = \sum_{w\in W} w\left(x^\lambda \prod_{\alpha\in R^+} \frac{1-t_\alpha x^{-\alpha}}{1-x^{-\alpha}}\right)$$

and

$$W_\lambda(t_\alpha) = \sum_{w\in W_\lambda}\Big(\prod_{\alpha\in R(w)} t_\alpha\Big),$$

where, as in (1.5), $R(w) = \{\alpha \in R^+ \mid w\alpha < 0\}$ is the inversion set of w. Then

(a) $R_\lambda(x;t_\alpha) = \sum_{\mu\in P^+} u_{\lambda\mu} s_\mu$,
with $u_{\lambda\mu} \in \mathbb{Z}[t_\alpha]$, $u_{\lambda\mu} = 0$ unless $\mu \leqslant \lambda$, and $u_{\lambda\lambda} = W_\lambda(t_\alpha)$.

(b) $P_\lambda(x;t) = \sum_{\mu\in P^+} c_{\lambda\mu} s_\mu$, *with $c_{\lambda\mu} \in \mathbb{Z}[t]$, $c_{\lambda\mu} = 0$ unless $\mu \leqslant \lambda$, and $c_{\lambda\lambda} = 1$.*

Proof (a) If $E \subseteq R^+$ let

$$t_E = \prod_{\alpha\in E} t_\alpha \quad \text{and} \quad \alpha_E = \sum_{\alpha\in E} \alpha,$$

and let a_μ be as defined in (2.4). Using the Weyl denominator formula, Proposition 2.3(a), and the second equality in (2.5),

$$\begin{aligned}
R_\lambda &= \sum_{w\in W} w\left(x^\lambda \prod_{\alpha\in R^+} \frac{1-t_\alpha x^{-\alpha}}{1-x^{-\alpha}}\right)\\
&= \sum_{w\in W} w\left(\frac{x^{\lambda+\rho}\prod_{\alpha\in R^+}(1-t_\alpha x^{-\alpha})}{x^\rho \prod_{\alpha\in R^+}(1-x^{-\alpha})}\right)\\
&= \frac{1}{a_\rho}\sum_{w\in W}(-1)^{\ell(w)} w\left(x^{\lambda+\rho}\prod_{\alpha\in R^+}(1-t_\alpha x^{-\alpha})\right)\\
&= \frac{1}{a_\rho}\sum_{w\in W}(-1)^{\ell(w)} w\left(\sum_{E\subseteq R^+}(-1)^{|E|} t_E x^{\lambda+\rho-\alpha_E}\right)\\
&= \frac{1}{a_\rho}\sum_{E\subseteq R^+}(-1)^{|E|} t_E a_{\lambda+\rho-\alpha_E} \;=\; \sum_{E\subseteq R^+}(-1)^{|E|} t_E s_{\lambda-\alpha_E},
\end{aligned}$$

which shows that R_λ is a symmetric function and $u_{\lambda\mu} \in \mathbb{Z}[t_\alpha]$.

By the straightening law for Weyl characters (2.7), $s_{\lambda-\alpha_E} = 0$ or $s_{\lambda-\alpha_E} = (-1)^{\ell(v)} s_\mu$ with

$$v \in W \text{ and } \mu \in P^+ \text{ such that } \quad \mu+\rho = v^{-1}(\lambda+\rho-\alpha_E).$$

Let E^c denote the complement of E in R^+. Since v permutes the elements of R^+,

$$\begin{aligned} v^{-1}(\lambda+\rho-\alpha_E) &= v^{-1}\lambda + v^{-1}\left(\tfrac{1}{2}\sum_{\alpha\in E^c}\alpha - \tfrac{1}{2}\sum_{\alpha\in E}\alpha\right) \\ &= v^{-1}\lambda + \left(\tfrac{1}{2}\sum_{\alpha\in F^c}\alpha - \tfrac{1}{2}\sum_{\alpha\in F}\alpha\right) = v^{-1}\lambda+\rho-\alpha_F, \end{aligned}$$

for some subset $F \subseteq R^+$ (which could be determined explicitly in terms of E and v). Hence

$$\mu = v^{-1}\lambda+\rho-\alpha_F-\rho = v^{-1}\lambda-\alpha_F \leqslant v^{-1}\lambda \leqslant \lambda. \tag{2.12}$$

This proves that $u_{\lambda\mu} = 0$ unless $\mu \leqslant \lambda$.

In (2.12), $\mu = \lambda$ only if $v^{-1}\lambda = \lambda$ and $\rho = \rho - \alpha_F = v^{-1}(\rho-\alpha_E)$ in which case

$$\rho - \alpha_E = v\left(\tfrac{1}{2}\sum_{\alpha\in R^+}\alpha\right) = \rho - \sum_{\alpha\in R(v)}\alpha \quad \text{and} \quad E = R(v).$$

Thus

$$u_{\lambda\lambda}(t_\alpha) = \sum_{v^{-1}\in W_\lambda} t_{R(v)}.$$

(b) Set $t_\alpha = t$ for all $\alpha \in R^+$. Applying (a) with $\lambda = 0$,

$$R_0(x;t) = \sum_{w\in W} w\left(\prod_{\alpha\in R^+}\frac{1-tx^{-\alpha}}{1-x^{-\alpha}}\right) = W_0(t). \tag{2.13}$$

Let W^λ be a set of minimal length coset representatives of the cosets in W/W_λ. Every element $w \in W$ can be written uniquely as $w = uv$ with $u \in W^\lambda$ and $v \in W_\lambda$ (see [1, IV §1 Ex. 3]). Let

$$Z(\lambda) = \{\alpha \in R^+ \mid \langle\lambda, \alpha^\vee\rangle = 0\},$$

and let $Z(\lambda)^c$ be the complement of $Z(\lambda)$ in R^+. Then $v \in W_\lambda$ permutes the elements of $Z(\lambda)^c$ among themselves and so

$$\begin{aligned} R_\lambda(x;t) &= \sum_{u\in W^\lambda} u\left(x^\lambda\Big(\prod_{\alpha\in Z(\lambda)^c}\frac{1-tx^{-\alpha}}{1-x^{-\alpha}}\Big)\sum_{v\in W_\lambda} v\Big(\prod_{\alpha\in Z(\lambda)}\frac{1-tx^{-\alpha}}{1-x^{-\alpha}}\Big)\right) \\ &= \sum_{u\in W^\lambda} u\left(x^\lambda\Big(\prod_{\alpha\in Z(\lambda)^c}\frac{1-tx^{-\alpha}}{1-x^{-\alpha}}\Big)W_\lambda(t)\right), \end{aligned}$$

where the last equality follows from (2.13). Thus there is an element $P_\lambda(x;t) \in \mathbb{F}[P]$ where $\mathbb{F}$ is the field of fractions of $\mathbb{Z}[t_\alpha]$ such that

$$R_\lambda(x;t) = W_\lambda(t) \sum_{u \in W^\lambda} u \left(x^\lambda \prod_{\alpha \in Z(\lambda)^c} \frac{1 - tx^{-\alpha}}{1 - x^{-\alpha}} \right) = W_\lambda(t) P_\lambda(x;t).$$

Since R_λ is a symmetric polynomial (an element of $\mathbb{Z}[t][P]^W$), $P_\lambda(x;t) \in \mathbb{F}[P]^W$. Since t only occurs in the numerators of the terms in the sum defining P_λ in fact P_λ is a symmetric polynomial with coefficients in $\mathbb{Z}[t]$. It follows that all the $u_{\lambda\mu}$ appearing in part (a) are divisible by $W_\lambda(t)$ and

$$P_\lambda(x;t) = \sum_{\mu \in P} c_{\lambda\mu} s_\mu, \quad \text{where } c_{\lambda\mu} = \frac{1}{W_\lambda(t)} u_{\lambda\mu}$$

are polynomials in $\mathbb{Z}[t]$ such that $c_{\lambda\lambda} = 1$ and $c_{\lambda\mu} = 0$ unless $\mu \leqslant \lambda$. □

Lemma 2.5 has the following interesting (and useful) corollary, see [13].

Corollary 2.6 *Let ρ and $\alpha^\vee$ be as in (1.8) and (1.1), respectively, and let $W_0(t)$ be as defined in (2.8).*

(a) $$\sum_{w \in W} w \left(\prod_{\alpha \in R^+} \frac{1 - tx^{-\alpha}}{1 - x^{-\alpha}} \right) = W_0(t).$$

(b) $$\prod_{\alpha \in R^+} \frac{1 - t^{\langle \rho, \alpha^\vee \rangle + 1}}{1 - t^{\langle \rho, \alpha^\vee \rangle}} = W_0(t).$$

Proof Part (a) follows from Lemma 2.5 (a) by setting $\lambda = 0$ and specializing $t_\alpha = t$ for all $\alpha \in R^+$.
(b) Applying the homomorphism

$$\begin{array}{ccc} \mathbb{Z}[t^{\pm 1}][P] & \longrightarrow & \mathbb{Z}[t^{\pm 1}] \\ x^\lambda & \longmapsto & t^{\langle -\rho, \lambda \rangle} \end{array}$$

to both sides of the identity in (a) for the root system $R^\vee = \{\alpha^\vee \mid \alpha \in R\}$ gives

$$W_0(t) = \sum_{w \in W} \prod_{\alpha \in R^+} \left(\frac{1 - t^{\langle \rho, w\alpha^\vee \rangle + 1}}{1 - t^{\langle \rho, w\alpha^\vee \rangle}} \right). \tag{2.14}$$

If $w \in W$, $w \neq 1$, and $w = s_{i_1} \cdots s_{i_p}$ is a reduced word for w then $w^{-1}(-\alpha_{i_1}) = (s_{i_1} w)^{-1} \alpha_{i_1} \in R(w)$ and so

there is an $\alpha \in R^+$ such that $w\alpha^\vee = -\alpha_{i_1}^\vee$.

Then

$$\prod_{\alpha\in R^+} \frac{1-t^{\langle\rho,w\alpha^\vee\rangle+1}}{1-t^{\langle\rho,w\alpha^\vee\rangle}} = \frac{1-t^{\langle\rho,-\alpha_{i_1}^\vee\rangle+1}}{1-t^{\langle\rho,-\alpha_{i_1}^\vee\rangle}} \prod_{\substack{\alpha\in R^+\\ w\alpha\neq-\alpha_{i_1}}} \frac{1-t^{\langle\rho,w\alpha^\vee\rangle+1}}{1-t^{\langle\rho,w\alpha^\vee\rangle}}$$

$$= \frac{1-t^{-1+1}}{1-t^{-1}} \prod_{\substack{\alpha\in R^+\\ w\alpha\neq-\alpha_{i_1}}} \frac{1-t^{\langle\rho,w\alpha^\vee\rangle+1}}{1-t^{\langle\rho,w\alpha^\vee\rangle}} = 0\,.$$

Thus the only nonzero term on the right hand side of (2.14) occurs for $w = 1$. □

Lemma 2.7 *For $\lambda \in P^+$ let $t_\lambda \in \tilde{W}$ be the translation in λ and let n_λ be the maximal length element in the double coset $Wt_\lambda W$. Let $M_\lambda = \mathbf{1}_0 x^\lambda \mathbf{1}_0$, as in (2.4). Then*

$$q^{-\ell(w_0)} W_0(q^2)\cdot \frac{W_0(q^{-2})}{W_\lambda(q^{-2})}\cdot M_\lambda = \sum_{x\in Wt_\lambda W} q^{\ell(x)-\ell(n_\lambda)} T_x,$$

in the affine Hecke algebra $\tilde{H}$.

Proof Let $\lambda \in P^+$. Let $W_\lambda = \mathrm{Stab}(\lambda)$ and let w_0 and w_λ be the maximal length elements in W and W_λ, respectively. Let m_λ and n_λ be the minimal and maximal length elements respectively in the double coset $Wt_\lambda W$. For each positive root α the hyperplanes $H_{\alpha,i}$, $1 \leqslant i \leqslant \langle\lambda,\alpha^\vee\rangle$, are between the fundamental alcove A and the alcove $t_\lambda A$ and so

$$\ell(t_\lambda) = \sum_{\alpha\in R^+} \langle\lambda,\alpha^\vee\rangle = 2\langle\lambda,\rho^\vee\rangle, \quad \text{where } \rho^\vee = \tfrac{1}{2}\sum_{\alpha\in R^+}\alpha^\vee. \tag{2.15}$$

Since $m_\lambda = t_\lambda(w_\lambda w_0)$ and $n_\lambda = t_{w_0\lambda} w_0$,

$$\begin{aligned} \ell(m_\lambda) &= \ell(t_\lambda) - \ell(w_0 w_\lambda) = \ell(t_\lambda) - (\ell(w_0) - \ell(w_\lambda)), \quad \text{and} \\ \ell(n_\lambda) &= \ell(t_\lambda) + \ell(w_0) = \ell(m_\lambda) + \ell(w_0) - \ell(w_\lambda) + \ell(w_0). \end{aligned} \tag{2.16}$$

For example, in the setting of Example 1.1, if $\lambda = 2\omega_2$ in type C_2, then $W_\lambda = \{1, s_1\}$, $w_\lambda = s_1$, $w_0 = s_1s_2s_1s_2$, $\ell(t_\lambda) = 6$, $\ell(m_\lambda) = 3$, and $\ell(n_\lambda) = 10$. Labeling the alcove wA by the element w, the 32 alcoves wA with $w \in Wt_\lambda W$

make up the four shaded diamonds.

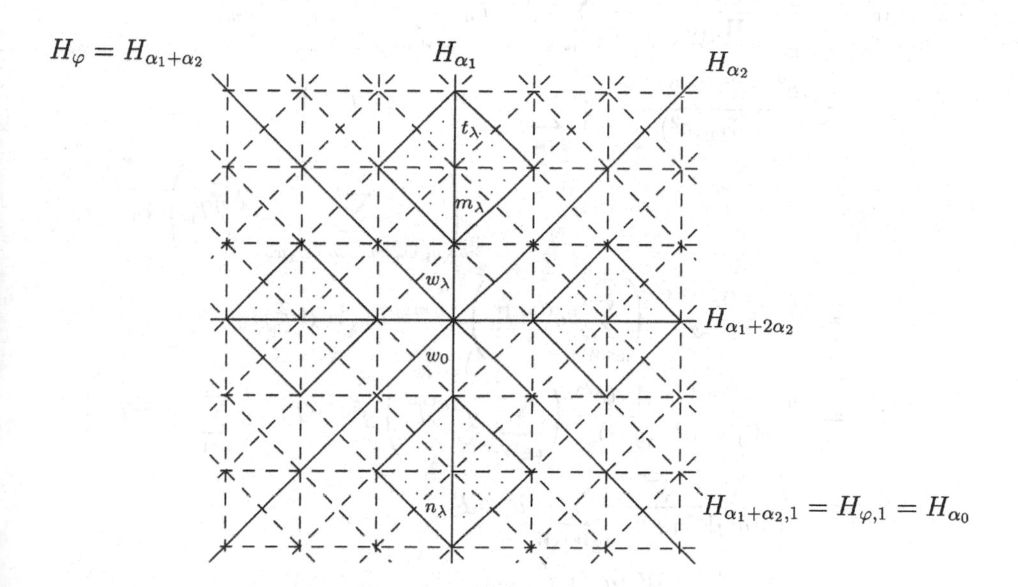

The double coset $Wt_\lambda W$

Then

$$\begin{aligned}
\mathbf{1}_0 x^\lambda \mathbf{1}_0 &= \mathbf{1}_0 T_{t_\lambda} \mathbf{1}_0 = \mathbf{1}_0 T_{m_\lambda w_0 w_\lambda} \mathbf{1}_0 = \mathbf{1}_0 T_{m_\lambda} T_{w_0 w_\lambda} \mathbf{1}_0 \\
&= q^{\ell(w_0)-\ell(w_\lambda)} \mathbf{1}_0 T_{m_\lambda} \mathbf{1}_0 \\
&= \frac{q^{\ell(w_0)-\ell(w_\lambda)-\ell(m_\lambda)}}{W(q^2)} \left(\sum_{w \in W} q^{\ell(w)} T_w \right) q^{\ell(m_\lambda)} T_{m_\lambda} \mathbf{1}_0.
\end{aligned}$$

Let W^λ be a set of minimal length coset representatives of the cosets in W/W_λ. Every element $w \in W$ has a unique expression $w = uv$ with $u \in W^\lambda$ and $v \in W_\lambda$. If $v \in W_\lambda$ then

$$vm_\lambda = vt_\lambda w_\lambda w_0 = t_\lambda v w_\lambda w_0 = m_\lambda (w_\lambda w_0)^{-1} v w_\lambda w_0 = m_\lambda (w_0^{-1} w_\lambda^{-1} v w_\lambda w_0).$$

Since conjugation by w_λ and conjugation by w_0 are automorphisms of W_λ and W respectively taking simple reflections to simple reflections,

$$\ell(v) = \ell(w_0^{-1} w_\lambda^{-1} v w_\lambda w_0)\ .$$

Thus

$$
\begin{aligned}
\mathbf{1}_0 x^\lambda \mathbf{1}_0 &= \frac{q^{\ell(w_0)-\ell(w_\lambda)-\ell(m_\lambda)}}{W_0(q^2)} \sum_{u\in W^\lambda} q^{\ell(u)}T_u \sum_{v\in W_\lambda} q^{\ell(v)}T_v q^{\ell(m_\lambda)}T_{m_\lambda}\mathbf{1}_0 \\
&= \frac{q^{2\ell(w_0)-2\ell(w_\lambda)-\ell(t_\lambda)}}{W_0(q^2)} \left(\sum_{u\in W^\lambda} q^{\ell(u)}T_u q^{\ell(m_\lambda)}T_{m_\lambda}\right) \\
&\qquad \cdot \left(\sum_{v\in w_0^{-1}w_\lambda^{-1}W_\lambda w_\lambda w_0} q^{\ell(v)}T_v\right)\mathbf{1}_0 \\
&= \frac{q^{-2\ell(w_\lambda)-\ell(t_\lambda)}}{W_0(q^{-2})} \left(\sum_{u\in W^\lambda} q^{\ell(u)}T_u\right) q^{\ell(m_\lambda)}T_{m_\lambda}W_\lambda(q^2)\mathbf{1}_0 \\
&= \frac{q^{-2\ell(w_\lambda)-\ell(t_\lambda)}W_\lambda(q^2)}{W_0(q^2)W_0(q^{-2})} \left(\sum_{u\in W^\lambda} q^{\ell(u)}T_u\right) q^{\ell(m_\lambda)}T_{m_\lambda}\left(\sum_{w\in W} q^{\ell(w)}T_w\right) \\
&= \frac{q^{-\ell(t_\lambda)}W_\lambda(q^{-2})}{W_0(q^2)W_0(q^{-2})} \sum_{x\in Wt_\lambda W} q^{\ell(x)}T_x \\
&= \frac{q^{-\ell(t_\lambda)+\ell(n_\lambda)}W_\lambda(q^{-2})}{W_0(q^2)W_0(q^{-2})} \left(\sum_{x\in Wt_\lambda W} q^{\ell(x)-\ell(n_\lambda)}T_x\right) \\
&= \frac{q^{\ell(w_0)}}{W_0(q^2)}\frac{W_\lambda(q^{-2})}{W_0(q^{-2})} \left(\sum_{x\in Wt_\lambda W} q^{\ell(x)-\ell(n_\lambda)}T_x\right).
\end{aligned}
$$

□

Lemma 2.8 *Let w_0 be the longest element of W and let $\lambda \in P$.*

(a) $\overline{x^\lambda} = T_{w_0}x^{w_0\lambda}T_{w_0}^{-1}$.

(b) $\overline{\mathbf{1}_0} = \mathbf{1}_0$ *and* $\overline{\varepsilon_0} = \varepsilon_0$.

(c) *If* $z \in \mathbb{Z}[P]^W$ *then* $\bar{z} = z$.

(d) $\overline{q^{-\ell(w_0)}A_{\lambda+\rho}} = q^{-\ell(w_0)}A_{\lambda+\rho}$.

Proof (a) If $\lambda \in P^+$ then $w_0t_\lambda = t_{w_0\lambda}w_0$, $\ell(w_0t_\lambda) = \ell(w_0) + \ell(t_\lambda)$ and $\ell(t_{w_0\lambda}w_0) = \ell(t_{w_0\lambda}) + \ell(w_0)$. Thus,

$$T_{w_0}T_{t_\lambda} = T_{w_0t_\lambda} = T_{t_{w_0\lambda}w_0} = T_{t_{w_0\lambda}}T_{w_0}, \quad \text{for } \lambda \in P^+.$$

Let $\lambda \in P$ and write $\lambda = \mu - \nu$ with $\mu, \nu \in P^+$. Since $-w_0\mu \in P^+$ and $-w_0\nu \in P^+$,

$$
\begin{aligned}
\overline{x^\lambda} &= \overline{T_{t_\mu}T_{t_\nu}^{-1}} = T_{t_{-\mu}}^{-1}T_{t_{-\nu}} = T_{w_0}T_{t_{-w_0\mu}}^{-1}T_{t_{-w_0\nu}}T_{w_0}^{-1} \\
&= T_{w_0}(x^{-w_0\lambda})^{-1}T_{w_0}^{-1} = T_{w_0}x^{w_0\lambda}T_{w_0}^{-1}.
\end{aligned}
$$

(b) For $1 \leqslant i \leqslant n$,

$$\overline{\mathbf{1}_0^2} = \overline{\mathbf{1}_0}^2 \quad \text{and} \quad T_i\overline{\mathbf{1}_0} = \overline{T_i^{-1}\mathbf{1}_0} = \overline{q^{-1}\mathbf{1}_0} = q\overline{\mathbf{1}_0},$$

$$\overline{\varepsilon_0^2} = \overline{\varepsilon_0}^2 \quad \text{and} \quad T_i\overline{\varepsilon_0} = \overline{T_i^{-1}\varepsilon_0} = \overline{-q\varepsilon_0} = -q^{-1}\overline{\varepsilon_0}.$$

These are the defining properties (before 2.1) of $\mathbf{1}_0$ and ε_0 and so $\overline{\mathbf{1}_0} = \mathbf{1}_0$ and $\overline{\varepsilon_0} = \varepsilon_0$.

(c) If $z = \sum_{\mu\in P} c_\mu x^\mu \in \mathbb{Z}[P]^W$, then, since $c_\mu \in \mathbb{Z}$, $\overline{c_\mu} = c_\mu$ and, by (a),

$$\overline{z} = \sum_{\mu\in P} \overline{c_\mu}\,\overline{x^\mu} = \sum_{\mu\in P} c_\mu T_{w_0} x^{w_0\mu} T_{w_0}^{-1} = T_{w_0}\left(\sum_{\mu\in P} c_\mu x^{w_0\mu}\right) T_{w_0}^{-1} = T_{w_0} z T_{w_0}^{-1},$$

since $z \in \mathbb{Z}[P]^W$ is W-invariant. Finally, since $\mathbb{Z}[P]^W \subseteq Z(\tilde{H})$, z is central, and $\overline{z} = T_{w_0} z T_{w_0}^{-1} = z$.

(d) By (a), (b) and the third equality in (2.5),

$$\begin{aligned}
\overline{q^{-\ell(w_0)}A_{\lambda+\rho}} &= q^{\ell(w_0)}\overline{\varepsilon_0 x^{\lambda+\rho}\mathbf{1}_0} = q^{\ell(w_0)}\varepsilon_0 T_{w_0} x^{w_0(\lambda+\rho)} T_{w_0}^{-1}\mathbf{1}_0 \\
&= q^{\ell(w_0)}(-q^{-1})^{\ell(w_0)}\varepsilon_0 x^{w_0(\lambda+\rho)}\mathbf{1}_0 q^{-\ell(w_0)} = (-q^{-1})^{\ell(w_0)} A_{w_0(\lambda+\rho)} \\
&= q^{-\ell(w_0)} A_{\lambda+\rho}.
\end{aligned}$$

□

The following theorem is due to Lusztig [10]. Part (a) was originally proved in a different formulation by Macdonald [12, (4.1.2)].

Theorem 2.9 *If $\mu \in P$ let W_μ be the stabilizer of μ and let $W_\mu(t)$ be as in (2.8).*

(a) *Let $\mu \in P$. Let $P_\mu(x;t)$ be the Macdonald spherical function defined in (2.9) and define $M_\mu = \mathbf{1}_0 x^\mu \mathbf{1}_0$ as in (2.4). In the affine Hecke algebra $\tilde{H}$,*

$$\frac{W_\mu(q^{-2})}{W_0(q^{-2})} \cdot P_\mu(x;q^{-2})\mathbf{1}_0 = M_\mu.$$

(b) *For $\lambda \in P^+$ let $t_\lambda \in \tilde{W}$ be the translation in λ and let n_λ be the maximal length element in the double coset $Wt_\lambda W$. Let s_λ be the Weyl character and let C'_{n_λ} be the Kazhdan–Lusztig basis element as defined in (2.6) and (1.26), respectively. In the affine Hecke algebra $\tilde{H}$,*

$$q^{-\ell(w_0)}W_0(q^2) \cdot s_\lambda \mathbf{1}_0 = C'_{n_\lambda}.$$

Proof (a) By Theorem 2.4 there is an element $\tilde{P}_\lambda \in \mathbb{K}[P]^W$ such that $\tilde{P}_\lambda \mathbf{1}_0 = \mathbf{1}_0 x^\lambda \mathbf{1}_0$. To find $\tilde{P}_\lambda$ first do a rank 1 calculation,

$$\begin{aligned}
(q^{-1}+T_i)x^\lambda \mathbf{1}_0 &= \left(q^{-1}x^\lambda + x^{s_i\lambda}T_i + (q-q^{-1})\left(\frac{x^\lambda - x^{s_i\lambda}}{1-x^{-\alpha_i}}\right)\right)\mathbf{1}_0 \\
&= \frac{1}{1-x^{-\alpha_i}}\left(\begin{array}{c} q^{-1}x^\lambda(1-x^{-\alpha_i}) + qx^{s_i\lambda}(1-x^{-\alpha_i}) \\ +qx^\lambda - qx^{s_i\lambda} - q^{-1}x^\lambda + q^{-1}x^{s_i\lambda}\end{array}\right)\mathbf{1}_0 \\
&= (1-x^{-\alpha_i})^{-1}(-q^{-1}x^{\lambda-\alpha_i} - qx^{s_i\lambda-\alpha_i} + qx^\lambda + q^{-1}x^{s_i\lambda})\mathbf{1}_0 \\
&= (1-x^{-\alpha_i})^{-1}(x^\lambda(q-q^{-1}x^{-\alpha_i}) + x^{s_i\lambda}(q^{-1}-qx^{-\alpha_i}))\mathbf{1}_0 \\
&= \left(\frac{q-q^{-1}x^{-\alpha_i}}{1-x^{-\alpha_i}}\cdot x^\lambda + \frac{x^{-\alpha_i}}{x^{-\alpha_i}}\cdot\frac{q^{-1}x^{\alpha_i}-q}{x^{\alpha_i}-1}\cdot x^{s_i\lambda}\right)\mathbf{1}_0 \\
&= (1+s_i)\left(\frac{q-q^{-1}x^{-\alpha_i}}{1-x^{-\alpha_i}}x^\lambda\right)\mathbf{1}_0.
\end{aligned}$$

Since $\mathbf{1}_0$ is a linear combination of products of T_i it can also be written as a linear combination of products of $q^{-1}+T_i$. Thus $\mathbf{1}_0 x^\lambda \mathbf{1}_0$ can be written as a linear combination of terms of the form

$$(1+s_{i_1})\left(\frac{q-q^{-1}x^{-\alpha_{i_1}}}{1-x^{-\alpha_{i_1}}}\right)\cdots(1+s_{i_p})\left(\frac{q-q^{-1}x^{-\alpha_{i_p}}}{1-x^{-\alpha_{i_p}}}\right)x^\lambda\mathbf{1}_0.$$

Thus

$$\mathbf{1}_0 x^\lambda \mathbf{1}_0 = \tilde{P}_\lambda \mathbf{1}_0, \quad \text{where} \quad \tilde{P}_\lambda = \sum_{w\in W} x^{w\lambda} w c_w,$$

and the c_w are some linear combinations of products of terms of the form $(q-q^{-1}x^\alpha)/(1-x^\alpha)$ for roots $\alpha \in R$. Since $\tilde{P}_\lambda$ is an element of $\mathbb{K}[P]^W$,

$$\tilde{P}_\lambda = \sum_{w\in W} w(x^{w_0\lambda}w_0c_{w_0}),$$

where w_0 is the longest element of W. The coefficient $w_0c_{w_0}$ comes from the highest term in the expansion of

$$\mathbf{1}_0 = \frac{1}{W_0(q^2)}(q^{2\ell(w_0)}T_{w_0} + \text{lower terms})$$

in terms of linear combination of products of the $(q^{-1}+T_i)$. If $w_0 = s_{i_1}\cdots s_{i_p}$ is a reduced word for w_0 then

$$\begin{aligned}
w_0c_{w_0} &= \frac{q^{\ell(w_0)}}{W_0(q^2)}s_{i_1}\left(\frac{q-q^{-1}x^{-\alpha_{i_1}}}{1-x^{-\alpha_{i_1}}}\right)\cdots s_{i_p}\left(\frac{q-q^{-1}x^{-\alpha_{i_p}}}{1-x^{-\alpha_{i_p}}}\right) \\
&= \frac{q^{\ell(w_0)}}{W_0(q^2)}s_{i_1}\cdots s_{i_p}\left(\frac{q-q^{-1}x^{-s_{i_p}\cdots s_{i_2}\alpha_{i_1}}}{1-x^{-s_{i_p}\cdots s_{i_2}\alpha_{i_1}}}\right)\left(\frac{q-q^{-1}x^{-s_{i_p}\cdots s_{i_3}\alpha_{i_2}}}{1-x^{-s_{i_p}\cdots s_{i_2}\alpha_{i_2}}}\right) \\
&\qquad\cdots\left(\frac{q-q^{-1}x^{-\alpha_{i_p}}}{1-x^{-\alpha_{i_p}}}\right) \\
&= \frac{q^{\ell(w_0)}}{W_0(q^2)}w_0\prod_{\alpha\in R^+}\frac{q-q^{-1}x^{-\alpha}}{1-x^{-\alpha}} = \frac{q^{2\ell(w_0)}}{W_0(q^2)}w_0\prod_{\alpha\in R^+}\frac{1-q^{-2}x^{-\alpha}}{1-x^{-\alpha}},
\end{aligned}$$

by Lemma 1.2 and the fact that $\ell(w_0) = \mathrm{Card}(R^+)$. Thus, since $q^{-2\ell(w_0)}W_0(q^2) = W_0(q^{-2})$,

$$\tilde{P}_\lambda = \frac{1}{W_0(q^{-2})} \sum_{w \in W} w \left(x^\lambda \prod_{\alpha \in R^+} \frac{1 - q^{-2}x^{-\alpha}}{1 - x^{-\alpha}} \right).$$

(b) Since $W_0(q^{-2}) = q^{-2\ell(w_0)}W_0(q^2)$, Lemma 2.8 gives

$$\overline{q^{-\ell(w_0)}W_0(q^2)s_\lambda \mathbf{1}_0} = q^{\ell(w_0)}W_0(q^{-2})\overline{s_\lambda}\mathbf{1}_0 = q^{-\ell(w_0)}W_0(q^2)s_\lambda \mathbf{1}_0.$$

By Lemma 2.5(b),

$$s_\lambda = \sum_{\mu \in P^+} K_{\lambda\mu}(t)P_\mu(x;t),$$

where $K_{\lambda\mu}(t) \in \mathbb{Z}[t]$, $K_{\lambda\mu}(t) = 0$ unless $\mu \leqslant \lambda$ and $K_{\lambda\lambda}(t) = 1$. Thus, by part (a) and Lemma 2.7

$$\begin{aligned} q^{-\ell(w_0)}W_0(q^2)s_\lambda \mathbf{1}_0 &= \sum_{\mu \in P^+} q^{-\ell(w_0)}W_0(q^2)K_{\lambda\mu}(q^{-2})P_\mu(x;q^{-2})\mathbf{1}_0 \\ &= \sum_{\mu \in P^+} \sum_{x \in Wt_\mu W} q^{\ell(x)-\ell(n_\mu)}K_{\lambda\mu}(q^{-2})T_x, \end{aligned}$$

where the polynomials $K_{\lambda\mu}(q^{-2}) \in \mathbb{Z}[q^{-2}]$ are 0 unless $\mu \leqslant \lambda$ and $K_{\lambda\lambda}(q^{-2}) = 1$. Hence $q^{-\ell(w_0)}W(q^2)s_\lambda \mathbf{1}_0$ is a bar invariant element of $\tilde{H}$ such that its expansion in terms of the basis $\{T_w \mid w \in \tilde{W}\}$ is triangular with coefficient of T_{n_λ} equal to 1 and all other coefficients in $q^{-1}\mathbb{Z}[q^{-1}]$. These are the defining properties (1.26)–(1.27) of C'_{n_λ}. □

3 Orthogonality and formulae for Kostka–Foulkes polynomials

Let $\mathbb{K} = \mathbb{Z}[t]$. If $f = \sum_{\mu \in P} f_\mu x^\mu \in \mathbb{K}[P]$ let

$$\bar{f} = \sum_{\mu \in P} f_\mu x^{-\mu}, \quad \text{and} \quad [f]_1 = f_0 = (\text{coefficient of 1 in } f). \tag{3.1}$$

Define a symmetric bilinear form

$$\langle\, ,\rangle_t : \mathbb{K}[P] \times \mathbb{K}[P] \to \mathbb{K} \quad \text{by} \quad \langle f, g\rangle_t = \frac{1}{|W|}\left[f\bar{g} \prod_{\alpha \in R} \frac{1 - x^\alpha}{1 - tx^\alpha} \right]_1. \tag{3.2}$$

"Specializing" t at the values 0 and 1 gives inner products

$$\langle\, ,\rangle_0 : \mathbb{K}[P] \times \mathbb{K}[P] \to \mathbb{K} \qquad \text{and} \qquad \langle\, ,\rangle_1 : \mathbb{K}[P] \times \mathbb{K}[P] \to \mathbb{K}$$

with

$$\langle f, g\rangle_0 = \frac{1}{|W|}\left[f\bar{g} \prod_{\alpha \in R}(1 - x^\alpha) \right]_1 \quad \text{and} \quad \langle f, g\rangle_1 = \frac{1}{|W|}[f\bar{g}]_1. \tag{3.3}$$

Proposition 3.1 *Let λ and $\mu \in P^+$. Then*

$$\langle m_\lambda, m_\mu\rangle_1 = \frac{1}{|W_\lambda|}\delta_{\lambda\mu}, \quad \langle s_\lambda, s_\mu\rangle_0 = \delta_{\lambda\mu}, \quad \textit{and} \quad \langle P_\lambda, P_\mu\rangle_t = \frac{1}{W_\lambda(t)}\delta_{\lambda\mu}.$$

Proof Letting $W\lambda$ denote the W-orbit of λ, the first equality follows from

$$|W_\lambda|\langle m_\lambda, m_\mu\rangle_1 = \frac{|W_\lambda|}{|W|}\sum_{\gamma\in W\lambda, \nu\in W\mu}[x^\gamma x^{-\nu}]_1 = \delta_{\lambda\mu}\frac{|W_\lambda|}{|W|}\sum_{\gamma\in W\lambda} 1 = \delta_{\lambda\mu}.$$

If $\lambda, \mu \in P^+$,

$$\begin{aligned}
\langle s_\lambda, s_\mu\rangle_0 &= \frac{1}{|W|}[\overline{a_\rho s_\lambda} a_\rho s_\mu]_1 = \frac{1}{|W|}[\overline{a_{\lambda+\rho}} a_{\mu+\rho}]_1 \\
&= \frac{1}{|W|}\sum_{v,w\in W}(-1)^{\ell(v)}(-1)^{\ell(w)}[x^{-v(\lambda+\rho)}x^{w(\mu+\rho)}]_1 \\
&= \delta_{\lambda\mu}\frac{1}{|W|}\sum_{v\in W}(-1)^{\ell(v)}(-1)^{\ell(v)} \;=\; \delta_{\lambda\mu},
\end{aligned}$$

giving the second statement.

By Lemma 2.5(b) the matrix K^{-1} given by the values $(K^{-1})_{\lambda\mu}$ in the equation

$$P_\lambda(x;t) = \sum_\mu (K^{-1})_{\lambda\mu}s_\mu,$$

has entries in $\mathbb{Z}[t]$ and is upper triangular with 1's on the diagonal, that is, $(K^{-1})_{\lambda\lambda} = 1$ and $(K^{-1})_{\lambda\mu} = 0$ unless $\mu \leqslant \lambda$. Since $P_\lambda(x;1) = m_\lambda$ the matrix k^{-1} describing the change of basis

$$m_\lambda = \sum_\mu (k^{-1})_{\lambda\mu}s_\mu,$$

is the specialization of K^{-1} at $t = 1$ and so k^{-1} has entries in $\mathbb{Z}$ and is upper triangular with 1's on the diagonal. Then the matrix $A = K^{-1}k^{-1}$ giving the change of basis

$$P_\lambda(x;t) = \sum_{\nu\leqslant\lambda} A_{\lambda\nu}m_\mu, \tag{3.4}$$

has $A_{\lambda\mu} \in \mathbb{Z}[t]$, $A_{\lambda\lambda} = 1$, and $A_{\lambda\mu} = 0$ unless $\mu \leqslant \lambda$.

Let Q^+ be the set of nonnegative integral linear combinations of positive roots. Then

$$\begin{aligned}
P_\mu(x;t)W_\mu(t)\Big(\prod_{\alpha\in R}\frac{1-x^\alpha}{1-tx^\alpha}\Big) &= \sum_{w\in W} w\left(x^\mu \prod_{\alpha\in R^+}\frac{1-x^\alpha}{1-tx^\alpha}\right) \\
&= \sum_{w\in W} w\left(x^\mu \prod_{\alpha\in R^+}(1+\sum_{r>0} t^{r-1}(t-1)x^{r\alpha})\right) \\
&= \sum_{w\in W} w\left(\sum_{\nu\in Q^+} c_\nu x^{\mu+\nu}\right) \\
&= \sum_{\nu\in Q^+} c_\nu\left(\sum_{w\in W} w x^{\mu+\nu}\right),
\end{aligned}$$

where $c_\nu \in \mathbb{Z}[t]$ and $c_0 = 1$. Hence

$$P_\mu(x;t)W_\mu(t)\prod_{\alpha\in R}\frac{1-x^\alpha}{1-tx^\alpha} = |W_\mu|m_\mu + \sum_{\gamma>\mu} B_{\mu\gamma}m_\gamma = \sum_{\gamma\geqslant\mu} B_{\mu\gamma}m_\gamma, \tag{3.5}$$

with $B_{\mu\gamma} \in \mathbb{Z}[t]$ and $B_{\mu\mu} = |W_\mu|$.

Assume that $\lambda \leqslant \mu$ if λ and μ are comparable. Then, by using (3.4) and (3.5),

$$\begin{aligned}
\langle P_\lambda, P_\mu\rangle_t &= \frac{1}{W_\mu(t)}\left\langle P_\lambda, P_\mu W_\mu(t)\prod_{\alpha\in R}\frac{1-x^\alpha}{1-tx^\alpha}\right\rangle_1 \\
&= \frac{1}{W_\mu(t)}\left\langle \sum_{\nu\leqslant\lambda} A_{\lambda\nu}m_\nu, \sum_{\gamma\geqslant\mu} B_{\mu\gamma}m_\gamma\right\rangle_1 .
\end{aligned}$$

Since $A_{\lambda\lambda} = 1$ and $B_{\mu\mu} = |W_\mu|$ the result follows from $\langle m_\lambda, m_\mu\rangle_1 = |W_\lambda|^{-1}\delta_{\lambda\mu}$. □

The following theorem shows that the spherical functions $P_\lambda(x,t)$ are uniquely determined by the triangularity in (3.4) and the orthogonality in the third equality of Proposition 3.1.

Theorem 3.2 *Let $\mathbb{K} = \mathbb{Z}[t]$. The spherical functions $P_\lambda(x;t)$ are the unique elements of $\mathbb{K}[P]^W$ such that*

(a) $P_\lambda = m_\lambda + \sum_{\mu<\lambda} A_{\lambda\mu}m_\mu$,

(b) $\langle P_\lambda, P_\mu\rangle_t = 0$ *if* $\lambda \neq \mu$.

Proof Assume that the P_μ are determined for $\mu < \lambda$. Then the condition in (a) can be rewritten as

$$P_\lambda = m_\lambda + \sum_{\mu<\lambda} C_{\lambda\mu} P_\mu,$$

for some constants $C_{\lambda\mu}$. Take the inner product on each side with P_ν, $\nu < \lambda$, and use property (b) to get the system of equations

$$0 = \langle m_\lambda, P_\nu\rangle_t + \sum_{\mu<\lambda} C_{\lambda\mu}\langle P_\mu, P_\nu\rangle_t = \langle m_\lambda, P_\nu\rangle_t + C_{\lambda\nu}\langle P_\nu, P_\nu\rangle_t.$$

Hence

$$C_{\lambda\nu} = \frac{-\langle m_\lambda, P_\nu\rangle_t}{\langle P_\nu, P_\nu\rangle_t}, \qquad \text{for each } \nu < \lambda,$$

and this determines P_λ. □

Remark 3.3 (a) The inner product $\langle\ ,\ \rangle_t$ arises naturally in the context of p-adic groups. Let $S^1 = \{z \in \mathbb{C} \mid |z| = 1\}$ and view the x^λ, $\lambda \in P$, as characters of

$$T = Hom(P, S^1) \quad via \quad \begin{array}{rccc} x^\lambda : & T & \longrightarrow & \mathbb{C}^* \\ & s & \longmapsto & s(\lambda). \end{array} \tag{3.6}$$

Let ds be the Haar measure on T normalized so that

$$\langle x^\lambda, x^\mu\rangle = \int_T x^\lambda(s)\overline{x^\mu(s)}ds = \delta_{\lambda\mu}. \tag{3.7}$$

Letting $\mathbb{Q}_p$ be the field of p-adic numbers, Macdonald [12, (5.1.2)] showed that the Plancherel measure for the p-adic Chevalley group $G(\mathbb{Q}_p)$ corresponding to the root system R is given by

$$d\mu(s) = \frac{W_0(p^{-1})}{|W|}\prod_{\alpha\in R}\frac{1 - x^\alpha(s)}{1 - p^{-1}x^\alpha(s)}. \tag{3.8}$$

The corresponding inner product is

$$W_0(p^{-1})\langle f, g\rangle_{p^{-1}} = \int_T f(s)\overline{g(s)}d\mu(s), \quad \textit{for } f, g \in C(T),$$

where $C(T)$ is the vector space of continuous functions on T.

(b) The inner product $\langle\ ,\ \rangle_t$ arises naturally in another representation theoretic context. The complex semisimple Lie algebra $\mathfrak{g}$ corresponding to the root system R acts on $S(\mathfrak{g}^*)$, the ring of polynomials on $\mathfrak{g}$, by the (co-)adjoint action. As graded $\mathfrak{g}$-modules the characters of $S(\mathfrak{g}^*)$ and the subring of invariants $S(\mathfrak{g}^*)^{\mathfrak{g}}$ are

$$\begin{aligned} \mathrm{grch}(S(\mathfrak{g}^*)) &= \left(\prod_{i=1}^r \frac{1}{1-t}\right)\left(\prod_{\alpha\in R}\frac{1}{1-tx^\alpha}\right) \quad \textit{and} \\ \mathrm{grch}(S(\mathfrak{g}^*)^{\mathfrak{g}}) &= \prod_{i=1}^r \frac{1}{1-t^{d_i}} = \frac{1}{W_0(t)}\prod_{i=1}^r\frac{1}{1-t}, \end{aligned} \tag{3.9}$$

where r is the rank of $\mathfrak{g}$ and $d_1, \ldots, d_r$ are the *degrees* of the Weyl group W. Let $\mathcal{H}$ denote the vector space of harmonic polynomials. An important theorem of Kostant [8, Theorem 0.2] gives

$$S(\mathfrak{g}^*) \cong S(\mathfrak{g}^*)^{\mathfrak{g}} \otimes \mathcal{H}, \quad \textit{and thus,} \quad \mathrm{grch}(\mathcal{H}) = W_0(t) \prod_{\alpha \in R} \frac{1}{1 - tx^{\alpha}}. \tag{3.10}$$

If $L(\lambda)$ denotes the finite dimensional irreducible $\mathfrak{g}$-module of highest weight $\lambda \in P^+$ then $L(\lambda)$ has character s_λ and using the notation of (3.2),

$$\begin{aligned} &\sum_{k \geqslant 0} \dim(Hom_{\mathfrak{g}}(L(\lambda), L(\mu) \otimes \mathcal{H}^k) t^k \qquad (3.11) \\ &= \left\langle s_\lambda, s_\mu W_0(t) \prod_{\alpha \in R} \frac{1}{1 - tx^{\alpha}} \right\rangle_0 \\ &= W_0(t) \left[s_\lambda \overline{s_\mu} \prod_{\alpha \in R} \frac{1 - x^{\alpha}}{1 - tx^{\alpha}} \right]_1 = W_0(t) \langle s_\lambda, s_\mu \rangle_t, \end{aligned}$$

where $\mathcal{H}^k$ is the vector space of degree k harmonic polynomials.

3.1 Formulae for Kostka–Foulkes polynomials

For $\lambda \in P$ let s_λ denote the Weyl character, as defined in (2.6). The *Kostka–Foulkes polynomials*, or *q-weight multiplicities*, $K_{\lambda\mu}(t)$, $\lambda, \mu \in P^+$, are defined by the change of basis formula

$$s_\lambda = \sum_{\mu \in P^+} K_{\lambda\mu}(t) P_\mu(x; t), \tag{3.12}$$

where the Macdonald spherical functions $P_\mu(x; t)$ are as in (2.9).

For each $\alpha \in R^+$ define the *raising operator* $R_\alpha \colon P \to P$ by

$$R_\alpha \lambda = \lambda + \alpha, \quad \text{and define} \quad (R_{\beta_1} \cdots R_{\beta_l}) s_\lambda = s_{R_{\beta_1} \cdots R_{\beta_l} \lambda}, \tag{3.13}$$

for any sequence $\beta_1, \ldots, \beta_l$ of positive roots. Using the straightening law for Weyl characters (2.7),

$$s_\mu = (-1)^{\ell(w)} s_{w \circ \mu}, \quad \text{where} \quad w \circ \mu = w(\mu + \rho) - \rho,$$

any s_μ is equal to 0 or to $\pm s_\lambda$ with $\lambda \in P^+$. Composing the action of raising operators on Weyl characters should be avoided. For example, if α_i is a simple root then (since $\langle \rho, \alpha_i^\vee \rangle = 1$) $s_{-\alpha_i} = -s_{s_i \circ (-\alpha_i)} = -s_{s_i(\rho - \alpha_i) - \rho} = -s_{-\alpha_i}$ giving that $s_{-\alpha_i} = 0$ and so

$$R_{\alpha_i}(R_{\alpha_i} s_{-2\alpha_i}) = R_{\alpha_i} s_{-\alpha_i} = R_{\alpha_i} \cdot 0 = 0, \text{ whereas } (R_{\alpha_i} R_{\alpha_i}) s_{-2\alpha_i} = s_0 = 1.$$

Let Q^+ be the set of nonnegative integral linear combinations of positive roots. Define the *q-analogue of the partition function* $F(\gamma;t)$, $\gamma \in P$, by

$$\prod_{\alpha\in R^+} \frac{1}{1-tx^\alpha} = \sum_{\gamma\in Q^+} F(\gamma;t)x^\gamma, \quad \text{and } F(\gamma;t)=0, \text{ if } \gamma \notin Q^+. \tag{3.14}$$

Theorem 3.4 *Let $\lambda,\mu \in P^+$. Let t_μ be the translation in μ as defined in (1.10) and let n_μ be the longest element of the double coset $Wt_\mu W$. Let $W_\mu(t)$ be as in (2.8), $P_\mu(x;t)$ as in (2.9) and let $\langle\ ,\ \rangle_t$ be the inner product defined in (3.2). For $y,w\in\tilde{W}$ let $P_{yw}\in\mathbb{Z}[t^{\pm\frac{1}{2}}]$ denote the Kazhdan–Lusztig polynomial defined in (1.26)-(1.27) and let $\rho^\vee = \frac{1}{2}\sum_{\alpha\in R^+}\alpha^\vee$.*

(a) $K_{\lambda,\mu}(t) = W_\mu(t)\,\langle s_\lambda, P_\mu(x;t)\rangle_t$.

(b) $K_{\lambda\mu}(t) =$ *coefficient of* s_λ *in* $\displaystyle\left(\prod_{\alpha\in R^+}\frac{1}{1-tR_\alpha}\right)s_\mu$.

(c) $\displaystyle K_{\lambda\mu}(t) = \sum_{w\in W}(-1)^{\ell(w)}F(w(\lambda+\rho)-(\mu+\rho);t)$.

(d) $K_{\lambda\mu}(t) = t^{\langle\lambda-\mu,\rho^\vee\rangle}P_{x,n_\lambda}(t^{-1})$, *for any* $x\in Wt_\mu W$.

Proof (a) This follows from the third equality in Proposition 3.1 and the definition of $K_{\lambda\mu}(t)$.

(b) Since

$$\begin{aligned}
&P_\mu(x;t)W_\mu(t)\prod_{\alpha\in R}\frac{1}{1-tx^\alpha}\\
&= \sum_{w\in W} w\left(x^\mu\prod_{\alpha\in R^+}\frac{1-tx^{-\alpha}}{1-x^{-\alpha}}\right)\prod_{\alpha\in R}\frac{1}{1-tx^\alpha}\\
&= \sum_{w\in W} w\left(x^{\mu+\rho}\frac{1}{x^\rho\prod_{\alpha\in R^+}(1-x^{-\alpha})(1-tx^\alpha)}\right)\\
&= \frac{1}{a_\rho}\sum_{w\in W}(-1)^{\ell(w)}w\left(\prod_{\alpha\in R^+}\left(\frac{1}{1-tx^\alpha}\right)x^{\mu+\rho}\right),
\end{aligned}$$

it follows that

$$\begin{aligned}
K_{\lambda\mu}(t) &= (\text{coefficient of } P_\mu(x;t) \text{ in } s_\lambda) = \langle s_\lambda, W_\mu(t)P_\mu(x;t)\rangle_t\\
&= \left\langle s_\lambda, W_\mu(t)P_\mu(x;t)\prod_{\alpha\in R}\frac{1}{1-tx^\alpha}\right\rangle_0\\
&= \text{coefficient of } s_\lambda \text{ in } \frac{1}{a_\rho}\sum_{w\in W}(-1)^{\ell(w)}w\left(\prod_{\alpha\in R^+}\left(\frac{1}{1-tx^\alpha}\right)x^{\mu+\rho}\right)\\
&= \text{coefficient of } s_\lambda \text{ in } \left(\prod_{\alpha\in R^+}\frac{1}{1-tR_\alpha}\right)s_\mu\,.
\end{aligned}$$

(c)

$$\begin{aligned}
K_{\lambda\mu}(t) &= \text{coefficient of } s_\lambda \text{ in } \frac{1}{a_\rho}\sum_{w\in W}(-1)^{\ell(w)}w\left(\prod_{\alpha\in R^+}\left(\frac{1}{1-tx^\alpha}\right)x^{\mu+\rho}\right) \\
&= \text{coefficient of } a_{\lambda+\rho} \text{ in } \sum_{w\in W}(-1)^{\ell(w)}w\left(\left(\sum_{\gamma\in Q^+}F(\gamma;t)x^\gamma\right)x^{\mu+\rho}\right) \\
&= \text{coefficient of } x^{\lambda+\rho} \text{ in } \sum_{w\in W}(-1)^{\ell(w)}w\left(\sum_{\gamma\in Q^+}F(\gamma;t)x^{\gamma+\mu+\rho}\right) \\
&= \sum_{w\in W}(-1)^{\ell(w)}F(w(\lambda+\rho)-(\mu+\rho);t),
\end{aligned}$$

since $w^{-1}(\gamma+(\mu+\rho))=\lambda+\rho$ implies $\gamma=w(\lambda+\rho)-(\mu+\rho)$.

(d) Let $\lambda\in P^+$. By Theorem 2.9 and Lemma 2.7

$$\begin{aligned}
\sum_{x\leqslant n_\lambda} q^{-(\ell(n_\lambda)-\ell(x))}P_{x,n_\lambda}(q^2)T_x &= C'_{n_\lambda} = q^{-\ell(w_0)}W_0(q^2)s_\lambda\mathbf{1}_0 \\
&= q^{-\ell(w_0)}W_0(q^2)\sum_{\mu\leqslant\lambda}K_{\lambda\mu}(q^{-2})P_\mu(x;q^{-2})\mathbf{1}_0 \\
&= q^{-\ell(w_0)}W_0(q^2)\sum_{\mu\leqslant\lambda}K_{\lambda\mu}(q^{-2})\frac{W_0(q^{-2})}{W_\mu(q^{-2})}M_\mu \\
&= \sum_{\mu\leqslant\lambda}K_{\lambda\mu}(q^{-2})\sum_{x\in Wt_\mu W}q^{\ell(x)-\ell(n_\mu)}T_x.
\end{aligned}$$

Hence, for $\mu\leqslant\lambda$ and $x\in Wt_\mu W$,

$$K_{\lambda\mu}(q^{-2}) = q^{\ell(n_\mu)-\ell(n_\lambda)}P_{x,n_\lambda}(q^2).$$

By (2.15) and (2.16),

$$\ell(n_\mu)-\ell(n_\lambda)=\ell(t_\mu)+\ell(w_0)-(\ell(t_\lambda)+\ell(w_0)) = 2\langle\mu,\rho^\vee\rangle-2\langle\lambda,\rho^\vee\rangle,$$

and the result follows on replacing q^{-2} by t. □

With the notation of Remark 3.3(b), it follows from Theorem 3.4(a) and $s_0=P_0(x;t)$ that

$$\begin{aligned}
K_{\lambda,0}(t) &= W_0(t)\langle s_\lambda, P_0(x;t)\rangle_t = W_0(t)\langle s_\lambda, s_0\rangle_t \\
&= \sum_{k\geqslant 0}\dim(\mathrm{Hom}_{\mathfrak{g}}(L(\lambda),\mathcal{H}^k)t^k.
\end{aligned} \tag{3.15}$$

This is an important formula for the Kostka–Foulkes polynomial in the case that $\mu = 0$.

Let $J \subset \{\alpha_1, \dots, \alpha_n\}$ be a subset of the set of simple roots and let

$$\mathfrak{h}_J^* = \mathbb{R}\text{-span}\{J\}, \qquad R_J = R \cap \mathfrak{h}_J^*, \qquad R_J^+ = R^+ \cap \mathfrak{h}_J^*, \tag{3.16}$$

$$W_J = \langle s_j \mid \alpha_j \in J \rangle \qquad \text{and} \qquad P_J^+ = P^+ \cap \mathfrak{h}_J^*, \tag{3.17}$$

so that R_J is a *parabolic subsystem* of the root system R, R_J^+ is the set of positive roots of R_J, W_J is the Weyl group of R_J, and P_J^+ is the set of dominant integral weights for R_J. Let $\mathfrak{h}_J^\perp$ be the orthogonal complement to $\mathfrak{h}_J^*$ with respect to the inner product $\langle\ ,\ \rangle$ so that

$$\mathfrak{h}^* = \mathfrak{h}_J^* \oplus \mathfrak{h}_J^\perp, \qquad \text{and write} \qquad \mu = \mu_J + \mu_J^\perp, \tag{3.18}$$

to denote the decomposition of an element $\mu \in \mathfrak{h}^*$ as a sum of $\mu_J \in \mathfrak{h}_J^*$ and $\mu_J^\perp \in \mathfrak{h}_J^\perp$.

Proposition 3.5 *Let J be a subset of the set of simple roots $\{\alpha_1, \dots, \alpha_n\}$. Then, with the notation of equations (3.16)–(3.18), and for $\lambda, \mu \in P^+$,*

$$K_{\lambda\mu}(t) = \textit{coefficient of } s_\lambda \textit{ in } \left(\prod_{\alpha \in (R^+ \setminus R_J^+)} \frac{1}{1 - tR_\alpha} \right) \sum_{\lambda_J \in P_J^+} K_{\lambda_J \mu_J}(t) s_{\lambda_J + \mu_J^\perp},$$

where the $K_{\lambda_J\mu_J}(t)$ are Kostka-Foulkes polynomials for the root system R_J.

Proof By the third equation in the proof of Theorem 3.4(b), $K_{\lambda\mu}(t)$ is the coefficient of $a_{\lambda+\rho}$ in

$$\begin{aligned}
&\sum_{w \in W} (-1)^{\ell(w)} w \left(\prod_{\alpha \in R^+} \left(\frac{1}{1 - tx^\alpha} \right) x^{\mu+\rho} \right) \\
&= \frac{1}{|W_J|} \sum_{w \in W} (-1)^{\ell(w)} w \sum_{v \in W_J} (-1)^{\ell(v)} v \\
&\qquad \left(\left(\prod_{\alpha \in (R^+ \setminus R_J^+)} \frac{1}{1 - tx^\alpha} \right) \left(\prod_{\alpha \in R_J^+} \frac{1}{1 - tx^\alpha} \right) x^{\mu_J + \rho_J} x^{\mu_J^\perp + \rho_J^\perp} \right) \\
&= \frac{1}{|W_J|} \sum_{w \in W} (-1)^{\ell(w)} w \left(\left(\prod_{\alpha \in (R^+ \setminus R_J^+)} \frac{1}{1 - tx^\alpha} \right) x^{\mu_J^\perp + \rho_J^\perp} \right. \\
&\qquad \left. \sum_{v \in W_J} (-1)^{\ell(v)} v \left(\prod_{\alpha \in R_J^+} \frac{1}{1 - tx^\alpha} \right) x^{\mu_J + \rho_J} \right) \\
&= \frac{1}{|W_J|} \sum_{w \in W} (-1)^{\ell(w)} w \left(\left(\prod_{\alpha \in (R^+ \setminus R_J^+)} \frac{1}{1 - tx^\alpha} \right) x^{\mu_J^\perp + \rho_J^\perp} \sum_{\lambda_J \in P_J^+} K_{\lambda_J \mu_J}(t) a_{\lambda_J + \rho_J} \right),
\end{aligned}$$

where the last equality follows from Theorem 3.4(b) applied to the root system R_J. Expanding $a_{\lambda_J+\rho_J}$ gives that $K_{\lambda\mu}(t)$ is the coefficient of $a_{\lambda+\rho}$ in

$$\frac{1}{|W_J|}\sum_{\lambda_J\in P_J^+} K_{\lambda_J\mu_J}(t)\sum_{w\in W}(-1)^{\ell(w)}w\left(\Big(\prod_{\alpha\in(R^+\setminus R_J^+)}\frac{1}{1-tx^\alpha}\Big)x^{\mu_J^\perp+\rho_J^\perp}\right.$$
$$\left.\sum_{v\in W_J}(-1)^{\ell(v)}v\,x^{\lambda_J+\rho_J}\right)$$
$$=\frac{1}{|W_J|}\sum_{\lambda_J\in P_J^+} K_{\lambda_J\mu_J}(t)\sum_{w\in W}(-1)^{\ell(w)}w\sum_{v\in W_J}(-1)^{\ell(v)}v$$
$$\left(\Big(\prod_{\alpha\in(R^+\setminus R_J^+)}\frac{1}{1-tx^\alpha}\Big)x^{\mu_J^\perp+\rho_J^\perp+\lambda_J+\rho_J}\right)$$
$$=\sum_{\lambda_J\in P_J^+} K_{\lambda_J\mu_J}(t)\sum_{w\in W}(-1)^{\ell(w)}w\left(\Big(\prod_{\alpha\in(R^+\setminus R_J^+)}\frac{1}{1-tx^\alpha}\Big)x^{\lambda_J+\mu_J^\perp+\rho}\right)$$

from which the desired formula follows on dividing by a_ρ and converting to raising operators (as in the proof of Theorem 3.4(b) above). □

4 The positive formula

In the type A case Lascoux and Schützenberger [9] have used the theory of column strict tableaux to give a positive formula for the Kostka–Foulkes polynomial. In this section we give a proof of this formula. Versions of this proof have appeared previously in [17] and in [2].

The starting point is the formula for $K_{\lambda\mu}(t)$ in Theorem 3.4(b). To match the setup in [11] we shall work in a slightly different setting (corresponding to the Weyl group W and the weight lattice of the reductive group $GL_n(\mathbb{C})$). In this case the vector space $\mathfrak{h}^*_\mathbb{R}=\mathbb{R}^n$ has orthonormal basis $\varepsilon_1,\dots,\varepsilon_n$, where $\varepsilon_i=(0,\dots,0,1,0,\dots,0)$ with the 1 in the ith coordinate, the Weyl group is the symmetric group S_n acting on $\mathbb{R}^n$ by permuting the coordinates, the weight lattice P is replaced by the lattice

$$\mathbb{Z}^n=\{(\gamma_1,\dots,\gamma_n)\mid\gamma_i\in\mathbb{Z}\}\quad\text{and}\quad\delta=(n-1,n-2,\dots,2,1,0)\tag{4.1}$$

replaces the element ρ. The positive roots are given by $R^+=\{\varepsilon_i-\varepsilon_j\mid 1\leqslant i<j\leqslant n\}$ and the Schur functions (defined as in (2.6)) are viewed as (Laurent) polynomials in the variables $x_1,\dots,x_n$, where $x_i=x^{\varepsilon_i}$ and the symmetric group S_n acts by permuting the variables. If $w\in S_n$ then $(-1)^{\ell(w)}=\det(w)$ is the *sign* of the permutation w and the straightening law for Schur functions (see (2.7) and [11, I paragraph after (3.1)]) is

$$s_\mu=(-1)^{\ell(w)}s_{w\circ\mu},\quad\text{where}\quad w\circ\mu=w(\mu+\delta)-\delta,\tag{4.2}$$

for $w \in S_n$ and $\mu \in \mathbb{Z}^n$. The set of *partitions*

$$\mathcal{P} = \{(\lambda_1, \dots, \lambda_n) \in \mathbb{Z}^n \mid \lambda_1 \geqslant \cdots \geqslant \lambda_n \geqslant 0\} \tag{4.3}$$

takes the role played by the set P^+. Conforming to the conventions in [11] so that gravity goes up and to the left, each partition $\mu = (\mu_1, \dots, \mu_n) \in \mathcal{P}$ is identified with the collection of boxes in a corner which has μ_i boxes in row i, where, as for matrices, the rows and columns of μ are indexed from top to bottom and left to right, respectively. For example, with $n = 7$,

$$(5,5,3,3,1,1,0) = \quad .$$

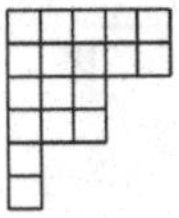

For each pair $1 \leqslant i < j \leqslant n$ define the *raising operator* $R_{ij}\colon \mathbb{Z}^n \to \mathbb{Z}^n$ (see (3.13) and [11, I §1 (1.14)]) by

$$R_{ij}\mu = \mu + \varepsilon_i - \varepsilon_j \quad \text{and define} \quad (R_{i_1j_1} \cdots R_{i_lj_l})s_\mu = s_{R_{i_1j_1}\cdots R_{i_lj_l}\mu}, \tag{4.4}$$

for a sequence of pairs $i_1 < j_1, \dots, i_l < j_l$. Using the straightening law (4.2) any Schur function s_μ indexed by $\mu \in \mathbb{Z}^n$ with $\mu_1 + \cdots + \mu_n \geqslant 0$ is either equal to 0 or to $\pm s_\lambda$ for some $\lambda \in \mathcal{P}$. Composing the action of raising operators on Schur functions s_λ should be avoided. For example, if $n = 2$ and s_1 denotes the transposition in the symmetric group S_2 then, by the straightening law, $s_{(0,1)} = -s_{s_1((0,1)+(1,0))-(1,0)} = -s_{(1,1)-(1,0)} = -s_{(0,1)}$ giving that $s_{(0,1)} = 0$ and so

$$R_{12}(R_{12}s_{(-1,2)}) = R_{12}s_{(0,1)} = R_{12} \cdot 0 = 0, \qquad \text{whereas}$$

$$(R_{12}^2)s_{(-1,2)} = s_{(1,0)} = x_1 + x_2.$$

With notation as in (4.2) and (4.4) we may define the *Hall–Littlewood polynomials* for this type A case by (see Theorem 3.4(b) and [11, III (4.6)])

$$Q_\mu = \left(\prod_{1 \leqslant i < j \leqslant n} \frac{1}{1 - tR_{ij}} \right) s_\mu, \quad \text{for all } \mu \in \mathbb{Z}^n, \tag{4.5}$$

and the *Kostka–Foulkes polynomials* $K_{\lambda\mu}(t)$, $\lambda, \mu \in \mathcal{P}$, by

$$Q_\mu = \sum_{\lambda \in \mathcal{P}} K_{\lambda\mu}(t) s_\lambda. \tag{4.6}$$

4.1 Insertion and Pieri rules

Let $\lambda, \mu \in \mathbb{Z}^n$ be partitions. A *column strict tableau of shape* λ *and weight* μ is a filling of the boxes of λ with μ_1 1s, μ_2 2s, …, μ_n ns, such that

(a) the rows are weakly increasing from left to right,

(b) the columns are strictly increasing from top to bottom.

If T is a column strict tableau write $\mathrm{shp}(T)$ and $\mathrm{wt}(T)$ for the shape and the weight of T so that

$$\begin{aligned} \mathrm{shp}(T) &= (\lambda_1, \dots, \lambda_n), \quad \text{where } \lambda_i = \text{number of boxes in row } i \text{ of } T, \quad \text{and} \\ \mathrm{wt}(T) &= (\mu_1, \dots, \mu_n), \quad \text{where } \mu_i = \text{number of } i\text{s in } T. \end{aligned}$$

For example,

$$T = \begin{array}{ccccccccc} 1&1&1&1&1&1&1&2&2 \\ 2&2&2&2&3&3&4 \\ 3&3&3&4&4&4&5 \\ 4&5&5&6 \\ 6&7 \\ 7 \end{array} \qquad \begin{array}{l} \text{has } \mathrm{shp}(T) = (9,7,7,4,2,1,0) \\ \text{and } \mathrm{wt}(T) = (7,6,5,5,3,2,2). \end{array}$$

For partitions λ and μ and, more generally, for any two sets $\mathcal{S}, \mathcal{W} \subseteq \mathcal{P}$ write

$$\begin{aligned} B(\lambda) &= \{\text{column strict tableaux } T \mid \mathrm{shp}(T) = \lambda\}, \\ B(\lambda)_\mu &= \{\text{column strict tableaux } T \mid \mathrm{shp}(T) = \lambda,\ \mathrm{wt}(T) = \mu\}, \qquad (4.7) \\ B(\mathcal{S})_\mathcal{W} &= \{\text{column strict tableaux } T \mid \mathrm{shp}(T) \in \mathcal{S},\ \mathrm{wt}(T) \in \mathcal{W}\}. \end{aligned}$$

Let λ and γ be partitions such that $\gamma \subseteq \lambda$ (as collections of boxes in a corner, that is $\gamma_i \leqslant \lambda_i$ for $1 \leqslant i \leqslant n$). The *skew shape* λ/γ is the collection of boxes of λ which are not in γ. The *jeu de taquin* reduces a column strict filling of a skew shape λ/γ to a column strict tableau of partition shape. At each step "gravity" moves one box up or to the left without violating the column strict condition (weakly increasing in rows, strictly increasing in columns). Once an empty box on the northwest side of the skew shape starts to move it must continue and exit the southeast border of the skew shape before another empty box can start its exit. The jeu de taquin is most easily illustrated by example:

$$\begin{array}{cccccc} \cdot&\cdot&1&1&2&2 \\ \cdot&1&2&3 \\ \cdot&3&4&4 \\ 2 \end{array} \longmapsto \begin{array}{cccccc} \cdot&\cdot&1&1&2&2 \\ \cdot&1&2&3 \\ 2&3&4&4 \end{array}$$

$$\longmapsto \begin{array}{cccccc} \cdot&\cdot&1&1&2&2 \\ 1&&2&3 \\ 2&3&4&4 \end{array} \longmapsto \begin{array}{cccccc} \cdot&\cdot&1&1&2&2 \\ 1&2&&3 \\ 2&3&4&4 \end{array}$$

$$\longmapsto \begin{array}{cccccc} \cdot&\cdot&1&1&2&2 \\ 1&2&3& \\ 2&3&4&4 \end{array} \longmapsto \begin{array}{cccccc} \cdot&\cdot&1&1&2&2 \\ 1&2&3&4 \\ 2&3&4 \end{array}$$

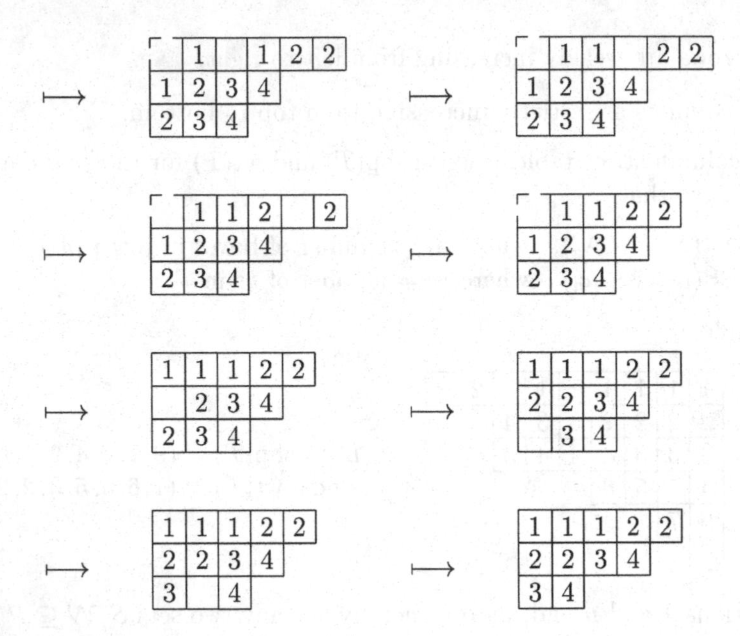

In this example $\lambda = (6,4,4,1)$, $\gamma = (2,1,1)$ and the resulting column strict tableau is of shape $(5,4,2)$. The result of the jeu de taquin is independent of the choice of order of the moves ([3, §1.2, Claim 2] which is proved in [3, §2 and §3].

The *plactic monoid* is the set $B(\mathcal{P})$ of column strict tableaux with product given by

$T_1 * T_2 =$ jeu de taquin reduction of

T_2

T_1

Because the result of the jeu de taquin is independent of the choice of the order of the moves this is an associative monoid.

If x is a "letter", that is, a column strict tableau of shape $(1) = \square$, then

$$\begin{aligned} &x * T \text{ is the } \textit{column insertion} \text{ of } x \text{ into } T, \quad \text{and} \\ &T * x \text{ is the } \textit{row insertion} \text{ of } x \text{ into } T. \end{aligned} \tag{4.8}$$

The shape λ of $P = T * x$ differs from the shape γ of T by single box and so if γ and P are given then the pair (T, x) can be recovered by "uninserting" the box λ/γ from P. The tableaux P and T differ by at most one entry in each row. The entries where P and T differ form the *bumping path of* x. The bumping path begins with x in the first row of P and ends at the entry in the box λ/γ. For example,

$$\begin{array}{|c|c|c|c|c|c|}\hline 1&1&1&1&2&2\\\hline 2&3&3&4\\\cline{1-4} 4&4&4&5\\\cline{1-4} 6\\\cline{1-1}\end{array} \quad * \quad \boxed{1} \quad = \quad \begin{array}{|c|c|c|c|c|c|}\hline 1&1&1&1&\mathbf{1}&2\\\hline 2&\mathbf{2}&3&4\\\cline{1-4} \mathbf{3}&4&4&5\\\cline{1-4} \mathbf{4}\\\cline{1-1} \mathbf{6}\\\cline{1-1}\end{array}\ ,$$

where the bold face entries form the bumping path.

The *monoid of words* is the free monoid B^* generated by $\{1, 2, \ldots, n\}$. The *weight* $\mathrm{wt}(w)$ of a word $w = w_1 \cdots w_n$ is

$$\mathrm{wt}(w) = \mathrm{wt}(w_1 \cdots w_n) = (\mu_1, \ldots, \mu_n) \quad \text{where} \quad \mu_i \text{ is the number of } i\text{'s in } w.$$

For example, $w = 3214566532211$ is a word of weight $\mathrm{wt}(w) = (3, 3, 2, 1, 2, 2)$. The *insertion* map

$$\begin{array}{ccc} B^* & \longrightarrow & B(\mathcal{P}) \\ w_1 \cdots w_n & \longmapsto & w_1 * \cdots * w_n \end{array} \tag{4.9}$$

is a weight preserving homomorphism of monoids.

A *horizontal strip* is a skew shape which contains at most one box in each column. The *length* of a horizontal strip λ/γ is the number of boxes in λ/γ. The boxes containing $\times$ in the picture

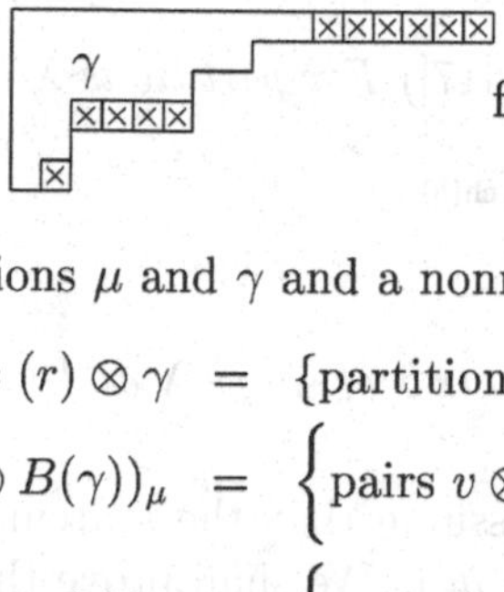

$\lambda =$ γ form a horizontal strip λ/γ of length 11.

For partitions μ and γ and a nonnegative integer r let

$$\begin{aligned} \gamma \otimes (r) = (r) \otimes \gamma &= \{\text{partitions } \lambda \mid \lambda/\gamma \text{ is a horizontal strip of length } r\}, \\ (B(r) \otimes B(\gamma))_\mu &= \left\{ \text{pairs } v \otimes T \;\middle|\; \begin{array}{l} v \in B(r), T \in B(\gamma) \\ \text{such that } \mathrm{wt}(v) + \mathrm{wt}(T) = \mu \end{array} \right\} \\ (B(\gamma) \otimes B(r))_\mu &= \left\{ \text{pairs } T \otimes v \;\middle|\; \begin{array}{l} v \in B(r), T \in B(\gamma) \\ \text{such that } \mathrm{wt}(v) + \mathrm{wt}(T) = \mu \end{array} \right\}, \end{aligned} \tag{4.10}$$

The following lemma gives tableau versions of the Pieri rule [11, I (5.16)]. The second bijection of the lemma is proved in [3, §1.1 Proposition], and the proof of the first bijection is similar (see also [2, Propositions 2.3.4 and 2.3.11]).

Lemma 4.1 *Let $\gamma, \mu, \tau \in \mathcal{P}$ be partitions and let $r, s \in \mathbb{Z}_{\geqslant 0}$. There are bijections*

$$\begin{array}{ccc} (B(r) \otimes B(\gamma))_\mu & \longleftrightarrow & B(\gamma \otimes (r))_\mu \\ v \otimes T & \longrightarrow & v * T \end{array} \quad \textit{and}$$

$$\begin{array}{ccc} (B(\gamma) \otimes B(s))_\tau & \longleftrightarrow & B(\gamma \otimes (s))_\tau \\ T \otimes u & \longrightarrow & T * u\ . \end{array}$$

4.2 Charge

Let $B(\mathcal{P})_{\geqslant} = \bigcup_{1\leqslant i\leqslant n} B(\mathcal{P})_{\geqslant i}$, where

$$B(\mathcal{P})_{\geqslant i} = \left\{ \text{column strict tableaux } b \;\middle|\; \begin{array}{l} \mathrm{wt}(b) = (\mu_1, \ldots, \mu_n) \text{ has} \\ \mu_1 = \cdots = \mu_{i-1} = 0 \text{ and} \\ \mu_i \geqslant \cdots \geqslant \mu_n \geqslant 0 \end{array} \right\}.$$

Let $i^k = \boxed{i}\boxed{i}\boxed{\cdots}\boxed{i}$ be the unique column strict tableau of shape (k) and weight $(0, \ldots, 0, k, 0, \ldots, 0)$, where the k appears in the i th entry. *Charge* is the function $\mathrm{ch}\colon B(\mathcal{P})_{\geqslant} \longrightarrow \mathbb{Z}_{\geqslant 0}$ such that

(a) $\mathrm{ch}(\emptyset) = 0$,

(b) if $T \in B(\mathcal{P})_{\geqslant(i+1)}$ and $T * i^{\mu_i} \in B(\mathcal{P})_{\geqslant i}$ then $\mathrm{ch}(T * i^{\mu_i}) = \mathrm{ch}(T)$,

(c) if $T \in B(\mathcal{P})_{\geqslant i}$ and x is a letter not equal to i then $\mathrm{ch}(x*T) = \mathrm{ch}(T*x)+1$.

The proof of the existence and uniqueness of the function ch is presented beautifully in [7].

Theorem 4.2 (Lascoux-Schützenberger [9], [17]) *For partitions λ and μ,*

$$K_{\lambda\mu}(t) = \sum_{b\in B(\lambda)_\mu} t^{\mathrm{ch}(b)},$$

where the sum is over all column strict tableaux b of shape λ and weight μ.

Proof The proof is by induction on n. Assume that the statement of the theorem holds for all partitions $\mu = (\mu_1, \ldots, \mu_n)$. We shall prove that, for all partitions $(\mu_0, \mu) = (\mu_0, \mu_1, \ldots, \mu_n)$, $Q_{(\mu_0,\mu)}$ has an expansion

$$Q_{(\mu_0,\mu)} = \sum_{p\in B(\nu)_{(\mu_0,\mu)}} t^{\mathrm{ch}(p)} s_\nu, \tag{4.11}$$

Beginning with the expression (4.5),

$$\begin{aligned} Q_{(\mu_0,\mu)} &= \left(\prod_{0\leqslant i<j\leqslant n} \frac{1}{1-tR_{ij}} \right) s_{(\mu_0,\mu)} \\ &= \left(\prod_{j=1}^{n} \frac{1}{1-tR_{0j}} \right) \left(\prod_{1\leqslant i<j\leqslant n} \frac{1}{1-tR_{ij}} \right) s_{(\mu_0,\mu)}. \end{aligned}$$

Proposition 3.5 shows that this particular product of raising operators can be composed and so, by applying the definition of the Kostka–Foulkes polynomials (4.6),

$$
\begin{aligned}
Q_{(\mu_0,\mu)} &= \left(\prod_{j=1}^{n} \frac{1}{1-tR_{0j}}\right) \sum_{\lambda\in\mathcal{P}} K_{\lambda\mu}(t) s_{(\mu_0,\lambda)} \\
&= \sum_{\lambda\in\mathcal{P}} K_{\lambda\mu}(t) \sum_{r\in\mathbb{Z}_{\geqslant 0}} t^r \sum_{\substack{k_1,\dots,k_n\in\mathbb{Z}_{\geqslant 0}\\ k_1+\cdots+k_n=r}} R_{01}^{k_1}\cdots R_{0n}^{k_n} s_{(\mu_0,\lambda)} \\
&= \sum_{\lambda\in\mathcal{P}} K_{\lambda\mu}(t) \sum_{r\in\mathbb{Z}_{\geqslant 0}} t^r \sum_{\substack{k_1,\dots,k_n\in\mathbb{Z}_{\geqslant 0}\\ k_1+\cdots+k_n=r}} s_{(\mu_0+r,\lambda-(k_1,\dots,k_n))}
\end{aligned}
$$

Let $\gamma = \lambda - (k_1, \dots, k_n)$ be such that λ/γ is not a horizontal strip (usually γ isn't even a partition). Let m be the first place a violation to being a horizontal strip occurs, that is,

$$\text{let } m \text{ be minimal such that } \lambda_m - k_m < \lambda_{m+1}.$$

For example, in the followng picture, $\gamma = \lambda - (3, 1, 2, 2, 1, 0)$ and $m = 3$.

$\lambda =$ 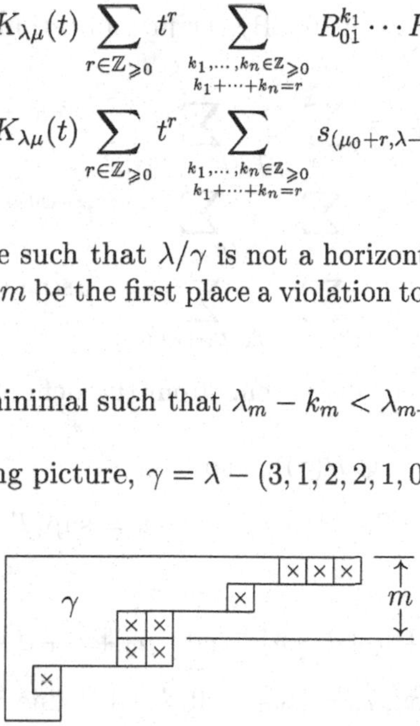

Let s_m be the simple transposition in the symmetric group which switches m and $m+1$ and define

$$\tilde{\gamma} = s_m \circ \gamma, \quad \text{so that} \quad s_{(\mu_0+r,\gamma)} = -s_{(\mu_0+r,\tilde{\gamma})}.$$

Then $\tilde{\gamma} = \lambda - (\tilde{k}_1, \dots, \tilde{k}_n)$ with $\lambda_i - \tilde{k}_i = \lambda_i - k_i$, for $i \neq m, m+1$, and

$$\lambda_m - \tilde{k}_m = \lambda_{m+1} - k_{m+1} - 1, \quad \text{and} \quad \lambda_{m+1} - \tilde{k}_{m+1} = \lambda_m - k_m + 1.$$

Thus $\tilde{\gamma}_m = \lambda_{m+1} - k_{m+1} - 1 < \lambda_{m+1}$ and so $\lambda/\tilde{\gamma}$ is not a horizontal strip. This pairing $\gamma \leftrightarrow \tilde{\gamma}$ provides a cancellation in the expression for $Q_{(\mu_0,\mu)}$ and thus

$$
\begin{aligned}
Q_{(\mu_0,\mu)} &= \sum_{\lambda\in\mathcal{P}} \sum_{r\in\mathbb{Z}_{\geqslant 0}} t^r K_{\lambda\mu}(t) \sum_{\substack{\gamma\in\mathcal{P}\\ \lambda\in\gamma\otimes(r)}} s_{(\mu_0+r,\gamma)} \\
&= \sum_{\gamma,r} \sum_{\substack{\lambda\in\mathcal{P}\\ \lambda\in\gamma\otimes(r)}} t^r K_{\lambda\mu}(t) s_{(\mu_0+r,\gamma)},
\end{aligned}
$$

where $\gamma \otimes (r)$ is as defined in (4.10). Then, by the induction assumption,

$$
\begin{aligned}
Q_{(\mu_0,\mu)} &= \sum_{\gamma,r} \sum_{\substack{\lambda\in\mathcal{P}\\ \lambda\in\gamma\otimes(r)}} \sum_{b\in B(\lambda)_\mu} t^r t^{\mathrm{ch}(b)} s_{(\mu_0+r,\gamma)} \\
&= \sum_{\gamma,r} \sum_{b\in B(\gamma\otimes(r))_\mu} t^{r+\mathrm{ch}(b)} s_{(\mu_0+r,\gamma)},
\end{aligned}
$$

with $B(\gamma \otimes (r))_\mu$ as in (4.10). By the first bijection in Lemma 4.1 this can be rewritten as

$$
\begin{aligned}
Q_{(\mu_0,\mu)} &= \sum_{\gamma,r} \sum_{v\otimes T\in(B(r)\otimes B(\gamma))_\mu} t^{r+\mathrm{ch}(v*T)} s_{(\mu_0+r,\gamma)} \\
&= \sum_{\gamma,r} \sum_{v\otimes T\in(B(r)\otimes B(\gamma))_\mu} t^{r+\mathrm{ch}(v*T*0^{\mu_0}))} s_{(\mu_0+r,\gamma)} \qquad (4.12) \\
&= \sum_{\gamma,r} \sum_{v\otimes T\in(B(r)\otimes B(\gamma))_\mu} t^{\mathrm{ch}(T*0^{\mu_0}*v))} s_{(\mu_0+r,\gamma)},
\end{aligned}
$$

where the last two equalities come from the defining properties of the charge function ch.

Let $v \otimes T \in (B(r) \otimes B(\gamma))_\mu$ and let

$$p = T * 0^{\mu_0} * v \quad \text{and} \quad \nu = \mathrm{shp}(T * 0^{\mu_0} * v).$$

Let d be such that

$$\mu_0 + r + d > \nu_d \quad \text{and} \quad \mu_0 + r + d - 1 \leqslant \nu_{d-1},$$

where, by convention, $\nu_0 = \mu_0 + r$. If $d > 1$ define $\tilde{\gamma}$ and $\tilde{r}$ by

$$\tilde{\gamma} = (\gamma_1, \ldots, \gamma_{d-2}, \mu_0 + r + d - 1, \gamma_d, \ldots, \gamma_n) \quad \text{and} \quad \mu_0 + \tilde{r} + d - 1 = \gamma_{d-1},$$

so that, if s_i denotes the transposition $(i, i+1)$ in the symmetric group, then $(\mu_0 + \tilde{r}, \tilde{\gamma}) = (s_0 \cdots s_{d-3} s_{d-2} s_{d-3} \cdots s_0) \circ (\mu_0 + r, \gamma)$, and

$$s_{(\mu_0+r,\gamma)} = (-1)^{2(d-3)+1} s_{(\mu_0+\tilde{r},\tilde{\gamma})} = -s_{(\mu_0+\tilde{r},\tilde{\gamma})}. \qquad (4.13)$$

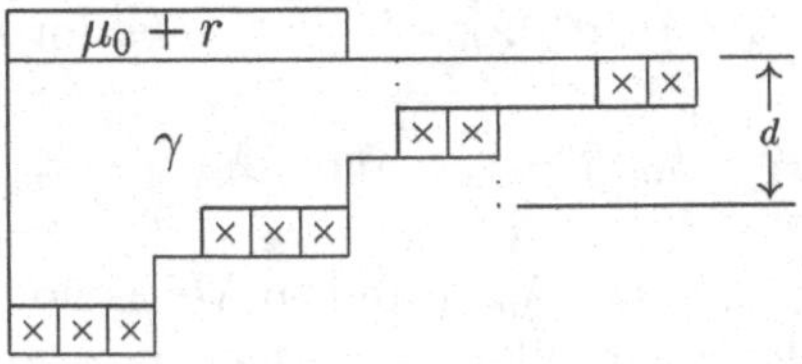

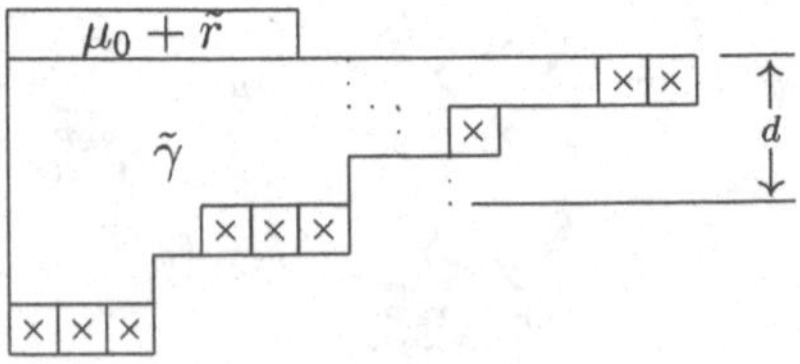

Note that $\tilde{\tilde{\gamma}} = \gamma$ and $\tilde{\tilde{r}} = r$.

Case 1: $d > 1$ and $(\mu_0 + r, \gamma) = (\mu_0 + \tilde{r}, \tilde{\gamma})$. In this case (4.13) implies $s_{(\mu_0+r,\gamma)} = 0$.

Case 2: $d > 1$ and $(\mu_0 + r, \gamma) \neq (\mu_0 + \tilde{r}, \tilde{\gamma})$. Then

$$\nu \in \gamma \otimes (\mu_0 + r) \qquad \text{and} \qquad \nu \in \tilde{\gamma} \otimes (\mu_0 + \tilde{r}).$$

Row uninserting the horizontal strips ν/γ and $\nu/\tilde{\gamma}$ from p, by using the second bijection in Lemma 4.1, produces pairs

$$T \otimes u = T \otimes (0^{\mu_0} * v) \in (B(\gamma) \otimes B(\mu_0 + r))_{(\mu_0,\mu)}$$

and

$$\tilde{T} \otimes \tilde{u} \in (B(\tilde{\gamma}) \otimes B(\mu_0 + \tilde{r}))_{(\mu_0,\mu)},$$

respectively. Consider the $\ell = \mu_0 + r$ bumping paths in the tableau p which arise from $T * u$. These begin with the letters $u_1 \leqslant \ldots \leqslant u_\ell$ of u and end at the boxes of the horizontal strip ν/γ. Similarly, there are $\tilde{\ell} = \mu_0 + \tilde{r}$ bumping paths in p arising from $\tilde{T} * \tilde{u}$. Note that

(a) since $u = 0^{\mu_0} * v$ begins with μ_0 0 s the leftmost μ_0 bumping paths in $T * u$ travel vertically, directly down the first μ_0 columns of p, and

(b) in rows numbered $\geqslant d$ the bumping paths for $\tilde{T} * \tilde{u}$ coincide exactly with the bumping paths for $T * u$, since the horizontal strips ν/γ and $\nu/\tilde{\gamma}$ coincide exactly in rows $\geqslant d$ and these paths are obtained by uninserting the boxes in this portion of the horizontal strip.

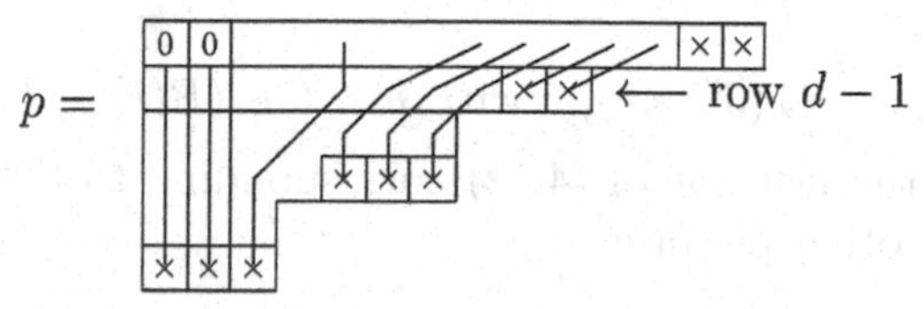

bumping paths in $T * u$

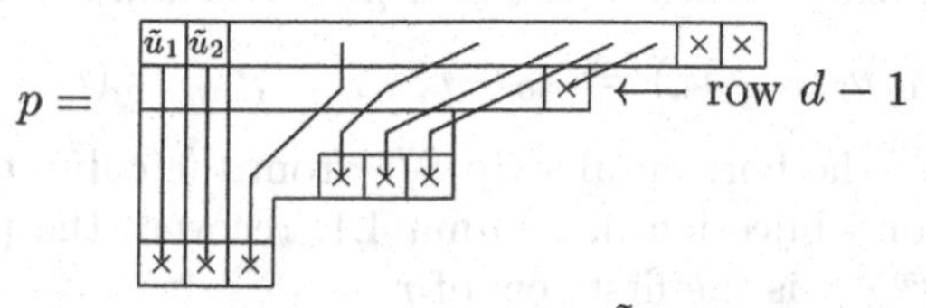

bumping paths in $\tilde{T} * \tilde{u}$

Suppose there are k bumping paths which end in rows $\geqslant d$. The picture above has $k = 6$ and corresponds to Case 2b below.

Case 2a: If $\mu_0 + \tilde{r} > \mu_0 + r$ then the k bumping paths which end in rows $\geqslant d$ are the same or slightly "more left" in $\tilde{T} * \tilde{u}$ than in $T * u$. Since the first μ_0 bumping paths cannot be any "more left" than vertical, this forces the first μ_0 entries of $\tilde{u}$ to be 0 so that $\tilde{u} = 0^{\mu_0} * \tilde{v}$ for some $v \in B(\tilde{r})$.

Case 2b: If $\mu_0 + \tilde{r} < \mu_0 + r$ then the k bumping paths which end in rows $\geqslant d$ are the same or slightly "more right" in $\tilde{T} * \tilde{u}$ than in $T * u$. We shall analyze how these k paths pass through row $d-1$ in $T * u$ and in $\tilde{T} * \tilde{u}$. Divide row $d-1$ into four disjoint regions, left to right:

Region 1: the leftmost μ_0 boxes of row d-1,
Region 2: the boxes which do not have a cross in them in $T * u$
(and are not in Region 1),
Region 3: the boxes which have a cross in them in $T * u$ but not in $\tilde{T} * \tilde{u}$,
Region 4: the boxes which have a cross in them in both $T * u$ and $\tilde{T} * \tilde{u}$.

Of the k bumping paths of $T * u$ which end in rows $\geqslant d$ the first μ_0 of these pass through Region 1 in $T * u$, and the others ($k - \mu_0$ of them) pass through Region 2. Since the total number of bumping paths (the number of crosses) in $\tilde{T} * \tilde{u}$ is $\mu_0 + \tilde{r}$ and there are some bumping paths of $\tilde{T} * \tilde{u}$ which end in row $d-1$ ($r - \tilde{r}$ of these), $k < \mu_0 + \tilde{r}$. Thus

$$k - \mu_0 < \mu_0 + \tilde{r} - \mu_0 < \mu_0 + \tilde{r} + (d-1) - \mu_0 = \mathrm{Card}(\text{Region } 2),$$

since Card(Region 1) $= \mu_0$ and Card(Region 1) + Card(Region 2) $= \mu_0 + \tilde{r} + d - 1$. Thus there must be a box in Region 2 of $T * u$ that does not have a bumping path passing through it. All the bumping paths of $T * u$ which pass through row $d-1$ to the left of this box remain the same as bumping paths for $\tilde{T} * \tilde{u}$ and the first μ_0 of these begin at an entry 0 in the first row of p. Thus, as in Case 2a, the first μ_0 entries of $\tilde{u}$ are 0 so that $\tilde{u} = 0^{\mu_0} * \tilde{v}$ for some $v \in B(\tilde{r})$.

So,

$$\tilde{T} \otimes \tilde{u} = \tilde{T} \otimes (0^{\mu_0} * \tilde{v}), \quad \text{with } \tilde{v} \otimes \tilde{T} \in (B(\tilde{r}) \otimes B(\tilde{\gamma}))_\mu,$$

and the terms in the last line of (4.12) corresponding to the pairs $v \otimes T$ and $\tilde{v} \otimes \tilde{T}$ cancel each other because

$$T * 0^{\mu_0} * v = \tilde{T} * 0^{\mu_0} * \tilde{v} \quad \text{and } s_{(\mu_0+r,\gamma)} = -s_{(\mu_0+\tilde{r},\tilde{\gamma})}.$$

Case 3: $d = 1$. Since $\mu_0 + r + 1 > \nu_1$ and $\nu \in \gamma \otimes (\mu_0 + r)$ the horizontal strip ν/γ has its boxes in each of the first $\mu_0 + r$ columns and

$$\nu = (\nu_0, \nu_1, \ldots, \nu_n) = (\mu_0 + r, \gamma_1, \ldots, \gamma_n) = (\mu_0 + r, \gamma).$$

Row uninsertion of the horizontal strip ν/γ from the column strict tableau p, i.e. using the second bijection in Lemma 4.1, recovers the pair $T \otimes (0^{\mu_0} * v)$ and shows that $0^{\mu_0} * v$ is the first row of p.

In conclusion, in the last line of (4.12) the terms corresponding to Case 1 vanish, the terms corresponding to Case 2 cancel, and the remaining Case 3 terms give formula (4.11), as desired. □

Acknowledgements

The research of A. Ram was partially supported by the National Science Foundation (DMS-0097977), the National Security Agency (MDA904-01-1-0032) and by EPSRC Grant GR K99015 at the Newton Institute for Mathematical Sciences. The research of K. Nelsen was partially supported by the National Science Foundation (DMS-0097977 and a VIGRE grant) and the National Security Agency (MDA904-01-1-0032).

References

[1] N. Bourbaki, *Groupes et algebres de Lie, Chapt. IV-VI*, Masson, Paris (1981).

[2] L.M. Butler, *Subgroup lattices and symmetric functions*, *Memoirs Amer. Math. Soc.*, 539, 112. (1994).

[3] W. Fulton, *Young tableaux: with applications to representation theory and geometry*, *London Math. Soc. Student Texts*, 35, Cambridge Univ. Press, Cambridge (1997).

[4] A. Joseph, G. Letzter and S. Zelikson, On the Brylinski-Kostant filtration, *J. Amer. Math. Soc.* **13** (2000), 945–970.

[5] S. Kato, Spherical functions and a q-analogue of Kostant's weight multiplicity formula, *Invent. Math.* **66** (1982), 461–468.

[6] D. Kazhdan and G. Lusztig, Representations of Coxeter groups and Hecke algebras, *Invent. Math.* **53** (1979), 165–184.

[7] K. Killpatrick, A combinatorial proof of a recursion for the q-Kostka polynomials, *J. Combin. Theory Ser. A* **92** (2000), 29–53.

[8] B. Kostant, Lie group representations on polynomial rings, *Amer. J. Math.* **85** (1963), 327–404.

[9] A. Lascoux and M.P. Schützenberger, Sur une conjecture de H.O. Foulkes, *C. R. Acad. Sci. Paris Sr. A-B* **286A** (1978), 323–324.

[10] G. Lusztig, Singularities, character formulas and a q-analog of weight multiplicities, in *Analysis and topology on singular spaces, II, III (Luminy, 1981) Astérisque*, 101-102, Soc. Math. France, Paris (1983), pp. 208–229.

[11] I.G. Macdonald, *Symmetric functions and Hall polynomials*, *Oxford Mathematical Monographs*, Oxford Univ. Press, New York (second edition, 1995).

[12] I.G. Macdonald, *Spherical functions on a group of p-adic type*, Publ. Ramanujan Institute No. 2, Madras (1971).

[13] I.G. Macdonald, The Poincaré series of a Coxeter group, *Math. Ann.* **199** (1972), 161–174.

[14] I.G. Macdonald, Affine Hecke algebras and orthogonal polynomials, *Séminaire Bourbaki, Vol. 1994/95*, *Astérisque* **237** (1996), 189–207.

[15] H. Nakajima, *Lectures on Hilbert schemes of points on surfaces*, *Univ. Lecture Series*, 18, Amer. Math. Soc.Providence, RI (1999).

[16] H. Nakajima, Quiver varieties and finite dimensional representations of quantum affine algebras, *J. Amer. Math. Soc.* **14** (2001), 145–238.

[17] M.P. Schützenberger, Propriétés nouvelles des tableaux de Young, *Séminaire Delange-Pisot-Poitou,* 19e *année 1977-1978, no. 26* Secréteriat Mathématique, Paris (1978),

[18] T. Shoji, Green functions of reductive groups over a finite field, *Proc. Symp. Pure Math.* **47** (1987), 289–301.

[19] T. Shoji, Green functions associated to complex reflection groups I and II, *J. Algebra* **245** (2001), 650–694, J. Algebra, in press.

[20] R. Steinberg, *Lectures on Chevalley groups*, Yale University, New Haven, CT (1968).

Department of Mathematics
University of Wisconsin, Madison
Madison, WI 53706 USA
kanelsen@students.wisc.edu
ram@math.wisc.edu

Printed in the United States
by Baker & Taylor Publisher Services